W0037053

DRUG DEPENDENCE
AND
EMOTIONAL BEHAVIOR

Neurophysiological
and
Neurochemical Approaches

DRUG DEPENDENCE
AND
EMOTIONAL BEHAVIOR
Neurophysiological
and
Neurochemical Approaches

Edited by
A. V. Valdman

Institute of Pharmacology
Academy of Medical Sciences of the USSR
Moscow, USSR

Translated by
L. R. Sandler

Translation Edited by
M. Sandler

Queen Charlotte's Hospital
London, England

CONSULTANTS BUREAU • NEW YORK AND LONDON

Library of Congress Cataloging in Publication Data

Drug dependence and emotional behavior.

Includes bibliographies and index.
1. Neuropsychopharmacology. 2. Drug abuse — Physiological aspects. 3. Psycho-
tropic drugs — Physiological effect. I. Val'dman, A. V. (Artur Viktorovich) II. Sandler,
Merton. [DNLM: 1. Behavior — drug effects. 2. Nervous System — drug effects. 3.
Substance Dependence. QV 76.5 D794]
RM315.D789   1986   615'.78   86-8127

ISBN-13: 978-1-4684-1658-9   e-ISBN-13: 978-1-4684-1656-5
DOI: 10.1007/978-1-4684-1656-5
This translation is published under an agreement with the Copyright Agency of the
USSR (VAAP)

© 1986 Consultants Bureau, New York
Softcover reprint of the hardcover 1st edition 1986
A Division of Plenum Publishing Corporation
233 Spring Street, New York, N.Y. 10013

All rights reserved

No part of this book may be reproduced, stored in a retrieval system, or transmitted
in any form or by any means, electronic, mechanical, photocopying, microfilming,
recording, or otherwise, without written permission from the Publisher

Contributors

Allikmets L. H.[2]

Andreev B. V.[3]

Avdulov N. A.[1]

Badyshtov B. A.[4]

Blednov Yu. A.[4]

Bondarenko N. A.[1]

Borisenko S. A.[1]

Burov Yu. V.[1]

Galustyan G. E.[3]

Garibova T. L.[1]

Ignatov Yu. D.[3]

Kampov-Polevoi A. B.[1]

Katkova E. B.[3]

Kharlamov A. N.[1]

Khodorova N. A.[1]

Klusha V. E.[5]

Kovalev G. I.[1]

Kozlovskaya M. M.[1]

Kudrin V. S.[1]

Maisky A. I.[1]

Mineyeva M. F.[1]

Mucinieze R. K.[5]

Poshivalov V. P.[3]

Rayevsky K. S.[1]

Rojanetz V. V.[1]

Shemanov A. Yu.[1]

Shevchenko N. M.[4]

Seredenin S. B.[4]

Valdman A. V.[1]

Vedernikova N. A.[1]

Voronina T. A.[1]

Yukhanov R. Yu.[1]

Zarkovsky A. M.[2]

Zhukov V. N.[1]

Zvartau E. E.[3]

1. Institute of Pharmacology, Academy of Medical Sciences of the USSR, Baltiyskaya str. 8, 125315 Moscow, USSR.
2. Department of Pharmacology, Tartu University, 202400 Tartu, Ulikooli 18, USSR.
3. Department of Pharmacology, Pavlov Medical Institute, Tolstoy str., 6/8, 197089, Leningrad, USSR.
4. Laboratory of Pharmacological Genetics, Second Medical Institute, Ostrovityanova str. 1, 117437 Moscow, USSR.
5. Institute of Organic Synthesis, Latvian Academy of Sciences, Ayzkraukles str. 21, 226006 Riga, USSR.

Foreword

English-speaking scientists start with one vast advantage: the bulk of the world's scientific transactions are conducted in English. There are many who would go further and say that any scientific work of importance is published in English. This book, which is, in effect, the tip of a large iceberg, gives them the lie! In the Soviet Union alone we have a vast wealth of expertise supported by a treasury of books and publications, but it is effectively cut off from Western scrutiny by the language barrier. It therefore seems timely to lift the curtain a little and put some of the best of it on display.

In this excellent compilation, Professor Valdman and Dr. Burov have assembled a cast list of leading Soviet scientists who provide us with a refreshingly different slant on a set of problems of concern to neuroscientists throughout the world. These scientific presentations are neither better, nor worse than but, rather, complementary to Western pharmacological thinking. Traditional Soviet approaches to animal psychology are here coupled with sophisticated latter-day neurochemistry and neurophysiology and, in the process, provide us with new insights into the molecular bases of animal responses to environment and to certain drugs. Apart from shedding new light on many contemporary problems, the findings reported here provide an important window on the thought processes of the foremost neuroscientists of the Soviet Union. This book cannot fail to be of interest to all who work in this expanding (and exciting) area.

M. Sandler

Contents

EMOTIONAL BEHAVIOR
(Edited by A.V. Valdman)

ix

DRUG DEPENDENCE
(Edited by Yu.V. Burov)

Emotional Behavior

Edited by A.V. Valdman

Psychophysiological Behavior and Pharmacological Aspects

Combined Method for Evaluating Emotional States and Individual Response of Cats Under Conditions of Social Interaction

M.M. Kozlovskaya and A.V. Valdman

Institute of Pharmacology, Academy of Medical Sciences
of the USSR, Moscow

1. REASONS AND PRINCIPLES BEHIND THE NEW METHODOLOGICAL APPROACH

Hinde aptly commented in 1970 on the gradual phasing out of the accepted boundaries between life sciences, with particular reference to animal behavior studies. This makes it difficult to decide on a yardstick for research into animal behavior. By the same token, workers in the field are making a contribution to the development of their own particular scientific discipline as well as the field of animal behavior as such.

Producing adequate methods of rating behavior – especially emotional response in animals, is extremely important to the psychologist, the physiologist and the ethologist, but even more so for the psychopharmacologist. Psychopharmacological agents do not produce selected effects on specific, recognized forms of mental illness, but affect the severity of certain syndromes which form part of the condition diagnosed. Emotional breakdown may be regarded as one such syndrome. Psychotropic agents are generally recognized to be "emotiotropic" substances first and foremost, in that they act on the target area of mental or emotional breakdown in man. The experimental psychopharmacologist is thus particularly concerned with the strength and specificity of these agents' effects on the range of emotions displayed by animals.

However, his work fails to provide a complete spectrum of a drug's psychotropic effects for the guidance of his clinical counterparts, a basic omission from their point of view. Nor does the existing battery of tests, i.e. primary and secondary screening, produce adequate syntheses illustrating any given aspect of a psychotropic drug's action.

Some pharmacologists, aware of the one-sidedness of this approach, advocated the propagation of methods for evaluating changes in animal behavior in a so-called free environment, with rats and cats as the experimental animals (Biossier 1965; Lat et al., 1965). Various alternatives for recording particular features of spontaneous animal behavior were suggested for the purposes of describing psychotropic action. Norton and De Beer (1954) described a method of analyzing spontaneous cat behavior which involved recording certain postures, how frequently these postures recurred, subsequent social interaction with the group and any other identifiable behavior patterns.

This method provided data on a range of psychotropic compounds (Norton, 1969). For the sake of objectivity, however, psychotropic action was only measured as and when changes in the frequency of certain motor effects occurred. A standard behavioral approach of this type excludes not only the psychological processes but also the physiological significance of behavior.

Valuable research was carried out on cats by Hoffmeister and Wuttke (1968), Wuttke and Hoffmeister (1968) and Wuttke (1971), differentiating between the effect of psychotropic substances in certain categories of individual behavior, such as sociability, pleasure or contentment, alertness, attentiveness and so forth. A range of psychotropic preparations were compared according to which moods they produce, suppress or fail to influence, and a whole spectrum of specific properties for the various substances tested emerged. These authors' behavioral research has a much more solid theoretical framework than that of workers using the Norton method. Some behavior patterns (German: Verhaltensweisen) are considered grouped rather than separate categories. The definition of an animal's behavioral state and how this state is displayed is founded not on mere assumption but on studies of cat behavior (Leyhausen, 1960; Tembrook, 1961).

Over the years we have been developing a new approach to assessing the action of psychotropic agent on animal behavioral states, including seemingly emotional ones, based on the "systemic principle" (Valdman et al., 1976, 1979; Valdman and Kozlovskaya, 1965, 1973, 1977; Kozlovskaya and Valdman, 1977). This system uses a combination of neurophysiological, physical, psychological and ethological approaches to animal group behavior, together with an analysis of social interaction in animals. It has broadened the scope of objective assessment of emotional response in animal behavior both in the normal and under model pathological conditions. Such an approach breaks new ground by promoting fuller preclinical definition of a psychotropic compound's range prior to clinical acceptance (Valdman, 1972, 1980).

Considering emotion as one of the brain's forms of mental activity, it would rank lower than consciousness from the evolutionary

standpoint. The primary or "vital" emotions are not the prerogative
of man but are inherited by him in the course of evolutionary development. It is these vital emotions which form the target area for
the action of the psychotropic agents used in psychopathology. However, an emotional state produced in an animal by external factors or
electrical stimulation of emotional centers in the brain may not be
measured by outward displays such as vocalization and movement. Any
description relating to emotional state seen through human eyes is
both subjective, and, inevitably, somewhat anthropomorphic. As
Delgado (1966) rightly observed, numerous judgements are confidently
made from the standpoint of personal feelings and perceptions. This
accounts for contradictions between existing facts and the abundance
of seemingly similar terms which are subject to contrasting inter-
pretations.

No one behavioral item suffices to characterize an emotional
state. Thus, the psychological state of an animal may not be
properly judged according to one striking behavioral display alone
(Valdman, 1972). However, our many years of research into emotion
and emotional behavior in animals have led us to conclude that a
subject's emotional state does, in fact, determine its behavior
(Valdman et al., 1976). Subjective imput such as emotional experience and emotional "tone" enters the functional progression
leading to performance of an action at the stage of afferent synthesis, and thus plays a substantial part in processes of decision
making, planned action and motivation. Experimentally-produced
emotional states can help to modify the structure of behavior, and
may considerably affect goal-directed behavior to the point of dis-
organization. This emerges from the pathology of emotional state
(psychopathology) in particular. One means adopted of evaluating the
emotional state of our experimental animals objectively was by gaug-
ing their effect on the behavior of other individuals of the same
animal species when interacting within an animal group - the newest
aspect of our approach. Displays of emotion in animals serve a
biological purpose as fine indicators of emotional state, and signal
specific information to other individuals.

An earlier description of ethologically meaningful postures and
behavior patterns in cats in a free environment was made by Leyhausen
(1960). We ourselves have worked extensively both on evolving useful
and objective criteria for defining emotional behavior in cats under
laboratory conditions and on methods of quantifying change in emotional behavior. Our cat behavior experiments, carried out under
conditions giving ample scope for wide-ranging emotionally triggered
behavior, showed both spontaneous and evoked behavior in these
animals to be highly individual. The set of outward manifestations
observed - movements displaying emotion - was basically limited, and
remained unaltered during prolonged contact with other cats or with
the experimentalist.

As a key feature of our approach, experimental animals were kept amongst others of their own species and interacted freely. Changes in emotional/behavioral response were produced either by stimulating the emotional centers of the brain, or by administering psychotropic substances, or by applying afferent aversive stimuli. The state of the experimental animal and the nature of the emotional change exhibited was assessed according to the reaction of other animals within the group and to modifications in the test animal's own responses to a series of challenges. The emotional/behavioral response pattern of the animals used in the experiment had previously been thoroughly investigated.

Detailed measuring tables on the lines of a rating scale were developed for quantifying and qualifying emotional and behavioral reactions in animals, awarding scores ranking their response according to a five-point system. Data obtained from the tables could be expressed in the form of a graph or as segments of a circle. The latter method of analysis could cover the whole range of emotional/behavioral response (see Figure 1). Each segment of the circle, as it traverses successive concentric rings, describes a scale from 0 – 5 moving outwards, with 0 (zero) as the center. Each subdivision of

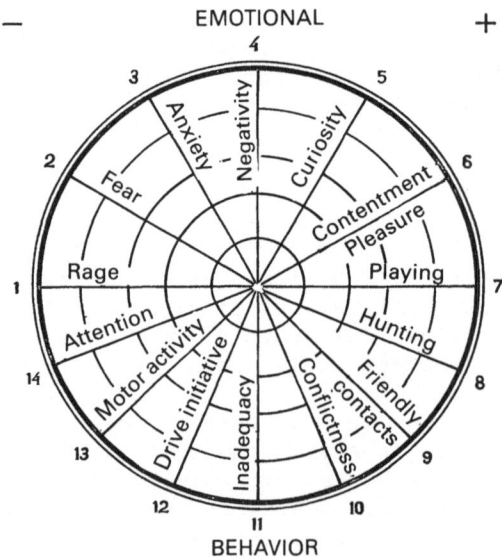

Fig. 1. Diagram showing range of emotional behavior in cats. Segments of a circle are sub-divided as the radii traverse 5 concentric rings. Each sub-division stands for a number of points scored on a 5-point scale from 0-5 moving outwards from the center (=0). Each segment stands for one category of behavior and points are scored for intensity of expression of the behavior pattern involved.

a segment represents points scored out of five for a given category
of behavior. The first eight segments of the circle stand for a
spectrum of emotionality. Various types of positive and negative
emotions are discerned, as follows: 1 = rage, 2 = fear, 3 = anxiety,
4 = negativity, 5 = curiosity, 6 = contentment or pleasure, 7 = playfulness, 8 = hunting; the two which follow, No. 9 - friendliness and
10 - conflict, apply to social interaction. Nos. 11 and 12 refer
respectively to behavioral inadequacy and drive or initiative as
shown in a set experimental situation. Of the two remaining segments, No. 13 expresses general level of motor activity and 14 -
orientation ability (alertness).

The figures obtained - or points scored on each count by each
animal group - were collected, the arithmetical mean was defined as
M, scatter as D and standard error of the mean as m. Subsequently a
correlated analysis of the structure of the spectrum of emotional
behavior was produced. This was based on a whole series of experiments for each type of emotion displayed (or each segment of the
circle) covering a wide range of values. The parameters obtained
were then paired and the correlation coefficient determined (details
given in Chapter II).

2. MATERIALS AND METHODS

This work aims to review the results of 10 years of cat behavior
observations and analytical research under laboratory conditions,
performed in open cages, vivariums or other experimental conditions.
The cat was chosen in preference to any other type of experimental
animal because it has a wide emotional range and displays groups of
emotions resembling some of those known to man, such as fear, aggressivity and pleasure or contentment. However, an animal's reaction to
a stimulus in experimental conditions is largely determined by its
individual make-up, experience drawn from life, acquired habit,
ability to assess a situation, how it relates towards the experimentalist and a number of other considerations. Research workers
observing animal behavior unconsciously interpret it subjectively,
thereby attaching human connotations to displays of affect in
animals. Emotional response observed in experimental conditions may
not truly reflect an animal's emotional state, moreover. For example, an animal may communicate its emotional state to its fellows
by pointing its ears in a particular direction - a sign which is
meaningful for animals of a certain species but which slips the
experimentalist's notice. Movements displaying emotion serve as
early warning signals conveying certain information to others within
a group. Hence, any type of emotional state, including fear, rage,
anxiety, repose, pleasure or contentment, which is usually accompanied by stereotyped voluntary and involuntary signs, has its effect
on relationships within the animal group.

The cat, having a fairly well developed neocortex, is sufficiently in control of its own behavior to match these signs of its (inner) state to the requirements of a given situation to some extent. In some animals the onset of aggression in a negatively predisposed individual may be provoked by a tiny movement made by a fellow-member of the group but not by any manipulation on the part of the familiar experimentalist.

All these factors should be borne in mind when establishing procedures for studying animal emotional states and response objectively, both where alteration of the emotional state is produced naturally (as far as experimental conditions will allow) or simulated, as with electrical stimulation for example, or destruction of the "emotional areas" of the hypothalamus.

Cross-bred male cats were used ranging from 2-3½ years old, and averaging 3-4.5 kg in weight. They were observed over extensive periods varying between 6 months and 2-3 years, and housed in a vivarium measuring 16 sq meters with an overall volume of 48 cu meters, having easy access to a 6 sq meters open-air cage. In the winter months daylight was replaced by artificial light until 6.30 pm. Six to twelve animals, having been in quarantine for 10-21 days, were kept in an isolation cubicle. The vivarium was fitted with shelving enabling the animals to exercise and to pick a favorite spot; netting was also provided for them to climb up and down. An underfoot grid served to record the distance kept between animals and any warning signs of impending confrontations, as well as feeding habits and any other type of observable activity relating to free behavior and group relationships. The animals were regularly given medical examinations and checks were made on their mucous membranes, eyes, coat, gastrointestinal tract, weight, body temperature, etc. They were fed twice daily, at 10 am and 3 pm. The daily ration consisted of a standard 100 g meat of fish with milk, vegetables and vitamins. The experiments were carried out for 10-16 hours daily from October to February.

The animals were transferred from the vivarium to the experimental enclosure in square containers measuring 35 x 20 x 25 cm.

The experimental enclosure itself measured 1.5 m x 1 m x 1 m and was lit by 500 lux. It was partitioned into 2 small compartments: the right-hand or "danger" compartment, fitted with an electrode floor for administering aversive electrical stimuli to the skin, and that on the left, for presenting the animal with different trials, whether in the form of stimuli or other testing devices. A number of other cats of varying standing within the group and varying emotional responsiveness were to be placed in the latter compartment; these were instrumental for analyzing relationships with fellow animals belonging to the group. The partition dividing the two compartments contained four openings, arranged in a chessboard pattern equidistant

from the floor, enabling animals to pass from one compartment to the
other. The openings were fitted with flaps which could be opened or
closed in any sequence by remote control. The enclosure also contained an electric bell and two niches, the first covered by a flap
where the animal could feed and the second where it could perch to
avoid "punishing" treatment, as well as a darkened corner where a
"punished" animal tended to withdraw. The floor of the chamber was
divided up into 70 squares each measuring 15 cm x 15 cm on which a
measuring grid was laid diagonally, stretching from corner to corner,
from front to back and from side to side, so as to measure the distance separating one animal from another, either when undisturbed or
during competitive interaction or some other activity.

Electrical stimulation was applied to the deep brain structures,
which produce emotionally-based behavioral responses in cats, by
means of a pliable cable, attached flexibly to a shelf within the
chamber. It was controlled from an electrode panel via a commutation
device. The panel was connected to the animal's skull, without
inhibiting its freedom of movement within the cage. Cutaneogalvanic
or galvanic skin reaction (CGR) further served to measure the intensity and duration of the emotional state produced in the animal. An
active electrode was attached to the pad on the cat's forepaw and
another (indifferent) to a shaved patch on its spine, so as to give
an unbroken CGR recording. The trace was made using standard Mishuka
equipment connected to an H-39 model automatic recorder. The electrodermal resistance was measured in kilohms, with any alterations
expressed as a percentage of the baseline level.

3. SEQUENCE OF EXPERIMENTS PERFORMED AND RATING THE INTENSITY OF EMOTIONAL/BEHAVIORAL REACTIONS

At least two observers took part in the experiment, working
separately. The first was to tape a running commentary of the trial
and perform the manipulations required, and the second to produce
standard recordings, strictly adhering to a set formula, for incor-
poration into a measuring table. By deciphering the magnetic tape
after the experiment a complete schedule could be reconstructed
covering all responses shown by the animals to trial situations or
stimuli. A cine film, a series of still photographs or a video
recording could also be made if required.

Comprehensive measuring tables or rating scales were used both
to describe meaningful response observed in the trial animals and
rate emotional and behavioral reactions. Scores were awarded for
response according to a five-point scale. These measuring tables,
which served as a back-up to the two monitoring systems described
above, could indicate:
- different emotions and behavior displayed
- typical postures

- position of certain parts of the body
- observable displays of voluntary action
- quality of vocalization
- active behavior towards other members within the group
- other group members' behavior towards the test animal

and a number of other factors (see Tables 1-2, 5-15, 17-18).

The experimentalist should always adhere to a strict sequence for introducing the test situations, stimuli and other testing devices used. He should also record the precise order and frequency of the testing devices or their effects, as well as the duration, regularity and intensity of any repetitions.

The research was carried out in several stages.

3.1.1. Cat Behavior in Changing Circumstances

(i)   In vivarium; in contact with experimentalist.
      Possible animal behavior patterns : indifference, approach,
      flight, defense.
(ii) Transfer from vivarium in portable cage.
      Possible animal behavior patterns : immobility, motor
      disturbances, vocal reaction, defecation.
(iii) 2-10 min after transfer from portable box to experimental
      chamber (depending on speed of adaption).

      Values were obtained for :

      - duration of the initial "frozen" attitude in secs.
      - position within the enclosure; whether central or close
        to exit or rear wall.
      - in which direction the animal first moves - around the
        periphery, to the center, the exit or the darkened corner of
        the enclosure.

The end of the adaptation period is marked by the animal adopting a relaxed sitting or lying posture and holding it for 2-3 min. These results were used in evaluating displays of anxiety or the emotional state of fear, as part of the range of emotional behavior (see Tables 11 and 13).

3.1.2. Spontaneous Motor Activity in an Experimental Chamber

After the adaptation period the animal is removed from the experimental enclosure for 3 min and put in a small dark cage. The number of squares it crosses spontaneously during the 5 min following its return to the cage is then counted.

The result are set out in tabular form (Table 1) giving measurements for degree of motor activity, expressed in predetermined units on a 5-point scale. Data relating to motor activity were likewise used for evaluating fear-related emotional states in Table 13.

3.1.3. Orientation Reaction or Alertness (ear-pricking) Produced by the Ring of a Bell as Afferent Stimulus

After the animal has adapted to its new surroundings in the experimental enclosure, an electric bell emitting a steady ring is set off. The animal's response is recorded - i.e. position of different parts of the body, quantity and force of voluntary and involuntary motor reactions. The results are set out in tabular form (see Table 2), with the intensity of the ear-pricking reaction expressed in pre-determined units, as scores on a five-point scale. The results, while showing the degree of alertness and orientation reaction displayed, also indicate the presence of anxiety - another facet of the range of emotional behavior.

3.2. DETERMINING EMOTIONAL RESPONSE TO A SERIES OF ASSORTED TESTS

3.2.1. Trials included presenting the animal with ten stimuli or testing devices, five of which caused it to react positively and five - negatively. Other factors were also considered, such as latent period of the first response, the speed and intensity of the emotional or behavioral reaction, the distance measured between presentation of the testing device and a particular behavioral reaction, how frequently a testing device was used or repeated and the duration of the ensuing stimulation.

Table 1. Measuring Table for Rating Spontaneous Motor Activity
          (Refers to segment 13 of the circle - see Figure 1)

Chief groups of signs	Points scored
Lethargy, prolonged holding of one posture, maximum of 1-3 squares crossed	1
Some mobility, 10-15 squares crossed, posture changed 2-3 times	2
Mobile, 10-15 squares crossed, swivelling, changes in posture, spontaneous combinations of motor activity shown	3
Increased motor activity, 25-50 squares crossed, cat often changes posture and frequents the periphery of the chamber	4
Over-active, constant aimless bustling about, over 50 squares crossed	5

Table 2.   Measuring Table for Rating Orientation Reaction
           (alertness). (Refers to segment 14 of the circle –
           see Figure 1)

Chief groups of signs	Main signs observed	Points scored
Slow reaction barely observable, completed in 10 secs.	Ears pricked, tail wound round body, slight turn of head and slightly faster breathing.	1
Calm response, noticeable orientation towards the stimulus, completed in 10 secs.	Adopts sitting posture, follows sound with ears, tail quivers, head turned towards the ringing sound, breathing considerably faster, pupils very dilated.	2
Rapid reaction, clearly-defined involuntary motor movements. Time taken: 3–5 secs.	Adopts posture on half-bent paws, ears pressed flat, tail between the legs, head turned towards the sound, breathing considerably faster, pupils very dilated.	3
Reaction over in a moment; no orientation towards the stimulus.	Poised for flight (which follows in 7–10 secs), ears pressed backwards, sharp twists of the head, quivering. Pupils fully dilated, with widened palpebral fissures.	4
Startle reflex type reaction, no orientation towards the stimulus.	Shuddering, leaping on all fours or jerking of hind legs, flight.	5

An animal's response, when tested in this way, consisted of displaying a combination of emotions, the extent and intensity of which depended on individual emotional pre-disposition. The results are set out in Tables 5-13, which describe the emotional/behavioral response obtained, and measure it in predetermined units (scores).

3.2.2. Prolonged research into how a certain response may increase in intensity was performed prior to devising the measuring tables. The response of each animal was duly recorded, including the position of different parts of the body, viz: the angle at which the ears were pointed, position of the tail, head, paws and flexing of the spinal column. Typical postures, already established as etho-logically significant in the cat, were also taken into account: positions of passive defense, paralyzing fear, aggressivity (or active defense response), submissiveness, playfulness, ritual move-ments, pleasure, breath-holding, etc. The nature of the animal's different vocal expressions, such as hissing, growling, snorting or

Table 3. Series of Reward Motivated Stimuli and Testing Devices
          for Cats

Order of tests	Stimulus or testing device	Number of repetitions of stimulus or testing device	Duration of test in secs	Distance between animal and testing device in cm
1	Stroking, scratching different parts of the body	6	30	
2	Moving toy (small ball)	6	30-60	15-5
3	Mouse:			
	outside window of chamber	3	10	15-20
	when brought into chamber	2-3	30	20
	when moving around the chamber	1	30	Not fixed
4	Food:			
	in food dispenser	3	15	20, 10, 0
	preferred food in food dispenser	3	15	20, 10, 0
	behind a 20 cm transparent barrier	6	45	15
	on a ledge behind a transparent flap	6	45	10
5	Various non-frightening objects:			
	from the vivarium	1	15	20
	unfamiliar	1	15	20

sniffing and mewing, as well as involuntary motor effects, e.g. on
the pupils of the eyes, breathing rate and piloerection on various
parts of the body were likewise included. The intensity and duration
of galvanic skin response was also recorded in a number of cases. By
synthesizing data obtained from a large sample of animals a series of
signs was produced representing graded levels of response and indicating how such responses are usually expressed. A five-point scale
was devised on this basis, each additional "point" corresponding to a
higher degree of intensity.

One such detailed measuring table referred to hunting – amongst
the most genetically determined behavioral reactions in the cat.
Table 5 gives a full list of factors used for evaluating cat response, awarding a score on a five-point scale for each.

The signs indicating a step-by-step rise in intensity of the
reaction were also graded (see Table 6).

A score of one was awarded for a minimal display of the hunting
reaction, where the cat was only in indirect contact with the mouse

Table 4. Series of Stimuli and Testing Devices with Negative
          (Punishing) Effect on Animals

Order of tests	Stimulus or testing device	No. of times each is reintroduced	Total duration of test in secs	Distance between animal and testing device in cm
1	Waving or clapping in front of snout	4	5	5-10
2	Poking with a stick	4	10	0
3	Presented with noisy moving object	4	45	50, 20, 10
4	Airstream from a hand-pump at the nose	6	30	10
5	Pressure mechanically exerted on the tail, 2 cm from the tip	2	15	0

and could well fail to catch it. Two points were awarded for a more
pronounced response, beginning as the mouse was first brought into
the enclosure, and intensifying as it began to move about. Although
the cat may claw the mouse repeatedly it does not usually grasp it
with the teeth at this stage. Typical hunting reaction with repeated
chasing and ritual playing with the mouse, ending by attacking it in
the neck - 3 points. Four points were awarded for a brisk response
(having once sighted the mouse through a window at a distance), displaying keen excitement, hot pursuit, pronounced involuntary movements and a sharp attack culminating in grasping and eating the
mouse. The fifth stage is characterized by abruptly, and sometimes
not altogether successfully, attacking the mouse, even if it is
separated by glass; the mouse is seized in an atypical way, either
by the tail or the back.

    Tables of the above type were produced for rage, fear, anxiety,
negativity or punishment, pleasure or contentment, playfulness and
all the other categories of emotional behavior identified. The
results of experimental data, analyzed according to this type of
measuring table, enabled us to pick out identifiable signs, for each
of which a score was awarded.

3.2.3. Simplified tables were used thereafter for evaluating displays of different types of emotional behavior and monitoring basic
groups of signs and their corresponding points score.

    The responses of cats to tests Nos. 1 and 2 in Table 3 for dis-
plays of rewarding playful or pleasurable emotion are further rated

Table 5.   Measuring Table for Rating Emotional/behavioral Response
           a) Hunting

Groups of signs recorded	Score out of five awarded for grade of response				
	1	2	3	4	5
Critical distance of response					
in direct contact	+				
at 15-10 cm		+			
at 20-30 cm			+		
at 70-80 cm (maximum)				+	+
Position of ears					
pricked	+				
pricked and strained		+			
turned towards testing device			+	+	
Position of tail					
tensed, tail waves rhythmically	+	+			
raised off floor, stretched			+	+	
Position of head					
usual position	+	+			
stretched forward, neck stiffened			+	+	
Posture					
forequarters slightly raised	+				
"lying in wait" - body inclined forward towards testing device and tensed		+			
"stealing up" attitude - body raised on paws and stretched forward			+		
makes a rush, with rapid forward movements of head and paws				+	
abrupt leap, absence of static posture					+
Movements leading up to seizing the mouse					
head sharply inclined forwards and lowered	+				
blow with paw and mauling of mouse		+			
turning mouse over repeatedly			+		
small bites followed by release (or gradual eating) of mouse			+		
snaps, sinking in teeth					+
How mouse is seized					
by the neck		+	+		
by the head			+	+	
in the middle of the back				+	+
in the area of the tail					+
Involuntary responses					
dilation of pupils		+	+	+	
widening of palpebral fissures				+	+
altered breathing		+	+	+	+

Note: with reference to the hunting reaction, it should be noted
       that not all domestic cats display this interspecific
       response, ending in killing or eating the mouse.

Table 6. Table for Rating Emotional/behavioral Response
          a) Hunting (refers to 8th segment of circle, see Figure 1)

Groups of signs recorded	Score
Brief alert reaction lasting > 15 sec prompted by mouse moving at short range	1
More prolonged state of alertness, lasting > 30 sec when mouse is sighted at 15-30 cm. Posture changes as mouse approaches; repeatedly touches mouse with paw	2
At a distance of > 50 cm quickly turns towards the mouse as it first appears in the chamber; various sets of motor reactions accompanying pursuit of, playing with and seizing the mouse with the paws and subsequently biting it in the neck	3
Darts towards the mouse as soon as it enters cat's field of vision, increasingly alert, forcefully attacks mouse, repeatedly biting it in the neck and head	4
Mouse grasped forcefully, killed outright and swallowed, i.e. standard reaction; "cold" attack	5

Table 7. Measuring Table Rating "Rewarding" States
          i) Pleasure or Contentment in Response to Stroking
          (refers to 6th segment of circle, see Figure 1)

Chief groups of signs displayed	Score
Resting position, stays in one spot, occasional blinking	1
Relaxed posture, narrows palpebral fissures, purrs, shakes head and neck	2
Actively displays contentment, shakes body and head, stretches claws of forelegs, mews	3
Marked and prolonged display of pleasure, shakes body and head, loud continuous mewing, rhythmic curling and uncurling of the distal parts of the forepaws and claws	4
Displays of pleasure arising spontaneously, rolling onto its back, mews loudly and uninterruptedly, continuous licking of fur	5

in measuring Tables Nos. 7 and 8. Similarly, measuring Tables Nos. 6 and 9 evaluate responses to tests Nos. 3 and 4, described in Table 3; these involve positive emotional behavior connected with hunting and feeding. The cat's behavioral reaction, when introduced to various unfamiliar test objects, as in test No. 5 of Table 3, was considered indicative of pursuit or curiosity, and is further graded in measuring Table No. 10.

Table 8. Measuring Table for Rating Playful Response when
          Introduced to a Toy (Small Ball)
          (refers to 7th segment of circle, see Figure 1)

Chief groups of signs displayed	Score
Sluggish and short-lived. Turns towards object at a distance of 5-6 cm, touches small ball once or twice when shown repeatedly	1
Brief reaction of under 15 secs when object is presented 3-4 times. Touches it with paw, but not followed up	2
Easily aroused by toy, even when still. Lasts over 30 secs. Chases object briefly	3
Reacts when shown object once, even at a distance. Reaction lasts up to 3 mins. Chases object repeatedly and seizes it with paws	4
Prolonged reaction lasting 5-10 min, occurring "spontaneously", set off even by non-moving object, rolling on its back, with loud continuous mewing	5

Intensity of goal-directed behavior, with a mouse as the goal, for instance, in the case of hunting, or food in the case of feeding behavior, comes under the combined heading of "behavior showing initiative" in Table 14.

An animal's responses to adverse stimuli and testing devices (see Table 4) served to evaluate the emotional state of a "punished" animal. Its emotional response may be one of fear-related passive defense, as in Table 11, or rage-related active defense (see Table 12), and varies according to the individual.

The intensity of emotion as displayed by cats, however, does not correspond to the degree of emotional tension actually experienced in many instances. Any conclusions reached should therefore take into account how other cats within the group relate to the test cat, in addition to the tests described in Table 4. These cats serve as a yardstick of emotional state (see Tables 17 and 18).

Anxiety was put into a special category (see Table 13). It could not merely be considered a state analogous to fear. This emotional state has no clear-cut set of physical signs. Nor is anxiety focused on any one object in the experimental environment. In fact, diffuse anxiety in the test cat produces virtually no re-action on other members of the group.

The test cat's anxiety state is communicated to others within the group. They never take up a lying posture; they keep a watch on the cat, without actually attacking it. A number of other factors

Table 9.   Measuring Table for Rating Food-seeking Behavior

Feeding rack placed in sight of animal	Food placed behind transparent barrier	Food placed behind a transparent flap on shelf
1. Feeds when shown food repeatedly, when feeding rack is put under snout	Jumps behind barrier once after being shown food 6 times	Obtains food (meat) once (max.), having been shown it repeatedly
2. Having sniffed food, settles down 10 cm away to feed	Jumps repeatedly (with breaks) behind barrier, having been shown food 2-3 times	Having been shown food 2-3 times, opens the flap with paw or head (inconsistent)
3. Actively approaches food placed 10 cm away	Jumps repeatedly behind barrier, having been shown food 2-3 times	Consistently opens the flap and gets food with paw, having been shown food 2-3 times
4. Actively approaches food placed 20 cm away	Jumps behind the barrier up to 10 times per 45 sec	Having been shown food 1-2 times, repeatedly gets food rapidly and more calmly
5. Actively approaches any visible food and follows the feeding rack around when it is moved	Hurried jumps with adventitious movements	Hurries to the flap, makes repeated food-seeking movements with paw, irrespective of availability

were also considered in rating anxiety; how animals adapted to new situations and speed of adaptation (see 3.1.1), degree of spontaneous motor activity (see Table 1), and how animals related within the group. A pure anxiety reaction is only to be seen extremely rarely, as, for example, when the "emotiogenic points" of the hypothalamus are electrically stimulated (see Section 5 of this chapter). Anxiety co-exists more commonly with fear where the signs displayed of these states oscillate between a 2- and 3-point score.

Initiative and inadequacy are additional criteria for rating animal behavior.

Initiative is a complex factor governing how an animal acts in an experimental situation, as well as its degree of drive. It is

Table 10. Measuring Table for Rating Positive Emotional States
           ii) Curiosity (exploration) (Refers to 5th segment of
           circle, see Figure 1)

Chief groups of signs	Score
Focusing attention onto object, turning head and body towards it	1
Altered posture, sniffs at objects, stretches neck, noticeably faster breathing	2
Rapid change of posture, approaches objects, sniffing them repeatedly	3
Fast, purposeful approach, examines objects, moves from one to another, attempts to shift them or turn them over, accelerated breathing	4
Hastily examines the empty chamber, prolonged sniffing of walls and corners, turns head, rears on hind legs, frequently switches attention between objects	5

rated according to how motivation is carried through into a set of
voluntary actions. Behavioral initiative is rated in Table 14, which
itemizes behavior and evaluates performance in tests for food-seeking
and hunting response, social interaction and so forth. Here initiative is taken as goal-directed behavior, whether stimulated by
reward or punishment, in pursuit of increasingly difficult aims.

Behavioral inadequacy is also difficult to rate. This was done
by analyzing :

- the breakdown of the most typical and common ways of displaying
  motivated behavior - food seeking reaction, hunting, flight from
  adverse situations or effects
- impaired ability to carry through planned action when faced with
  obstacles
- altered emotional response to standard test procedures
- loss of ability to adjust its behavior towards other animals
  within the group (or in a pair) on the basis of ethologically
  recognizable signs displayed by its fellow animals.

Rating and descriptions of behavioral inadequacy may vary according to prevailing experimental conditions. The basic scale is
outlined in Table 15.

3.3. ANIMAL GROUP INTERACTION

Animal group relationships were analyzed according to observations made on the behavior of cats in a vivarium, in an open cage -

Table 11.   Measuring Table for Rating Negative Emotional States
a) Fear (passively defensive)   (Refers to 2nd segment
of circle, see Figure 1)

Chief groups of signs	Score
Huddling, immobility, shifts away without taking flight, ears pressed flat briefly, retreats. May display different defensive signs if unable to avoid contact	1
Quivering, tail between legs, paws tensed, head stretched out, ears pointing sideways, occasional narrowing of the palpebral fissures, breath holding. Submissive posture when coming in contact with other members of the group	2
Tip of tail curled tightly under. Hugs the floor, closes eyes, irregular breathing. Refrains from initiating any action. Paralyzed by fear on coming into contact with other members of the group	3
Hugs the floor, completely motionless. Frozen catatonic posture. Completely refrains from any type of activity. Attacked when interacting within the group	4
State of stupor, catatonia maintained for long periods, repeatedly attacked when interacting within the group	5

where interaction could develop more freely amongst the animals, and
in an experimental chamber, using groups of 2-3 cats selected by the
experimentalist.

The test cat was the main "performer" in these group experiments. The experimentalist had already fully established the cat's
behavioral traits and emotional response. As a rule it had spent up
to 2 years in a vivarium environment, but was kept isolated from
other cats. This animal would have the most even temperament and the
most "adequate" behavior of the cats and would usually occupy a
leading position within the group.

Interaction involving the test cat in a pair was investigated by
testing emotional state under standard experimental conditions. The
scope of the relationships defined could be extended by including
other male cats from within the group or a female, depending on the
purpose of the experiment.

The test procedures used are given in Table 16.

Test 1 covered both the distance separating the test cat from
the other members of the group and the stance it adopted in relation
to another cat - whether head-to-head, side-to-head, side-to-side,
head-to-back or back-to-back. Test 2 determined distance at which
vying for food or a mouse begins; it also measures the latency

Table 12.   Measuring Table for Rating Negative Emotional States
            b) Rage (aggression or active defense reaction)
            (Refers to 1st segment of the circle, see Figure 1)

Chief groups of signs	How displayed	Score
On the alert, turns towards object	Ears turned slightly outwards, growling, tail quivers, pupils briefly dilated	1
Changes in posture, different displays of active defense (e.g. striking out with paw) on coming into contact with the object, isolated motions of the head	Ears pointed and laid back, rhythmic tail-swinging, front paws tensed, swallowing motions, hissing, pupils dilated, piloerection of the tail, one paw raised	2
Turns abruptly towards the object, striking out repeatedly, various reactions of the head. Bouts of fighting on coming into contact with other members of the group; threatening motions at a distance of 20-25 cm	Ears laid right back, flailing movements of the tail, extremities rigid, growling, hissing, pupils dilated continuously, piloerection on tail and along the spine. Strikes out repeatedly with forepaws	3
Approaches testing device or fellow-member of the group actively and rapidly with intent to attack, a distance of 25-30 cm. Repeated attacks or bouts of fighting. Advance threatening signals made at a distance of over 25 cm	Entire body stiffened, with rigidity of extremities, back either straight or arched. Generalized piloerection, pupils dilated continuously. Darts forward abruptly. Bites opponent in the neck during fight	4
Unprovoked and aggressive attack made on any object	Attacks angrily without forewarning	5

Table 13. Table for Rating Negative or "Punishing" Emotional
           Responses. c) Anxiety (refers to 3rd segment of the
           circle, see Figure 1)

Chief groups of signs	Score
General cautiousness, slight muscular tension, isolated movements of the head, adapts slowly in "new" situation (20 mins). The ring of a bell produces a strong (2 point) alerting reaction	1
Quivers, looks round anxiously, ears pricked and turned outwards, sitting position adopted, muscular tension. Takes over 20 mins to adapt; alerting reaction (3 points). Isolated reactions of the head, attempts to hide	2
Motor unrest, frequent movements of the head, muscular tension, slow adaptation process. Afferent stimuli produce panic reaction. Other members of the group behave anxiously	3
Continuous reactions of the head, muscular tremor, attempts to escape from the chamber, produces no reaction on other members of the group	4
Increased muscle tonus reaches catatonic state, continuous vocal reactions, bursts of flustered movement around the chamber	5

period between the aggressive cat's taking up its position in front
of the exit and actually leaving the "danger" compartment of the
enclosure. Test 3 consisted of slowly drawing a member of the group
on a moving platform towards the test animal and measuring the distance at which the latter began showing threatening or defensive
behavior; it may advance, retreat or stay frozen to the spot. Test 4
showed the critical distance for producing confrontation, fight or
flight when another animal rushes up forcefully or spontaneously.

Findings from monitoring mutually-determined behavior were used
for rating additional forms of animal group contacts, such as friendliness or conflict (see Tables 17 and 18) as well as emotional states
a, b and c (fear, anxiety and rage or aggression, see Tables 11, 12
and 13 respectively).

4. MAIN PATTERNS OF EMOTIONAL/BEHAVIORAL RESPONSE IN CATS

Findings from reviewing and analyzing experimental material
accumulated over our 10 years' work on cats enabled us to identify a
number of basic response patterns. Our observations of individual
animals over extensive periods lasting 6 months to 2-2½ years con-

Table 14. Measuring Table for Rating Behavioral Initiative
            (refers to segment 12 of the circle, see Figure 1)

Chief groups of signs	Score
Approaches feeding rack from a distance of 15-20 cm and feeds. Examines isolated objects. Leaves enclosure by open exit, 20 cm away. Flight from "danger" compartment through familiar open door	1
Gets food off shelf from behind closed flap, having been shown it 2-3 times. Approaches objects and examines them. Leaves chamber by open exit 50 cm away. Flight from "danger" compartment through any one of 4 exits	2
Repeatedly opens flap and jumps barrier to get food. Attempts to open the closed exit from the enclosure. Pursues mouse. Actively competes with other members of the group. Actively seeks an exit from the "danger" section	3
Active behavior - approaching objects, examining them, repeated attempts to open the closed exit. Prevails against an assertive fellow-animal, i.e. obstacle from within the animal group, when leaving the enclosure or the "danger" compartment. Fast and effective response to all testing devices and situations entailing drive	4
Very active behavior, carrying out its intentions with unnecessary haste and behavioral drive. Inactive social interaction with group. Behavioral reactions and intentions rapidly and energetically followed through	5

vinced us that such behavior stays mainly within the same categories, whether spontaneous, produced by stimuli in an experimental enclosure, or occurring within a stable group. This means that cats may be classed according to which emotional/behavioral response each animal most commonly displays.

Firstly, there are male cats showing a "desirable" range of emotional/behavioral response. Their behavior displays initiative and purpose. They show leadership or leadership potential within the group. Their dominating position is recognized by other members of the group. Experiments show that they are easily trained and quick to make decisions to meet the contingencies of an experimentally-created situation. They initially react to provocation by adopting a defensively warning attitude, and then respond by actively attacking if the noxious influence continues. Cats of this type take part in conflict within the group. They often tend towards behavior involved with positive emotions.

Table 15.   Measuring Table for Rating Behavioral Inadequacy
            (refers to 11th segment of circle, see Figure 1)

Chief groups of signs	Score
Responses geared to the nature and intensity of the provocation, but are unduly slow. Some links in the behavioral chain missing	1
Not all responses geared to the nature and intensity of the provocation	2
Decreased accuracy in assessing situations and carrying through complex series of actions requiring behavioral drive	3
More severe breakdown of ability to assess situations, deteriorating emotional response	4
Complete breakdown of ability to assess situations. Decline or exacerbation of responses	5

Next come the animals whose emotional/behavioral response is dominated by the negative state of anxiety or fear. Other areas of response are thereby partially reduced; these cats do not show contentment either spontaneously or when stroked, for instance. They have a low level of initiative exploratory activity and sense of orientation. They have to struggle to cope in a stressful situation, to which they respond by adopting a passively defensive attitude. Any noxious agent induces a sustained state of fear. These animals do not dominate within a group, their behavior pattern being predominantly passive.

Aggressive cats constitute a third group. Their range of emotional/behavior response is deficient in display of positive (rewarding) emotions and therefore narrower. They neither play nor respond to petting nor show any spontaneous signs of pleasure or contentment. Their emotional behavior pattern is overt. They achieve leadership within the group on the strength of their aggressivity.

The next article gives a fuller statistical analysis of the psychophysiological emotional/behavioral responses of type I and II animals.

5.   EVALUATING CHANGES IN CATS' EMOTIONAL STATE PRODUCED BY ELECTRICAL STIMULATION OF THE BRAIN

The term "emotion" is generally recognized as having the dual meaning of emotion as experienced (or emotional state) and emotion as displayed. Where naturally-occurring emotions are concerned, experi-

Table 16.   Test Procedure for Interaction Between Two or More Cats

Order of performance	Testing device	No. of repeats	Duration of test in secs
1	Test cat or one of its "partners" introduced into chamber and mutual, spontaneous reactions recorded	1	60-180
2	Vying for food following moderate food deprivation lasting 15-18 hours, competing over hunting a mouse and for precedence in leaving enclosure and "danger" compartment	3	180
3	"Partner" slowly approaches the test cat until early signs of defensive behavior or submission appear	1	60
4	Interaction triggered by partner making vigorous and frequent approaches	5-6	180-240

encing and display of emotion are inter-linked.   The biological purposes served by movements expressing emotion are seen to be con- veying certain information under conditions of social interaction; they serve as a fine indicator of emotional state, signalling to other individuals of the same species.

Electrical stimulation of the subcortical structures or sub- strata (the so-called "emotion circle") produces various well co- ordinated involuntary observable signs in animals, generally referred to as emotional reactions.   However, displays of emotion resulting from electrical stimulation of the emotiogenic areas of the brain, which has been a preferred method for reproducing, simulating and studying emotions experimentally, by no means always correspond qualitatively and quantitatively to the emotion's true nature.   In a number of cases, they may not be accompanied by alterations in emotional state, as we have frequently pointed out (Valdman and Kozlovskaya, 1972; Valdman et al., 1976).   All the experimentalist has to go by are isolated fragments of the spectrum of emotional behavior (Valdman, 1972).   The changing picture of interaction within an animal group, however, provides an informative and fairly objec- tive criterion for evaluating the types of emotional state produced by electrical stimulation.

Our experiments were performed on cats whose behavior had al- ready been investigated.   The experimental situation is shown in

Table 17. Measuring Table for Rating Mutually-determined Behavior.
           a) Friendliness (refers to 9th segment of circle, see
           Figure 1)

Pattern of interaction	Main indications	Score
Friendliness, fostered by non-aggressive behavior in other members of group	Distance between partners varies. Does not join competitive interaction, gives up food it has acquired. Defensive behavior displayed at a distance of under 30 cm	1
Passive friendliness, irrespective of fellow animals' behavior	Avoids conflict at a distance of under 30 cm from partners. Side-to-side or side-to-back position adopted. Does not compete for food. Passive defense shown at a distance of 15-20 cm. Does not enter into fights	2
More communicative, drawn towards the group	Does not enter into conflicts. Tries to limit distance between self and fellows to 20-15 cm. Reacts to approaching partner by paying attention (not on the defensive). Will not fight, even after repeated provocation	3
Initiates contacts	Takes up position close to other group members, edges closer towards them. Avoidance of any form of conflict and rivalry	4
Craves contact, irrespective of fellow animals' reaction	Hungry for contact, follows other members of the group around assiduously, impervious to their threatening behavior	5

Figure 2. The group consisted of 2-3 other cats in addition to
the test cat, serving as reference animals for evaluating the test
cat's behavior and emotional state. The so-called "observer" or
control cat was an essential member of the group; its emotional
display patterns had been fully investigated previously. The hypothalamic structures were implanted with 2-3 nichrome electrodes of

Fig. 2.   Position of experimental animals in control situation, no
          electrical stimulation.

150-180 μ diameter, according to stereotaxic coordinates of the
Jasper and Ajmone-Marsan (1961) atlas.   These were insulated along
their entire length, except at the tip.   The implanting operation was
performed using standard universal stereotaxic equipment under sterile conditions.   The animals were anesthetized by a dose of 40 mg/kg
of 2% Nembutal solution injected intraperitoneally.   The electrodes
were attached to the surface of the skull by dental phosphate cement.
The loose ends of the electrodes were connected up to a small panel
attached to the bone surface, to serve as commutator.

     No experiments involving stimulation were performed for at least
a week after the operation.   The amplitude of the electric impulses
was monitored oscillographically.   The stimulation, lasting 30 sec,
consisted of square impulses (frequency: 300 per sec) of 0.5 msec
duration and a current of 0.2-1 mA.

     Stimulation was applied according to two different schedules.
The first was a weak threshold activation with current amplitude
0.2-0.5 mA.   This stimulation, when the electrode made contact with
the "active points", produced immediate minimal voluntary or emotion-
ally-related involuntary outward signs.   Changes in emotional state
were observed and evaluated (see Figures 3 and 4) after stimulation
of brain sites and could be observed from altered response to test
stimuli and changed interaction between the test animal and other
members of the group.   The method used is described in Section 3.
A stable decrease in electrodermal resistance, lasting for 5-15 min
after the stimulation had ceased, was demonstrated by an uninterrupted galvanic skin response (see Figure 5).   The changes in galvanic skin response correlate with the behavior displayed by the
trial animal when its emotional behavior during interaction with
other members of the group was tested.

Table 18. Measuring Table for Rating Mutually-determined Behavior
           in an Animal Group. b) Conflict (refers to 10th segment
           of circle, see Figure 1)

Pattern of interaction	Main indications	Score
Minimal conflict even when repeatedly provoked	With animals 20-30 cm apart head-to-back position adopted. No rivalry over food. Gives way in contacts with fellow animals. Lifts paw once under repeated provocation without threatening behavior	1
Conflict in response to provocation from fellows	Head-to-side position at a distance of 20 cm. Rivalry over food at a distance of 20 cm. Impending threatening behavior; single attacks on coming into contact with fellow animal	2
Passively hostile, avoidance of contact when fellows display friendly behavior	Head-to-head position when animals are 50-80 cm apart. Food rivalry at a distance of 30-40 cm. Impending threatening behavior at a distance of 30-50 cm. Displays of defensiveness; attacks fellow animals	3
Actively hostile avoidance of contacts	Maximum distance kept between animals. Sitting posture, head-to-head position. Always prevails in rivalry over food. Threatening behavior at a distance of over 50 cm. Positive advances; repeated bouts of fighting	4
Actual aggression and conflict	Threatening behavior shown even at a distance from fellow-animals. Positive advances; repeatedly in prolonged fights. Fight sometimes continues beyond adoption of submissive posture by opponent	5

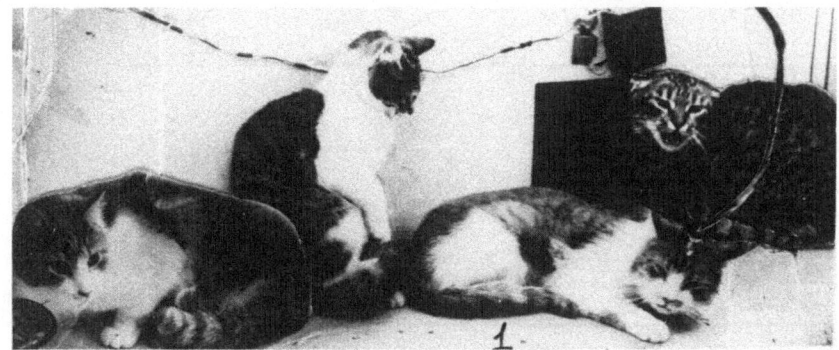

Fig. 3.   Passively defensive fear response develops in the test cat
          (1) resulting from electrical stimulation of the hypothal-
          amus and behavioral response of other members of the group
          (containing different types).

Fig. 4.   Rage response develops in the test cat (1) resulting from
          electrical stimulation of the hypothalamus; fellow-animals'
          response of fear and rejection (2).

        Changes in state produced by weak activation of emotiogenic
brain sites which are too sporadic or too faint to attract the
experimentalist's particular attention cannot be used to gauge shifts
of emotional emphasis. These are, however, meaningful to other
animals in the group. It may thus be seen that the level of display
of affect resulting from electrical stimulation of the hypothalamus
often fails to correspond to the level of emotion as experienced.
This fact is clearly detected by other group members, which adjust
their behavior accordingly (see Figures 6 and 7).

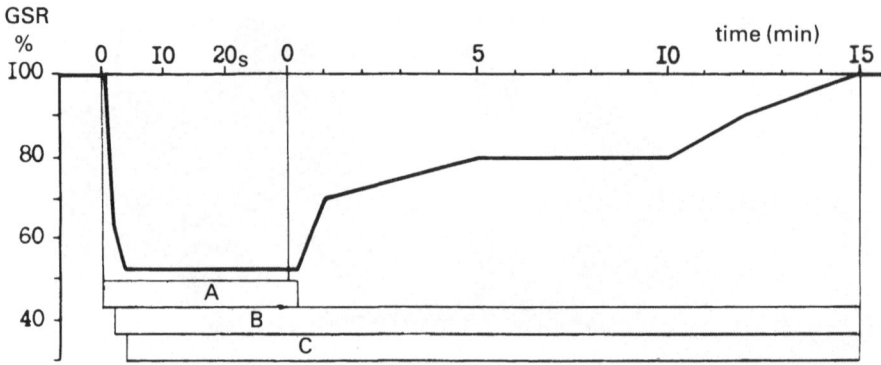

Fig. 5. Dynamics of galvanic skin response following threshold
         stimulation of the hypothalamus. GSR = Galvanic skin
         response expressed as a percentage of the baseline,
         taken as 100. The figure shows a period of hypothalamic
         stimulation of 30 sec duration and a post-stimulation
         period of 15 min duration. The boxes show intensity
         of response at baseline (A = caution), and changes in
         emotional state during and after stimulation (B =
         negativity; C = anxiety/fear).

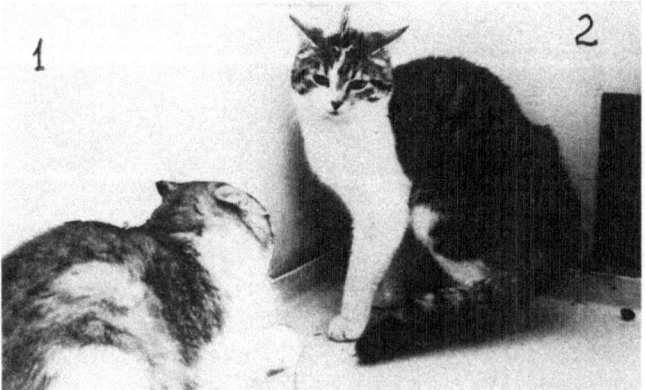

Fig. 6. Fear and passive defense displayed by partner (1) upon
         signs of impending rage and threat in the test cat (2)
         post-stimulation.

        Figure 8 shows how relationships within a group of three cats
change following electrical stimulation of emotiogenic areas of the
hypothalamus. Two males and one female cat were used in this experiment. The female had a conflicting relationship with cat No. 1 and
was friendly towards cat No. 2 originally. Electrical stimulation of

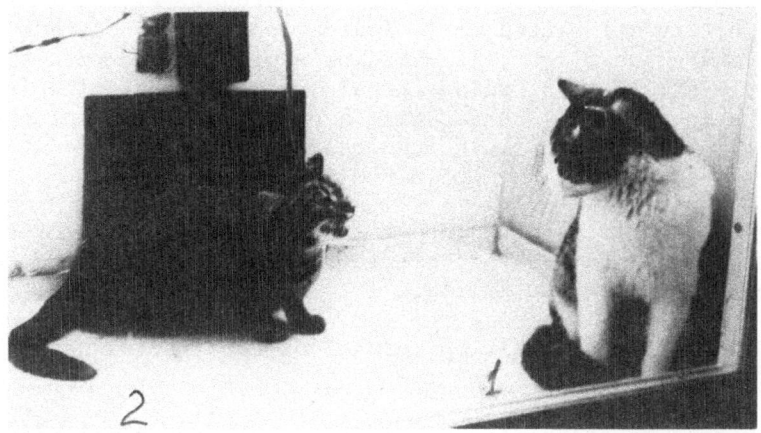

Fig. 7. Partner's response: cautious and on the defensive at
minimal signs of rage in the test cat (2) produced by
electrical stimulation of the hypothalamus.

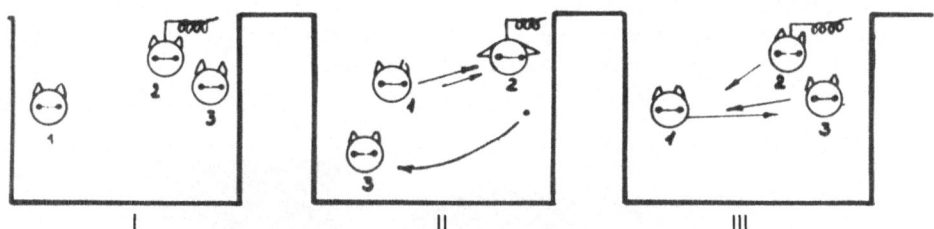

Fig. 8. Altered relationship with group partners following
electrical stimulation of the hypothalamus, where
2 = electrically stimulated male cat; 1 = subordinate
male cat; 3 = female cat. I - before stimulation;
II - when fear is displayed by cat 2, post-stimulation;
III - when cat 2 is restored to its original emotional
state.

the hypothalamus of cat No. 2 (as previously tested out using the
method described in Section 3) potentiated a state of fear, which
in turn provoked the weaker and normally submissive cat No. 1 to
attack. This altered relationship between the two male cats changed
the female cat's tactics; she went to take up a position nearer the
leader. As the bout of induced fear passed and the test cat began
to return to its original emotional state, the female went back to
cat No. 2. Characteristically, both cats attacked and otherwise
showed hostility towards cat No. 1, once the original group pattern
had been re-established. The fainter the outward signs of emotion
experienced noted by the experimentalist, the more meaningful they

appear to be for other animals within the group. Figure 9 illustrates this; the stimulated cat's indications of its current
emotional state are slight; it makes no expressive movements and
keeps fairly still. Its fellow animals, however, unfailingly recognize the increasingly threatening and angry attitude of the test
cat through barely perceptible ethological signs, and adjust their
behavior accordingly.

General negativity was another factor serving to illustrate the
adverse state induced by hypothalamic stimulation. Results for this
state are given below and in Table 19:

- prolonged post-stimulation refusal of food, notwithstanding
  deprivation.

- Suppressed reactions to stimuli and testing devices normally
  producing rewarding response such as pleasure, playfulness,
  hunting and curiosity.

Negativity follows a different pattern of development from rage
or anxiety/fear. It more closely resembles the "punished" state.
Development of negativity may be assessed by an increasingly inhibited response to rewarding test devices to the point of complete
rejection during the post-stimulation period.

Fig. 9. Fear and submissive posture develop in a fellow member
         of the group when the test cat shows minimal signs of
         threatening behavior and aggression resulting from elec-
         trical stimulation of the hypothalamus. 1 - test animal
         immediately (30 secs) after stimulation of the hypothalamus;
         2 - a fellow member of the group - type II according to our
         classification.

Table 19.   Table for Rating General Negativity Produced by a)
            Stimulation of Emotionally Negative "Points" of the
            Hypothalamus, b) Experimentally-produced Aversive
            Situations   (refers to 4th segment of the circle, see
            Figure 1)

Chief groups of signs	Score
Increased latent period between sighting food at a distance of up to 5 m and feeding following 24 h food deprivation; attenuated pleasurable response to stroking (scoring up to 1); muted "playing" response	1
Refuses food, muted response to stroking and exploratory response inhibited. Resists being placed in chamber	2
Actually withdraws from food and avoids stroking, all signs of positive (pleasurable) emotions inhibited. Isolated long drawn out vocalization. Escape from enclosure	3
Rejects food placed in its mouth, defensive reaction against stroking. Actively resists being placed in enclosure	4
Same as above; clearly on the defensive	5

        Negativity may be induced by the repetition of aversive stimuli
appled within the experimental enclosure as well as by electrical
stimulation. Here, the experimental environment itself is blamed as
the cause of the punishment-induced negative emotional state. The
animal therefore reacts defensively when attempts are made to place
it in the experimental enclosure and endeavors to escape at the
earliest opportunity.

1.2. A stronger current of 0.5-1 mA was applied to facilitate further evaluation of the emotional/behavioral response produced by
stimulating a particular "point". Any reaction occurring during
stimulation assumed the character of affect. Three basically different situations could then arise:

  (i) The usual range of outer emotional expression displayed by the
      animal under natural conditions remains unchanged in the case
      of artificially-induced emotion.
 (ii) The cat may display expressive movements which differ from its
      normal range, and may, for example, rush about, leap up, wave
      its paws and sometimes actually fling itself, unawares, at the
      window of the enclosure. Such atypical patterns of emotional
      expression are not deemed indicators of threat, impending attack
      or rage by its fellows. They respond cautiously to behavior
      patterns of this type, remaining attentive without growing
      either defensive or afraid (see Figure 10).

Fig. 10. Displays of emotion simulating "rage" response, but not
          corresponding to any experienced emotional state.

(iii)Activation of a series of hypothalamic "points" may also provoke
     pronounced motor-affective displays which serve no purpose
     whatsoever. Although the movements made by the animal do imi-
     tate extreme states of rage/aggression the animal may, in fact
     be handled when actually under stimulation; its aggression is
     not focused against the aggravating stimulus (see Figure 11).
     This is not followed by any alteration in emotional state. Such
     displays of "false rage" are not looked upon as threatening by
     other members of the group.

     Of course, it would be quite wrong to make judgements on the
nature or intensity of animal emotion solely on the basis of motor
display, whether produced by electrical stimulation of the brain or
occurring spontaneously. The styles of motor activity displayed by
animals, although seemingly varied, are in fact fairly standard, and
may be reduced to such notions as flight, walking, lying, stretching,
etc. However, behind this motor activity the emotion actually being
experienced may vary from one individual of the species to another.
It is wrong for the experimentalist to define such animal movements
arbitrarily as displays of rage, anxiety, fear and so forth. Viewing
an induced emotional state from the animal's own standpoint helps to
avoid the pitfalls of subjectivity and anthropomorphism.

     The method described in Section 3 was used for qualifying and
quantifying emotional state produced in animals by hypothalamic
electrical stimulation. Different "emotional indicators" were
awarded a score or rating for intensity and represented in the form
of segments of a circle.

     Figure 12 depicts the behavior of two animals as an example;
I = before and II = immediately after threshold stimulation of
various points of the hypothalamus. It will be seen by contrasting
the two diagrams illustrating the cats' range of emotional behavior
prior to the experiment that they broadly resembled each other.

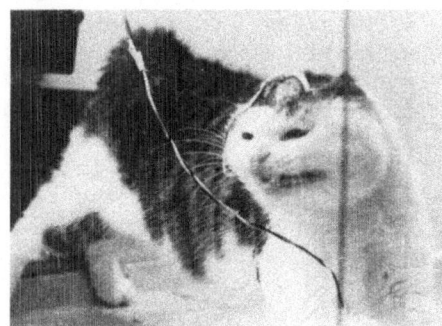

Fig. 11.   Forceful display of motor affect, simulating aggression,
           but not goal-directed.

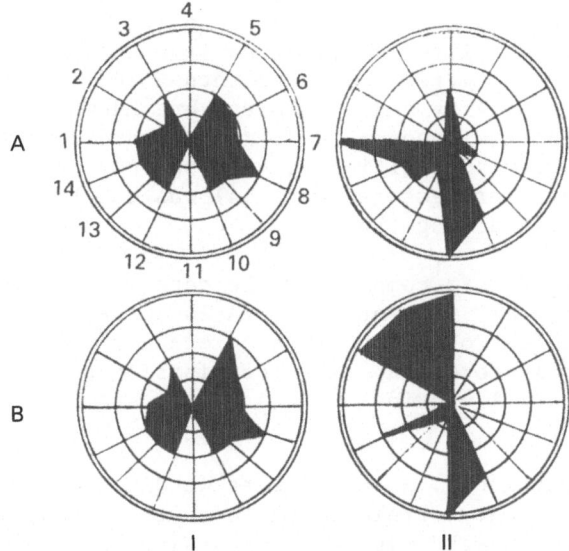

Fig. 12.   Changes of emphasis in emotional behavior; (I) before and
           (II) immediately after stimulation of "points" of the
           hypothalamus (see text).

Activating the hypothalamus produced pronounced discrepancies in
their range of emotional activity.

     A state of frustrated rage resulted from stimulation of the
ventromedial sections of the hypothalamus in one animal (A);   this
upsurge of rage is illustrated in the 1st segment of the circle,
accompanied by almost total disappearance of displays of positive
emotions, such as curiosity, pleasure or contentment, playfulness
and hunting, represented by the 5th, 6th, 7th and 8th segments of
the circle respectively.   Segment 10, standing for conflict in a

social situation, increases, while capacity for reacting to contact
with other animals in a friendly way is eliminated (cf. segment 9).
The activity of the animal in its changed emotional state is clearly
aggression orientated; it is inadequate, however, to meet the requirements of a situation; nor does it match fellow-animals' behavior
patterns (segment 11). The animal's motor activity increases along
with rage (segment 13); but this effect is only visible from its
behavior as and when tests are carried out. Alertness, represented
by segment 14, is sharply aroused, has distinct emotional overtones
and alternates abruptly with displays of aggression.

Activating the dorso-lateral areas of the hypothalamus, as in
the case of cat B, produced marked changes of emphasis in the cat's
emotional behavior. The emotional states of fear (segment 2) and
anxiety (segment 3) not only inhibited the display of all positive
emotions (segment 5, 6, 7 and 8) but also prevented friendly contacts
being established between the test animal and its fellows (segment
9). This was due to a reinforcement of passive unfriendliness and
fear rather than actual aggression. Segments 4 and 11 show respectively pronounced negativity and a complete breakdown of the capacity
to assess a set of circumstances and to react to various tests. The
animal's motor activity and initiative also sharply decreased (segments 12 and 13) but alertness (segment 14) was heightened due to
increased tension felt during the trials.

Once observations were completed, the cats were put to sleep by
a large dose of Nembutal. A constant 5 mA 20-30 sec current was
passed through the electrodes to act as an electrolytic marker and
the brain was fixed in formalin. The localization of the electrodes
was determined in serial sections of the brain, stained by Nissl's
stain.

It should be mentioned, however, that we did not manage to
demonstrate any clear-cut correlation between the localization of the
tip of the electrode and the type of emotional/behavioral response
produced. The set of physical and mental signs developing, firstly
during stimulation and secondly - and more importantly - maintained
following stimulation, are caused by activation of both the brain
area in close proximity to the electrode and of connected pathways.
In work involving electrical stimulation of the hypothalamus it is
advisable to implant the cat with several electrodes and then select
for further stimulation the "point" which has given the most definite
reaction.

The method we have described for making an objective assessment
of ethological emotional/behavioral response in animals involved in
group activity (in normal individuals, as well as those subjected to
electrical stimulation of the brain) offers new prospects for studying the psychopharmacological characteristics of different pharmaceutical agents. Results of these analyses are given in the articles
starting on pages 79 and 101.

REFERENCES

Boissier, J. R., 1965, Situation libre et psychotropes, in: "Pharm-
        acology of Conditioning, Learning and Retention," M. Michelson
        and V. Longo, eds., Pergamon Press, Oxford - London -
        Edinburgh - New York - Paris - Frankfurt, pp.25-46.
Delgado, J., 1966, Emotions, Self-selection psychology textbook,
        Iowa.
Hinde, R. A., 1970, "Animal Behavior. A Synthesis of Ethology and
        Comparative Psychology," Second ed., McGraw-Hill Book Co., New
        York.
Hoffmeister, F., and Wuttke, W., 1969, On the action of psychotropic
        drugs on the attack and aggressive-defensive behaviour of mice
        and cats, in: "Aggressive Behaviour," S. Garatini and E. B.
        Sigg, eds., Excerpta Medica, Amsterdam, pp.273-283.
Kozlovskaya, M. M., and Valdman, A. V., 1977, Brain stimulation
        provoked emotional reactions of cats in a social situation,
        Act.Nerv.Super.(Praha), 19:221-222.
Lat, J., 1965, The spontaneous exploratory reactions as a tool for
        psychopharmacological studies, in: "Pharmacology of Condition-
        ing, Learning and Retention," M. Michelson and V. Longo, eds.,
        Oxford, London, Edinburgh, New York, Paris, Frankfurt, Prague,
        Pergamon Press, pp.47-66.
Leyhausen, P., 1960, Verhaltenstudien an Katzen, Berlin-Hamburg.
Norton S., 1969, The effects of psychoactive drugs on cat behaviour,
        Ann.N.Y.Acad.Sci., 158:915-927.
Norton, S., and De Beer, E. J., 1956, Effects of drugs on the
        behavioural patterns of cats, Ann.N.Y.Acad.Sci., 65:249-257.
Tembrook, G., 1961, Verhaltensforschung, Jena.
Valdman, A. V., 1972, Essential problems of studying emotional behav-
        iour on animals, in: "The Experimental Neuropsychology of
        Emotions, A. V. Valdman, ed., Nauka, Leningrad, pp.6-26.
Valdman, A. V., 1980, Preclinic prediction of the spectrum of psycho-
        tropic activity of tranquilizers, Pharmacol.Res.Commun.,
        12:225-266.
Valdman, A. V., and Kozlovskaya, M. M., 1965, The influence of
        neurotropic drugs upon various types of diencephalic behav-
        ioural reactions in relation to intensity and localization of
        the stimuli, in: "Pharmacology of Conditioning, Learning and
        Retention," M. Michelson and V. Longo, eds., Pergamon Press,
        Oxford - London - Edinburgh - New York - Paris - Frankfurt,
        pp. 327-337.
Valdman, A. V., and Kozlovskaya, M. M., 1972, Psychopharmacological
        analysis as a method of analytical neurophysiology of
        emotions, in: "The Experimental Neuropsychology of Emotions,"
        A. V. Valdman, ed., Nauka, Leningrad, pp. 211-243.
Valdman, A. V., Zwartau, E. E., and Kozlovskaya, M. M., 1976, Psycho-
        pharmacology of Emotions, Meditsina, Moscow.
Valdman, A. V., and Kozlovskaya, M. M., 1977, Psychotropic drug
        effects on the emotional state of cats in a social situation,
        Act.Nerv.Super. (Praha), 19:222-224.

Valdman, A. V., Kozlovskaya, M. M., and Medvedev, O. S., 1979,
        Pharmacological Regulation of Emotional Stress, Meditsina,
        Moscow.
Valdman, A. V., Zvartau, E. E., and Kozlovskaya, M. M., 1975, Exper-
        imental study of the action of psychotropic drugs on emotions,
        motivations and social behaviour of animals, in: "CNS and
        Behaviour Pharmacology," Vol,. 3, Proc. 6th Intern. Congr.
        Pharmacol., M. Airaksinen, ed., pp. 297-211.
Wuttke, W., 1971, Das Spontanverhalten der Katze und seine -
        Beeinflussung durch Psychopharmaka, Arzneimittel-Forsch.,
        21:2059-2068.
Wuttke, W., and Hoffmeister, F., 1968, Veranderungen des Spontanver-
        haltens der Katz durch Chlordiazepoxid and Chlorpromazin,
        Naunyn-Schmiedeberg's Arch.Pharmak.Exp.Path., 260:221-222.

Factor Analysis of Emotional/Behavioral Reactions in Cats of Different Types

A.V. Valdman and E.B. Katkova

Institute of Pharmacology, Academy of Medical Sciences
of the USSR, Moscow

Tranquilizers are not found to exert any specific anxiolytic effect, even when dealing with overall trends in large groups of up to 200 healthy volunteers (Di Mascio, 1969). However, by dividing the whole population under investigation into sub-groups, according to anxiety level, a clear-cut tranquilizing effect may be seen in highly anxious individuals (De Mascio, Barrett, 1965). Investigations into the influence of tranquilizers on a number of physical and mental parameters showed (Nakano et al., 1978) that they increase speed of accomplishing a task while the number of mistakes made drops with highly anxious volunteers; the level of task fulfillment falls, however, in those with a low anxiety level. These and similar findings clearly indicate the part played by personal characteristics and individual level of emotional response in observed psychopharmacological effect. Hence the importance of choosing the right group of animals at the experimental stage of pre-clinical evaluation of psychotropic drugs; they should be physically and mentally suited to illustrate a specific effect. The averaging out of experimental data obtained in an entire animal group often conceals one of the unknown effects of a pharmacological compound. Many years of experience in experimental psychopharmacology have convinced us of the need for preliminary classification of the animals into more or less homogeneous sub-groups (Valdman et al., 1976). Even in the case of genetically pure lines, e.g. of rats and mice, some individual animals show quite considerable variations in their emotional and behavioral response and their stability under stress. This applies even more to large laboratory animals, such as cats and rabbits, where individuality is of paramount importance.

Animal behavior is determined partly by genetic factors, which contribute the basic repertoire of the animal's actions and postures,

39

and are common to the strain as a whole, and partly by individual
factors, such as the animal's history, accommodation, learning ex-
periences, etc. These are peculiar to each animal and help to govern
its responses, etc. That is why a full investigation of an animal's
initial style of emotional/behavioral response is essential for
psychopharmacological research. Wuttlke (1971) emphasized the need
to make distinctions between the individual behavior patterns
dominant in each cat.

By reviewing the wealth of experimental material accumulated
over 10 years of work on cats, Kozlovsky succeeded in dividing these
animal's individual and group emotional/behavioral response patterns
into a number of basic types (see the first article). Further
analysis was required, however, in order to answer the questions:
how real are the differences between the various types of emotional
response and how are they produced? How does the inherent pattern
of emotional behavior influence the "rank" of various types? The
statistical factor analysis method was adopted for continuing work
in the field.

1. CORRELATION OF EMOTIONAL/BEHAVIORAL RESPONSE IN TYPE I AND II CATS

The material used for statistical analysis was provided by the
measuring tables given in the first article. These scales showed the
intensity of various types of emotional behavior displayed in 20 type
I and 20 type II cats (see Tables 1 and 2 respectively). Male cats
aged 2-3 years were used, which have spent a prolonged period of 2-3
years under observation in a vivarium. A score was awarded to rate
particular features of individual and group behavior by the method
given in the first article.

The results for each group of animals were pooled and the mean
determined. The symbol M represented the mean and m - the standard
error of the mean. Table 3 presents values for both groups of
animals combined. The column headed "positive emotions" gives total
scores awarded for displays of pleasure or contentment, playfulness
and exploratory behavior (curiosity). The only real quantitive
discrepancies observed between type I and type II animals' responses
lie in the differing intensity of the "positive" and "negative"
emotions displayed (negative = fear and anxiety) as well as level of
motor activity.

The correlation analysis (determining the correlation coefficient by paired comparison of individual indices of emotional/
behavioral response) was carried out using a CORR program (Bolch,
Huang, 1974) on an EC 1020 computer.

The correlation analysis brought out a number of common points
in the responses of both types of cat. Positive correlations were

Table 1. Intensity of Different Types of Emotional Behavior Displayed by Type I Animals in Control Observations

No. of experiment	Animal No.	Aggression	Fear	Anxiety	Positive emotions*	Hunting	Friendliness	Conflict	Motor activity	Alertness (orientation)
1	K-8	0.5	0.5	1.0	6.0	3.0	2.0	1.0	2.5	2.0
2	K-5	3.5	0.5	1.0	2.0	1.0	0	4.0	2.0	3.0
3	K-17	2.5	0.5	0.5	4.0	3.0	0	3.0	2.0	2.0
4	K-10	1.0	1.0	1.5	5.0	2.0	2.0	1.0	2.0	2.0
5	K-5	3.0	1.0	1.0	3.0	1.0	0	3.0	2.0	3.0
6	K-9	0.5	0.5	1.0	9.0	0	3.0	0.5	2.0	1.5
7	K-8	0.5	0.5	0.5	6.0	3.0	2.0	1.0	1.5	1.5
8	K-17	2.5	0.5	0.5	4.0	3.0	1.0	3.0	2.0	2.0
9	K-7	0.5	0.5	1.0	6.0	2.0	3.0	1.0	3.0	2.0
10	K-21	1.5	1.5	1.5	3.0	3.0	1.0	1.5	2.0	2.0
11	K-15	1.0	1.0	2.0	3.0	1.0	1.0	1.0	3.0	3.0
12	K-24	3.0	0.5	0.5	6.0	3.0	1.0	3.0	2.0	2.0
13	K-19	0	1.0	1.0	3.0	2.0	1.0	0.5	2.0	2.0
14	K-21	1.5	1.0	1.5	8.0	3.0	2.0	1.5	4.0	3.0
15	K-25	0	0.5	1.0	5.0	0	2.0	2.0	2.0	2.0
16	K-28	1.0	1.0	2.0	4.0	2.0	1.0	1.0	3.0	3.0
17	K-12	1.0	0.5	1.0	8.0	3.0	3.0	1.0	3.0	2.0
18	K-33	3.5	0.5	0.5	3.0	1.0	0	3.5	2.0	2.0
19	K-3	0	1.0	1.0	4.0	1.0	1.0	0.5	2.0	2.0
20	K-18	1.0	0	0	6.5	3.0	2.0	1.0	3.0	2.0
M		1.4	0.7	1.0	4.9	2.0	1.4	1.6	2.4	2.2
D		1.38	0.12	0.26	3.8	1.2	0.98	1.38	0.37	0.25
m		0.26	0.08	0.11	0.44	0.24	0.22	0.26	0.14	0.11

Figures are the average of scores awarded for emotional behavior displayed by an individual animal versus a set of controlled observations (details of method given in the first article).
* = total score for exploratory behavior, pleasure/contentedness and playfulness.
M = arithmetical mean; D = deviation; m = standard error of the mean.

Table 2. Intensity of Different Types of Emotional Behavior Displayed by Type II Animals in Control Observations

No. of experiment	Animal No.	Aggression	Fear	Anxiety	Positive emotions*	Hunting	Friendliness	Conflict	Motor activity	Alertness (orientation)
1	K-13	2.5	1.0	1.5	0	0	0	3.5	1.0	2.0
2	K-3	1.0	3.0	2.0	1.0	1.0	1.0	2.0	1.0	2.0
3	K-10	0.5	3.0	3.0	1.0	0	1.0	1.0	3.0	3.0
4	K-3	1.0	3.0	3.0	1.0	0	1.0	2.0	1.0	3.0
5	K-14	0.5	2.0	2.0	2.0	1.0	1.0	1.0	2.0	2.0
6	K-10	0.5	3.0	3.0	1.0	0	1.0	1.0	3.0	3.0
7	K-6	0	3.0	3.5	1.0	0	1.0	1.0	1.0	3.0
8	KK-2	2.0	0.5	1.0	3.0	3.0	1.0	3.0	3.0	2.0
9	K-4	0	1.0	1.0	1.0	0	1.0	0	1.0	1.5
10	K-14	0	2.0	2.0	3.0	1.0	1.0	1.0	2.0	2.0
11	K-Mp	0	0.5	1.0	0	0	1.0	0	1.0	1.5
12	K-20	0	2.0	2.0	2.0	1.0	1.0	1.0	1.0	2.0
13	KK-2	2.0	1.0	1.0	2.0	3.0	1.0	3.0	3.0	3.0
14	K-22	0	3.0	3.5	0	1.0	1.0	0	2.0	3.0
15	K-38	2.0	0.5	1.0	2.0	3.0	0	3.0	2.0	2.0
16	K-4B	2.0	1.0	1.0	1.0	0	0	3.0	2.0	2.0
17	K-34	2.0	1.0	1.0	2.0	3.0	0	3.0	2.0	3.0
18	K-30	0	0.5	1.0	0	0	1.0	0	1.0	2.0
19	K-31	0	2.0	2.0	1.0	1.0	1.0	1.0	2.0	3.0
20	K-32	0	2.5	2.5	0	0	1.0	0	2.0	3.0
	M	0.8	1.8	1.9	1.2	0.9	0.8	1.5	1.8	2.4
	D	0.85	0.01	0.83	0.91	1.36	0.17	1.5	0.59	0.33
	m	0.21	0.22	0.20	0.21	0.26	0.09	0.27	0.17	0.13

Figures are the average of scores awarded for emotional behavior displayed by an individual animal versus a set of control observations (details of method given in the first article).

* = total score for exploratory behavior, pleasure/contentedness and playfulness.

M = arithmetical mean; D = deviation; m = standard error of the mean.

Table 3. Correlation Coefficient between the Emotional/behavioral Response Patterns Displayed by Types I and II cats

Type of Response	Ag-gression	Fear	Anxiety	Positive emotions	Hunting	Friend-liness	Conflict	Motor activity	Alert-ness (orien-tation)
I M	1.4	0.7	1.0	4.9	2.0	1.4	1.6	2.4	2.2
±									
n=20 m	0.26	0.08	0.11	0.44	0.24	0.22	0.26	0.14	0.11
II M	0.8	1.8*	1.9*	1.2*	1.0	0.8	1.5	1.8*	2.4
±									
n=20 m	0.21	0.22	0.20	0.21	0.36	0.17	0.27	0.17	0.13

M = mean value of score obtained for intensity of display.

m = standard error of the mean.

* = difference from type I (significant where $p < 0.05$).

n = No. of animals.

revealed for both types: between anxiety and both fear and alertness, between alertness and motor activity and also between hunting
and the positive emotions. Negative correlations are shown between
conflicting and friendly relationships within the group. Table 4
conveys the numerical significance of the correlation coefficients of
responses displaying emotional behavior in types I and II.

Apart from such overall interconnections between disparate
signs, specific correlation clusters also become apparent, peculiar
to only one of the two types. A characteristic response pattern for
type I animals would be: a positive correlation between positive
(pleasurable) emotions and friendly intraspecies contact (r = +0.839)
and a negative correlation between positive emotions and aggression
(r = -0.385) or conflict (r = -0.445). Conflict correlates positively with aggression (r = +0.977); fear is negatively correlated
with friendliness.

A negative correlation between anxiety and fear on one hand and
aggression and hunting on the other characterize type II animals. A
positive correlation is shown between fear and friendliness (r =
+0.459) as well as hunting and aggressivity (r = +542). It is significant that in these animals positive emotions have no particular
correlation with friendliness (friendly contacts made within the
group) while they do correlate closely with conflict (r = +0.442).

Correlation analysis thus serves to bring out marked differences
in how particular mental and physical phenomena of individual and
group behavior interrelate. The different emotional/behavioral
responses described in the first article are now shown to have a
factual basis.

2. FACTOR ANALYSIS OF CORRELATION MATRICES FOR CATS WITH
TYPES I AND II EMOTIONAL/BEHAVIORAL RESPONSE PATTERNS

Factor analysis enabled us to present the pattern of correlation
matrices in the form of factors or components determining behavior
patterns. This helps us to understand the nature of emotionally-
related behavioral status amongst various animals. Using factor
analysis we were able to divide into three headings the unrelated
factors which form the kernel of type I and II emotional response
patterns and account for up to 80% of total deviation. These factors
are given in Tables 5 and 6, together with factor loadings for all
displays of emotional behavior.

The following factors characterize emotional response in type I:

Factor 1 includes displays of positive emotions, friendliness
and alertness marked with a plus, and aggression or conflict preceded
by a minus sign. This helps to demonstrate the dominance of positive

Table 4. Correlation Coefficient between the Emotional/Behavioral Response Patterns Displayed by Types I and II Cats

Type I cats (n=20)

Type II cats (n=20)	Ag-gression	Fear	Anxiety	Positive emotions	Hunting	Friend-liness	Conflict	Motor activ-ity	Alert-ness (orien-tation)
Ag-gression		-0.111	+0.261	-0.385	0.131	-0.683**	+0.977**	-0.168	+0.350
Fear	-0.447*		**+0.753****	-0.390	-0.072	-0.249	-0.151	+0.025	+0.373
Anxiety	-0.478*	**+0.935****		-0.157	-0.238	-0.052	-0.327	+0.379	+0.567**
Positive emotions	+0.287	-0.170	-0.249		+0.511*	+0.836**	-0.445*	+0.397	-0.388
Hunting	+0.542*	-0.401	-0.431	+0.683**		+0.098	+0.125	+0.241	-0.147
Friend-liness	-0.736**	+0.459*	+0.436	+0.027	+0.264		-0.711**	+0.365	-0.446*
Conflict	+0.957**	-0.334	-0.390	+0.442	+0.568**	-0.690**		+0.235	+0.301
Motor activity	+0.274	+0.041	+0.045	+0.448*	+0.447*	-0.033	+0.245		**+0.495***
Alertness (orien-tation)	-0.039	+0.595**	+0.631**	-0.106	+0.063	+0.134	+0.015	**+0.449***	

$* < 0.05$; $** < 0.01$; **Bold type** = correlations common to both types of cat.

Table 5. Results of Factor Analysis on Correlation Matrices for
          Animals with Type I Emotional Response Pattern

Emotional behavior displayed	Factors		
	1	2	3
	Weight		
	43.4%	23.3%	13.8%
Aggression	−0.641*	+0.401	+0.588*
Fear	−0.194	−0.782*	−0.118
Anxiety	+0.037	−0.950*	+0.058
Positive emotions	+0.940*	+0.184	+0.221
Hunting	+0.230	+0.328	+0.337
Friendliness	+0.945*	−0.058	−0.100
Conflict	−0.675*	+0.520*	+0.458*
Motor activity	+0.413	−0.570*	+0.628*
Alertness (orientation)	+0.564*	−0.589*	+0.566*

* = factor loadings significant where $p < 0.05$.

Table 6. Results of Factor Analysis on Correlation Matrices for
          Animals with Type II Emotional Response Pattern

Emotional behavior displayed	Factors		
	1	2	3
	Weight		
	43.4%	23.3%	13.8%
Aggression	+0.772*	−0.273	+0.481*
Fear	−0.626*	+0.604*	+0.311
Anxiety	−0.676*	+0.561*	+0.326
Positive emotions	+0.783*	+0.531*	−0.250
Hunting	+0.897*	+0.104	+0.167
Friendliness	−0.549*	+0.537*	+0.450*
Conflict	+0.794*	−0.123	+0.468*
Motor activity	+0.382	+0.618*	+0.298
Alertness (orientation)	−0.262	+0.510*	+0.718*

* = factor loadings significant where $p < 0.05$.

emotions, with a factor loading of 0.940, and friendliness (0.945) and
the negative correlation between aggression, conflict and positive
emotions. The low factor loading for negative emotions stresses
their independence from other emotions displayed. This factor is
general for type I and represents 43.4% of total deviation. We have
defined this factor as "positive emotion-dominated".

Factor 2 lists fear, anxiety, motor activity and alertness with a minus sign. It characterizes the independent but secondary significance of negative emotions and their correlation with displays of motor activity and alertness. This factor accounts for 23.3% of the total deviation.

Factor 3 includes aggression, conflict and alertness, marked with a plus sign; this indicates their interrelatedness; i.e. that animals of this type show an actively defensive type of aggression, and accounts for 13.8% of total deviation.

The following factors characterize emotional response patterns in type II cats:

Factor 1 includes displays of fear, anxiety and friendliness marked with a minus sign and displays of positive emotions, hunting, conflict and aggression with a plus. This indicates a) the high interdependence between positive and negative emotions, b) the predominance of negative emotions and c) the inadequate readjustments in relationships between positive emotions, friendliness, aggression and conflict. This factor is general for type II emotional response and may be regarded as dominated by negative emotions and inadequacy of emotional response. It accounts for 43.4% of total deviation.

Under factor 2 both negative and positive emotions are preceded by a plus sign, together with friendliness, motor activity and alertness. In view of the generally low level of activity shown by these animals' behavior, this factor may indicate a trend towards a passive type of emotional reaction, expressed by withdrawal and paralyzing fear, as well as lowered differentiation between emotions displayed. It accounts for 21.8% of total deviation.

Factor 3 - aggression, conflict and alertness - is marked with a plus. The independence of displays of conflict is shown, and the absence of any correlation between the latter and motor activity may indicate that the aggression displayed is of a passively defensive type.

To recap, the above method indicates the pattern of correlations of emotional behavior displayed by type I and II animals' emotional response patterns. The main factors which emerge from factor analysis bring out the chief qualitative differences between the two types:

Type I is marked by the independence of the main types of emotion displayed and the predominance of positive type emotions in the pattern of correlations.

Type II is marked by the predominant role of negative emotions in the pattern of correlation matrices and by a certain inadequacy of

emotional response as well as a tendency towards passively defensive
behavior patterns.

The above results show that the emotional behavior originally
ascribed to an animal is an incomplete indicator of its emotional/
behavioral rank. A thorough statistical analysis led us to recognize
the possibility of and the need for classifying animals into types.
The following guidelines for performing psychopharmacological trials
at the pre-clinical stage may help to demonstrate a tranquilizer's
range of emotiotropic activity:

1. Animals exhibiting type II emotional/behavioral response pattern
   are the most suitable subjects for investigating the emotiotropic action of tranquilizers.
2. Correlation analysis of interrelatedness of displays of emotional behavior aids objective differentiation between animals
   according to their emotional response pattern.
3. Correlation analysis may be found a suitable method for use in
   evaluating the effect of pharmacological compounds on the structure of animals' innate emotional/behavioral "status".
4. The emotiotropic or therapeutic effect of tranquilizers should
   be judged by comparing and contrasting the readjusted correlations amongst (treated) type II animals against the correlation matrix of those with type I emotional/behavioral
   response patterns.

REFERENCES

Bolch, B. W., and Huang, C. J., 1974, "Multivariate statistical
        methods for business and economics", Prentice-Hall Inc.,
        Englewood Cliffs, New Jersey.
Di Mascio, A., 1969, The use of "normal" in predicting clinical
        utility of psychotropic drugs, in: "The Psychopharmacology of
        the Normal Human," W. O. Evans and N. S. Kline, eds., N.Y.,
        pp. 114-125.
Di Mascio, A., and Barrett, Y., 1965, Comparative effects of oxazepam
        in high and low anxious volunteers, Psychosomatics, 6:298-303.
Nakano, S., Ogawa, N., Kawasu, Y., and Osata, E., 1978, Effects of
        anti-anxiety drugs and personality on stress inducing psycho-
        motor performance test, J.Clin.Pharmacol., 18:125-130.
Uberla, K., 1977, Faktorenanalyse, New York.
Valdman, A. V., Zvartau, E. E., and Kozlovskaya, M. M., 1976,
        "Psychopharmacology of Emotions," Meditsina, Moscow.
Wuttke, W., 1971, Das Spontanverhalten der Katze und seine Beein-
        flussung durch Psychopharmaka, Arzneimit-Forsch, 24:2059-2068.

Hereditary Traits of Animals with Different Types of Reaction to Stress and Benzodiazepine Tranquilizers

S.B. Seredenin, Yu.A. Blednov, B.A. Badyshtov
and N.M. Shevchenko

Laboratory of Pharmacological Genetics, Moscow

1. INTRODUCTION

Individual reactions to benzodiazepines have been widely discussed in the literature since these tranquilizers were first accepted into clinical practice (Aleksandrovsky, 1973, 1976). Neither pharmacokinetics nor other pharmacological approaches already clinically approved for other groups of medicaments have provided a reliable means of predicting their effects to date (Dasberg, 1974; Ghoneim et al., 1981). This state of affairs appears to be due to benzodiazepines' broad pharmacological profile, hampering attempts to account for their varying effects. Further research is obviously required to improve our understanding of them.

Nowadays it is accepted that individual reactions to drugs are closely bound up with genetic factors (Vogel and Motulsky, 1979; Seredenin et al., 1982). Genetically-produced differences in pharmokinetics and drug metabolism are the most commonly studied aspects (Vesell, 1971). A number of drugs have side-effects due to hereditary deficiency of certain enzymes, potentiating their pharmacodynamic effect. The majority of pharmacogenetic phenomena described involve monogenic control (WHO, 1973). It is much more difficult to demonstrate hereditary traits influencing an observed pharmacological effect where polygenic control is concerned. There is no doubt, however, that differences in polygenically controlled mechanisms, potentiating the pharmacodynamics and pharmacokinetics of medicaments, are all-important for determining the end-response to drug treatment. There are thus ample grounds for taking genetic predisposition as a factor in the action of benzodiazepine tranquilizers, since it is known that reaction to stress (their target area of action) (Cook and Sepinwall, 1975) is genetically determined in various animal species (Borodin et al., 1976; Markel and Borodin, 1978; Marple, et al., 1972; Lankin et al., 1979, 1980; Broadhurst, 1975). Variations on the same theme of individual reaction to stress have also been described in man (Frankenhauser, 1970).

This means that laboratory studies of the genetic variations in the biochemical indicators of stress reactions may be investigated for their applicability to stress response in man (see Vavilov, (1920) on general biological law of homologous series in genetic variation). Indications drawn from cases where pharmacological correction of stress is shown to be governed specifically by genetic variations serve as predictors of a trial drug's effects.

The present work therefore aims:

(i) to show how the action of benzodiazepine derivatives on inbred animals reacting differently to stress is genetically determined.

(ii) to seek out genetic differences in the biochemical indicators of stress. The biological transformation and the specific binding of benzodiazepine derivatives were also studied genetically, with a view to evaluating the part these processes play in observed variations in tranquilizer effect.

2. METHODS

2.1. EXPERIMENTAL ANIMALS

Our experiments were performed on male mice of strains C57BL/6 (= B6), BALB/c (= c) from the "Stolbovaya" Acad. Med. Sci. breeding center (USSR) and reciprocal hybrids F_1, F_2 and F_1 x c obtained in the laboratory, weighing 18-20 g. The animals were kept on a standard diet on a 12-hour day-night cycle with 10 animals to a cage, and housed in a separate vivarium for one month prior to each set of experiments. The research was carried out from 10 h to 14 h daily from October to January. During the intervening period the controls revealed no real differences in the parameters being studied. The results obtained for the F_1 reciprocal hybrids were similar (data had been presented for the B6 x c combination).

2.2. MATERIALS

The following tranquilizers were used in the experiments: phenazepam, trihydroxyphenazepam, ^{14}C-phenazepam (activity 1 Ci/mole) produced by the Odessa Institute of Physical Chemistry, Acad. Sci. (Ukraine), diazepam from Gideon Richter (Hungary), ^{3}H-diazepam (2-methyl-^{3}H, 71 Ci/mmole) from Amersham, England.

2.3. OPEN FIELD METHOD

Animal behavior was investigated using the "open field" method
described by Borodin et al. (1976) Levels of general, horizontal,
peripheral, central and vertical locomotor activity were determined.
Drugs were given intraperitoneally, in an emulsion with Tween-80:
phenazepam in doses of 0.05, 0.075 and 0.1 mg/kg, and diazepam in
doses of 5.0, 10.0 and 20.0 mg/kg.

2.4. INVESTIGATION OF THE ANTICONVULSANT ACTION OF TRANQUILIZERS

Anticonvulsant action was estimated by the pentylenetetrazole
titration test. Doses of 0.875, 1.75, 3.5, 7.0 and 14.0 mg/kg phenazepam were given in suspension with Tween-80. After an interval of
30 min pentylenetetrazole was administered into the caudal vein and
the dosage producing convulsions was noted.

2.5. INVESTIGATING THE METABOLISM OF ^{14}C-PHENAZEPAM

The metabolites of ^{14}C-phenazepam were determined radiochromato-
graphically, using methods described previously (Golovenko et al.,
1979).

2.6. INVESTIGATING THE BIOCHEMICAL INDICATORS OF STRESS

Certain biochemical indicators of stress were noted in the
plasma of the animals, viz. ACTH by a radioimmunoassay (Berson and
Yalow, 1968), using reagents from CIS Intern. of France, Corticosterone, by Murphy's (1967) method, cyclic AMP by the method of Brown
et al. (1974) using biochemical kits from Amersham and cyclic GMP by
a radioimmunoassay method using reagents from the same source
(Steiner et al., 1972). Concentrations of hormones and nucleotides
in the animals' plasma were measured a) at varying intervals after
the open field experiments, b) straight after they were removed from
the cage and decapitated and c) in the controls - after all handling
involved in the experiment except when put in the open field.

2.7. INVESTIGATION OF SPECIFIC BINDING OF ^{3}H-DIAZEPAM

^{3}H-diazepam binding was investigated in homogenates of the
cerebral cortex, mid-brain, diencenphalon, cerebellum and brain stem
using the Mohler and Okada (1977) method. Tissues were homogenized
in 50 mM tris-HCl buffer, pH 7.4. Concentration of the homogenates:
10 mg/ml. Final concentration of the added ^{3}H-diazepam - 3 nM.
Volume of incubation mixture: 1 ml. Incubation took place at 0°C and
lasted 30 min. Unlabelled diazepam in a final concentration of 3 µM
was added to estimate non-specific binding in the incubation mixture.

After trial incubation was complete the mixture was filtered through glass filters GF/B (Whatman), and washed twice with 5 ml of cold buffer solution. The filters were then placed in scintillation fluid consisting of toluene:Triton X-100, 2:1; PPO - 0.4%; POPOP - 0.01% and left in the dark for 16 h. Radioactivity was measured on an LS-100 (Beckman) liquid scintillation counter.

2.8. STATISTICAL EVALUATION

The validity of the results obtained was estimated according to Student and Fisher tests and the number of genes was calculated using the Serebrovsky formula (Rokitsky, 1978). The dispersion analysis method (Seredenin et al., 1981) served to estimate interstrain differences in the speed of ^{14}C-phenazepam metabolism.

3. RESULTS

3.1. THE EFFECTS OF PHENAZEPAM AND DIAZEPAM ON THE BEHAVIOR OF B6 and c MICE IN "OPEN FIELD" TESTS

Data obtained for mouse behavior in open field test show that B6 mice respond actively to exposure to stress, whereas c mice respond by a freezing reaction. Clear-cut inter-strain differences were noted in their response to tranquilizers. Phenazepam produced a dose-dependent reduction in locomotor activity in B6 mice, while c mice displayed biphasic activity; a dose of 0.05 mg/kg of the drug exerted a stimulating effect on their behavior but their locomotor activity level was again reduced by a 0.1 mg/kg.

Diazepam exercized a similar effect, only differing in the wider dose range at which the tranquilizer increased locomotor activity in c mice (see Table 2).

The behavior of F hybrids in an open field was comparable to that of the parent strain B6; similarly, their reaction to phenazepam was an inherited trait. The tranquilizer produced a dose-dependent decrease in locomotor activity in F hybrids as it did in B6 mice (see Table 3).

It is thus evident that the action of benzodiazepines is not identical in animals of different genetic strain, but depends on their pattern of reaction to stress. If B6 are crossed with c mice, active behavior in initial open field trials is dominant and the same type of tranquilizer effect is seen as with B6 mice (see Table 3).

These investigations enabled us to determine the minimum number of genes influencing the shaping of initial response to a stress

Table 1. Effect of Phenazepam on the Behavior of Mice of C57BL/6 and BALB/c Strain in the "Open Field" Test (M±SEM)

Strain of Mouse	Locomotor activity	Control	Dose given (mg/kg)			
			0.05	0.075	0.1	
C57BL/6	General	126.4±7.8 (25)	111.5±7.9* (15)	92.4±6.7** (15)	58.5±9.8*** (15)
	Horizontal	112.7±7.3	101.1±7.2*	84.1±8.6**	55.6±8.4***
	Peripheral	83.8±3.8	87.2±6.9	72.5±7.9*	49.6±7.7***
	Central	28.9±3.8	16.1±2.4**	10.6±2.3**	8.5±1.7**
BALB/c	General	24.5±3.6 (27)	39.5±6.9* (27)	32.7±11.7 (15)	14.8±1.8** (15)
	Horizontal	24.5±3.0	39.5±6.9*	32.7±11.7	14.8±1.8**
	Peripheral	23.5±4.0	38.9±7.1*	32.4±11.6	14.4±1.9
	Central	0.9±0.5	0.5±0.4	0.2±0.1	0.4±0.2

* $p < 0.05$.
** $p < 0.01$.
*** $p < 0.001$.
In brackets - number of animals.

Table 2. Effect of Diazepam on the Behavior of C57BL/6 and BALB/c Strain in the "Open Field"
          Test (M±SEM)

Strain of mouse	Locomotor activity	Dose given (mg/kg)			
		Control	5.0	10.0	20.0
C57BL/6	General	118.2±13.3 (15)	113.3±11.9 (15)	70.2±10.3** (15)	65.2±11.2** (15)
	Horizontal	106.8± 9.7	107.8±11.6	69.3± 9.9**	63.9±10.8**
	Peripheral	93.9± 6.8	94.9±10.6	66.7± 9.6*	49.1±10.5**
	Central	12.9± 3.0	12.9± 2.1	2.6± 0.9**	4.9± 1.0*
BALB/c	General	17.6± 3.6 (17)	18.2± 3.5 (17)	28.5± 3.9* (17)	28.1± 4.0 (21)
	Horizontal	17.5± 3.5	18.2± 3.5	28.5± 3.9*	28.1± 4.0*
	Peripheral	17.5± 3.5	18.2± 3.5	28.2± 3.7*	27.9± 3.8*
	Central	0.06±0.08	0.1± 0.1	0.3± 0.2	0.2± 0.2

* p < 0.05.
**p < 0.01.
In brackets - number of animals.

Table 3. Effect of Phenazepam on Behavior of F_1 and F_2 Hybrids Obtained by Crossing C57BL/6 with BALB/c Mice in the "Open Field" Test

Strain of mouse	Locomotor activity	Control	Dose given (mg/kg)		
			0.05	0.075	0.1
C57BL/6 x BALB/c F_1 (12)	General	106.2±12.9	68.1±16.3	60.9±14.4*	41.7±10.8***
	Horizontal	99.4±11.2	64.7±13.1	58.4±13.8*	40.3± 9.9***
	Peripheral	86.7± 9.8	57.7±12.3	49.9± 9.6*	34.0± 7.6***
	Central	12.7± 3.3	7.0± 2.8	8.4± 4.2	6.1± 2.3
	Vertical	8.4± 1.7	3.4± 1.2*	2.5± 1.2*	1.3± 0.7***
F_2 (15)	General	95.8±10.8	57.4±14.3*	40.5± 9.5**	43.1± 6.4**
	Horizontal	91.5±10.0	55.3±13.4	39.1± 8.9**	42.0± 6.3**
	Peripheral	74.7± 9.2	49.1± 9.6*	32.7± 8.5**	36.7± 5.8**
	Central	16.8± 4.8	7.3± 3.7	6.5± 3.3	5.9± 2.3
	Vertical	3.7± 1.6	2.1± 0.9	1.4± 0.6	1.1± 0.2

* $p < 0.05$.
** $p < 0.01$.
***$p < 0.001$.
In brackets – number of animals.

situation in an open field. It was shown, using the Serebrovsky
formula, that the response is polygenically controlled, and that the
number of genes in the group is less than three.

3.2. INVESTIGATIONS OF THE ANTICONVULSANT ACTION OF PHENAZEPAM AND
      DIAZEPAM ON B6 AND c MICE AND THEIR HYBRIDS

    In the following series of experiments, conditions were created
such as to isolate the baseline action against which tranquilizers'
effect was measured, irrespective of genotype. Pentylenetetrazole
(i.v.), which produced convulsions in B6 and c mice at the same
dosage, was used for this purpose. Phenazepam had an anticonvulsant
action on both strains, but this tranquilizer consistently worked
more effectively in B6 than in c mice (see Table 4).

    As may be seen in Table 5, diazepam exerted a protective effect
both in B6 and c mice. Quantitative variations in sensitivity to
phenazepam were thus demonstrated by experiments involving intravenous pentylenetetrazole.

    Similar experiments were performed on F_1 and F_2 hybrids, to
investigate how their reaction was inherited.

    As predicted, the dosage of pentylenetetrazole inducing con-
vulsions in the hybrids did not differ noticeably from that producing
the same effect in the original strain. The protective action of
phenazepam was similar to that observed in c mice; this finding
indicates genetically dominant low sensitivity to the anticonvulsant
action of the tranquilizer (Figure 1). Calculation of the number of
genes influencing this effect showed it to be polygenically controlled; minimum number of genes = 3.

    If the results of investigations into the influence of
phenazepam on the behavior of inbred mice in an open field are
examined together with those obtained for this drug's anticonvulsant
effect, it may be noted that genetic factors determine a) initial
pattern of reaction to stress and tranquilizer action (involving
qualitative variations) and b) sensitivity to the drug's anticonvulsant effect (involving quantitative differences only).

    The reaction as inherited also tends towards the dominant for
the values studied, in that mice of different strains bear dominant
genes. This shows polygenic control shaping the tranquilizer's
compound effect and leads us to predict the existence of a minimum of
two groups of genes, each containing not less than three. One of
these controls the type of stress reaction and how the drug's action
deals with it; another group regulates sensitivity, at least to the
drug's anticonvulsant effect.

Table 4. Anticonvulsant Action of Phenazepam on Mice of C57BL/6 and BALB/c Strains Following Pentylenetetrazole

Strain	Control	Dose (mg/kg)				
		0.875	1.75	3.5	7.0	14.0
BALB/c	99.9±4.7*	221.3±20.4	239.9±14.2	355.2±14.2	335.2±16.4	465.6±18.4
C57BL/6	93.3±3.7	284.4±10.4	415.1±20.5	499.5±22.7	532.4±21.1	603.7±15.5

$p < 0.001$, $p < 0.05$, $p < 0.001$, $p < 0.001$, $p < 0.002$, $p < 0.002$, $p < 0.002$ (BALB/c comparisons)

$p < 0.001$, $p < 0.001$, $p < 0.01$, $p < 0.01$, $p < 0.01$ (C57BL/6 comparisons)

*Dose levels of pentylenetetrazole inducing tonic convulsions.

Table 5. Anticonvulsant Action of Diazepam on Mice of C57BL/6 and BALB/c Strains Following
Pentylenetetrazole

Strain	Control	Dose (mg/kg)				
		2.5	5.0	10.0	15.0
BALB/c	99.9±4.7*	117.7±7.4	239.3±13.7	288.4±17.9	333.96±25.2
		↕ p < 0.05	↕ p < 0.001		
C57BL/6	93.3±3.7	129.9±7.3	204.8±14.0	264.8±19.5	316.9 ±18.03
		↕ p < 0.001	↕ p < 0.01		

*Dose levels of pentylenetetrazole inducing tonic convulsions.

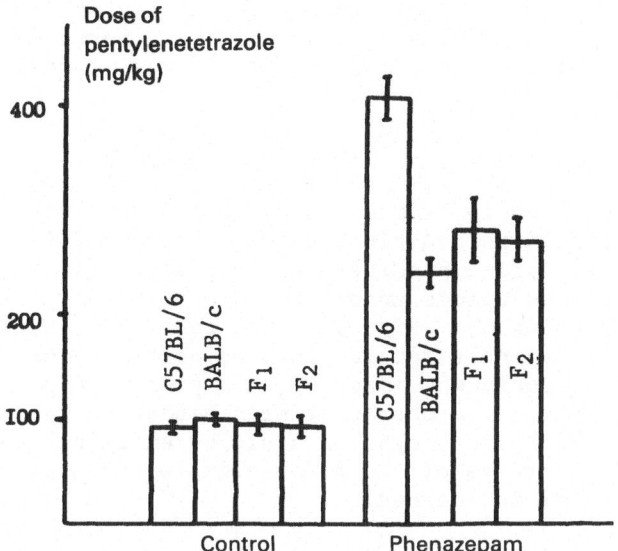

Fig. 1.   Analysis of the anticonvulsant action of phenazepam in
          hybrids at a dose of 1.75 mg/kg.

        Genetic variations in the parameters of the stress reaction and
mechanisms governing quantitative variations in sensitivity towards a
drug may therefore prove significant for predicting the effects of
phenazepam. Guided by this, we proceeded to investigate these traits
amongst the biochemical indicators of stress and capacity for specific binding of benzodiazepine derivatives. Our study included
pharmacokinetic characteristics and metabolism of phenazepam in
animals with differing sensitivity to the drug.

3.3.   INVESTIGATIONS OF GENETIC VARIATIONS IN THE BIOCHEMICAL
       INDICATORS OF STRESS REACTION

        Emotionally stressful stimuli are currently believed to trigger
a whole range of neuroendocrine changes in the organism, activating
the sympathetic and parasympathetic nervous systems and stimulating
hypothalamo-adrenocortical function. These changes are accompanied
by an increase in noradrenaline release from the sympathetic nerve
endings, of adrenaline from the adrenals and of hormones from the
anterior pituitary (gonadotropic, thyrotropic, adrenocorticotropic)
leading in turn to increased hormone secretion for the glands it
controls (Levy, 1970).

        We investigated ACTH and corticosterone concentrations in plasma
both before and after exposure to stress, in order to evaluate
genetic variations in these processes. Levels of cyclic nucleotides

(cyclic AMP and cyclic GMP), which may be considered indicators of
activity of the sympathetic and parasympathetic nervous system
(Kunitada et al., 1978; Honma and Ui, 1978), were likewise measured.

3.3.1. ACTH

The results given in Figure 2 and Table 6 show that baseline
concentration of ACTH is genetically controlled, and is inherited
following the basic principles set for Rokitsky's (1978) "quantitative signs". Calculations according to Serebrovsky minimum gene
count (Π), influencing the "sign", showed that $\Pi = 0.6$, i.e. ≈ 1;
thus a monogenically inherited initial concentration of ACTH in blood
plasma may well be found when mice of the strains under study are
crossed. The absence of a characteristic Mendelian ratio in F_2 and
the offspring of back cross hybrids could be explained by the unbroken distribution range of the above sign in these strains under
strong environmental influences.

Changes in ACTH concentrations in the plasma of B6 and c mice
after exposure to stress are given in Figure 3. In B6 mice, handling
produced no substantial deviation from the baseline hormone level
during a one-hour period. In the open field, however, ACTH concentration increased sharply immediately after the end of the experiment. The next substantial increase occurred 15 min later (Figure
3).

A different pattern was observed for c mice. A continuous drop
in ACTH concentration, beginning from the first minutes of handling,
was observed in the control animals, only starting to rise again one
hour later.

Placing the c mice in an open field produced a brief increase in
hormone level straight after the end of the experiment. This was
followed, however, by a steady decrease in concentration, as in the
control animals (see Figure 3).

Investigation of ACTH content in F_1 hybrids revealed changes
basically similar to those noted in strain B6. The baseline hormone
level and the time course of subsequent rises in its concentration in
plasma were also similar (see Table 7). During exposure to an open
field ACTH concentration in B6 x c mice is thus inherited in the same
way as the main behavioral parameters are genetically determined.

3.3.2. Corticosterone

It may be seen from information provided in Table 8 that the
initial corticosterone concentration in c mice is barely half the
B6 level.

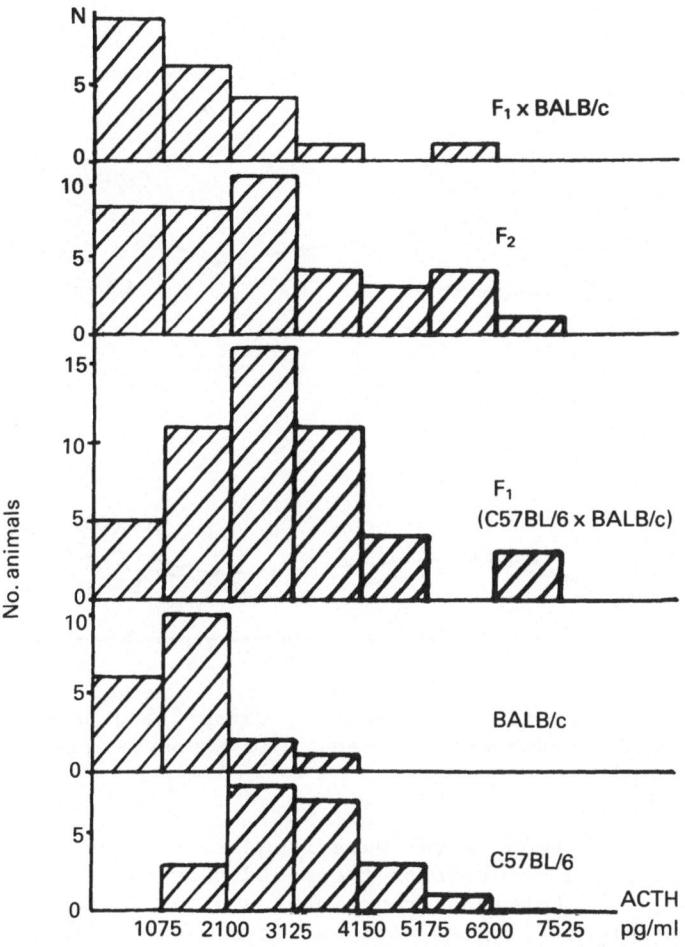

Fig. 2.   Inherited initial ACTH concentration in plasma when C57BL/6
          and BALB/c mice are crossed.   Horizontal axis – concentra-
          tion of ACTH (pg/ml).   Vertical axis – number of animals.

Table 6.   Indices of Baseline ACTH Concentrations in the Plasma in
           B6 and c Strains and F_1, F_2 and F_1 x c Hybrids

Generation	No. of animals	Mean M ± SEM		Deviation σx
B6	22	3100±215	p < 0.001 vs. c	1007
c	19	1350±162		707
F_1	50	2747±214	p < 0.001 vs. c	1511
F_2	38	2636±284		1749
F_β (F_1 x c)	21	1682±351	p x 0.05 vs. B6	1608

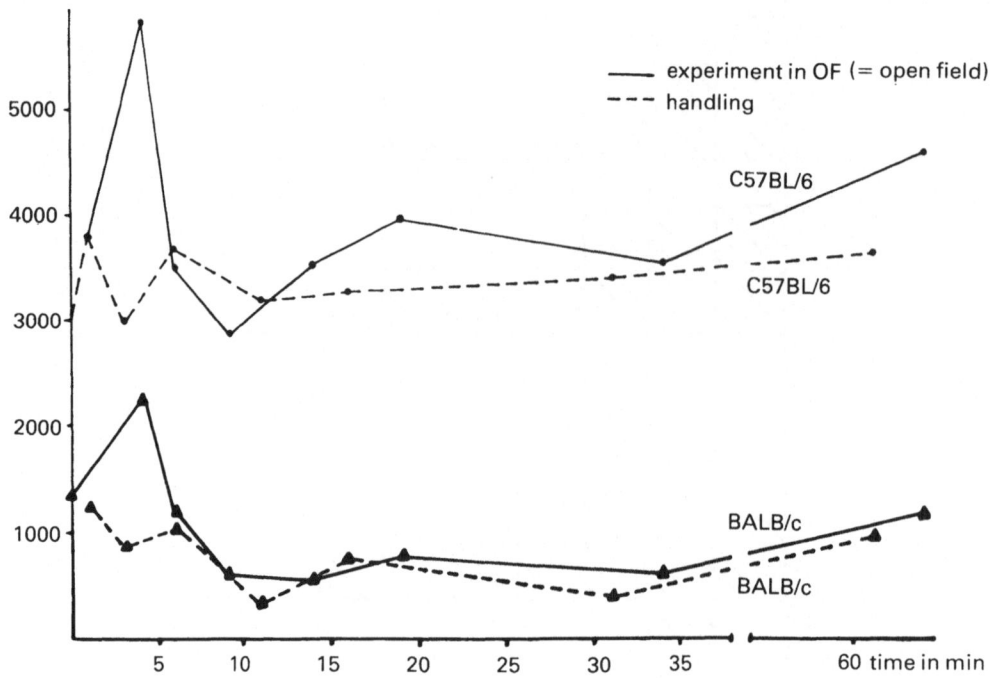

Fig. 3. ACTH concentration in plasma in C57BL/6 and BALB/c mice
         following stress during open field exposure and after
         handling.

No changes in hormone level were found in B6 mice 10 min after
handling, while the concentration increased sharply in c animals.
The concentration of corticosterone in plasma increased equally in
both strains after handling as well as after open field exposure (see
Table 8). The corticosteroid reaction thus occurs more readily and
with greater intensity in c mice than in the B6 strain.

3.3.3. Cyclic nucleotides

Analysis of initial levels of cyclic nucleotides in plasma
showed that cyclic AMP concentration is almost twice as high in B6
mice as in c, whole no interstrain differences were seen in the case
of cyclic GMP (Figure 4). Cyclic AMP concentration in B6 mice was
reduced by handling, with an even sharper reduction following trials
in an open field. No substantial changes in the cyclic GMP levels of
B6 were observed.

In c mice cyclic AMP concentration increased by 20% after handling and 50% after open field exposure. Cyclic GMP level rose by
only 20%.

Table 7.   ACTH Concentration in the Blood Plasma of F_1 Hybrids
           C57BL/6 x BALB/c Following Exposure to Stress in an Open
           Field (OF + t)* and after Handling (-OF + t)

| Experiment | No. of animals | ACTH concentration (mg/ml) | Significantly different | |
			a) from initial level	b) post-handling level
Initial concentration	50	2747±214	from BALB/c*** $p < 0.001$	from C57BL/6 $p > 0.1$
1st min in OF**	13	3604±462	$p > 0.05$	
OF + 0	30	2565±144	$p > 0.1$	$p > 0.1$
- OF + 0	30	2602±196	$p > 0.1$	
OF + 2	8	3463±362	$p > 0.1$	$p > 0.1$
- OF + 2	8	3675±188	$p > 0.1$	
OF + 5	8	2581±397	$p > 0.1$	$p > 0.1$
- OF + 5	9	2844±344	$p > 0.1$	
OF + 10	9	5672±667	$p < 0.001$	$p > 0.1$
- OF + 10	9	5594±630	$p < 0.001$	
OF + 15	9	4494±513	$p < 0.01$	$p > 0.1$
- OF + 15	6	3775±304	$p > 0.1$	
OF + 30	8	3513±455	$p > 0.1$	$p > 0.1$
- OF + 30	8	3000±355	$p > 0.1$	
OF + 60	9	2222±474	$p > 0.1$	$p > 0.1$
- OF + 60	9	2944±1118	$p > 0.1$	

*   time since 3-min experiment in open field.
**   ACTH concentration found 1 min after placing in open field.
*** information on significance of differences compared with initial
     level of the parental strain.

The initial nucleotide level in F_1 hybrids resembled that observed in B6 mice. Changes in cyclic AMP also more closely resembled
those occurring in B6, although the level did not drop substantially
after handling, while cyclic GMP did rise in F_1 after handling and
open field experiments, a reaction broadly resembling that of c mice.

The changes in cyclic AMP/cyclic GMP ratio in F_1 hybrids were
closer to those observed in B6 animals. It follows that initial
levels of nucleotide concentration and their modification following
exposure to stress differ in animals of different genotypes with
different patterns of response under stress.

The F_1 hybrids obtained by crossing B6 and c mice, whose behavior most resembled parental strain B6, showed a) initial levels of
cyclic nucleotide concentration and b) ratios between these levels
after stress again resembling the B6 strain.

Table 8.   Corticosterone Concentration in the Plasma of C57BL/6 and
           BALB/c Mice after "Open Field" Experiment

Experimental conditions	Strain of mouse			
	c		B6	
Initial level	1.37±0.18 (n=8)		3.5±0.23 (n=10)	
-OF + 10	5.37±0.94 (n=8)	$p < 0.001$	4.8±0.69 (n=10)	$p_1 > 0.05$
+OF + 10	4.37±0.43	$p_1 < 0.001$ $p_2 > 0.05$	7.9±0.4	$p_1 < 0.001$ $p_2 < 0.01$
-OF + 20	7.0 ±0.64 (n=8)	$p_1 < 0.001$	7.1±0.4 (n=10)	$p_1 < 0.001$
+OF + 20	8.12±0.37 (n=8)	$p_1 < 0.001$ $p_2 > 0.05$	6.4±0.49 (n=10)	$p_1 < 0.001$ $P_2 > 0.05$

In brackets - number of animals.
p_1 - significance compared with initial level.
p_2 - significance compared with corresponding control.

3.4.   ANALYSIS OF SPECIFIC BINDING OF ^{3}H-DIAZEPAM IN B6 AND c MICE
       AND THEIR F_1 HYBRIDS

     The results summarized in Table 9 show that the degree of spe-
cific ^{3}H-diazepam binding in the cerebral cortex of B6 mice is higher
than in c animals. No substantial differences were found in the
other areas of benzodiazepine receptor binding investigated. The
hybrids had inherited a lower capacity for specific binding of the
radioactive ligand characteristic of c mice (see Table 9).

3.5.   GENETIC DIFFERENCES IN THE PHARMACOKINETICS AND METABOLISM OF
       ^{14}C-PHENAZEPAM

     A great deal of research has been devoted to showing that a
varying degree of pharmacological response may be due to the pharmacokinetics of a given medicament (Golovenko, 1981; Soloviev et al.,
1980). Vesel's (1977) work showed that the distribution and metabolism of drugs is genetically controlled. Guided by these findings,
we studied the biotransformation of ^{14}C-phenazepam in B6 and c mice
and their F_1 hybrids, which showed differing sensitivity to the anticonvulsant action of tranquilizers.

     Radiochromatography of chloroform extracts of biological samples
enabled us to establish the presence of:

  i) phenazepam, the starting material

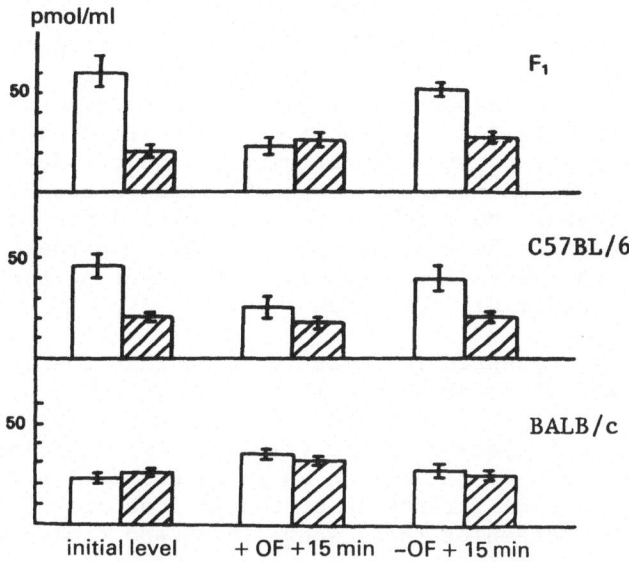

Fig. 4.   □ Cyclic AMP and ⊠ cyclic GMP concentrations in the plasma
         of inbred mice and their hybrids after stress during open
         field exposure.

Table 9.   Specific Binding of ^{3}H-diazepam in C57BL/6 and BALB/c Mice
          and their F_1 Hybrids (in counts/min per 10 mg/tissue)

Area of brain	C57BL/6	BALB/c	F_1
Cerebral cortex	58740±2426 n=10	51491±1127* n=10	50417±2053* n=7
Cerebellum	32740±849 n=5	35877±1803 n=5	–
Mid-brain and diencephalon	48228±1262 n=5	44079±1287 n=5	–
Brain stem	32714±4511 n=5	26629±688 n=5	–

n – Number of animals.
* – $p < 0.05$ compared with strain C57BL/6.

 ii) a considerable quantity of its trihydroxy derivative
iii) the sum of the metabolites formed from oxidation of the aromatic
      rings of molecule 1.

    A portion of the radioactive material was bound to protein
(Seredenin et al., 1981). No qualitative differences in the
formation of metabolites were found in mice of separate strains and

their hybrids. Quantitative differences in the speed of metabolism
of compound (i) were shown, however; these were most pronounced at
the stage of hydroxylation of the molecule in position 3 (see Table
10). This process is characteristic of the biotransformation of
benzodiazepine tranquilizers in the mouse (Seredenin et al., 1981).
Compound (ii) appears to be an intermediate product, derived from
subsequent metabolic change. Its concentration increases as the
experiment progresses, to reach a constant level after a set time.
Monitoring changes in the concentration of (ii) can therefore be used
as an effective and sensitive criterion for comparing rate of metabolism of (i) in mice of different genotype.

The results, presented in Table 10, show that the speed of
oxidation is lower in B6 than in c mice.

Investigation of metabolism of (i) in the hybrids indicated
the presence of the same metabolism as in B6 and c mice. As shown
in Tables 10-12, the difference in concentrations of (i) and (ii)
between F_1 and B6 resembles those shown in B6 and c mice in most
cases.

The index indicating the degree of influence exerted by hereditary factors η^2_A on the concentration of compound (i) in the brain
$\frac{1}{2}$ to 1 hour later in B6 and c was 27-37%, 36-49% in F_1 and B6, and
37-38% for F_1, B6 and c mice. No substantial differences were seen
between the hybrids and c mice. The level of compounds (ii) found
in the brain one and two hours after administering the drug was even
more heavily dependent on genotype: η^2_A varied within the limits 48-
77% for B6 and c, 53-64% for F_1 and B6, and 43-68% for F_1, B6 and c.
No differences were found between F_1 and c over these periods of time.

In the liver – the site of transformation of compound (i) into
compound (ii) – the compounds were present in similar concentrations
for c and F_1 mice, but throughout the experiment their level varied
substantially from that observed in B6.

The concentration of compounds (i) and (ii) in the plasma of the
hybrids $\frac{1}{2}$, 1 and 6 hours later were midway between those of the
parental strains; the η^2_A for F_1 and c were lower, however, than for
F_1 and B6.

It was thus shown that the transformation of compound (1) into
(ii) in the organism of F_1 hybrids takes place largely as in c.

Contrasting the results of investigations on the anticonvulsant
action of (1) and its metabolism, a reciprocal relationship may be
observed between genetically determined sensitivity to the substance
and its speed of oxidation. Due to the tranquilizer's higher rate
of metabolism, it is probably present in lower concentrations in
the organism of c and F_1 than of B6; this accounts for its reduced

Table 10. Distribution of Compounds (i) and (ii) (in counts/min$\cdot 10^{-3}$) in the B6 and c Mouse (n=5) after Administration of 14 mg/kg Dose of Phenazepam (M±SEM)

Organ	Time in hours	Compound (i)			Compound (ii)		
		B6 mice	c mice	t_d	B6 mice	c mice	t_d
Liver	1/2	147.7±38.1	264.7±37.5	2.19	12.7±3.6	51.4±6.4	5.22**
	1	241.9±14.7	229.8±33.5	0.33	21.3±8.6	54.3±9.2	2.62*
	2	185.6±12.1	162.8±14.6	1.20	38.5±4.6	58.1±6.4	2.47*
	3	153.8±22.7	133.5±13.2	0.77	51.0±6.7	57.9±4.0	0.88
	6	122.5±9.2	40.4±11.2	5.67**	61.6±9.3	65.2±7.2	0.31
Brain	1/2	64.9±13.2	101.9±10.9	3.19*	3.6±0.9	7.1±1.3	2.28
	1	122.5±13.9	101.9±10.9	2.31*	6.3±0.9	12.1±0.6	5.48**
	2	75.8±4.6	75.5±7.0	0.04	8.8±1.0	17.3±2.4	3.22*
	3	70.4±7.8	51.4±6.7	1.86	14.4±2.2	17.4±2.4	0.90
	6	56.7±12.1	19.2±3.9	2.96*	18.7±3.0	19.1±4.2	0.88
Plasma	1/2	36.0±4.8	131.1±14.1	6.33**	3.1±0.7	22.4±4.3	4.44**
	1	82.1±4.1	50.9±5.1	4.79**	3.2±0.3	13.0±1.0	9.12**
	2	-	43.6±1.7	-	7.5±1.1	14.9±0.6	5.84**
	3	52.3±8.9	37.3±4.1	1.53	14.6±2.6	16.3±2.0	0.48
	6	33.0±1.8	9.2±1.1	11.05**	21.2±1.7	12.2±1.7	3.77**

t_d - Student's t statistic of significance of differences between strain B6 and c mice.

* - $p < 0.05$, ** - $p < 0.01$

Table 11. Concentration of Compound (i) (in counts/min g·10^{-3}) in the Tissues of F1 Hybrids (n=5) after Administering a 14 mg/kg Dose of Phenazepam (contrasted with parental strain)

Organ	Time in hours	Compound (i) F$_1$ (M±SEM)	^1t_d	^2t_d	A F_ϕ	A $\eta_A^2,\%$	B F_ϕ	B $\eta_A^2,\%$	C F_ϕ	C $\eta_A^2,\%$	D F_ϕ	D $\eta_A^2,\%$
Liver	1/2	217.5±14.2	1.17	1.72	-	-	-	-	-	-	-	-
	1	188.0±19.1	1.09	2.39*	-	-	-	-	-	-	-	-
	2	146.0±6.9	1.04	2.84*	-	-	8.08	44.0	-	-	-	-
	3	111.2±8.3	1.42	1.93	-	-	-	-	-	-	-	-
	6	57.2±3.7	1.51	6.58**	-	-	43.4	83.5	25.2	77.6	32.08	77.5
Brain	1/2	111.3±5.3	0.78	3.25*	-	-	9.73	49.2	5.82	38.2	4.39	27.3
	1	83.7±7.1	0.06	2.48	-	-	6.11	36.2	5.02	36.6	6.22	36.7
	2	61.2±5.1	1.63	2.14	-	-	-	-	-	-	-	-
	3	56.6±5.2	0.59	1.48	-	-	-	-	-	-	-	-
	6	32.5±2.1	3.07*	1.57	9.02	47.1	-	-	6.53	44.1	8.70	46.1
Plasma	1/2	73.1±3.6	3.98**	6.04**	15.89	62.3	37.01	80.3	29.02	80.0	40.25	81.3
	1	47.4±4.9	0.52	5.43**	-	-	29.50	76.0	16.38	68.7	22.73	70.7
	2	35.6±2.9	1.15	-	-	-	-	-	-	-	-	-
	3	33.3±1.6	0.18	2.09	-	-	-	-	-	-	-	-
	6	21.4±2.0	5.26**	4.31**	28.57	75.6	18.59	61.1	50.30	87.6	127.3	93.3

Note: ^1t_d= Student t statistic of significance between B6 and F$_1$ animals; ^2t_d= Student t statistic of significance between c and F$_1$ animals; A = index of deviation analysis of differences between c and F$_1$ animals; B = index of deviation analysis of differences between B6 and F$_1$; C = index of deviation analysis of differences between B6, c and F$_1$ animals; D = index of deviation analysis of differences between B6 and c animals. *p<0.05; **p<0.01.

Table 12. Concentration of Compound (ii) (in counts/min g·10^{-3}) in the Tissues of F_1 Hybrids (n=5) after Administering a 14 mg/kg Dose of Phenazepam (contrasted with parental strain)

Organ	Time in hours	Compound (ii) F_1 (M±SEM)	$1t_d$	$2t_d$	A F_ϕ	A η_A^2,%	B F_ϕ	B η_A^2,%	C F_ϕ	C η_A^2,%	D F_ϕ	D η_A^2,%
Liver	1/2	29.7±2.8	3.08*	3.71**	9.65	49.0	13.89	58.9	18.28	71.2	15.28	61.7
	1	45.8±4.6	0.81	2.53*	–	–	6.31	37.1	4.90	35.8	6.87	39.5
	2	57.3±3.1	0.11	3.48**	–	–	11.49	53.8	9.90	48.8	13.89	58.9
	3	59.1±6.1	0.17	1.52	–	–	–	–	–	–	–	–
	6	45.8±4.6	1.13	1.95	–	–	–	–	–	–	–	–
Brain	1/2	6.1±1.0	0.57	1.95	–	–	–	–	–	–	–	–
	1	10.6±0.5	0.66	4.22**	–	–	17.44	64.6	20.76	68.4	31.74	77.4
	2	14.8±1.5	0.88	3.32**	–	–	11.08	52.8	6.35	43.3	10.67	48.2
	3	20.9±5.1	0.62	0.16	–	–	–	–	–	–	–	–
	6	19.2±3.9	0.03	0.11	–	–	–	–	–	–	–	–
Plasma	1/2	9.5±0.8	2.95*	5.75**	8.70	46.1	36.25	79.7	14.78	32.7	19.63	67.4
	1	9.0±0.8	4.91*	6.99**	10.4	53.0	37.70	82.1	37.59	85.9	88.11	92.6
	2	12.0±1.8	1.49	2.09	–	–	–	–	8.67	52.3	34.88	79.0
	3	15.4±1.7	0.35	0.23	–	–	–	–	–	–	–	–
	6	14.7±1.4	1.11	2.97*	–	–	8.45	45.3	8.17	45.3	13.70	58.5

Note: $1t_d$ = Student t statistic of significance between c and F_1 animals; $2t_d$ = Student t statistic of significance between B6 and F_1 animals; A = index of deviation analysis of differences between c and F_1 animals; B = index of deviation analysis of differences between B6 and F_1; C = index of deviation analysis of differences between B6, c and F_1 animals; D = index of deviation analysis of differences between B6 and c animals. *p<0.05; **p<0.01.

effect. This presumption is supported by experiments investigating
the anticonvulsant activity of the trihydroxy-metabolite of phenazepam, which provided less protection than the original compound when
pentylenetetrazole was given intravenously (see Figure 5).

4. DISCUSSION

Investigations into the behavior of B6 and c mice in an open
field corroborated earlier findings on genetically determined types
of stress response and the involvement of polygenic control
(Borodin et al., 1976; Archer, 1973). The similarity between
F_1 B6 x c hybrids and the B6 parental strain would make B6, c and F_1
hybrids suitable models for use in pharmacological experiments investigating the genetics of behavior. It should be borne in mind, however, that B6 and c mice differ in learning capacity (Will, 1977);
thus the information obtained from initial exposure to stress in
experiments analyzing emotional stress response in these strains
should be regarded as the most trustworthy.

Our pharmacogenetic investigations of phenazepam and diazepam
have enabled us to demonstrate that their effect is genetically
determined and the pharmacological signs relating to the drug's
action are under polygenic control. In spite of our present lack

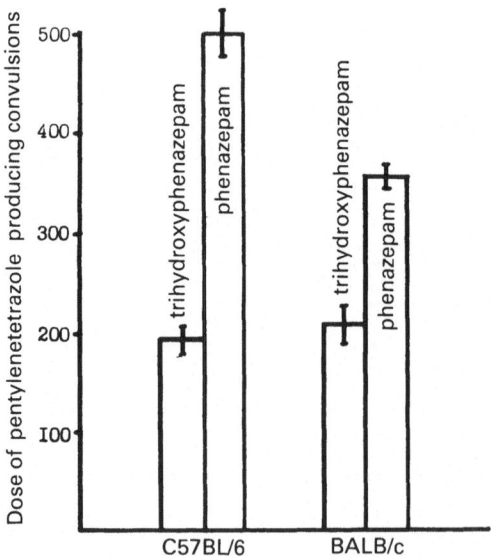

Fig. 5. Anticonvulsant effect of phenazepam (3.5 mg/kg) and
         trihydroxyphenazepam (3.5 mg/kg) in C57BL/6 and BALB/c mice
         "titrated" with pentylenetetrazole.

of information on the mechanisms producing differences in the effects
of benzodiazepines, the inherited traits responsible for stress and
metabolic potential suggest a lead to follow in seeking a set of
predictors for their action; they may also prove useful in the subsequent deciphering of these mechanisms.

Experiments on human volunteers have provided results confirming
the dependence of benzodiazepines' effect on patterns of reaction to
stress (Valdman and Martynikhin, 1982).

Genetic variability in response to stress may also be followed
up biochemically. The data obtained on ACTH concentration in plasma
of B6 and c mice support the results of earlier investigations into
hormone level in the pituitary of mice of the same strains (Crabbe
et al., 1981). The analysis of ACTH concentration in hybrids gives
ample grounds for postulating that this parameter may be monogenically controlled. We have been unable to discover other information
on this theme in the current literature; our results do, however,
corroborate current ideas on the molecular mechanisms of hormone
synthesis (Fox, 1981; Barchas and Sullivan, 1982).

The results obtained show the presence of genetically controlled
variants in patterns of adaptation. This is demonstrated by the
different changes in ACTH level observed in B6 and c mice in response
to such stressful stimuli as handling and open field exposure. The
lower concentration of ACTH observed in c mice than in B6 mice following open field exposure tallies with c animals' considerably increased corticosterone level. An increase in the concentration of
this hormone occurs in c mice within 10 min of handling, while a
similar but less pronounced reaction is only produced within this
time in B6 mice following open field exposure. These results indicate that the hormonal adaptation systems of c mice are more responsive; they also correspond with data on their behavior under stress.
A contrasting rise in corticosterone level, inversely proportional to
its initial concentration under stress, was found in c and B6 mice,
and was comparable with the results obtained from investigations into
cortisol secretion after ACTH administration to healthy volunteers
(Kukreja and Williams, 1981).

Our investigation of nucleotides in inbred mice was founded on
research showing their role as cell mediators of the adrenergic and
cholinergic systems, performing essential functions in the organism's
adaptation processes (Dorofeev et al., 1978). Many authors have
shown that the altered levels of cyclic AMP and cyclic GMP in the
tissues are associated with corresponding changes in the plasma
(Broadus et al., 1970; Strange and Mjos, 1975). The literature,
however, points to the compound nature of cyclic nucleotide control.
Many hormones and nucleotides, secretion of which changes under
stress, may influence their concentration in target tissues (Dorofeev
et al., 1978). It thus appears that cyclic nucleotide concentration

in blood plasma should be viewed as an integral parameter, showing
many homeostatic alterations, resulting from stress. It is nonetheless interesting that changes occurred in cyclic AMP after open
field exposure along different lines in B6 and c mice with different
response patterns to stress, while those occurring in the behavior-
ally alike B6 and F_1 hybrids were similar. The reasons for the
differences seen evidently require further investigation.

The relationship between open field behavior and biochemical
parameters established for B6 and F_1 hybrids mice would appear to be
the principal results obtained from investigation of the biochemical
indicators of stress. This correspondence between the hereditary
factor in behavioral and biochemical signs brings out the reason for
the interconnection between the animals' behavior pattern and the
pattern of hormonal and nucleotide response observed, and this, in
its turn, determines its usefulness as a predictor of benzodiazepine
effects.

The overall results showing genetic control over these indices -
taking account of those data on the biochemical action of the oligo-
peptides which go to make up ACTH molecules, may help us to understand such physiological and behavioral variations between B6 and c
mice as a) emotionality (Archer, 1973), b) ability to develop conditioned reflexes and memory formation (Bovet and Oliverio, 1973), c)
alcohol consumption (McClean and Rodgers, 1959) and a number of other
factors brought out by the extensive work published on these strains
of animal.

Another reason for benzodiazepine sensitivity and reaction to
stress following different patterns could be connected with the
genetic heterogeneity of benzodiazepine receptors. The specific
binding of ^{3}H-diazepam in B6 and c mice and their hybrids was studied
to shed light on this question. The findings supported those of
Robertson, who has shown that receptor binding of benzodiazepines in
the CNS was greater in B6 than in c (Robertson, 1979). At first
glance these results appear to support the existence of an endogenous
ligand, providing stronger protection against stressful stimuli in B6
than c animals. No conclusions, however, can be drawn from analyzing
specific binding data in F_1 hybrids, since benzodiazepine binding in
B6 x c animals resembled that of c, whereas pattern of response to
stress correction of this reaction by benzodiazepines was inherited
according to B6 parental strain. The findings may be used to explain
the quantitative differences in anticonvulsant action of phenazepam
observed in B6 and c mice and their hybrids. This is the conclusion
of a number of papers showing that level of specific benzodiazepine
binding correlates with their anticonvulsant effect rather than their
influence on behavior in a conflict situation (Mackerer et al., 1978;
Speth et al., 1978). The contradiction arising from the absence of
inter-strain differences in the protective action of diazepam against
graded doses of pentylenetetrazole is partially solved by analyzing
the metabolic activity of hepatic microsomes of B6 and c animals.

Research into the biodegradation of phenazepam showed that the formation of its major metabolite – a trihydroxy derivative – occurs more intensively in the organism of c mice and F_1 hybrids than in B6 animals. Its speed of metabolism in c and F_1 animals is therefore in inverse proportion to its anticonvulsant action in these animals.

Experiments on phenazepam in vivo confirmed the results we had obtained earlier in vitro, showing that the metabolic activity of hepatic microsomes and the concentration of cytochromes P 450 in c mice is higher than in B6 (Seredenin, 1979). The absence of inter-species variations in the anticonvulsant action of diazepam may be explained in the light of these results. Diazepam would appear to be metabolized more actively in c mice, resulting in increased output of desmethyldiazepam, which, according to the literature, has a stronger anticonvulsant effect in mice than the parent compound (Marcucci et al., 1968); this levels out differences in specific binding.

Combining all findings obtained on the specific binding and metabolism of benzodiazepine tranquilizers, we are led to conclude that in the case of both processes, inherited differences are responsible for the drugs' quantitatively different action.

The phenotypes shown of "poor" and "extensive" oxidation in B6 and c mice and the genetic shaping of this capacity corroborated the data of Idle, Smith, Kalow and Eichelboum, who found similar pheno-types in a human population in the course of experiments with model preparations with antipyrine, sparteine and debrizoxine. These authors proposed that the oxidation reaction might be monogenically controlled (Inabe et al., 1980; Idle and Smith, 1979; Eichelboum et al., 1979). Another source deals with the possibility of using antipyrine to predict the pharmacokinetics of benzodiazepines in connection with differences in oxidation (Greenblatt et al., 1982).

Pharmacogenetic experimentation has thus enabled us to show how the effect of benzodiazepines is genetically determined. Such ef-fects are believed to be directly or indirectly controlled by a minimum of two groups of genes. Findings show that benzodiazepine predictors should be sought using evidence from both stress patterns and pharmacokinetic indicators. Needless to say, the most varied combinations of known characteristics may be found in the heterozy-gote population, resulting from polygenic control of the processes being discussed. Under natural conditions, they may also be modified by environmental factors. Even so, showing how the pharmacological effects of benzodiazepines under model conditions vary, depending on genetic laws, should be an essential prerequisite for further inves-tigations with a cross-bred population and for studying the influence of environmental factors.

The fact that this work corroborates a number of other findings on genetic determination of benzodiazepine tranquilizers' effects and

clinical results on individual sensitivity to these substances
(Valdman and Martynikhin, 1982) augurs well for our approach.

Abbreviations

ACTH   - adrenocroticotrophic hormone
cyclic AMP - cyclic 3',5'-adenosine monophosphate
cyclic GMP - cyclic 3',5'-guanosine monophosphate
CNS   - central nervous system
B6   - C57BL/6 strain mice
c   - BALB/c strain mice
PPO   - 2,5-diphenyloxazole
POPOP   - 1,4-di[2-(5-phenyloxazolyl)]benzene.

REFERENCES

Aleksandrovsky, Yu. A., 1973, The Clinical Pharmacology of Tran-
        quilizers, Moscow, "Meditsina".
Aleksandrovsky, Yu. A., 1976, Psychic Desadaptation and Its
        Compensation, Moscow, "Nauka".
Archer, J., 1973, Test for emotionality in rats and mice. A review,
        Anim.Behav., 21:205-229.
Barchas, J. D., and Sullivan, S., 1982, Opioid peptides as neuro-
        regulators: potential areas for the study of genetic-
        behavioural mechanisms, Behav.Genet., 12:69-91.
Berson, A. A., and Yalow, R. C., 1968, Radioimmunoassay of ACTH in
        plasma, J.Clin.Invest., 47:2725-2751.
Borodin, P. M., Schuler, L., and Belyev, D. K., 1976, Problems in
        stress genetics. I. Genetic analysis of the behavior of mice
        in stressful situations, Genetics (USSR), 12:62-71.
Bovet, D., and Oliverio, A., 1973, Pharmacogenetic aspects of
        learning and memory, in "Pharmacology and Future of man,
        Proc., 5th Int. Congr. Pharmacology," San Francisco, 1972,
        4:18-28.
Braestrup, C., and Squires, R. F., 1978, Pharmacological character-
        ization of benzodiazepine receptors in the brain, Eur.J.
        Pharmacol., 48:263-270.
Broadhurst, P. L., 1975, The Maudsley reactive and nonreactive
        strains of rats: survey, Behav.Genet., 5:299-319.
Broadus, A. E., Kaminsky, N. I., Northcutt, R. C., Hardman, J. G.,
        Sutherland, E. W., and Liddle, G. W., 1970, Effects of
        glucagon on adenosine 3',5'-monophosphate and guanosine
        3',5'-monophosphate in human plasma and urine, J.Clin.
        Invest., 49:2237-2245.
Brown, B. L., Albano, J. D., Barnes, G. D., and Ekins, R. P., 1974,
        The saturation assay of adenosine 3',5'-cyclic monophosphate
        in tissues and body fluids, Biochem.Soc.Trans., 2:10-12.
Cook, L., and Sepinwall, J., 1975, Behavioral analysis of the effects
        and mechanisms of action of benzodiazepines, in: "Mechanism of

Action of Benzodiazepines; Advances in Biochemical Psychopharmacology," E. Costs and P. Greengard, eds., Raven Press, New York, 14:1-28.

Crabbe, J. C., Allen, R. G., Gaudette, N. D., Young, E. R., Kosobut, A., and Stack, Y., 1981, Strain differences in pituitary endorphin and ACTH content in inbred mice, Brain Res., 219:219-223.

Dasberg, H., 1974, The effect of daily oral dosage of diazepam plasma concentrations and metabolic clearance of diazepam and des-methyldiazepam on various constituents of the acute clinical anxiety syndrome, Psychother.Psychosom., 24:113-118.

Dorofeev, G. I., Kozhemjakin, L. A., and Ivashkin, V. T., 1978, Adaptation and Cyclic Nucleotides, Leningrad, "Nauka".

Eichelboum, M., Spannbrucker, N., Steinche, B., and Denger, H. I., 1979, Defective N-oxidation of sparteine in man: a new pharm-acogenetic defect, Eur.J.Clin.Pharmacol., 16:183-187.

Fox, Y., 1981, Some nervous system genes curiously complex, Chem. and Eng.News., 59:26-27.

Frankenhauser, M., 1967, Some aspects of research in physiological physiology, in: "Emotional Stress, Physiological and Psycho-logical Reactions; proceedings," L. Levy, ed., Basel - New York, 16-26.

Ghoneim, M. M., Korttila, K., Chlang, C. K., Yacobs, L., Schoenwald, R., Mewaldt, S. P., and Kayle, R. O., 1981, Diazepam effects and kinetics in Caucasians and orientals, Clin.Pharmacol. Ther., 29:749-756.

Golovenko, N. Ya., The Mechanisms of Metabolic Reactions of Xenobiotics in Biological Membranes, 1981, Kiev, "Naukova Dumka".

Golovenko, N. Ya., Zinkovsky, V. G., Bogatsky, A. V., Andronatti, S. A., and Yakubovskaja, L. N., 1979, The comparative study of phenazepam metabolites after single and long-term treatment of rats, Chem-Pharm.J.(USSR), 13:21-26.

Greenblatt, D. J., Divoll, M., Abernethy, D. R., Harmatz, J. S., and Shader, R. I., 1982, Antipyrine kinetics in the elderly: prediction of age-related changes in benzodiazepine oxidizing capacity, J.Pharmacol.Exp.Ther., 220:120-126.

Honma, M., and Ui, M., 1978, Plasma cyclic GMP: response to cholin-ergic agents, Eur.J.Pharmacol., 47:1-10.

Idle, J. R., and Smith, R. L., 1979, Polymorphism of oxidation at carbon centers and clinical significance, Drug Metab.Rev., 9:301-307.

Inabe, T., Otton, S. V., and Kalow, W., 1980, Deficient metabolism of debrisoquine and sparteine, Clin.Pharmacol.Ther., 27:547-549.

Kaminsky, N. I., Broadus, A. E., Hardman, J. G., Jons, D. J., Ball, J. H., Sutherland, E. W., and Liddle, G. W., 1970, Effects of parathyroid hormone on plasma and urinary adenosine 3',5'-monophosphate in man, J.Clin.Invest., 49:2387-2395.

Kukreja, S. C., and Williams, G. A., 1981, Corticotropin stimulation test: inverse correlation between basal serum cortisol and its response to corticotropin, Acta Endocr., 97:522-524.

Kunitada, S., Honma, M., and Ui, M., 1978, Increases in plasma cyclic
     AMP dependent on endogenous catecholamines, Eur.J.Pharmacol.,
     48:159-169.
Lankin, V. S., Stakan, G. A., and Naumenko, E. V., 1979, Domestic
     behaviour of sheep. I. The relation of the functional state of
     the hypothalamic-pituitary-adrenal system to behaviour and age
     in thin-fleeced sheep, Genetics (USSR), 15:891-900.
Lankin, V. S., Stakan, G. A., and Naumenko, E. V., 1979, Domestic
     behaviour of sheep. II. A. study and analysis of differences
     in behaviour characteristics in thin-fleeced sheep, Ibid.,
     15:901-911.
Lankin, V. S., Stakan, G. A., and Naumenko, E. V., 1980, Domestic
     behaviour of sheep. III. Hereditary determination of the
     endocrine function of the pituitary-adrenal system in
     thin-fleeced sheep with different behaviour type, Ibid.,
     16:1088-1095.
Levy, L., 1967, Endocrine reactions during emotional stress, in:
     "Emotional Stress, Physiological and Psychological Reactions;
     Medical, Industrial and Military Implications; proceedings,"
     L. Levy, ed., Basel - New York, 109-113.
Mackerer, C. R., Kochman, R. L.,Bierschenk, B. A., and Bremner, S.
     S., 1978, The binding of ^{3}H-diazepam to rat brain homogenates,
     J.Pharmacol.Exp.Ther., 206:405-413.
Marcucci, F., Guaitani, A., Kretina, J., Mussini, E., and Garattini,
     S., 1968, Species difference in diazepam metabolism and anti-
     convulsant effect, Eur.J.Pharmacol., 4:467-470.
Markel, A. L., and Borodin, P. M., 1978, Stress phenomenon: genetical
     and evolutional approach, Int. Congr. of Genetics, Moscow,
     21-30 Aug. 1978, Plenary sessions symposia, Abstr., Moscow,
     Nauka, p.106.
Marple, D. N., Aberly, E. D., Forrest, J. C., Blake, W. H., and
     Yudge, M. D., 1972, Endocrine responses of stress-susceptible
     and stress-resistant swine to environmental stress, J.Anim.
     Sci., 35:576-579.
McClean, C. E., and Rodgers, D. A., 1959, Differences in alcohol
     preference among inbred strains of mice, Quart.J.Stud.Alcohol,
     20:691-695.
Mohler, H., and Okada, T., 1977, Properties of ^{3}H-diazepam binding to
     benzodiazepine receptors in rat cortex, Life Sci.,
     20:2101-2110.
Murphy, B. E. P., 1967, Some studies of the protein-binding of
     steroids and their application to the routine micro and
     ultra-micro measurement of various steroids in body fluids by
     competitive protein-binding radioassay, J.Clin.Endocrinol.
     Metab., 27:973-990.
Robertson, H. A., 1979, Benzodiazepine receptors in emotional and
     "non-emotional" mice: comparison of four strains, Eur.J.
     Pharmacol., 56:163-166.
Rokitsky, P. F., 1978, Introduction to Statistical Genetics, Minsk,
     "Higher School".

Seredenin, S. B., Blednov, Yu. A., and Badyshtov, B. A., 1982,
Hereditary dependence of drug reactions, in: "Results of
Science and Engineering, Problems of Human Ecogenetics,"
6:90-143.

Seredenin, S. B., Zinkovksy, V. G., Badyshtov, B. A., Golovenko, N.
Ya., and Rybina, I. V., 1981, Study of genetic differences in
the anticonvulsant effect and metabolic rate of phenazepam,
Bull.Exp.Biol.Med., No.10, 450-452.

Seredenin, S. B., Zinkovsky, V. G., Golovenko, N. Ya., and Rybina, I.
V., 1981, Study of genetic differences of metabolism and
distribution of phenazepan-^{14}C, Chem.Pharm.J.(USSR), 9, 23-26.

Solovjev, V. N., Firsov, A. A., and Fylov, V. A., 1980, Pharmacokinetics, Moscow, "Meditsina."

Speth, R. C., Wastek, G. J., Johnson, P. C., and Yamamura, H. I.,
1978, Benzodiazepine binding in human brain: characterization
using ^{3}H-flunitrazepam, Life Sci., 22:859-866.

Steiner, A. L., Pagliara, A. S., Chase, L. R., and Kipnis, D. M.,
1972, Radioimmunoassay for cyclic nucleotides. II. Adenosine
3',5'-monophosphate and guanosine 3',5'-monophosphate in
mammalian tissues and body fluids, J.Biol.Chem.,
247:1114-1120.

Strange, R. C., and Mjos, O. D., 1975, The sources of plasma cyclic
AMP; studies in the rat using isoprenaline, nicotinic acid and
glucagon, Eur.J.Clin.Invest., 5:147-152.

Valdman, A. V., and Martynikhin, A. V., 1982, The study of the action
of psychotropic drugs on psychophysiological characteristics
of operating activity under emotional stress, in: "Pharmacological Normalization of Fatigue," Yu. G. Bobkov, ed., Moscow,
83-97.

Valdman, A. V., Kozlovskaja, and M. M. Medvedev, O. S., 1979,
Pharmacological Regulation of Emotional Stress, Moscow,
"Meditsina".

Vavilov, N. I. 1922, The law of homology series in variation,
J.Genetics., 12:47-49.

Vessel, E. S., ed., 1971, Drug metabolism in man, Ann.N.Y.Acad.Sci.,
179:1-773.

Vessel, E. S., 1977, Genetic and environmental factors affecting drug
disposition in man, Clin.Pharmacol.Ther., 22:659-679.

Vogel, F., and Motulsky, A. G., 1979, Human Genetics. Problem and
Approaches, Springer-Verlag, Berlin - Heidelberg.

Wehmann, R. E., Blonde, L., and Steiner, A. L., 1974, Sources of
cyclic nucleotides in plasma, J.Clin.Invest., 53:173-179.

WHO Tech. Rep., Pharmacogenetics, 1973, Ser. No.524, 1-40.

Will, B. E., 1977, Neurochemical correlates of individual differences
in animal learning capacity, Behav.Biol., 19:143-171.

Qualitative and Quantitative Assessment of Benzodiazepine Action on Emotional/ Behavioral Responsiveness

A.V. Valdman, E.B. Katkova and M.M. Kozlovskaya

Institute of Pharmacology, Academy of Medical Sciences
of the USSR, Moscow

Psychopharmacologists tend to divide pharmacological compounds into groups according to structural resemblances or some aspect of their effect. This leads to the supposition that, therapeutically speaking, all those substances belonging to one group have a similar action. Many clinical pharmacologists subscribe to the view that one benzodiazepine is as good as another for treating phobias (McCurdy, 1980) and indeed, most controlled trials have not shown any real difference between their clinical efficacy (Greenblatt and Shader, 1978). These drugs have been regarded mainly in the light of their long-term action on anxiety-related syndromes. After the discovery of benzodiazepine receptors, it was found that the degree of inter-action with benzodiazepine binding sites often provides a good yard-stick for a particular product's clinical action. If the size of the recommended 24 h dose of the tranquilizer is an index of its clinical potency, there is likely to be a good correlation with the degree of binding in the human cerebral cortex (Möhler and Okada, 1978). An equally clear correlation has been established between receptor affinity and the corresponding potency of the substance in pharmaco-logical tests – i.e. antagonism with pentylenetetrazole and muscle relaxant action (Braestrup and Nielsen, 1980). These facts indicate that few real differences exist between the pharmacological and clinical profile of benzodiazepines. Differences in the pharmaco-kinetics of particular benzodiazepines do exist; these are reflected in the characteristics of their clinical actions. Lader (1980), however, having performed clinical trials on benzodiazepines, con-siders that these compounds differ pharmacokinetically and pharmaco-dynamically from each other. Individual benzodiazepines have their own pharmacological properties apart from their shared psychotropic effect. Some compounds in the group work better as anti-convulsants or muscle relaxants than anxiolytics. There is no doubt that certain

characteristics of their action which emerge more clearly from single
dose administration are smoothed over by repeated clinical use.

Iversen (1980), writing on the subject of separate identity of
tranquilizer action, headed one section of his paper with the
rhetorical question "Are the properties of the benzodiazepines
dissociable?" He emphatically answers the question in the affirm-
ative and points to the urgent need to develop specific animal behavior tests providing an accurate measure of benzodiazepines'
anxiolytic, sedative and anticonvulsant actions.

The fact was stressed by Iversen (1980) that hardly any systematic research has been performed on different types of behavioral
displays and cognitive performance. Behavioral tests as used in
recent years have illustrated the similarity of different benzodiazepines' effects but, to quote, "no one used a wide range of
behavioral tests to seek differences among them" (p. 873).

Thus the question of distinctions between the pharmacological
effects of different benzodiazepines on emotional behavior remains
unsolved.

2. COMPARATIVE STUDY OF BENZODIAZEPINES' ACTION ON THE INITIAL RANGE OF EMOTIONAL/BEHAVIORAL RESPONSE

We compared the effect of a number of tranquilizers on the
behavior and emotional response characterizing type I and II response
patterns (see the second article), using the method of evaluating
emotional and group behavior in cats outlined in the first article.
We also attempted to identify any specific "remedial" effect of
these compounds in experimental models of neurotic breakdown. All
substances were given orally; tablets were ingested by the cat
together with a lump of meat. The effect of the following drugs was
investigated:

diazepam ("Seduxen") from Gedeon Richter:
30 test doses given to 9 cats.

chlordiazepoxide ("Elenium") from Polfa:
35 test doses given to 12 cats.

oxazepam ("Tasepam") from Polfa:
24 test doses given to 8 cats.

nitrazepam ("Eunoctin") from Gedeon Richter:
24 test doses given to 8 cats.

lorazepam ("Ativan") from Wyeth:
27 test doses given to 9 cats.

phenazepam from the Institute of Pharmacology,
Acad.Med.Sci.(USSR):
27 test doses given to 9 cats.

Table 1. Effect of Diazepam on the Range of Emotional Behavior of
          Different Types of Animal

Type I cats (n = 4)						Type II cats (n = 5)		
Diazepam (mg/kg)			Control	Range of emotional behavior	Control	Diazepam (mg/kg)		
2-3	1-1.5	0.5				0.5	1-1.5	2-3
0*	0*	0.2*	0.8	Fear	2.5	1.0*	0.5*	0*
0*	0.2*	0.6	1.0	Anxiety	2.7	1.2*	1.0*	0.5*
0.3*	0.5*	0.8	1.4	Aggression	1.0	1.4	1.2	1.2
0.6*	0.6*	0.6*	1.6	Conflict	2.0	1.3*	1.0*	1.0*
3.5*	3.6*	3.0*	2.0	Pleasure	0.8	1.2*	2.0*	2.5*
0.5	2.0*	2.0	1.0	Playing	0	0.4	0.8*	0.5
1.0	2.4	2.5	2.0	Hunting	0.6	0.8	1.4	0.7
3.0*	2.3*	2.4*	1.4	Friendliness	0.8	1.6*	1.5*	1.5*
1.3*	2.0	2.0	2.1	Alertness	2.3	2.0	2.0	1.8
1.5	3.2*	2.8	2.5	Curiosity	1.3	1.8	2.0*	1.3
2.0	3.0*	2.6	2.2	Motor activity	2.2	1.8	2.4	2.3

Figures are the mean of displays graded for intensity on a fixed
scale (see the first article). * = change significant according to
Student's t-test, p < 0.05. Figures for 9 cats. n = 30.

Each benzodiazepine's mode of action was analyzed separately for
each type of animal. Patterns of different types of display were
rated according to intensity, expressed as the mean of the points
scored. Mean values were compared by Student's t-test, using a
significance level of p < 0.05. Findings are given in Tables 1, 2,
3, 4, 5 and 6.

Diazepam. Displays of the emotional state of fear-and-alarm in
cats were attenuated by a dose of 0.5-1 mg/kg of diazepam and
virtually suppressed by 2.0 mg/kg. At the same time, the animals'
behavior was activated to show, for example, increased initiative and
more exploratory and hunting activity. Passively defensive responses
to adverse and threatening influences became actively defensive.
They showed pleasure in response to fondling. Most became playful.
Changes were seen in the pattern of relationships within the group:
the male test cat, which had formerly been subordinate, would not
tolerate attacks from fellow-animals following diazepam. It began to
initiate friendly contacts and show rivalry.

A 0.5-1.0 mg/kg dose of diazepam raised motor activity and
exploratory behavior in type I animals. Conflict and aggression were
either less forcefully displayed or actually suppressed, and the cat

Table 2. Effect of Chlordiazepoxide on the Range of Emotional
          Behavior of Different Types of Animal

Type I cats (n = 6)					Type II cats (n = 6)		
Chlordiazepoxide (mg/kg)			Control	Range of emotional behavior	Control	Chlordiazepoxide (mg/kg)		
8-9	5-6	3-4				3-4	5-6	8-9
0*	0.2*	0.5	0.9	Fear	2.3	1.3*	0.8*	0.6*
0*	0.2*	0.5	1.0	Anxiety	2.4	1.7*	1.0*	0.6*
1.0	1.4	1.2	1.5	Aggression	1.0	1.2	1.0	1.0
1.2*	1.4	1.8	2.0	Conflict	1.6	1.0	1.0	1.2
2.5*	2.6*	2.0	1.6	Pleasure	0.8	1.6*	2.0*	1.9*
1.0	0.4	1.0	1.0	Playing	0.4	0.4	0.4	0.7
1.0	1.4	1.8	1.3	Hunting	0.8	1.4	1.0	1.3
2.2*	1.3	1.0	1.0	Friendliness	0.8	1.0	1.5*	2.0*
1.0*	1.2*	1.8	2.4	Alertness	2.6	1.8	1.2	1.0
0.5*	1.0*	2.6	2.0	Curiosity	1.0	1.2	1.2	0.7
1.0*	1.2*	2.6	2.4	Motor activity	1.8	2.0	1.3	1.2

Figures are the mean of displays graded for intensity on a fixed
scale (see the first article). * = change significant according to
Student's t-test, p < 0.05. Figures for 12 cats. n = 30.

stubbornly refused to get involved in fights, only attacking in
self-defense after continued provocation. Displays of "positive" or
pleasurable emotions increase in scale.

    Chlordiazepoxide. Displays of fear-and-alarm were mitigated by
a 3-4 mg/kg dose and virtually eliminated by 6-8 mg/kg in type II
animals. Actively defensive behavior gradually began to replace fear
as a response to adverse stimuli or contacts with aggressive fellow-
animals. "Positive" or pleasurable emotions were increasingly
displayed.

    A dose of 6-8 mg/kg heightened external displays of pleasure in
type I animals. They made more friendly contacts with fellow-
animals. Upon raising the dose to 9 mg/kg, both types of animal
became ataxic, showed reduced motor activity and became less aware,
showing general sedation and muscular relaxation.

    Oxazepam. Displays of fear-and-alarm were alleviated by a dose
of 1-1.5 mg/kg and almost disappeared at a dose of 4-6 mg/kg; the
animals also began to react more. For example, cats reacted to
fellow animals' aggressive or adverse stimuli with somewhat aggressive and marked actively defensive response.

Table 3. Effect of Oxazepam on the Range of Emotional Behavior of
          Different Types of Animal

Type I cats (n = 4)						Type II cats (n = 4)		
Oxazepam (mg/kg)			Control	Range of emotional behavior	Control	Oxazepam (mg/kg)		
4-6	2-3	1-1.5				1-1.5	2-3	4-6
0*	0.3*	0.5	1.2	Fear	2.6	1.5*	0.9*	0.6*
0*	0.3*	1.0	1.5	Anxiety	2.7	1.6*	1.2*	1.0*
0.5*	1.0	0.8	1.4	Aggression	0.6	0.8	1.4*	0.8
1.0*	1.5	1.5	1.8	Conflict	1.0	1.3	1.9*	1.4
2.2	2.2	2.0	1.6	Pleasure	0.8	1.6	1.8*	1.4
1.0	1.0	1.7	1.4	Playing	0	0.5	0.5	0.8
1.8	2.0	2.0	1.6	Hunting	1.0	1.2	1.4	1.4
1.6	1.8*	1.6	1.0	Friendliness	1.0	1.4	1.4	1.4
2.0	2.0	2.0	2.3	Alertness	2.3	2.0	1.8	1.8
2.0	1.8	2.4*	1.6	Curiosity	1.0	1.2	1.8*	1.4
3.3*	3.2*	2.8	2.3	Motor activity	2.1	2.4	2.9*	2.2

Figures are the mean of displays graded for intensity on a fixed
scale (see the first article). * = change significant according to
Student's t-test, p < 0.05. Figures for 8 cats. n = 30.

Oxazepam promoted motor activity in type I animals at a dose of
2-3 mg/kg, causing ataxia when the dose was raised to 4-6 mg/kg. The
level of motor activity remained high but seemed to lack purpose.
Displays of "positive" (pleasurable) emotions were not increased.

Nitrazepam. Displays of fear-and-alarm were suppressed in both
types I and II. The animals became ataxic. They tried to detach
themselves from fellow animals as far as possible, turning their
backs on them. Exploratory and motor activity was inhibited. Response to provocation remained passively defensive. A sedative
effect occurred at a higher dose.

Lorazepam. Displays of fear-and-alarm decreased in both types
of animals at a dose of 0.05 mg/kg and virtually disappeared at an
increased dose of 0.15 mg/kg. Pleasure was increasingly shown, as
was exploratory behavior in type II animals; they began to assert
themselves within the group. A dose of 0.25 to 0.3 mg/kg produced
ataxia. They became generally less active.

Phenazepam. Phenazepam potentiated motor activity, especially
in type II animals. Displays of fear-and-alarm were considerably
reduced, especially for type II. The animals' motor and exploratory
activity dropped and they became ataxic. Type II animals showed more

Table 4. Effect of Nitrazepam on the Range of Emotional Behavior of
          Different Types of Animal

Type I cats (n = 4)					Type II cats (n = 4)			
Nitrazepam (mg/kg)			Control	Range of emotional behavior	Control	Nitrazepam (mg/kg)		
4-6	2-3	1-1.5				1-1.5	2-3	4-6
0*	0.3*	0.6	1.0	Fear	2.4	1.4*	1.0*	0.8*
0.5*	0.3*	0.8	1.3	Anxiety	2.5	1.8	1.3*	0.5*
0.8	1.0	0.8	1.3	Aggression	0.7	1.0	0.6	0.6
1.0	1.2	1.2	1.5	Conflict	1.5	1.6	1.4	0.6
0.5*	2.0	2.3	2.2	Pleasure	1.0	1.1	1.2	0.5
0	0.5	1.0	1.0	Playing	0	0	0	0
0.5*	1.5*	2.2	2.4	Hunting	1.0	1.2	0.9	0.5
1.2	1.2	1.0	1.0	Friendliness	0.8	1.0	1.0	1.0
1.0*	1.0*	2.0	2.3	Alertness	2.5	2.2	1.8	1.0*
0.5	1.3	1.3	2.3	Curiosity	1.5	0.5	0.5	0.5
1.4*	1.8	2.0	2.3	Motor activity	2.3	3.6*	2.0	1.8

Figures are the mean of displays graded for intensity on a fixed
scale (see the first article). * = change significant according to
Student's t-test, $p < 0.05$. Figures for 8 cats. n = 30.

"positive" emotions. Phenazepam noticeably restricted all displays
of emotional/behavioral response. The animals' behavior became
inadequate and they grew increasingly ataxic. The sedative effect of
the drug predominated.

It was thus seen that the benzodiazepines changed the behavior
patterns of normal animals in varying ways according to the subjects'
"character" type. Sometimes a substance actually produced different
effects on different occasions. All the tranquilizers suppressed
fear-and-alarm, but at the same time they produced readjustments in
the individual range of emotional/behavioral response. This fact was
reflected in a number of physical and mental indices of individual
and reciprocal behavior. Our method of evaluating the emotiotropic
action of tranquilizers enabled us to grade the various properties of
certain drugs' action.

It was obviously difficult to obtain precise levels or figures
from our population of cross-bred cats. They had been divided up
into two groups somewhat arbitrarily, since some sub-sets fell between the divisions. But notwithstanding these reservations, our

Table 5. Effect of Lorazepam on the Range of Emotional Behavior of
          Different Types of Animal

Type I cats (n = 4)					Type II cats (n = 5)			
Lorazepam (mg/kg)			Control	Range of emotional behavior	Control	Lorazepam (mg/kg)		
0.25-0.3	0.15-0.2	0.05-1				0.05-0.1	0.15-0.2	0.25-0.3
0*	0.2*	0.2*	1.0	Fear	2.4	1.0*	0.4*	0*
0*	0.2*	0.6	1.3	Anxiety	2.8	1.2*	0.8*	0*
0.6*	0.8	1.0	1.4	Aggression	0.6	1.2	1.0	1.0
0.8*	1.0*	1.2	1.8	Conflict	1.2	1.4	1.2	1.2
2.6*	2.4*	1.7	1.4	Pleasure	0.8	1.5*	1.8*	2.0*
0	0	0.8	0.9	Playing	0.3	1.0	0.8	0
1.4	1.0*	1.5	1.8	Hunting	1.2	1.2	0.6	0.6
2.0*	1.2	1.0	1.0	Friendliness	1.0	1.0	1.4	1.6
1.6*	1.8	2.0	2.2	Alertness	2.3	2.0	2.0	2.0
1.2*	1.4*	2.0	2.0	Curiosity	1.4	2.2*	2.2*	1.0*
1.3*	1.5*	1.9	2.4	Motor activity	2.0	2.6	1.8	1.8

Figures are the mean of displays graded for intensity on a fixed
scale (see the first article). * = change significant according to
Student's t-test, $p < 0.05$. Figures for 9 cats. n = 30.

findings show notable differences in the range of different tranquilizers. The same differences apply to their action in animals of
inherently varying emotional response.

    Research demonstrating that the effect of benzodiazepines on
inbred animals responding differently to exposure to stress is genetically determined is described in the third article. The effect of
benzodiazepines varies from one animal to another according to
genetic type and is controlled by a minimum of two groups of genes
(Seredenin et al.)

3. COMPARISON BETWEEN THE ACTION OF BENZODIAZEPINE TRANQUILIZERS IN MODEL NEUROSIS STATES

    The first state to be observed during the onset of neurosis are
changes in the more complex forms of adaptive behavior, together with
a collapse of relatively stereotyped response to the environmental
situation. This is common to both man and animals, and occurs at the
borderline between normal and pathological. Breakdown of adaptational processes is more clearly displayed in emotionally charged situ-

Table 6. Effect of Phenazepam on the Range of Emotional Behavior of
          Different Types of Animal

Type I cats (n = 4)						Type II cats (n = 5)		
Phenazepam (mg/kg)			Con-trol	Range of emotional behavior	Con-trol	Phenazepam (mg/kg)		
0.25-0.3	0.15-0.2	0.05-0.1				0.05-0.1	0.15-0.2	0.25-0.3
0*	0.3*	1.0	1.3	Fear	2.5	2.2	1.0*	0.5*
0*	0.5	1.0	1.0	Anxiety	2.7	2.4	1.0*	0.5*
0.3*	1.0	1.0	1.2	Aggression	1.0	1.2	1.2	1.0
0.6	1.0	0.9	1.2	Conflict	1.2	1.5	1.2	1.0
1.3	1.2	1.3	1.3	Pleasure	0.7	0.5	1.5*	1.7*
0	0	0.5	1.0	Playing	0.3	0.3	0	0
0.5*	1.0*	1.5	2.4	Hunting	1.0	0.5	0.3*	0.5*
1.2	1.0	1.0	1.0	Friendliness	0.8	0.6	0.6	1.0
1.3*	1.5	2.0	2.1	Alertness	2.3	2.4	1.6	1.2*
0.5*	0.5*	1.8	2.5	Curiosity	1.3	1.0	0.3*	0.5
1.2	2.0	2.8*	2.0	Motor activity	2.2	2.4	1.6	1.2*

Figures are the mean of displays graded for intensity on a fixed
scale (see the first article). * = change significant according to
Student's t-test, $p < 0.05$. Figures for 9 cats. n = 30.

ations. Neurosis-like conditions were therefore simulated in
animals, involving fear and emotional tension, which were reinforced
and increased as the same circumstances recurred. For further details see Valdman, Kozlovskaya and Medvedev (1979).

     Direct electrical stimulation of the emotiogenic zones of the
hypothalamus by the method described in the first article induced a
state of fear-and-alarm, especially when a number of "points" in the
dorso-lateral hypothalamus were stimulated; this caused a distinct
change in the range of emotional/behavioral response (see Figure 1).
The "negative" or noxious emotional state produced by negative stimulation (see segments 2, 3 and 4 of the circle) suppressed all displays of positive or pleasurable emotion (see segments 5, 6 and 7)
and eliminated friendly contacts between the test cat and its fellow
animals (see segment 9) without an accompanying reduction in aggression or conflict (segments 11 and 10) or increase in passivity
and fear. Motor activity (see segment 12) dropped, while alertness,
represented by segment 13, was heightened, this being an indication
of an increasingly anxious and frightened response to afferent influences.

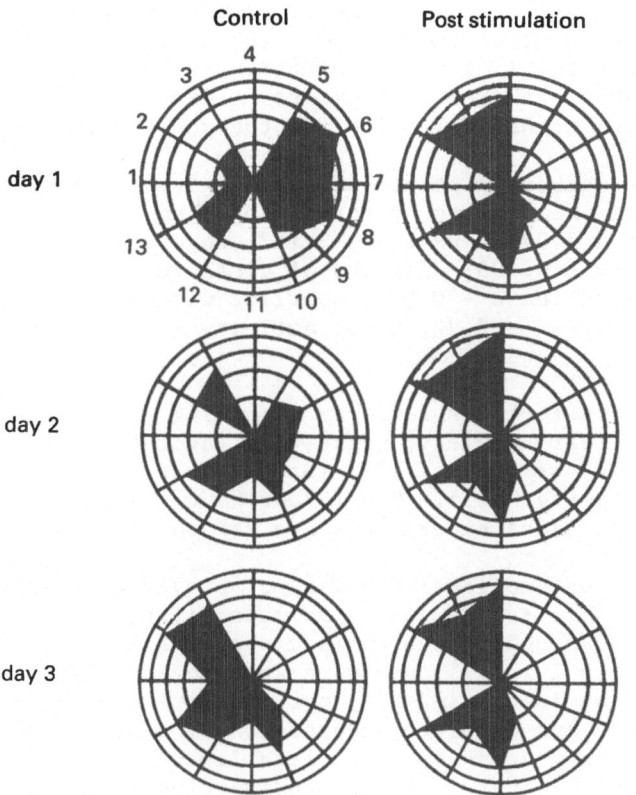

Fig. 1. Neurosis-like state sets in with repeated stimulation of the
         "negative points" of the hypothalamus.

An animal's behavior within the group changed following "negative" or adverse hypothalamic stimulation. Conflict could develop,
but not on the part of the test cat. It was often the victim of
attack; this in turn promoted fear.

Post-stimulation fear-and-alarm grew in duration and intensity
with repeated noxious hypothalamic stimulation over a number of days.
A neurosis-like state developed. The very placing of the animal in
the experimental chamber triggered displays of anxiety through a
conditioned reflex mechanism, and further "punishing" electrical
stimulation noticeably emphasized its "negative" emotional condition.

A neurotic state was easily induced experimentally by repeated
adverse electrical stimulation in type II animals with an inherent
fear-dominated response pattern. In type I animals the neurotic
state occurred in two stages. Initially type I behavior patterns
were changed by 2-3 experiments involving brain stimulation to resemble those of type II animals. The anxiety-related emotional

response increased with each successive experiment until it occurred
not just after stimulation but whenever the animal is in the environment of the experimental chamber. Differences between inborn types
of emotional response therefore began to be ironed out; it was thus
possible to obtain combined results for the whole animal group.

The average number of points scored for induced fear-and-alarm
response over the whole group of animals studied, where n = 55, were
as follows: fear - 2.975; anxiety - 3.183; negativity - 3.025; exploratory activity - 0.105; displays of pleasure - 0.05; playing - 0;
hunting - 0; friendliness - 0.558; conflict within the group - 1.25;
inadequacy of behavior - 2.92; motor activity - 1.41; alertness -
3.016.

Tranquilizer effect was rated by combining the points scored by
the responses of fear, anxiety, negativity (average: 9.18) and those
scored for displays of the positive responses of pleasure, playing
and hunting (averaging 0.05). The latter have a virtually zero
score. Results of these experiments are set out in Table 7.

The spectrum of emotional behavior displayed where a neurotic
response is artificially induced in animals differs substantially
from the control throughout the whole group of animals (n = 55). On
this occasion types I and II cats were kept together. A neurosis-
like state could be identified by a noticeable increase in "negative"
emotions displayed and complete suppression of displays of "positive"
emotion. Exploratory behavior also virtually disappeared. Anxiety
led to heightened alertness. All benzodiazepine tranquilizers
produced a dose-related reduction in displays of induced emotionally
negative states (fear-and-alarm). The variations between the effect
of different benzodiazepines are clear-cut as far as the display of
positive emotions is concerned.

Diazepam and chlordiazepoxide, as well as reducing the fear
response, restore previously suppressed responses to fondling, as
well as playful and hunting response. Diazepam reinforces attempts
at establishing friendly contacts within the group and restores
exploratory behavior, while potentiating motor activity to some
extent. Chlordiazepoxide also intensifies the display of positive
emotions and the friendly contacts made within the group, but acts
as a sedative at higher doses. Oxazepam restores behavioral displays connected with positive (or pleasurable) emotional states and
promotes motor activity. Nitrazepam did not reinstate such positive displays as exploratory behavior; the animals grew passive.
Phenazepam likewise failed to restore exploratory behavior and promoted the return of displays of positive emotions to a lesser extent. Lorazepam was less successful than diazepam in both restoring displays of positive emotions and potentiating exploratory behavior.

Table 7. Comparative Action of Benzodiazepine Tranquilizers on Range of Emotional/Behavioral Response in the Neurosis Model

Drug	Dose (mg/kg)	Emotions displayed			Group behavior				
		Negative emotions	Positive emotions	Ag-gression	Friend-liness	Con-flict	Motor ac-tivity	Explo-ration	Alert-ness
Control (n=55 initially)		3.5	3.2	1.2	1.0	1.5	2.2	1.6	2.3
Induced neurosis-like fear-and-alarm response (n=55)		9.2+	0+	0.9	0.6	1.3	1.4+	0+	3.0
Diazepam (n=9)	0.5	4.6*	1.3*	0.8	1.5	1.0	1.2	0.9*	2.2
	1.5	2.0*	3.6*	0.9	1.8*	0.8	1.7	1.5*	2.0
	3.0	0.9*	3.0*	0.7	1.8*	0.8	2.0*	1.3*	1.7*
Chlordiazepoxide (n=12)	3.0	6.2*	1.9*	1.1	1.0	1.0	1.1	0.8	2.2
	6.0	2.3*	2.7*	0.8	1.4	1.0	1.0	1.0*	1.6*
	9.0	2.4*	3.0*	1.0	2.0*	1.0	1.0	0.6	1.2*
Oxazepam (n=8)	1.0	6.3*	0.5	0.9	1.0	1.0	1.5	0	2.4
	3.0	3.3*	2.1*	0.6	1.2	1.2	1.8	0.7	2.0
	6.0	2.1*	2.9*	0.5	1.4*	1.1	1.8	1.1*	2.0
Nitrazepam (n=8)	1.0	6.1*	0	1.1	1.0	1.1	1.0	0	2.4
	3.0	3.5*	0.5	0.5	1.0	1.0	1.4	0.3	1.5*
	6.0	2.7*	0.4	0.4	1.0	1.0	1.2	0	1.3*
Lorazepam (n=9)	0.05	5.3*	0.4	0.4	1.0	0.6	1.2	0.3	2.3
	0.15	2.4*	1.8*	0.8	1.0	0.8	1.2	1.0*	2.2
	0.30	0.7*	2.3*	0.9	1.1	1.0	1.1	1.0*	1.8
Phenazepam (n=9)	0.05	6.0*	0	0.5	0.8	0.8	1.2	0	2.5
	0.15	4.1*	0.3	0.7	0.7	0.7	1.2	0	2.0
	0.30	2.4*	1.0	0.5	1.0	0.5	1.0	0	1.7*

Baseline shown as the mean for all cats prior to induction of neurosis, without subdivision into groups. + = significant difference between neurosis-type reaction and control; * = difference, post-tranquilizer from original level of neurosis, significant at $p < 0.05$ using Student's t-test; n = number of animals.

Each benzodiazepine tranquilizer thus differs in emphasis within
its range of emotiotropic action on induced fear-and-alarm or
neurosis-like reaction. In this context, however, the drugs act
mainly as an antidote to fear; they have much less influence on other
physical and mental effects (excluding the positive emotions) displayed by these animals than on behavior of other animals unaffected
by neurosis.

4. STUDY OF CORRELATIONS OF THE INFLUENCE OF BENZODIAZEPINE
    TRANQUILIZERS ON THE PHYSICAL AND MENTAL PATTERN OF EMOTIONAL
    RESPONSE IN THE ANIMAL NEUROSIS MODEL

Changes were produced by benzodiazepine tranquilizers within the
fear-and-alarm range of the emotional/behavioral response spectrum by
hypothalamic stimulation of neurotically affected animals; by comparing these changes certain quantitative differences between drugs
were revealed. This type of analysis is uninformative, however,
regarding the nature of a tranquilizer's "healing" effect, i.e. of
readjusting an animal's emotional response pattern. Analysis by
correlation, as described in the second article, was the method used
to study how tranquilizers influence the underlying pattern of a
neurotically affected animals' responses.

A correlation matrix of emotional/behavioral response was produced by determining the correlation coefficient with paired comparison of separate indices. As shown in the second article, the
study of correlations among non-pre-treated animals revealed typical
sets of relationships which characterize the emotional/behavioral
response of both types 1 and II (See Table 4, the second article) and
indicate resemblances and differences between them.

The correlation matrix of model emotional/behavioral response
known as "fear-and-alarm", as evoked by hypothalamic electrical
stimulation of neurotically affected animals, had much in common with
that relating to type II cats. Exposure to neurosis-promoting agents
had increased and reinforced correlations peculiar to type II
emotional response pattern. The sort of response found of type I
animals obviously comes closer to the physiological norm. The benzodiazepines' effects were therefore rated by comparing and contrasting
the readjustments of correlations which they had produced with the
correlation matrix of type I animals' response patterns (see the
second article for a description of the two types of emotional/
behavioral response patterns). The disappearance of correlations
peculiar to type II response pattern and the appearance of those
characteristic to type I (i.e. the normal) and the maintaining of
common (non-specific) correlations may be considered an indication of
optimum "healing" tranquilizer action.

The effects of various benzodiazepines are given in Tables 8-13. No numerical value is given for the correlation coefficient in the correlation matrix as in the format of the second article. The correlation signs (+) or (-) only are used for clarity's sake; correlations are shown either as maintained or phased out (shaded and X'ed out respectively). Correlations typical of type I response pattern - those changes indicating the normalizing or "healing" effects of the drug - are boxed. The correlations which reappear and are atypical for "normal" but specific for a certain drug are ringed. Only significant correlation coefficients are considered (p < 0.05).

Table 8 gives the readjusted pattern of correlations between emotional/behavioral displays under the influence of diazepam: the disappearance of a number of correlations typical of type II response, produced by the high level of negative emotions. A set of relationships typical for type I emotional/behavioral response (i.e. for the physiologically normal state) also appeared. Quite new correlations also made their appearance, which did not belong to the mainstream types of emotional response: negative correlations between motor activity and the negative emotions of anxiety and fear, positive ones between motor activity, the positive emotions and friendliness. These correlations may be viewed as indications of a drug's inhibiting or potentiating action on motor and general behavioral activity. The switch from the initial negative correlation of aggression with negative emotions to a positive one is seen in the qualitative changes of the fear response, which takes on an actively-defensive character, as well as reflecting the somewhat anti-aggressive effect of the drug.

Clear-cut reciprocal relationships between the positive emotions and a) aggression and b) the negative emotions, occurring with diazepam and apparently connected with its euphoric action, go beyond the point of adequacy (interdependence) typical of type I emotional response. Diazepam's range of effect thus includes a series of adjustments in the pattern of correlations produced by its psychotropic or emotiotropic effects as well as its tranquilizing action.

Chlordiazepoxide (see Table 9) destroyed both type II correlations and to some degree, the relationships typical of type I between a) negative emotions and conflict, and b) aggression and positive emotions. Raising the level of positive emotions produced a positive correlation between positive emotions and friendliness typical of type I. The inhibition of motor activity and alertness occurring with chlordiazepoxide caused the relationship between aggression and motor activity to disappear and potentiated a negative correlation between aggression and alertness. This type of readjustment of correlations demonstrates the strong tranquilizing action of chlordiazepam accompanying its sedative effect.

Table 8. Outline of Readjustment of correlations produced by Diazepam (n=9)

Emotional behavior displayed	Aggression	Negative emotions	Positive emotions	Friendliness	Conflict	Motor activity	Alertness
Aggression		-		-	+		
Negative emotions	⊕				-		
Positive emotions	[-]	⊖			-		
Friendliness	-		[+]				
Conflict	x	x	[-]	-			
Motor activity		⊖	⊕	⊕			+
Alertness			⊕			+	

Key: [ ] = type 1 correlations appears.
    ○ = atypical correlation (but specific for this tranquilizer).
    = correlation maintained.
   x = correlation disappears.
  +, - = positive or negative correlation.

Table 9. Outline of Readjustment of Correlations Produced by Chlordiazepoxide (n = 12)

Emotional behavior displayed	Aggression	Negative emotions	Positive emotions	Friendliness	Conflict	Motor activity	Alertness
Aggression							
Negative emotions	x	-	-	▒	+ -	-	+
Positive emotions	x ▒ +	x	⊞	▒	▒		+
Friendliness							
Conflict	+	x					
Motor activity	x						+
Alertness	○	+				+	

Key: [ ] = type 1 correlations appears.
○ = atypical correlation (but specific for this tranquilizer).
▒ = correlation maintained.
x = correlation disappears.
+, - = positive or negative correlation.

Type II correlations also disappeared with oxazepam and relationships typical of type I emotional response were produced (Table
10). The clear-cut activating effect of the drug was shown by the
positive correlation appearing between a) motor activity and aggression and b) positive emotions and friendliness.

Nitrazepam (see Table 11) phased out not only type II correlations, but also a few non-specific relationships of a more or less
stable nature. This would appear to indicate sedative action. The
positive correlation arising between negative emotions and motor
activity demonstrate that fear response maintains its active quality.
It is significant that positive relationships between the positive
emotions and a) aggression and b) conflict were produced by nitrazepam which are atypical of other tranquilizers. Taken together with
nitrazepam's tranquilizing and sedative action this may be interpreted as diffusely inhibiting the positive and negative reinforcement systems. The failure of animals to respond adequately to test
stimuli is an outward expression of this phenomenon.

Lorazepam (see Table 12) optimally readjusted the correlation
matrix and brought it closer to type I emotional response pattern.
The positive correlation arising between the positive emotions and
motor activity tallied with potentiation of exploratory behavior and
playfulness. The switch from a positive correlation of friendliness
and negative emotions to a negative one is peculiar to lorazepam.

Phenazepam (see Table 13) like nitrazepam, produced a positive
correlation between aggression and a) positive emotions and b)
friendliness. Unlike nitrazepam, it did not break down non-specific
correlations (or links) and no observable damage to the adequacy of
an animal's responses occurred. With diazepam both positive emotions
and aggression showed a positive correlation with motor activity;
this points to the active nature of these displays.

To conclude, we were able, using the above method of analyzing
emotional response in animals and adopting current methods of multi-
factor analysis, to demonstrate the separate emotiotropic and "healing" properties of benzodiazepine tranquilizers' action and to assess
these on a qualitative and quantitative basis. Our method may be
used to investigate new psychotropic compounds with a view to
predicting their emotiotropic action at the pre-clinical stage. Our
experimental mode reflects short-term shifts in the pattern of
emotional/behavioral response - here, fear-and-alarm - and the potent
effect exerted by benzodiazepine tranquilizers. It is doubtless more
valuable to investigate these relationships in animals by prolonging
the model psychopathological state and by chronic tranquilizer
treatment. Needless to say, findings in animals cannot be directly
extrapolated to man; nonetheless our approach does provide further
opportunities for thorough investigations of benzodiazepines and
their psychotropic action.

Table 10. Outline of Readjustment of Correlations Produced by Oxazepam (n=8)

Emotional behavior displayed	Aggression	Negative emotions	Positive emotions	Friendliness	Conflict	Motor activity	Alertness
Aggression		—					
Negative emotions	x			▓ +	▓ +		
Positive emotions	[-]			▓ +	▓ —		
Friendliness	▓ +	x	[+]				
Conflict	▓ +	x	[-]	⊕			
Motor activity	⊕		⊕		▓▓		▓ +
Alertness						▓▓	

KEY:
    [ ] = type 1 correlations appears.
    ◯ = atypical correlation (but specific for this tranquilizer).
    ▓ = correlation maintained.
    x = correlation disappears.
    +, - = positive or negative correlation.

Table 11. Outline of Readjustment of Correlations Produced by Nitrazepam (n=8)

Emotional behavior displayed	Aggression	Negative emotions	Positive emotions	Friendliness	Conflict	Motor activity	Alertness
Aggression		-		-	+	-	
Negative emotions	x			+	-		+
Positive emotions	⊕				-		
Friendliness	-	x					
Conflict	+	⊕	⊕	x			
Motor activity	x	+					+
Alertness			⊖			x	

KEY: ◯ = atypical correlation (but specific for this tranquilizer).

       = correlation maintained.

  x = correlation disappears.

  +, - = positive or negative correlation.

Table 12. Outline of Readjustment of Correlations Produced by Lorazepam (n=9)

Emotional behavior displayed	Aggression	Negative emotions	Positive emotions	Friendliness	Conflict	Motor activity	Alertness
Aggression							
Negative emotions	x	-	+	▨	▨		▨
Positive emotions	[-]			▨ +	- + ▨		
Friendliness	▨ -	○	[+]	▨	▨		
Conflict	+	x	x				
Motor activity			⊕			▨	
Alertness	+					+	+

KEY: [ ] = type 1 correlations appears.
○ = atypical correlation (but specific for this tranquilizer).
▨ = correlation maintained.
x = correlation disappears.
+, - = positive or negative correlation.

Table 13. Outline of Readjustment of Correlations Produced by Phenazepam (n=9)

Emotional behavior displayed	Aggression	Negative emotions	Positive emotions	Friendliness	Conflict	Motor activity	Alertness
Aggression		-		-	+	+	
Negative emotions	x		-	+	-		+
Positive emotions	⊕	x					
Friendliness	-	x			-		+
Conflict	+	x	⊕				
Motor activity	+		⊕	-			
Alertness		+		x		+	

KEY: ◯ = atypical correlation (but specific for this tranquilizer).

  = correlation maintained.

x = correlation disappears.

+, - = positive or negative correlation.

REFERENCES

Braestrup, C., and Nielsen M., 1980, Benzodiazepine receptors,
    Arzneim.Forch., 30:852-857.
Greenblatt, D. J., and Shader, R. I., 1978, Pharmacotherapy of
    anxiety with benzodiazepines and β-adrenergic blockers, in:
    "Psychopharmacology: A Generation of Progress," M. Lipton,
    A. Di Mascio, and K. Killian, eds., Raven Press, New York,
    pp.1381-1390.
Iversen, L. L., 1980, The present status of benzodiazepines in psy-
    chopharmacology, Arzneim.-Forsch., 30:907-910.
Iversen, S. D., 1980, Animal models of anxiety and benzodiazepine
    actions, Arzneim.-Forsch., 30:862-868.
Lader, M., 1980, The present status of the benzodiazepines in psy-
    chiatry and medicine, Arzneim.-Forsch., 30:910-913.
McCurdy, L., 1980, The short-term use of benzodiazepines, Arzneim.
    Forsch., 30:895-897.
Mohler, H., and Okada, T., 1978, The benzodiazepine receptor in
    normal and pathological human brain, Br.J.Psychiat.,
    133:261-268.
Valdman, A. V., Kozlovskaja, M. M., and Medvedev, O. S., 1979,
    Pharmacological Regulation of Emotional Stress, Moscow,
    Meditsina.

Psychophysiological and Neurochemical Analysis of Benzodiazepine Tranquilizers' Activating Effect

A.V. Valdman and B.V. Andreev

Institute of Pharmacology, Academy of Medical Sciences
of the USSR, Moscow

1. INTRODUCTION

It is only anxiolytic drugs, as distinct from other groups of sedatives, which heighten behavioral response in animals over a wide dose range. The increased motor and exploratory activity in rats and mice in an open field facilitate operative behavior, increase the incidence of behavioral response under negative reinforcement and potentiate self-stimulation. The potentiating and anxiolytic effects are to some extent linked. If fear is eliminated from passive behavior, this leads to disinhibition and consequently to potentiating certain behavior. The potentiating effect, however, is bound up with the animal's emotional and behavioral response patterns. In a number of cases aggression actually increased as a result of tranquilizer action (see Lader, 1980).

The methods normally used for assessing behavioral depression experimentally are: response-contingent punishment, fear conditioning, aversion, novelty and reduction of reward. It was the psychophysiological pattern of tranquilizer potentiating effect that we set out to investigate under laboratory conditions. We relied on appropriate behavioral methods enabling us to assess for the first time the whole range of emotional/behavioral response.

Our next aim was to analyze the neurochemical mechanisms potentiating benzodiazepine action and how they are involved in the anxiolytic and sedative functions of these compounds - a topic of special interest.

2. PSYCHOPHYSIOLOGIGAL AND BEHAVIORAL DISPLAYS POTENTIATING TRANQUILIZER ACTION IN CATS

Range of emotional/behavioral response was assessed in male cats both individually and in a group under experimental conditions. The method used is described in the first article.

Behavioral potentiation is only considered to have taken place when some type of behavioral display has previously been suppressed or inhibited which appears appropriate to the experiment in hand (see the works of Haefely, 1978). When working with animals such as cats and dogs, however, each subject's own psychological make-up should be taken into account, whether moulded by, for example, animal history, genetic factors or housing. Tranquilizer action may vary considerably from one animal to another for these reasons. Individual, i.e. personality factors, fashion tranquilizer effects observed in man to an even greater extent (Alexandrovski, 1981). The fact that the same dosage of a certain benzodiazepine tranquilizer produces aggression in one individual and a hypnotic effect in another (Lader, 1980) illustrates this point; such differences cannot be entirely explained by individual pharmacokinetic variation. Differences in tranquilizer effect shown in animals under experimental conditions may be occasioned either by their acting on innate forms of behavior or on behavior patterns which may have been suppressed during their own developmental period or related to certain environmental factors.

Details were given of how extensively cats differ in their emotional and behavioral response pattern in the second article. The changing pattern of some elements of their emotional behavior produced by increasing doses of tranquilizers were compared for two contrasting types of cat, described in the same article. Table 1 shows diazepam's effect on the emotional/behavioral range of response of these two types.

Behavioral displays of fear-and-alarm are observed to fall off in both groups of animals. This anxiolytic effect is dose-dependent. A reduction in aggression and conflict appeared most noticeably in group I. A section of group II animals showed increased aggression due to a growing tendency towards dominance within the group and their exploratory activity was potentiated. Diazepam heightened the display of positive emotions and friendly contacts within the group. This was especially noticeable in type I animals.

Figure 1 gives an example of chlordiazepoxide's effect on one type I and one type II cat. The drug acted differently, according to the animal's innate response pattern. Displays relating to positive emotions and friendliness, behavioral initiative and motor activity were all increased in type I animals. Displays of fear-and-alarm disappeared in the case of type II animals. Their displays of behavior patterns relating to positive emotions hardly changed. Ag-

Table 1. Effect of Increasing Dosage of Diazepam on the Emotional/Behavioral Response Patterns of Type I and II Cats

Diazepam (mg/kg) (Type I cat)			Control	Emotional behavior	Control	Diazepam (mg/kg) (Type II cat)		
2-3	1-1.5	0.5-1				0.5-1	1-1.5	2-3
0	0	0.2*	0.8	Fear	2.5	1.0*	0.5*	0
0	0.2*	0.6	1.0	Anxiety	2.7	1.2*	1.0*	0.5*
0.3*	0.5*	0.8	1.4	Aggression	1.0	1.4	1.2	1.2
0.6*	0.6*	0.6*	1.6	Conflict within the group	2.0	1.3*	1.0*	1.0*
3.5*	3.6*	3.0*	2.0	Pleasure	0.8	1.2*	2.0*	2.5*
0.5	2.0*	2.0*	1.0	Playfulness	0	0.4	0.8*	0.5
3.0*	2.3*	2.4*	1.4	Friendly contacts within the group	0.8	1.6*	1.5*	1.5*
1.3*	2.0	2.0	2.1	Alertness	2.3	2.0	2.0	1.8
1.5	3.2*	2.8	2.5	Exploratory activity	1.3	1.8	2.0*	1.3*
2.0	3.0*	2.6	2.2	Motor activity	2.2	1.8	2.4	2.3

Figures are the mean score (M ± SEM) from a group of 9 cats.
n = 30.
* $p < 0.05$. See the first two articles for details of rating system.

CONTROL   CHLORDIAZEPOXIDE mg/kg

                              3   6   9

CAT TYPE-I

CAT TYPE-II

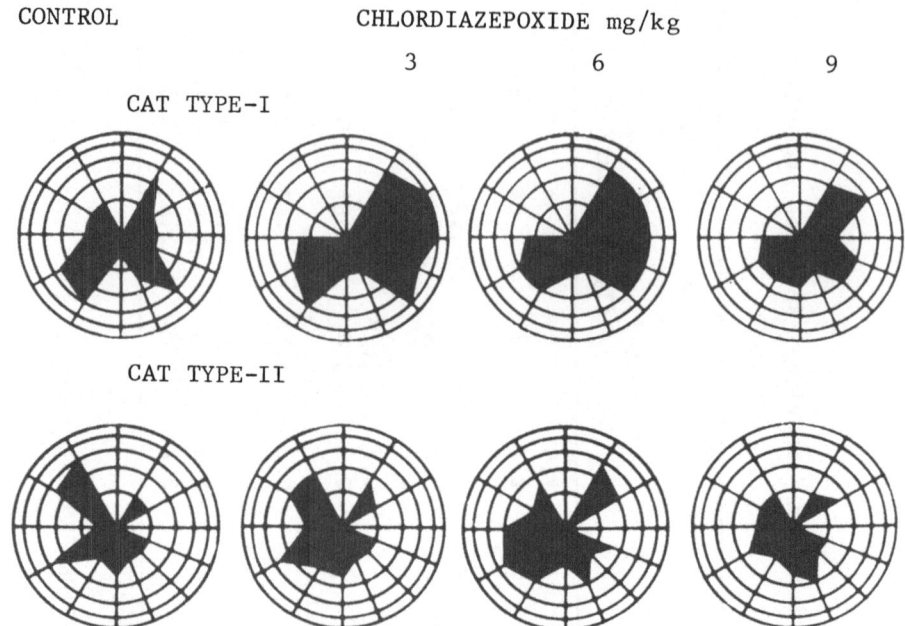

Fig. 1.   A comparison of chlordiazepoxide's effect on the pattern of
          emotional/behavioral response in one type I and one type II
          cat.

gression was displayed and motor activity increased. Different
tranquilizers of the benzodiazepine group, however, varied in the
extent of their potentiating effect. Combined findings relating to
type II for the whole set of animal experiments are given in diagrammatic form (see Figure 2).

A 0.1 to 0.15 mg/kg oral dose of lorazepam has a marked potentiating effect, especially on contesting leadership and dominant
behavior patterns shown within the group, as well as goal-directed
activity. An oral dose of 1-2 mg/kg of diazepam, while suppressing
fear-and-alarm, heightened (sometimes purposeless) motor activity and
reinforced displays related to the positive emotions. Oxazepam
increased behavioral initiative at roughly sub-anxiolytic doses of
0.5-1.5 mg/kg. An oral dose of chlordiazepoxide, while increasing
displays of positive emotion, failed to potentiate behavior patterns;
in fact its anxiolytic action was accompanied by a fall in motor
activity, exploration and initiative, as well as reduced contacts
within the group. Neither nitrazepam, at a 0.5-5 mg/kg dose, nor a
6-12 mg/kg dose of grandaxin (= tofizopam) showed any potentiating
effect.

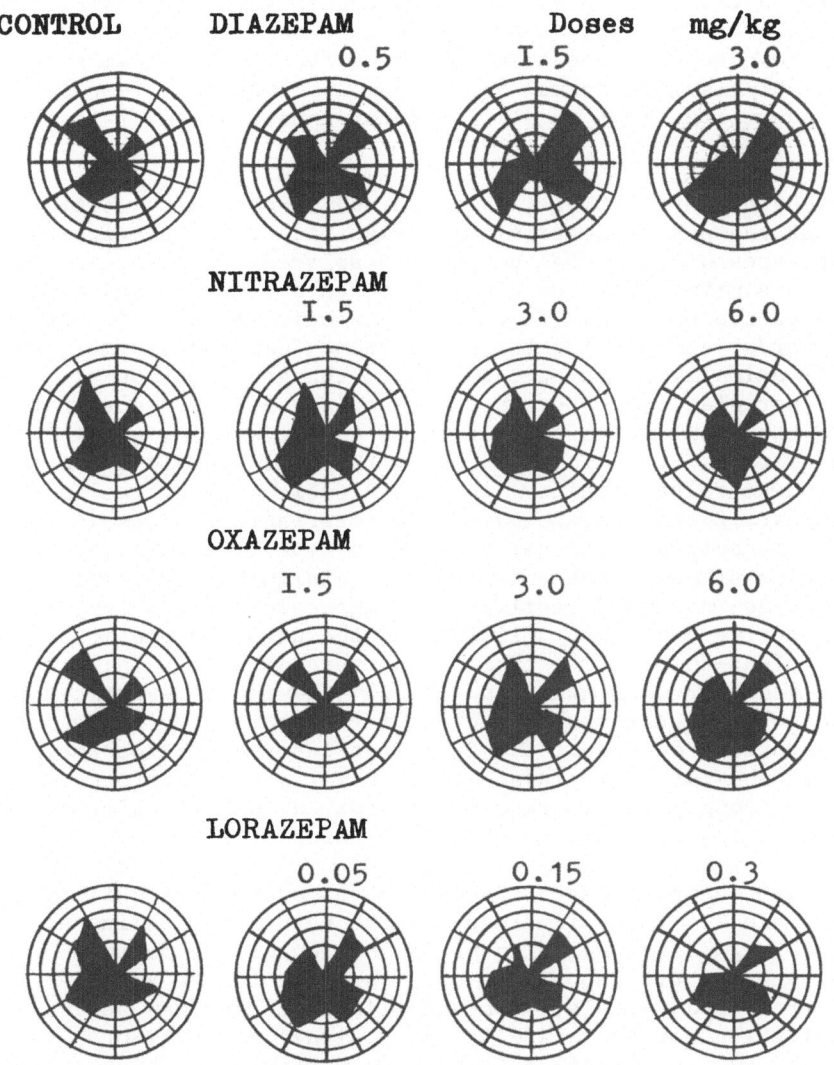

Fig. 2. Varying tranquilizer action on the spectrum of emotional/
behavioral response in type II cats.

A tranquilizer's potentiating effect is thus reflected by a
number of interrelated physical and mental factors. Changes of
emphasis occur within the range of behavioral response; behavior
patterns such as play, hunting and exploration, which are connected
with positive emotions, are reinforced. In type II animals, which
tend towards passively defensive behavior and displaying anxiety, it

is actively defensive behavior, with a tendency to aggression, which
is potentiated. They also show an increase in motor activity. It is
important to note that the potentiating effect of diazepam operates
in type I animals with normal behavioral and emotional response
patterns as well as in type II with their innate state of fear-and-
alarm.

A study of correlations, as explained in the second article,
showed that the set of statistically significant correlations between
different types of emotional behavior displayed, inherent in type II
cats, depends principally on the predominance of fear- and anxiety-
related negative emotions affecting these animals. Stabilization,
in other words, eliminating the predominant fear-and-alarm factor
in type II animals, either attenuates or changes the emphasis of a
number of correlations. But the fear-and-alarm factor does not
influence the positive correlation between aggression, conflict and
positive emotions; this fact indicates the solid physical and mental
basis of these relationships. It is obviously a reduction in the
fear-and-alarm factor which causes characteristic changes in the
emotional/ behavioral response pattern in type II animals. These are
observable as signs of the potentiating (disinhibiting) tranquilizer
action such as increased conflict and aggression plus a very slight
rise in positive emotions. Eliminating the negative emotions - the
element of fear-and-alarm - from the pattern of type I animals'
emotional behavior does not substantially change the existing correlations, whereas the stabilization of the dominating element of
positive emotions decreases the degree of interrelatedness between
emotional/ behavioral displays. It is clearly the reinforcement of
positive emotions which is responsible for the potentiating action of
tranquilizers on the behavior of type I animals.

The main observed psychotropic effects of the tranquilizers are
given in Table 2. These findings are the result of pooling experimental data regarding benzodiazepines' action on the pattern of
emotional/behavioral response in cats (see the fourth article)
obtained from group trials. A tranquilizing effect may be exhibited
as a combination between anxiolytic and either sedative or potentiating action. But the mental and physical pattern of action of the
potentiating component itself depends on the subject's inherent
emotional/behavioral response pattern.

3. THE INFLUENCE OF TRANQUILIZERS ON CAT BEHAVIOR AND
   SYSTEMS OF REWARD AND PUNISHMENT

Our study of correlations between different emotional/behavioral
displays in cats (see the second article) showed that the prevalent
type of behavior was connected with dominance in the cat's emotional
make-up; positive for type I and negative for type II cats. The way
which tranquilizers produce shifts of emphasis in the pattern of an

Table 2. Illustrating the Psychotropic Action of Benzodiazepine
          Tranquilizers (trials on individual and group behavior in
          cats).

I Anxiolytic Action

- attenuation of fear and anxiety behavior
- normalization of impaired adaptation to the changing
  exigencies of the experimental situation
- disappearance of fear-and-alarm behavior elicited by
  electrical stimulation of hypothalamus

II Sedative Action

- drop in motor activity
- drop in exploratory activity
- drop in alertness
- drop in emotional response
- drop in rate of self-stimulation of the brain
- drop in interaction within the group

III Activating Effects

- increase in behavioral initiative, goal-directed activity
- increase in alertness and exploratory activity
- tendency to dominance in the group
- conflict under provocation or when frustrated in its aims
- increase in evoked or spontaneous manifestations of positive
  emotional behavior (pleasure, playing, etc.)
- increased rate of self-stimulation of the positive
  reinforcing brain sites

$$\text{TRANQUILIZING EFFECT} = \frac{\text{anxiolytic} + \text{sedative action}}{\text{anxiolytic} + \text{activating effect}}$$

---

animal's emotional/behavioral response is connected with their effect
on the reinforcement systems of the brain. Reinforcement is the
final and decisive step in every action taken. Thus, a drug's psychopharmacological effect depends in many ways on how it works on
reward and punishment systems. A detailed review of our investigations in this area can be found in the monographs by Valdman,
Zwartau and Kozlovskaya (1976); and Valdman, Kozlovskaya and Medvedev
(1979).

Our experiments were performed on cats, using unipolar electrodes, implanted in the hypothalamus according to the coordinates in
the stereotaxic atlas (Jasper, Aimone-Marsane, 1961), see the first
article. Stimulation was applied; parameters - square pulses,
100/sec; duration, 0.5-1.0 msec; amplitude, 0.2-7 V at 0.1-1.0 mA.
The positive reinforcement effects were measured by the pedal self-
stimulation

method (SS). No limits were imposed on the duration of the stimulation. Our previous investigations had shown that duration of pedal- or lever-pressing is a better indicator of the noxious element in the stimulation than frequency; it is the total duration of stimulation which properly reflects the degree of reinforcement (Valdman, Zwartau, Kozlovskaya, 1976; Zwartau, and Patkin, 1974). The method of locomotor SS in a shuttlebox was also used. An experimental chamber measuring 170 x 150 x 110 cm was divided in two by a barrier 25 cm high. Brain stimulation was set off by the cat jumping over the barrier into the "active" portion of the chamber and switched off by its jumping back again. Each session of SS lasted 10 min. If the animal failed to switch off the current during 30 sec of stimulation, both the area stimulated and the reaction itself were taken as plainly positive. If, after electrical stimulation, the cat did not return to the active compartment, both the area stimulated and the response elicited were considered negatively reinforcing or punishing. Should the animal keep shifting backwards and forwards from the active or rewarding portion of the chamber into the negative or punishing section, the site stimulated was considered equivocal. Here the latency of escape and return reflected the balance between the rewarding and punishing elements in central stimulation (see also Patkina and Lapin, 1976a,b).

The tranquilizer noticeably decreased SS response in the case of both plainly positive and equivocal points, as seen by the increased frequency of pedal pressings. The overall time spent under current increased either almost imperceptibly, or in step with the frequency of pressings. The average duration of one pressing was barely changed. Diazepam produced a similar effect.

After administering 1-3.5 mg/kg of chlordiazepoxide an increase in the animals' activity was observed, together with an urge to move towards the experimental chamber, where the SS experiments had taken place. If an attempt is made to hold the animal back or distract it, the cat reacts by hissing and snarling, a self-protective reaction.

Table 3. The Effects of 3-5 mg/kg chlordiazepoxide on Pedal Self-stimulation. (% of initial values M ± SEM)

Type and number of stimulation points	Changes in:		
	No. of pressings	Total time remaining under current	Mean duration of one pressing
Plainly positive	+65.0± 7.3*	+69.0±11.0*	+12.0±3.0*
Equivocal	+38.0±14.8*	+41.0±12.0*	+11.2±8.5

* $p < 0.05$

The cat behaves aggressively to the "top cat" of the group if it blocks its way to the pedals.

The effect produced by this tranquilizer during locomotor SS is shown in Figure 3. Here, the stimulated site caused an equivocal response. Chlordiazepoxide noticeably accelerated the animal's return to the active half of the chamber. As the overall number of returns increased, the latent period between them diminished; this shows that the tranquilizer reinforced the rewarding quality of the stimulation. The time when the current was switched off either remained unchanged, or diminished. The fact that the escape reaction was maintained (or even facilitated) shows that the tranquilizer did not suppress the punishing quality of stimulation.

"Punished" response was also investigated. Cats deprived of food for a 12-hour period were placed in the chamber, where electrical stimulation was applied to the negative reinforcement points of the hypothalamus whenever the animal took any meat or milk. (The negativity of this stimulus had been established earlier by the

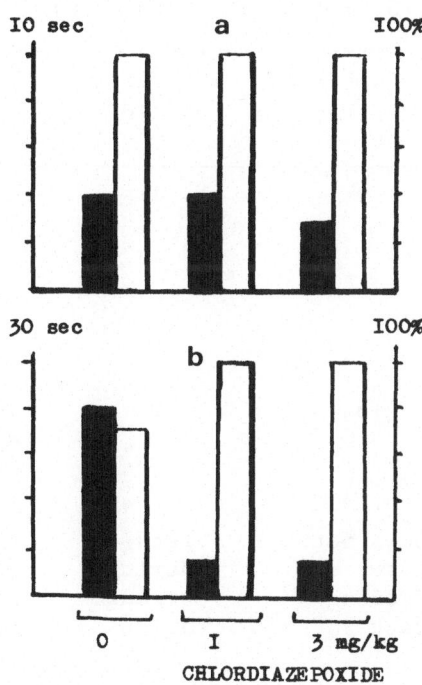

Fig. 3.   Effect of chlordiazepoxide on locomotor self-stimulation in cats. a) Reaction of turning off current; b) return to active compartment. Along ordinate: on right, latent period (shaded column); on left, No. of reactions counted (as percentage of total number of stimulations).

shuttle-box arrangement). The behavior of the cat was assessed as a
whole, including the number of approaches made and incidence of
food-taking.

One concrete instance of these experiments is shown in Table 4.
In the control, the animal refrained from approaching the food for
some time after application of punishing stimulation. When low doses
of tranquilizer were given the number of approaches increased without
the animal actually taking any food. When the dose of chlordiazepoxide was increased, the punishing agent began to affect the
animal's behavior differently. As stimulation was applied the cat
responded by looking round threateningly and hissing each time the
current was switched on; this did not, however, stop it from approaching the feeding rack and feeding. These findings show that the
noxious (aversive) experience is registered and has an effect on its
defense mechanisms. At the same time the tranquilizer eliminated the
emotionally negative response to stimulation which had produced a
prolonged passive avoidance reaction in the control.

Table 4.   Influence of Chlordiazepoxide on the Punishing Effects of
           Hypothalamic Stimulation (in cat M-6)

Compound	Intensity of stimu-lation	Number of approaches per 5 min	Incidence of food taking	Behavior patterns
Control (physio-logical saline solution)	1	9	1	Leaps away from food. Sits in corner of chamber. Attempts to escape.
Chlor-diazepoxide 1 mg/kg	1	13	0	Repeatedly approaches food, without taking any. Aroused.
3 mg/kg	1-2	Does not leave feeding rack	Feeds during stimul-ation	On application of stimulation: briefly adopts frozen position, looks around, hisses threateningly. Moves to rack and feeds. Persistent growling and hissing. Feeds without interruption.

It became apparent that a distinction must be made between two
elements of negative reinforcement, the perceptual and the emotional.
This was realized from results of past and present investigations
performed in our laboratories (Valdman et al., 1975, 1976; Zwartau,
1977) using a number of psychotropic agents, including barbiturates,
amphetamine and haloperidol. The perceptual element of the system of
punishment lies in the initial registering and interpretation of the
stimulus as negative or aversive; this triggers a behavior pattern
which seeks to avert negative (punishing) stimulation. The emotional
element is seen in an animal's relating emotionally (subjectively) to
the noxious agent. It may be identified by the duration of passive
avoidance in a punishing or conflict situation.

Tranquilizers of the benzodiazepine group in a fixed dose range
do not change the perceptual element in the system of punishment.
Adverse stimulation continues to convey the message which produces a
corresponding behavioral response. The cause of the tranquilizer's
disinhibiting effect on goal-directed behavior is likewise attenuation of the negative emotional state. This is why no conflicting
motives are seen, such as a clash between eating and drinking motivations, based on reward and punishment. The switching-off response
does not occur in the shuttle-box environment, however. It follows
that a dissociation of these two elements of the punishment system is
taking place; this indicates that the perceptual and emotional elements are functionally distinct. At the same time benzodiazepines
potentiate SS based on rewarding stimuli. Our findings on the initial action of tranquilizers on the punishment system differed from
those of Panksepp et al. (1970); we observed potentiation of SS with
stimulation of equivocal points as well as with plainly positive
stimulation.

4. ANALYSIS OF TRANQUILIZER ACTION IN INTERACTION BETWEEN
   POSITIVE AND NEGATIVE REINFORCEMENT SYSTEMS OF THE BRAIN

Complex neurophysiological relationships connect the brain
systems of punishment and reward. The potentiating effect of benzodiazepines on SS response may occur as the result of either potentiating positive systems or suppressing negative ones. The inhibition
of the adverse element of electrical stimulation of equivocal points
is thought by Panksepp et al. (1970) to produce reciprocal reinforcement of reward mechanisms. Direct potentiating of reward systems is
not excluded. Interaction between reinforcement systems was investigated in most research previously described, by combining central
rewarding and peripheral punishing stimuli. The effects of tranquilizers have not been assessed, however, on the balance between the
opposing systems of reinforcement combined with central stimulation.

We assessed the functional level of the system of encouragement
with central negative reinforcing stimulation. When the systems of

reward and punishment are stimulated together a model conflict situation between emotion and involvement on one hand and punished behavior on the other is produced.

Unipolar insulated 180 μm diameter electrodes were first implanted in male Wistar rats weighing 250-310 g under nembutal anesthesia, according to the coordinates of König and Klippel's atlas (1963). The indifferent electrode was placed in the region of the frontal sinuses. The electrodes were implanted in the region of the lateral hypothalamus, mainly in the area traversed by the medial forebrain bundle (MFB) and dorsal portion of the central gray (CG). Stimulus was applied, with the parameters square pulse 100 imp/sec, 1 msec; duration of train of impulses 0.20 sec. The electrodes used had been chosen according to their known ability to produce an exclusively rewarding or punishing response. It was considered rewarding only when the number of SS pedal pressings in a Skinner chamber did not fall below 60. The response of escaping into the other half of a shuttle-box served as a measure of aversive stimulation of the CG. Latency period for switching off the current was measured in one trial and then the average time over 10 responses was taken. Three different levels of intensity (in mA) for stimulating "punishing" brain sites were used, inducing an escape reaction with stepwise reduction of the range of latency; with J_1 the latent period lasted 10-20 sec, with J_2, 5-10 sec, and with J_3, 1-2 sec.

The rats, already adapted to laboratory conditions, were housed in a chamber measuring 40 x 30 x 30 cm. It was permanently divided into two sections, one "active" and the other "neutral". A 1 min stimulation of the negative point was started up during steady SS of 3 min duration. This induced the animal either to escape into the neutral zone, thus shutting off the noxious influence, whereupon the latent period was measured, or else remain in the active compartment, either continuing or breaking off SS. The degree of SS response was noted, as shown by the number of lever-pressings, together with "positive time" (or the influence of the time spent under current on the number of lever-pressings) and "negative time" (or time spent in the neutral, stimulation-free compartment).

SS response was inhibited at the more intense negative stimulation (J_2 and J_3), during background stimulation of the positive zone, and the balance shifted in favor of the system of negative reinforcement (see Table 5). But at the starting level of negative stimulation (J_1), the frequency of SS does not change substantially; in some cases, it actually increases. The rise in SS rate may be interpreted as a sign of adaptation, compensating for the boost in negative reinforcement system activity. During the first minute post negative stimulation SS frequency substantially exceeds the control level. This resembles the familiar physiological phenomenon of "rebound".

Table 5. Changes in SS Response During and After Stimulation of the
Negative Reinforcing Brain Sites

SS level	Intensity of punishing stimuli		
	J_1 $<$	J_2 $<$	J_3
Control SS (no punishing)	93.1±2.7	92.4±3.5	92.2±3.0
During punishing stimulation	88.2±4.2	55.4±6.7*	7.8±4.0*
Post-stimulation	107.3±2.0*	106.1±2.2*	104.2±2.4

Number of pedal-pressings per min (M±SEM); n = 35; * p < 0.01.

The results of experiments on the effect of tranquilizers on
a conflict situation involving the interaction of central systems
of punishment and reward and escape response are given in Tables 6
and 7.

Various drugs in the benzodiazepine group acted differently on
the escape response triggered by central stimulation of the negative
reinforcement sites, as shown in Table 6. Diazepam and chlordiazepoxide failed to increase the latent period of switching off the
negative stimulation at all three strengths. Escape response did
decrease, however. This fact shows that the perceptual element of
"punishment" is not suppressed, and confirms our data on cats (see
Section 2). Phenazepam and nitrazepam both prolong the latency of
escape response, indicating that the animals are more indifferent to
negative stimulation of the brain. We may therefore conclude that
the perceptual as well as the emotional element in negative reinforcement stimulation is inhibited by these tranquilizers at the
above doses.

Distinctions emerged between the actions of different tranquilizers on the model conflict situation created by simultaneous
stimulation of the systems of regard and punishment.

Chlordiazepoxide, diazepam and nitrazepam all increased positive
time in a conflict situation as compared with the control. Diazepam
and chlordiazepoxide, but not nitrazepam, shifted the balance in the
system of reinforcement in favor of the system of reward. They increased positive time while shortening the time during which the rat
is adversely stimulated (i.e. the negative time). This follows the
same lines as their action on the escape response: without diminishing – and sometimes while sharpening – perception of a noxious agent,
these tranquilizers did not prevent the animals' managing to escape.
Nitrazepam failed to shift the balance between the action of the

Table 6.   Influence of Varying Doses of Tranquilizers on Latent
           Period (LP) of the Escape Reaction

Drug	Dose (mg/kg)	Latency of escape response in sec (M ± SEM) Intensity of negative stimulation		
		J_1 <	J_2 <	J_3
Diazepam	C	14.8±0.9	5.6±0.7	1.6±0.1
	0.5	10.2±1.9*	4.2±0.7	2.0±0.2
	2.5	11.0±2.6	3.6±0.5	1.2±0.1
Chlordiazepoxide	C	12.4±1.0	6.0±0.7	2.4±0.8
	5	10.8±2.0	4.2±0.2	1.4±0.2
	10	8.6±2.1*	4.4±0.4	1.6±0.5
Nitrazepam	C	12.2±0.9	6.0±0.2	1.6±0.2
	0.5	16.6±2.8*	8.6±3.0	3.0±1.2
	2.5	17.2±2.0*	10.4±4.0	4.2±0.3*
Phenazepam	C	18.5±0.1	6.6±0.2	1.7±0.9
	0.5	20.0±0.5	8.3±0.8	5.0±0.4*
	1	22.0±0.6	18.5±0.3*	3.3±0.7

C - control (physiological saline).
* $p < 0.01$.

reinforcement systems, but increased the length of time under negative stimulation with no SS; this conforms with its inhibiting effect
on the perception of negative stimulation (increasing the latency of
the escape reaction). As the above doses of phenazepam imperceptibly
potentiate SS and inhibit escape reaction from punishing stimulation,
so negative time, when intense currents of J_2 and J_3 were applied,
rose steeply when conflict was set up between the two reinforcement
systems. The fact that phenazepam considerably increases negative
time means that when noxious stimulation is being applied the animal
neither accomplishes avoidance response, by retreating into the safe
compartment for a while before returning to the active area, nor
presses the pedal in order to obtain positive reinforcement. Phenazepam also prolongs the latency of escape reaction in a shuttle box,
which indicates that the animals are becoming more indifferent to
negative or punishing brain stimulation (diazepam shortens this
period). We may therefore conclude that phenazepam inhibits the
emotional element while considerably diminishing the perceptual
component of negative reinforcement. Phenazepam exercises a weaker
activating effect on the system of positive reinforcement than diazepam, judging by the frequency of self-stimulation. Hence, motivation towards positive or rewarding SS does not prevail in the balancing of the two opposing reinforcement systems.

     In her review "Animal models of anxiety and benzodiazepine
action" S. Iversen (1980) sets herself the question: Do benzodiaze-

Table 7. Influence of Varying Doses of Tranquilizers on "Positive", "Negative" and "Neutral" Times During Simultaneous Stimulation of the Systems of Positive and Negative Reinforcement

Drug	Dose (mg/kg)	Mean time in sec								
		Positive			Negative			Neutral		
		$J_1 <$	$J_2 <$	J_3	$J_1 <$	$J_2 <$	J_3	$J_1 <$	$J_2 <$	J_3
Diazepam	C	53.2	45.1	6.0	0	1.2	10.3	6.8	13.7	43.7
	0.5	53.7	44.0	17.7	0	4.3	9.4	6.3	11.7	32.9
	1	57.5	49.3	15.0	0	0.6	5.6	2.5	10.1	39.4
Chlordiazepoxide	C	56.9	49.3	13.7	3.1	6.5	20.4	0	4.2	25.9
	5	60.0	51.6	21.6	0	2.2	11.6	0	6.2	26.8
	10	60.0	60.0	32.5	0	0	17.1	0	0	10.4
Nitrazepam	C	53.5	15.3	6.0	0	8.8	9.1	6.7	35.9	44.9
	0.5	51.0	21.4	10.8	0	18.3	12.7	9.0	20.3	36.5
	2	56.8	35.0	20.1	3.2	16.1	19.7	0	8.9	20.5
Phenazepam	C	60.0	55.3	12.7	0	0.7	20.3	0	4.0	26.0
	0.5	51.3	20.2	9.8	0	18.0	23.9	8.7	21.8	26.3
	1	59.0	30.0	7.1	0	21.4	28.3	1.0	8.6	24.6

C - control (physiological saline).
$J_1 < J_2 < J_3$ - increasing intensity of punishing stimuli.
Total time - 60 sec.
"Positive" time - duration of SS during period of punishing stimulation.
"Negative" time - period of punishing stimulation without SS.
"Neutral" time - post-escape; free of both rewarding and punishing stimulation.

pines affect behavior unconnected with anxiety? The author believes
that benzodiazepines have a shared capacity for reinforcing motor
output and heightening response not controlled by anxiety-inducing
factors. The findings of different authors fail to agree, since they
have been operating under differing experimental conditions.

Wuttke and Kelleher (1970) incline towards a non-specific benzodiazepine action on response, in that benzodiazepines tend to raise a
low response rate rather than exert any specific action on behavior
suppressed by punishment. According to the findings of Tye et al.
(1977), chlordiazepoxide releases punished response much more than
either positively reinforced behavior or behavior free of positive
and negative influences.

Our own findings have shown the overriding importance of taking
the following into account when assessing benzodiazepines' effect on
behavior: 1) behavior patterns inherent in an individual, 2) the
varying action of different benzodiazepines on the positive and
negative reinforcement systems and ultimately on their perceptual and
emotional elements.

5.   THE POTENTIATING EFFECT OF DIAZEPAM AND MODULATION OF
THE CENTRAL NORADRENERGIC SYSTEMS

A great many experimental findings support the hypothesis of
noradrenergic (NA) brain systems' involvement in mediating reinforcement of behavior (Stein and Wise, 1970; German and Bowden, 1974;
Ritter and Stein, 1974). According to Cytawa and Trojnia (1979),
NA-containing neurones are crucial in mediating some effects of
reward on behavior, while DA-containing neurones are responsible for
potentiating positive reinforcement-motivated drive. Many data fail
to confirm the involvement of the NA system deriving from the locus
ceruleus (LC) in the rewarding effects produced by activation of
these sites (Corbett and Wise, 1979). Those opposing the NA-rein-
forcing hypothesis maintain that the cerulocortical NA system is
involved in behavior induced by noxious stimuli and plays a decisive
part in generating anxiety (Lader, 1974), and that benzodiazepines'
anxiolytic effect also works via this system (Gray et al., 1975).
Stress increases NA turnover; benzodiazepines forestall this effect
but destruction of the cerulocortical NA systems releases inhibition
of behavior produced by non-reward (Corrodi et al., 1971; Mason and
Iversen, 1977).

Guided by these premises, we investigated the involvement of NA
brain systems in the potentiating or disinhibiting effect observed
with diazepam.

Wistar male rats, each weighing 200-250 g, were used in these
experiments, with electrodes implanted in the MFB positive reinforcement area in the region of the lateral hypothalamus (see Section 3).

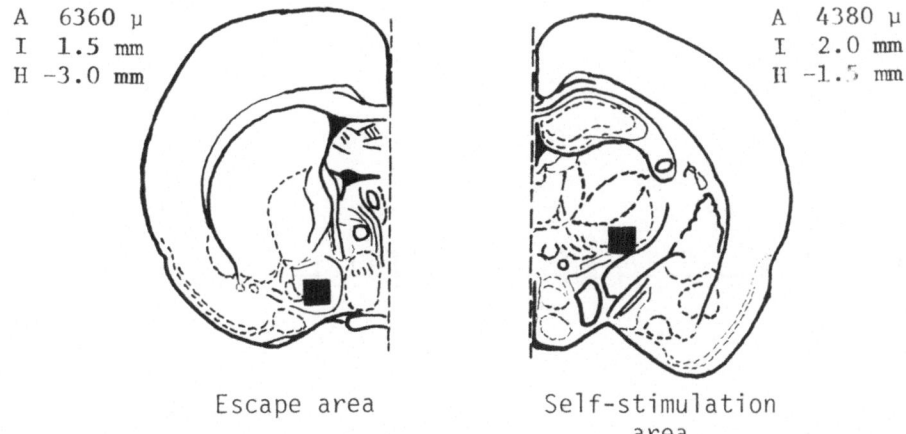

Escape area   Self-stimulation
                          area

Fig. 4. Diagram of frontal sections of rat brain marking location of
stimulating electrodes, according to König and Klippel's
(1963) atlas of the rat brain.

Figure 4 gives an illustration of how the electrodes were placed.
The frequency of pedal- or lever-pressings was measured in a Skinner
chamber. The same rate of reinforcement was maintained: sets of
0.25 sec square 1 msec impulses of negative polarity at a frequency
of 100 impulses/sec. The current used was that which produced
400-700 pressings per 10 min SS session.

Since changes in the amount of SS can be brought about by the
degree of alertness and general behavioral activity overall, as well
as by changes in the actual reinforcement system, the following
measurements were also made in the chamber where the SS took place:

- the number of squares crossed, each square measuring 5 cm^2
  within the 40 x 30 x 30 cm chamber,
- number of rearings,
- incidence and duration of grooming over the 10 min test period.

All parameters were registered on the tape of a recording machine.
Pressing the pedal switched on the lever-pressing counter and the
cumulative device for recording total SS time. Behavior patterns
were recorded by means of a keyboard system (see Figure 5).

A conflict situation was reproduced as follows. After 48 h food
deprivation rats were housed in a 40 x 50 x 55 cm chamber containing
a shelf. A recording was made of:

- the number of approaches made towards the shelf,
- incidence of water-taking,
- overall motor activity (how many 5-cm^2 squares were crossed, and
  number of rearings).
- overall incidence and total time of grooming.

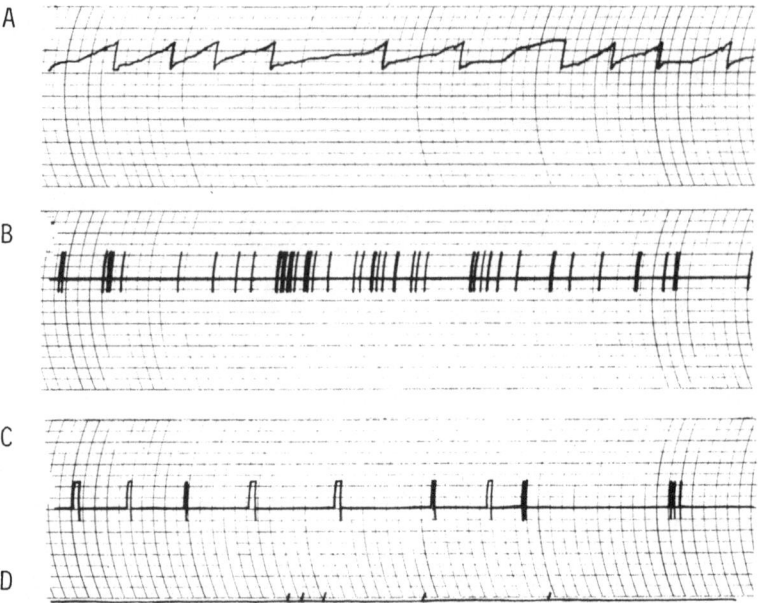

Fig. 5. Recording of one experiment involving self-stimulation with
         regular reinforcement. A: self-stimulation. 36 pedal-
         (lever-)pressings produce biggest waves; B: locomotion
         (number of 5 cm^2 squares crossed); C: incidence and
         duration of grooming; D: rearing.

The animals were given 1 mg/kg of diazepam intraperitoneally in
ampoules ("Seduxen", Gedeon Richter). Its effect on test models may
be seen in Table 8. Diazepam had been administered 30 min prior to
testing and effects were compared to the control which had been given
physiological saline.

    Tests were performed twice on both sets of rats - those given
diazepam and the control - i.e. a pre-treatment and a post-treatment
session. Changes in values were calculated as a percentage of the
untreated control. The dopamine-β-hydroxylase (DBH) inhibitor,
fusaric acid (FD-008 from Nippon Kagaku Co., Japan), was used to
reduce the physiological activity of central NA systems. FD-008 per
os, at a dose of 100 mg/kg, reduces the level of rat brain NA by 15%
without significant alteration of DA and 5-HT concentration (Ishii et
al., 1975). The animals' spontaneous behavior hardly changed and no
signs of sedative effect were observed two hours after ingesting 100
mg/kg FD-008 in excipient. Frequency of SS did drop, however, to
71.5% of the control level, where $p < 0.001$, as shown in Table 8.
Diazepam increased the frequency of SS when its action followed
treatment with the DBH inhibitor, in contrast with the latter
compound used alone. The rise in the number of pressings as a

Table 8. The Effects of FD-008, Nisotoxetin and Diazepam, Used Separately and in Combination, on SS Pattern

| Drug | Dose (mg/kg) | Level of SS | Locomotor activity | | Grooming | |
			Ambulation	Rearing	Counts	Total time
FD-008	100	71.5±3.9*	95.2±2.1	92.6±3.6	93.2±4.5	112.2±3.7*
Diazepam	1	119.4±7.8*	75.5±11.4	76.3±18.0	82.3±2.2	34.5±1.4
FD-008 +	100	88.0±7.4*	89.0±10.7	54.2±10.4*	94.5±6.1	76.2±2.2
Diazepam	1					
Nisoxetine	10	118.0±4.2*	96.2±11.8	102.6±9.1	85.6±11.2	87.2±8.0
Diazepam	1	130.5±10.2*	76.2±7.2*	80.3±17.2	88.9±14	91.4±15.7
Nisoxetine +	10	124.4±5.6*	79.2±17.9	93.7±17.9	73.2±14.2	89.0±12.7
Diazepam	1					

Number of responses per 10 min as a percentage of control level, M ± SEM.
* $p < 0.05$ (vs. control).
n = 6 throughout.

percentage of the starting level with diazepam used alone (vs.
control) or diazepam combined with FD-008 (vs. effect of FD-008
alone) was roughly the same - 119.4 and 114.4% respectively.

Our findings show that FD-008 inhibits SS when sites of the
substantia nigra (where DA-containing neurones are concentrated) are
stimulated, as well as when electrodes are placed in the region of
the lateral hypothalamus. Franklin et al. (1976) also found SS to
be inhibited when another DBH inhibitor (FLA-63) was given to an
animal with electrodes implanted in the substantia nigra. These
findings show that NA mechanisms connected with positive reinforcement may be activated otherwise than by stimulating the projections
of NA pathways or NA-containing neurones of the LC.

We also investigated the effect of the NA reuptake inhibitor
nisoxetine, which acts more selectively (Wong et al., 1975). No
substantial change in the animals' general behavior was noted 30 min
after administering 10 mg/kg nisoxetine. It exercised a potentiating
effect on SS, however; frequency of pedal-pressing rose by 18.0 ±
4.2%. Diazepam following pre-treatment with nisoxetine did not
potentiate the activating effect produced by diazepam on SS.

The activating effect of diazepam on SS is thus also observed
when there is a drop in NA concentration, as well as when its reuptake is inhibited.

In conflict situations, 10 mg/kg of i.p. nisoxetine produced no
disinhibitory effect. The number of approaches made to the trough,
and motor activity in general, decreased even more when this drug was
used alone than in the case of the control on saline. Diazepam
following pre-treatment with nisoxetine did not noticeably counter
conflict. Although the incidence of both water taking and approaches
to the trough when diazepam and nisoxetine were combined was somewhat
higher than that observed after treating with nisoxetine alone, these
differences were not statistically significant.

Table 9. The Effect of Diazepam and Nisoxetine, Used Separately and
          in Combination, on Behavior in a Conflict Situation

Drug	Dose (mg/kg)	Incidence of:		Locomotor activity	
		Approaches	Water taking	Ambulation	Rearing
Diazepam	1	142.0±26.9*	227.1±40.6	53.8±8.5	64.1±8.6
Nisoxetine	10	29.2±6.7*	29.2±7.3*	38.0±5.4	9.0±1.5*
Diazepam + nisoxetine	1 10	42.8±7.2	54.9±19.6	44.4±8.4	9.6±1.3

No. of responses per 10 min as a persentage of control.
* p < 0.01.

It should also be noted that the general behavioral changes produced by nisoxetine were not reversed by diazepam.

Our investigations have thus shown how the level of activation and disinhibition after diazepam are to some extent dependent on the functioning of the NA system. It should not be assumed that it is the NA systems which are decisive in determining diazepine action; degree of general activity and drive do, however, depend on these systems. The hypothesis of the involvement of the ascending NA system in generating anxiety was not confirmed by File et al. (1979). They used three behavioral models on which benzodiazepines exert a disinhibitory effect: the social interaction anxiety test, anxiosoif test and the home-cage intruder test. Bilateral destruction of the LC NA neurones by administering 6-hydroxydopamine (6-OH DA) failed to produce a disinhibitory effect in any of the tests.

If the disinhibitory effect of benzodiazepines were connected with a decrease in central NA system activity, other compounds would likewise show an anti-punishing effect. However, neither the α-noradrenergic antagonist, phentolamine, nor propranolol (the β-noradrenergic antagonist) showed any anti-conflict effect on punished behavior in rats (Stein, 1980). However, intragastric 1-noradrenaline exerted a disinhibitory effect similar to that of benzodiazepines and increased the anxiolytic or releasing effect of oxazepam (Stein, Wise and Berger, 1973).

6. INVESTIGATING THE INVOLVEMENT OF SEROTONINERGIC SYSTEMS IN THE POTENTIATING (DISINHIBITORY) EFFECT OF DIAZEPAM

According to Stein's hypothesis, serotoninergic (5-HT) mechanisms are responsible for the suppression of punishment-induced behavior.   The anti-serotonin compounds methysergide (Stein et al., 1973), sianserin (Cook and Sepinwall, 1975) and the inhibitor of 5-HT biosynthesis, parachlorophenylalanine (PCPA) (Robichaud and Sledge, 1969; Wise, Berger and Stein, 1973) exert no activating effect on behavior in a conflict situation. Intragastric administration of 5-HT reverses the activating effect of oxazepam in such a situation (Stein et al., 1975). This and similar findings raise the possibility of the existence of serotonin "punishing" systems which inhibit behavior and whose activity may be reduced by benzodiazepines or other anxiolytics (Stein, 1980).

Findings on 5-HT involvement in reward systems fail to agree. Wise et al. (1973) observed inhibition of SS by intragastric 5-HT. In Stark and Fuller's (1972) experiments PCPA, the 5-HT biosynthesis inhibitor, suppressed SS patterns; Black and Cooper (1970) observed a falling off of SS 3-7 days after administering PCPA. More research is required on the whole question of serotonin involvement in the activating effect of benzodiazepines. We used two compounds affect-

ing activity of the 5-HT system: citalopram (Lu-10-171), a selective
5-HT reuptake inhibitor (Hytell, 1977) and parachloramphetamine, an
inhibitor of 5-HT synthesis (PCA, Sanders-Bush et al., 1977). The
effects of PCA (Sigma Chemical Co., USA) were measured 24 hours and
those of citalopram 45 min after intraperitoneal injection - i.e. the
time required for each to achieve maximum effect. The former reduced
brain 5-HT concentration by 48 and 76% at doses of 5 and 15 mg/kg
respectively. Diazepam was given 30 min later, and 15 min later when
used in combination with citalopram.

A 10 mg/kg dose of citalopram reduced the frequency of SS by
14.4 ± 4.9% (p < 0.05) during steady reinforcement via an electrode
planted in the lateral hypothalamic region. When a fixed ratio
schedule of 6:1 was adopted, and reinforcement was only obtained for
one pedal pressing out of every six, citalopram produced a drop in SS
of 72.4 ± 6.4% compared with the control (p < 0.001, as shown in
Figure 6). At the same time, an increase in rearing and total duration of grooming was observed in the SS chamber (Table 10). Fluoxetine, another 5-HT uptake inhibitor, produced a similar decrease in
SS (Fuller et al., 1974). A dose of 1 mg/kg diazepam, superimposed
on a background of citalopram action, restored the frequency of SS,
which had been reduced by citalopram to below the control value. It
also suppressed displays of such behavior as grooming and rearing,
which citalopram had promoted. The difference between citalopram
used separately and combined with diazepam is statistically significant (p < 0.001).

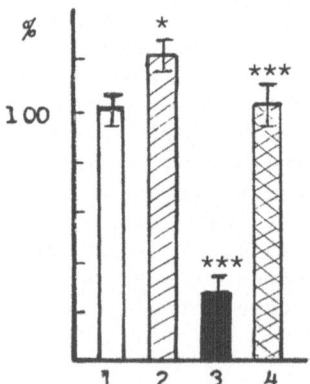

Fig. 6. Effect of diazepam and citalopram, used separately and in
         combination, on frequency of self-stimulation on a "fixed"
         6:1 ratio schedule. Ordinate - average frequency of pedal-
         (lever-)pressings. 1: control; 2: diazepam (1 mg/kg); 3:
         citalopram (10 mg/kg); 4 citalopram (10 mg/kg) + diazepam
         (1 mg/kg). * p < 0.05, *** p < 0.001 (for 4, compared to
         effect 3). Student's t test.

Table 10. The Effect of Citalopram and Diazepam, Used Separately and in Combination, on SS Behavior

Drug	Dose (mg/kg)	SS (6:1)	Motor activity		Grooming	
			Ambulation	Rearing	Counts	Total time
Diazepam	1	122.4± 6.3*	89.2±13.0*	72.6±11.0*	93.4± 8.2	59.2±18.4*
Citalopram	10	27.6± 6.4*	41.3± 8.1*	183.3± 3.6*	108.3±10.2	512.1±79.4*
Citalopram + Diazepam	10 1	93.9±11.4**	86.1±24.2	8.4±12.2**	16.6±18.1**	24.4±10.5**

Number of responses per 10 min as percentage of control level, M ± SEM. n = 7.
* p < 0.01 vs. control; ** p < 0.001 vs. effect of citalopram alone.

Parachloramphetamine (PCA): The rats were given a 5 mg/kg dose
of PCA followed by a second (10 mg/kg) dose 24 hours later, making a
total dose of 15 mg/kg. The animals were tested 24 hours after each
dose. Considerably impaired behavior developed 40 min after administering 5 mg/kg PCA, with stereotypy, muscle relaxation and
heightened reactivity. SS patterns were inhibited or completely
suppressed. Behavior seemed to have largely returned to normal 24 h
later; however, grooming and rearing decreased and frequency of SS
was similar to the control.

A 5 mg/kg dose of PCA given 24 h later failed to produce a
substantial change of SS frequency under the regular reinforcement
schedule, although a tendency towards increased pedal-pressing was
noted (see Figure 7). A 1 mg/kg dose of diazepam following pre-
treatment with PCA increased the frequency of SS, but the activating
effect of diazepam was not potentiated. The frequency of SS rose by
24.5% ± 9.5% compared with the level produced by PCA (p < 0.05). But
the difference between the effects of PCA used separately and in
combination with diazepam was not statistically significant. Increasing the dose of PCA to 15 mg/kg did not bring about any real
change in SS frequency. Diazepam given after pre-treatment with this
dose of PCA also activated the SS pattern – by 52.4 ± 16.5% – p <
0.05 – as compared with the effect of PCA alone. A slight tendency

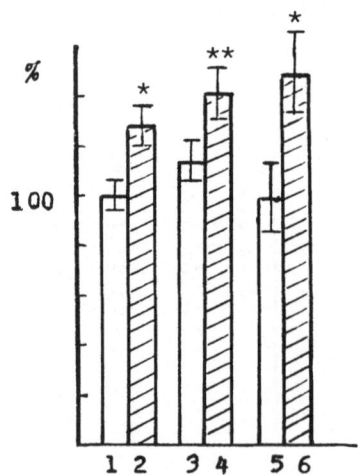

Fig. 7.   Effect of diazepam and parachloramphetamine, given
          separately and in combination. 1: control; 2: diazepam
          (1 mg/kg); 3: parachloramphetamine (5 mg/kg); 4: para-
          chloramphetamine (5 mg/kg) + diazepam (1 mg/kg); 5:
          parachloramphetamine (15 mg/kg); 6: parachloramphetamine
          (1 mg/kg) + diazepam (1 mg/kg); * p < 0.05 (Student's
          t-test); ** p < 0.05 (Wilcoxon's rank sum test).

towards potentiating the effect of diazepam when given following 15 mg/kg PCA did not prove to be statistically significant (p < 0.05).

Citalopram did not appear to counter "punishing" effect in a conflict situation. In fact it actually reinforced inhibition of the number of approaches and incidence of water-taking in the conflict situation as against the control (receiving saline). The combined effect of citalopram and diazepam was to increase both these activities, but at levels barely distinguishable from the effects of diazepam alone. Incidence of grooming fell. Other 5-HT-active compounds, such as zimelidine, a 5-HT uptake inhibitor (15 mg/kg), 5-hydroxytryptophan and quipazine, at doses of 40 and 7 mg/kg respectively, displayed no disinhibiting effect in a similar conflict situation.

Doses of 5 and 15 mg/kg PCA failed to produce any disinhibiting effect in a conflict situation; in fact, the number of approaches made to the trough actually dropped. Diazepam administered after pre-treatment with 5 mg/kg PCA did have a releasing action and the incidence of water-taking more than doubled (x 2.1). The effect could not be distinguished from that of diazepam used alone. A 15 mg/kg dose of diazepam following pre-treatment with PCA released punishment-suppressed behavior very effectively. But even here, the potentiating effect of diazepam was substantially activated (see Table 11). The rats' motor activity also rose.

Both citalopram and other selective inhibitors of serotonin uptake, such as fluoxetine and zimelidine, without any substantial influence on NA uptake, raise 5-HT concentration in the post-synaptic area; these drugs lower 5-HT turnover while reducing the concentration of HIAA and either raise or otherwise alter that of brain 5-HT. The drop in 5-HT turnover and release is produced by a decrease in the rate of discharge of 5-HT-containing neurones from the raphe nucleus because of either activation of the post-synaptic autoreceptors (feedback reinforcement) or else stimulation of the pre-synaptic receptors (uptake inhibitation; Waldmeier et al., 1979; Christensen et al., 1977; Hytell, 1977; Fuller et al., 1974). PCA produces a long-lasting reduction of 5-HT, 5-HIAA and tryptophan hydroxylase activity within the brain (Sanders-Bush et al., 1975). A 5 mg/kg dose reduces 5-HT concentration to 0.08 ± 0.01 µg/g (control 0.64 ± 0.03 µg/g) and is maintained at a similar level for up to 10 days. PCA, however, possesses an amphetamine-like action (Steranka et al., 1977), which may explain the activation of grooming and motor activities observed.

Our study showed that diazepam serves to potentiate SS patterns, both against a background of functionally suppressed reward systems (an effect produced by such 5-HT uptake inhibitors as citalopram) and by reduced functional activity of the level of brain 5-HT caused by PCA; here, the initial level of SS is indistinguishable from that of the control. The state of the 5-HT system of the brain is apparently

Table 11. The Effect of Diazepam Used Separately and in Combination
          with PCA and Citalopram on Behavior Patterns in a Conflict
          Situation

Drug	Dose (mg/kg)	Incidence of:		Locomotor activity	
		Approaches	Water taking	Ambulation	Rearing
Diazepam	1	142.0 ± 26.9*	227.1 ± 40.6*	53.8 ± 8.5	64.1 ± 8.6
PCA+diazepam	15	$327.7 \pm 64.1^{+}$	$800.0 \pm 135.7^{++}$	$126.4 \pm 27.2^{+}$	46.3 ± 6.6
Citalopram+ diazepam	10	141.0 ± 24.9*	261.9 ± 41.2*	45.1 ± 12.1	$22.0 \pm 1.8^{+}$

No. of responses per 10 min as percentage of control M ± SEM.
* $p < 0.01$ (vs. control); + $p < 0.05$; ++ $p < 0.001$ (vs. effect of
diazepam alone).

not decisive in diazepam's potentiation of behavior induced by
rewarding stimuli. The disinhibiting effect of diazepam in a conflict situation (when behavior is restrained by punishing stimuli)
occurs both against a background of 5-HT uptake blockade, with a rise
in 5-HT concentration and a drop in its turnover (using citalopram),
and with a decrease in 5-HT biosynthesis and concentration (using
PCA); that is, either by raising or reducing the functional level of
the 5-HT system.

     The activating component of diazepam action, consisting of
potentiating stimuli-rewarding SS and disinhibiting punishment-
induced behavior, is thus seen to operate independently of the functional state of the serotinergic brain system.

     The fact that citalopram reverses diazepam's inhibitory effect
on SS points to the capacity of benzodiazepines to activate reward-
related behavior previously suppressed by an excess of 5-HT. This is
not proof, however, that the tranquilizer acts directly on the 5-HT
brain system, or possesses an anti-serotonin effect.

     Many experimental observations show that 5-HT-containing neur-
ones of the nucleus raphe dorsalis, with their massive serotoninergic
projections into the forebrain, are involved in the restrained behavior induced by punishing stimuli (Stein, 1980). Both an increase
in brain 5-HT concentration and in activity of the 5-HT system reinforce the suppression of punished behavior. Citalopram reinforced
the restraint on behavior in a conflict situation in our experiments,
in comparison with a control administration of saline. We were able
to show that benzodiazepines' non-punishing or disinhibiting action
on behavior was not observed if 5-HT pathways had been destroyed (Tye
et al., 1977), or 5-HT mechanisms potentiated (Simon and Soubzié,
1979). However, in our experiments involving an inhibitor of 5-HT

reuptake – i.e. when the 5-HT system is potentiated – diazepam
counters punishing effects. Inhibition of 5-HT synthesis or
destruction of 5-HT terminals (5,6-dihydroxytryptamine-DHT) is associated with a releasing effect on punishment-suppressed behavior
(Wise et al., 1973; Cook and Sepinwall, 1975; Tye et al., 1977,
1979). In our investigations, however, neither large nor small doses
of PCA, while considerably reducing functional activity of the 5-HT
system, appeared able to counteract conflict, which the addition of
diazepam clearly did. Cook and Sepinwall (1979) noted how chlordiazepoxide brought about a reduction in conflict during their experiments, in which the 5-HT inhibitor, PCA, was used.

Chlordiazepoxide, microinjected into the nucleus raphe dorsalis,
releases behavior suppressed by threat of punishment without altering
non-punished behavior (Thiébot et al., 1980). GABA microinjected
into the nucleus raphe dorsalis produces a similar disinhibitory
effect. The functional activity of 5-HT neurones of the raphe nucleus is inhibited by the influence of GABA-ergic projections
(Gallager, 1978). Benzodiazepines' modulating effect on the 5-HT
system mediating punishment may therefore operate indirectly via
GABA-ergic influences.

7.   INVOLVEMENT OF THE GABA-ERGIC SYSTEM ON THE POTENTIATING EFFECT OF DIAZEPAM

The present state of our knowledge of the functional interaction
between the benzodiazepines and GABA is derived from a wealth of
neurochemical and neurophysiological data (see the reviews of Costa
and Guidotti, 1979, and Haefely, 1978). From the psychopharmacological viewpoint, however, it is not quite clear for exactly which
elements in the psychotropic range of benzodiazepines GABA receptors
play such a decisive role. The fact that benzodiazepine tranquilizers' anticonvulsant and muscle relaxant effects are interlinked
with GABA mechanisms appears undisputed. The connection between
benzodiazepines' anxiolytic and (particularly) their activating
effect on the one hand, and the GABA system on the other, is less
apparent, however.

Cook and Sepinwall (1975) did not find that the GABA-T inhibitor, aminohydroxyacetic acid (AHAA) countered conflict; nor did
it potentiate the inhibitory effect of diazepam in a conflict situation. The GABA receptor agonists, muscimol and THIP, failed to
potentiate benzodiazepines' suppression of conflict. Muscimol had
little effect in a conflict situation; THIP was inactive (Sullivan et
al., 1978; Cook and Sepinwall, 1979). The investigations of Thiébot
et al. (1979) did not support the hypothesis that GABA mechanisms are
involved in the control of behavioral depression and the releasing
effect of benzodiazepines. According to their findings, muscimol did
not release rat behavior inhibited by either novelty, punishment or

non-reward; nor did it potentiate diazepam effect. At the same time
it was shown that the GABA antagonists picrotoxin, bicuculline and
thiosemicarbazide reduce the anti-conflict effect of benzodiazepines
(Zakusov et al., 1977).

An aqueous solution of a) the GABA-mimetic, muscimol, b) bicuculline - a specific GABA receptor blocker, in a 1 M HCl solution,
subsequently titrated with 1 M NaOH solution,* c) picrotoxin (Sigma,
USA), d) thiosemicarbazide, a GABA biosynthesis inhibitor and e) the
irreversible GABA transaminase (GABA-T) inhibitor, γ-acetylenic-GABA
(Merell International, France), were all used to investigate GABA
mechanisms. All the drugs were given intraperitoneally; diazepam and
muscimol 30 min before testing, bicuculline - 5 min before, and, in
combination with diazepam - 25 min after it. Picrotoxin was administered 20 min before testing and 10 min after diazepam when the two
compounds were used in combination. γ-acetylenic-GABA (GA-GABA) was
given 4 hours before testing and when used in combination with
diazepam, the latter was given 3 h 40 min after GA-GABA.

A 2 mg/kg dose of bicuculline did not substantially change
frequency of SS or ambulation in rats. The incidence of rearing
increased. Diazepam's potentiating effect on SS did not change
significantly (see Table 12). Neither doses of 0.5 nor 1 mg/kg
muscimol noticeably changed the animals' general behavior. SS frequency hardly changed, although a tendency was noted towards fewer
pedal-pressings. A 2 mg/kg dose impaired adequate response to external stimuli in rats; stereotyped and circular movements began to
appear, together with signs of catalepsy and muscle relaxation. A
sharp drop in SS of 93.3 was observed. A dose of 0.5 mg/kg muscimol
combined with 1 mg/kg diazepam brought the activating effect on SS to
the control level. No potentiation of this effect was observed.
Sedation was seen in mice, however, when these two compounds were
used in combination; motor activity and grooming both decreased.
Biggio et al. (1977) had noted that muscimol potentiated inhibiting
effect on motor activity in mice.

A dose of 3 mg/kg thiosemicarbazide, the GABA synthesis inhibit-
or, given 1 h later, and 2 mg/kg picrotoxin, a non-specific GABA
receptor blocker, produced a marked SS inhibition, reducing SS by
45.8 and 92.8% respectively. This effect was accompanied by signs of
inhibition of other forms of behavior, however: a decrease in spontaneous motor activity (frequent adoption of a "frozen" posture),
heightened response to external stimuli and general tension (pre-
convulsion state).

Thiosemicarbazide does not exert an inhibitory effect on SS when
combined with diazepam. SS frequency equals 97.7% ± 13.0% of the

---

*These first two compounds through kind courtesy of Dr. W. Haefely,
 Basel.

Table 12. The Effect of Diazepam Used Separately and in Combination with GABA-ergic Compounds on SS

Compound	Dose (mg/kg)	n	Locomotor activity			Grooming	
			Frequency	Ambulation	Rearing	Counts	Total duration(s)
Diazepam	1	8	123.5± 5.9*	78.1±14.6	74.2±13.6	80.0±22.4	46.5±13.8*
Bicuculline	2	6	102.2± 6.2	105.3± 7.5	225.0±18.7*	104.5±11.0	67.4±10.3*
Bicuculline + diazepam	2	6	127.0±11.2* **	94.3± 5.2	41.4± 8.1* **	64.6±12.0* **	48.7±10.2*
Muscimol	0.5	6	96.4±14.4	75.6±18.2	88.7± 7.2	82.0±11.0	107.6± 4.8
	1	9	93.1±10.6	67.5±10.0*	66.0±15.3*	119.6±12.4	126.4±10.3*
	2	5	6.7± 5.0*	67.1±11.8*	84.9±17.2	18.4±15.2**	17.8± 9.0*
Muscimol + diazepam	0.5, 1	8	130.9±10.5* **	55.1±14.0*	76.3±18.4	27.2±10.2* **	25.5± 6.1* **
Thiosemicarbazide	5	6	52.4± 7.9*	34.2±18.2*	26.3±19.4*	43.8±16.4*	40.1±26.0*
Thiosemicarbazide + diazepam	3, 1	6	97.7±13.0**	69.4±22.1**	76.2±14.6**	75.4±29.6	59.4±31.4
Picrotoxin	2	5	7.2± 4.3*	56.4±19.0*	102.4± 6.2	87.2±14.7	79.2± 8.0*
Picrotoxin + diazepam	2, 1	5	66.5±14.0* **	78.2±29.0	85.6±14.2	54.7±12.1*	48.7±10.2*

Number of responses as a percentage of control (M ± SEM).
* $p < 0.05$ vs. control.
** $p < 0.05$ vs. effect of test substance alone.

control, a consistent increase in rate of 80.3% compared with the
effect of thiosemicarbazide alone (p < 0.05). Diazepam also countered the inhibitory effect of thiosemicarbazide on associated forms
of behavior. SS frequency equalled 66.5% of control when diazepam
and picrotoxin were given in combination, a substantially higher
(p < 0.001) frequency of SS than that produced by picrotoxin alone.

The irreversible GABA-T inhibitor, GA-GABA, does not significantly change the frequency of SS (see Table 13) when combined with a
threshold dose of 0.5 mg/kg diazepam; the potentiating effect of
diazepam is not observed with GA-GABA. When a 2.5 mg/kg dose of
diazepam is given in combination with GA-GABA, diazepam's activating
effect on SS is not reinforced. The difference between diazepam
given separately and combined with GA-GABA is not statistically
significant.

Little work has been done on GABA's involvement in the "reward"
system. Nasegara (1976) did not observe any clear-cut influence on
SS - in this case, on stimulation of the substantia nigra in rats -
when GABA was given intracisternally.

Kent and Fedinets (1976) noted that picrotoxin and bicucculine
inhibited SS in rats. According to our findings, the inhibiting
effect of GABA-active compounds on SS occurs as and when other forms
of behavior are also inhibited. This illustrates the non-specific
way in which these compounds act - whether as sedatives or proconvul-
sants, as in the case of thiosemicarbazide and picrotoxin. The fact
that the GABA-mimetic, muscimol, and the GABA receptor blocker
bicuculline, in doses producing no drastic behavioral changes, failed
to affect frequency of SS shows that GABA-ergic systems are not

Table 13. The Effect of Diazepam and γ-Acetylenic GABA (GA-GABA)
Used Separately and in Combination on SS

Compound	Dose (mg/kg)		Change in incidence of SS (as percentage of control; M ± SEM)
Diazepam	0.5	4	+19.0±6.5
	1.0	8	+23.5±5.9*
	2.5	6	+33.4±6.4**
GA-GABA	10	6	− 3.8±4.3
GA-GABA + diazepam	10 0.5	4	+20.4±12.4
GA-GABA + diazepam	10 2.5	6	+43.8±7.8**

GA-GABA given 4 h before testing.
Diazepam given 3.5 h after GA-GABA and 30 min before testing.
* p < 0.05; ** p < 0.05 (Student's t-test).

playing a decisive part in producing the "reward" effect accompanying central stimulation. Our investigations indicated that a change in activity of central GABA-ergic mechanisms is not essential for producing activation of diazepam's effect on SS.

Muscimol itself failed to produce release in a conflict situation at doses which did not affect motor activity level (see Table 14). In fact, the incidence of approaches made and water-taking somewhat decreased. Diazepam's disinhibiting action was not observed after pre-treatment with muscimol, possibly due to an increase in sedative effect, since the rats' motor activity had decreased.

In view of evidence of muscimol's rapid metabolism and poor penetration into the brain (Maggi and Enna, 1979; Cananzie et al., 1980), the compound was administered intraventricularly. It was then found that both muscimol and THIP diminished inhibition of rat behavior induced by a punishing (conflict) situation. Muscimol was shown to counter conflict in a narrow dose range of 150-200 ng. When given at higher dosage, motor activity becomes inhibited and no disinhibiting effect is observed. The GABA receptor agonist, isoguvacine, failed to exert a disinhibiting effect, however. In the experiments of Thiébot et al. (1980) muscimol microinjected into the nucleus raphe dorsalis did not release suppressed conditioned reflex operative behavior. Chlordiazepoxide introduced into this area brought about a typical disinhibiting effect, however. The potentiating effect of chlordiazepoxide was not potentiated when this benzodiazepine was given in combination with muscimol.

Bicucculine actually increased the effect of "punishment" on behavior in a conflict situation, and the incidence of approaches made and of water-taking dropped substantially. Diazepam antagonized the effects of pre-treatment with bicuculline. The incidence of water-taking in a conflict situation did increase, but less than when diazepam was given separately. Motor activity rose concurrently. Ostrovskaya and Voronina noted that diazepam and bicuculline appeared to compete in a similar test conflict situation (Zakusov et al., 1977).

The effect of diazepam alone and diazepam following pre-treatment with GA-GABA were also studied, using rats implanted with electrodes in the medial lemniscus region, as in Figure 4. The method used is described in Section 4. The animals were first trained to switch off a central aversive stimulus of 100 Hz, 1 ms, 70-250 μA by running across to the opposite section of a shuttle box. The latency of response was measured (<3-5 sec in the control).

A 0.5 mg/kg dose of diazepam produced no change in latency prior to turning off the central aversive stimulation as the animal made its escape. A 2.5 mg/kg dose, however, extended this time lapse. Latency of escape showed a slight tendency to increase but was

Table 14. The Effect of Diazepam Used Separately and in Combination with Muscimol and Bicuculline on Behavior in a Conflict Situation

Compound	Dose (mg/kg)	n	Conflict situation incidence of:		Motor activity		Grooming	
			Water taking	Approaches to trough	No. squares crossed	No. rearings	Counts	Total duration (sec)
First testing	–	10	10.9±2.8	27.2±4.0	42.9±5.4	24.5±3.6	5.6±0.6	36.8± 4.8
Second testing: Saline (control)	–	10	4.0±1.2	21.1±3.8	31.4±4.2	16.6±2.2	6.4±0.8	97.4±18.4
Diazepam	1	10	21.8±3.9*	50.7±9.6*	39.2±6.2	17.8±2.4	6.9±1.0	16.1± 2.1
Muscimol	0.5 1	9 10	2.8±0.4 2.1±0.2	7.2±0.9* 5.4±0.8*	22.4±3.5 21.9±7.3	8.7±0.8 1.3±0.2	9.6±1.1 4.0±0.2	64.9±10.1 83.4±18.4
Muscimol + diazepam	0.5 1	9	1.8±0.2**	5.1±1.7**	20.7±3.2	2.3±0.6**	3.7±0.8	38.0± 4.7
Muscimol + diazepam	1 1	10	1.6±0.2**	2.0±0.2**	5.0±0.9**	0.5±0.1**	3.0±0.8	25.6± 3.8
Bicuculline	2	10	1.6±0.2*	4.8±0.3*	11.6±2.0*	0.8±0.5*	2.4±0.9	21.0± 4.2
Bicuculline + diazepam	1 2	9	8.3±2.2***	3.8±1.4	34.3±5.7***	6.8±1.2***	7.0±1.7	27.5± 4.4

Average number of behavioral displays (M ± SEM).

* p < 0.01 vs. control (on saline).

** p < 0.001 vs. effect of diazepam.

*** p < 0.01 vs. effect of bicuculline (Student's t test).

scarcely changed by a dose of 10 mg/kg GA-GABA. The diazepam effect
was not potentiated by administering this drug following pre-
treatment with the GABA-T inhibitor (GA-GABA). In fact, the diazepam
effect was substantially reduced in a proportion of the animals
during 7 out of the experiments involving 5 stimulation points,
although in these cases the latency of active escape between comparable groups of animals had been close to start with (4.9 ± 0.7 and
4.5 ± 0.6 sec). Taking the increase in latency of active escape
after treatment with diazepam as +73.7 ± 16.6% (p < 0.01) the effect
of diazepam following pre-treatment with GA-GABA did not exceed 13.9
± 15.9% (p < 0.05).

Diazepam and GA-GABA substantially inhibit the action of GABA-T;
a dose of 2.5 mg/kg diazepam and 10 mg/kg GA-GABA has an almost
identical effect in this respect (Galustyan, 1979). Equivalent doses
of diazepam and GA-GABA inhibiting GABA-T, however, influence behavior differently: a dose of 2.5 mg/kg diazepam exerts a disinhibiting
effect in a conflict situation and increases the latency of active
escape, whereas 10 mg/kg GA-GABA was shown to be ineffectual in these
tests. A statistically significant increase in time spent under
central aversive stimulation is only observed following a dose of 100
mg/kg GA-GABA, when the activity of GABA-T drops by 63.8%. The two
compounds combined do not potentiate the anxiolytic effect of diazepam. GA-GABA neither activates SS nor influences diazepam's activating effect. A well-substantiated hypothesis has been put forward
explaining the anxiolytic and sedative effect exerted by benzodiazepines based on their action on various types of benzodiazepine receptors (Lippa et al., 1979). GABA-ergic mechanisms are involved,
according to these authors, in producing the sedative but not the
anxiolytic effects of benzodiazepines (Klepner et al., 1979; Lippa
et al., 1978).

In this context, it should be mentioned that Andreev and Ignatov
(1982) found that picrotoxin, but not bicuculline, attenuates the
general sedative action of diazepam given at a large dose (5 mg/kg),
potentiating a marked disinhibitory effect. Picrotoxin likewise
reduces the effect of a high (100 mg/kg) dose of GA-GABA, but no
releasing effect is observed, as in the case of diazepam.

Picrotoxin impairs the function of type 2 benzodiazepine receptors by blocking the passage through the chloride channels (Lippa
et al., 1979), thereby eliminating the sedative element masking
diazepam's anxiolytic effect. The sedative action of GA-GABA is also
attenuated. The separate activation of GABA mechanisms is not
responsible for the potentiating effect, however.

According to Braestrup and Nielsen (1980), the benzodiazepine
receptor may not necessarily be a link between GABA receptors and the
chloride channel. Complete GABA agnoists such as muscimol alter the
conformation of the GABA receptor and open up the Cl^- channel. At

the same time the affinity of the benzodiazepine receptor is
increased. Partial GABA agonists with rigid molecules, such as THIP
and isoguvacine, however, induce smaller conformational alterations
so that the benzodiazepine receptor is not completely activated,
although the Cl^- channel is opened up.

Neither the GABA-T inhibitor, GA-GABA, nor the GABA synthesis
inhibitor, thiosemicarbazide, exerts a selective effect on GABA
mechanisms in the synaptic compartment. GABA concentration in the
non-synaptosomal region increases, especially in the glial elements
(Iadrola and Gale, 1979). Valproic acid (n-dipropylacetate, n-DPA)
is a more selective GABA-positive compound. This substance also
inhibits GABA-T (Fowler et al., 1975). But an increase in GABA level
takes place, mainly in the nerve terminals (Iadrola and Gale, 1979).
Succinic semialdehyde dehydrogenase is inhibited by n-DPA; the polyaldehyde of succinic acid accumulates and the transformation of
GABA into succinic polyaldehyde is blocked via the mechanism of
end-product inhibition (Van der Laan et al., 1979). This process
takes place mainly at GABA-ergic terminals. Valproic acid exerts a
tranquilizing effect here (Lal et al., 1979, 1980; Rayevsky and
Kharmalov, 1980; Kharmalov et al., 1981).

The psychopharmacological range of valproate's tranquilizing
action differs from that of diazepam, however. This is clearly seen
from investigations in the cat (Kharmalov et al., 1981). These were
carried out on 9 animals with a complete range of emotional/behav-
ioral response, belonging to the 1st category described in the second
article. Valproic acid from Germed (GDR) was administered i.p. in
a 2% sterile vehicle at a dose of 50 and 200 mg/kg. A dose of 0.5-2
mg/kg of diazepam was given orally. The changing pattern of a)
initial spectrum of emotional/behavioral response (as described in
the first article) and b) emotional tension and fear response to
repeated exposure to punishing stimulation were both evaluated.

Table 16 shows the results of these experiments pooled. Valproate changes the initial spectrum by reducing exploratory activity,
motor reactions and initiative in motor behavior. Insignificant
displays of fear-and-alarm caused by the experimental environment
were virtually eliminated. Behavioral displays of positive emotions
were attenuated and food-seeking behavior reduced. The status of the
animal within the group is unchanged; it keeps its control over
territory, although reciprocal conflict behavior is decreased. The
fear response built up by repeated punishing stimulation was marked
by heightened anxiety, self-protective and vocal reactions, adynamia,
muscular hypertonus, loss of a) exploratory activity, b) alertness
and c) competitive interaction with fellow animals. Valproate
completely suppressed display of the fear-and-alarm response;
actively self-defensive reactions were retained, however. Both
exploratory and motor activity remained unchanged. Displays of
positive emotions were not increased. Findings on diazepam's action
under similar experimental conditions are given in Table 1.

Table 15.   Change in Latent Period of Active Escape Produced by
            Diazepam and GA-GABA Given Separately and in Combination

Compound	Dose (mg/kg)	n	Change in latency of escape as percentage of control (= 100)
Diazepam[1]	0.5	6	– 5.0
	2.5	10	+54.8*
GA-GABA[2]	10	7	+13.9
GA-GABA[3] + diazepam	10 0.5	7	+11.6
GA-GABA[3] + diazepam	10 2.5	10	+27.7

[1] 30 min before testing; [2] 4 h before testing; [3] GA-GABA 3 h 40 min before diazepam; * $p < 0.01$.

A comparison of valproate and diazepam shows differences in their effects. Figure 8 illustrates one experiment where both valproate and diazepam exhibit a clear-cut anti-anxiety effect. In this context, the pharmacological effect of both n-DPA and diazepam could be described as tranquilizing. With a certain dose range of diazepam, however, an activating effect of motor and exploratory activity and motivated behavior previously inhibited by fear is observed; displays of positive emotions are also reinforced. A 50 mg/kg dose of n-DPA produces no such activating effect. When 200 mg/kg is given the sedative component of the drug's tranquilizing action prevails - i.e. a drop in emotional response and alertness occurs. Valproate hardly alters the negativity induced by punishment, which is substantially suppressed by diazepam. This is the main effect which distinguishes valproate from diazepam.

Our findings and those of Lal et al. (1980) concur. In rats involved in a conflict situation, both valproate and diazepam reinstated bar pressing milk reinforcing behavior repressed by punishing stimulus. Valproate, however, also inhibits non-conflict response, while diazepam exerts no inhibiting effect on behavior. According to Ticky and Davis's (1980) findings, valproate reinforces inhibitory GABA responses acting on the picrotoxin site, extending the life of GABA-receptor-stimulated ^{3}H-diazepam binding to rat brain membranes.

CONCLUSION

The physical and mental mechanisms of the action of benzodiazepine tranquilizers and their connection with certain neurochemical brain processes have yet to be fully investigated. Nothwithstanding the enormous body of research performed in many laboratories through-

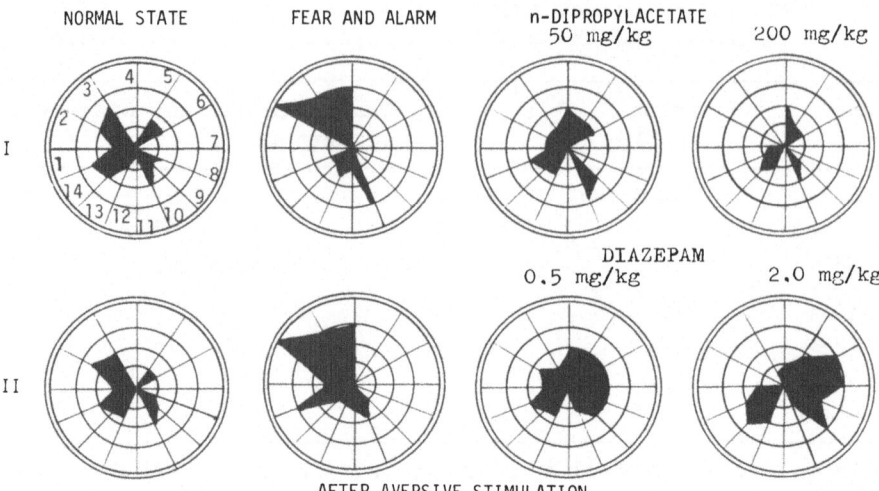

Fig. 8. Changing pattern of the tranquilizing effect of n-DPA and
         diazepam compared and contrasted on a model of fear-and-
         alarm. Spectra of emotional/behavioral response, a: in-
         itially; b: after treatment with n-DPA: I: in doses of 50
         and 200 mg/kg respectively, and with diazepam; II: in doses
         of 0.5 and 2 mg/kg respectively.
         Segments of circle: 1: aggression; 2: fear; 3: anxiety; 4:
         negativity; 5: exploration; 6: pleasure or contentment; 7:
         playfulness; 8: hunting; 9: friendliness; 10: conflict; 11:
         inadequate responses; 12: initiative; 13: motor activity;
         14: alertness.

out the world, contributing to the study of benzodiazepine effects
on, for instance, behavior, receptors (binding sites) and brain
membranes, the key to their anxiolytic action has not yet been discovered. The activating effect of tranquilizers, together with their
anti-anxiety action, are major factors in their psychotropic application, and of considerable importance for their therapeutic usefulness.

     Our detailed study of tranquilizer action on behavior clearly
indicates the differing effects of these compounds according to the
particular animal's response patterns and the characteristic of its
physical and mental status. Benzodiazepines' activating effect
should not always be regarded as a consequence of their anxiolytic
action. By means of the correlations introduced in the second
article we were able to pinpoint independent factors representing key
physical and mental characteristics of various types of emotional
response in animals. The activating effect of tranquilizers viewed
from this angle may be regarded as a readjusting of correlations
between separate physical and mental indicators which go to make up
the complete range of emotional/behavioral response. Further
research along these lines is doubtless required, but the tran-

quilizing effect of benzodiazepines should clearly not be regarded
as a unimodal category.

Psychologically speaking, the specific psychotropic effect of
benzodiazepine tranquilizers largely arises from two associated
processes: reduced function and activation of (respectively) the
punishment reinforcement and reward reinforcement brain systems. It
is significant that different benzodiazepine tranquilizers produce
contrasting effects on the perceptual and emotional components of the
punishment system and the balanced interaction between the punishment
and reward systems. Since the subject's emotional state reflects the
activity of these two basal reinforcement systems, and the emotional
state of the individual in turn determines how the functional system
shapes all behavior (Valdman et al., 1976, 1979) the influence of
benzodiazepines on these processes produces a fundamental readjustment of emotional behavior patterns. Apart from activating the
simpler types of display, such as exploration and self-stimulation of
the brain, the more complex forms of animal group behavior are also
potentiated, involving tendency towards dominance within the group
and signs of heightened (actively self-defensive) aggression, which
may set a puzzle for tranquilizer use.

The analysis we performed using a number of neurochemical
markers leads us to conclude that diazepam's activating effect on the
reward system occurs with both reinforcement and inhibtion of the NA,
5-HT and GABA brain systems. The activation of the reward system
produced by benzodiazepines obviously takes place away from these
systems. Anti-serotonin action potentiated the disinhibitory effect
of diazepam in a conflict situation. Level of emotional response to
punishment is apparently bound up with the 5-HT system. According to
Gray et al.'s (1980) hypothesis, anxiolytics mediate behavior by
modulating the 5-HT and NA-ergic input into the septo-hippocampal
system, where the 5-HT system is more closely connected with punishment and NA with non-reward. Regarding GABA mechanisms, diffuse
activation of this system is not the best background for these tranquilizers to exercise their tranquilizing effect upon; it has a
closer involvement with their sedative action. One aspect of the
subject of the neurochemical bases of benzodiazepines' disinhibitory
and activating effects requiring further study is that of the two
types of benzodiazepine receptors (Klepner et al., 1979; Lippa et
al., 1980). There is also some evidence that the disinhibitory
effect of diazepam in a punishing situation is partially potentiated
by endogenous opiate peptides (Soubrié et al., 1980; Duka et al.,
1980).

REFERENCES

Alexandrovski, J. A., 1982, Mit den Augen des Psychiatres, MIR,
     Moscow, S. Hirzel, Leipzig.

Biggio, Y., Della Bella, D., Frigeni, V., and Guidotti, A., 1977, Potentiation of morphine analgesia by muscimol, Neuro-pharmacology, 16:149-150.

Black, W. C., and Cooper, B. R., 1970, Reduction of electrically-rewarded behaviour by interference with monoamine synthesis, Physiol.Behav., 5:1415-1409.

Braestrup, C., and Nielsen, M., 1980, Benzodiazepine receptors, Arzneim.-Forsch., 30:852-857.

Cananzi, A., Costa, E., and Guidotti, A., 1980, Potentiation by intraventricular muscimol of the anticonflict effect of benzodiazepines, Brain Res., 196:447-453.

Christensen, A. V., Fjalland, B., Pedersen, V., Danneskjold-Samsof, P., and Swesen, O., 1977, Pharmacology of a new phthalane (Lu 10 171) with specific 5-HT uptake inhibitory properties, Eur.J.Pharmacol., 41:153-162.

Cook, L, and Sepinwall, J., 1975, Behavioural analysis of the effects and mechanisms of action of benzodiazepines, in: "Mechanism of Action of Benzodiazepines," E. Costa and P. Greengard, eds., Raven Press, New York, pp.1-28.

Cook, L., and Sepinwall, J., 1979, The relationship of GABA to the anxiolytic effects of benzodiazepines, Brain Res.Bull., 4:692.

Corbett, D., and Wise, R. A., 1979, Intracranial self-stimulation in relation to the ascending noradrenergic fiber systems of the pontine tegmentum and caudal midbrain: a moveable electrode mapping study, Brain Res., 169:423-436.

Corrodi, H., Fuxe, R., Lindbrink, P., and Olson, L., 1971, Minor tranquilizers, stress and central catecholamine neurons, Brain Res., 29:1-16.

Costa, E., 1980, Benzodiazepines and neurotransmitters, Arzneim. Forsch., 30:858-861.

Costa, E., Guidotti, A., and Toffano, G., 1978, Molecular mechanisms mediating the action of diazepam on GABA receptors, Brit.J. Psychiat., 133:239-248.

Cytawa, J., and Trojniar, W., 1979, The pleasure system of the brain and its neurotransmitters, Pol.J.Pharmacol.Pharm., 31:283-292.

Duka, T., Wuster, M., and Herz, A., 1980, Benzodiazepines modulate striatal enkephalin levels via a GABA-ergic mechanism, Life Sci., 25:771-776.

File, S. E., Deakin, J. F., Longden, A., and Crow, T. J., 1979, An investigation of the role of the locus coeruleus in anxiety and agonistic behaviour, Brain Res., 169:411-420.

Fowler, L. J., Beckford, J., and John, R. A., 1975, Analysis of the kinetics of the inhibition of rabbit brain gamma-aminobutyrate aminotransferase by sodium n-dipropylacetate and some other simple carboxylic acids, Biochem. Pharmacol, 24:1267-1270.

Franklin, K. B., Stephens, D. N., and Herzberg, L. J., 1976, Even "dopaminergic" self-stimulation requires noradrenergic acitivity, in: "Brain-Stimulation Reward," Elsevier, Amsterdam, pp.264-266.

Fuller, R. W., Perry, K. W., and Molloy, B. B., 1974, Effect of an uptake inhibitor on serotonin metabolism in the rat brain:

studies with 3-(p-trifluoromethylphenoxy)-N-methyl-3-phenyl-
propylamine (Lilly 110 140), Life Sci., 15:1161-1172.

Gallager, D. W., 1978, Benzodiazepines: potentiation of a GABA inhibiting response in the dorsal raphe nucleus, Eur.J.Pharmac.,
49:133-143.

German, D. S., and Bowden, D. M., 1974, Catecholamine systems as the
neural substrate for intracranial self-stimulation: a hypothesis, Brain Res., 73:381-419.

Gray, J. A., McNaughton, N., James, D. T., and Kelly, P. H., 1975,
Effect of minor tranquilizers on hippocampal theta rhythm is
mimicked by depletion of forebrain noradrenaline, Nature
(Lond.), 258:424-425.

Gray, J. A., Davis, N., Feldon, J., Rawelins, J. N., and Owen, S. R.,
1981, Animal models of anxiety, Prog.Neuro-Psychopharmacol.,
5:143-157.

Haefely, W. E., 1978, Behavioural and neuropharmacological aspects of
drugs used in anxiety and related states, in: "Pharmacology: a
Generation of Progress," M. A. Lipton, A. Di Mascio, and K. G.
Killam, eds., Raven Press, New York, pp.1359-1374.

Hasegawa, K., 1976, Changes in the substantia nigra self-stimulation
behavior by intraventricular injection of norepinephrine and
GABA, Fol.Pharmacol.Japon., 72:985-990.

Hyteel, J., 1977, Neurochemical characterization of a new potent and
selective serotonin uptake inhibitor, Lu 10-171, Psychopharmacology, 51:225-233.

Iadrola, M. J., and Gale, K., 1979, Evaluation of increases in nerve-
terminal dependent vs nerve-terminal-independent compartments
of GABA in vivo: correlation with anticonvulsant effects of
GABA-T inhibition, Brain Res.Bull, 4:686.

Iadrola, M. J., and Gale, K., 1979, Dissociation between drug-induced
increases in nerve terminal and non-nerve terminal pools of
GABA in vivo, Eur.J.Pharmacol., 59:125-129.

Ignatov, Yu, Galustyan, G. E., and Andreev, B. V., 1982, The role of
GABAergic mechanisms in stress-protective effect of benzo-
diazepine tranquilizers, A. Valdman, ed., Institute of
Pharmacology Press, Moscow, pp.118-126.

Ishii, Y., Homma, M., Yoshikawa, A., and Umezawa, H., 1975, Pharmacological action of FD-008, a new dopamine-β-hydroxylase inhibitor, Arzneim.-Forsch., 25:383-385.

Iversen, S. D., 1980, Animal models of anxiety and benzodiazepines
action, Arzneim.-Forsch., 30:862-868.

Jasper, H. J., and Ajmone-Marsan, C., 1961, Diencephalon of the cat,
in: "Electrical Stimulation of the Brain," Univ. of Texas
Press, pp.203-232.

Kent, E. W., and Fedinets, P., 1976, Effect of GABA blockade on
lateral hypothalamic self-stimulation, Brain Res.,
197:628-632.

Kharlamov, A. N., and Rayevsky, K. S., 1980, Tranquilizing effect of
n-dipropylacetate and other GABAergic substances in conditions
of conflict situation, Bull.Exp.Biol.Med., 89:35-37.

Klepner, C. A., Lippa, A. S., Benson, D. I., Sano, M. C., and Bier,

B., 1979, Resolution of two biochemically and pharmacologi-
    cally distinct benzodiazepine receptors, Pharmacol.Biochem.
    Behav., 11:457-462.

König, H. F. R., and Klippel, R. A., 1963, The rat brain, Williams
    and Wilkins, Baltimore.

Kozlovskaya, M. M., Kharlamov, A. N., Rayevsky, K. S., and Valdman,
    A. V., 1981, The GABAergic link in the development of
    tranquilizing effect in cats, Bull.Exp.Biol.Med., 91:45-48.

Lader, M., 1980, The present status of the benzodiazepines in psy-
    chiatry and medicine, Arneim.-Forsch., 30:910-913.

Lal, H., Shearman, G. T., Fielding, S., Dunn, R., Kruse, H., and
    Theurer, K., 1979, Effect of valproic acid on anxiety related
    behaviours in the rat, Brain Res.Bull., 4:711.

Lal, H., Shearman, G. T., Fielding, S., Dunn, R., Kruse, H., and
    Theurer, K., 1980, Evidence that GABA mechanisms mediate the
    anxiolytic action of benzodiazepines: a study with valproic
    acid, Neuropharmacology, 19:785-789.

Lippa, A. S., Klepner, C. A., Young, L., Sano, M., Smith, W. V.,
    and Beer, B., 1978, Relationship between benzodiazepine
    receptors and experimental anxiety in rats, Pharmacol.
    Biochem.Behav., 9:853.

Lippa, A. S., Nash, P. A., and Greenblatt, E., 1979, Pre-clinical
    neuropsychopharmacological testing procedures for anxiolytic
    drugs, in: "Anxiolytics," S. Fielding and H. Lal, eds., Future
    Publ. Co. New York, pp.41-81.

Lippa, A. S., Critchett, D., Sano, M. C., Klepner, C. A., Greenblatt,
    E. N., Coupet, J., and Beer, B., 1979, Benzodiazepine recep-
    tors: cellular and behavioural characteristics, Pharmacol.
    Biochem.Behav., 10:831-843.

Maggi, A., and Enna, S., 1979, Characteristics of muscimol
    accumulation in mouse brain after systemic administration,
    Neuropharmacology., 18:361-366.

Mason, S. T., and Iversen, S. D., 1977, Effects of selective fore-
    brain noradrenaline loss on behaviour inhibition in the rat,
    J.comp.physiol.Psychol., 91:165-173.

Panksepp, J., Gandelman, R., and Trowill, J., 1970, Modulation of
    hypothalamic self-stimulation and escape behaviour by chlor-
    diazepoxide, Physiol.Behav., 5:965.

Patkina, A., and Lapin, I. P., 1976, Effect of serotoninergic drugs
    on positive and negative reinforcing systems in cats,
    Pharmac.Biochem.Behav., 5:247-252.

Patkina, N. A., and Lapin, I. P., 1976, Effect of catecholaminergic
    drugs on systems of reward and punishment in experiments on
    cats, Pharmac.Behav., 5:247-252

Poshel, B. P., and Ninteman, F. W., 1971, Intracranial reward and
    forebrain's serotonergic mechanisms, Physiol.Behav., 7:39-46.

Ritter, S., and Stein, L., 1974, Self-stimulation in the mesen-
    cephalic trajectory of the ventral noradrenergic bundle,
    Brain Res., 81:145-157.

Robichaud, R. C., and Sledge, K. L., 1969, The effects of p-chloro-

        phenylalanine in experimentally induced conflict in the rat,
        Life Sci., 8:965-969.
Sanders-Bush, E., Bushing, J., and Sulser, F., 1975, Long-term
        effects of p-chloramphetamine and related drugs on central
        serotonergic mechanisms, J.Pharmacol.Exp.Therap., 192:33-41.
Sanders-Bush E., Massari, V. J., 1977, Actions of drugs that deplete
        serotonin, Fed.Proc., 36:2149-2153.
Simon, P., and Soubrié, P, 1979, Behavioural studies to differentiate
        anxiolytic and sedative activity of the tranquilizing drugs,
        in: "Modern Problems in Pharmacopsychiatry, vol. 14, Differen-
        tial Psychopharmacology of Anxiolytics and Sedatives," J. R.
        Boissier, ed., Karger, Basel, pp.99-143.
Soubrie, P., Jobert, A., and Thiebot, M. H., 1980, Differential
        effects of naloxone against the diazepam-induced release of
        behaviour in rats in three aversive situations, Psychopharma-
        cology, 69:101-105.
Soubrie, Ph., Thiebot, M., and Simon, P., 1979, Enhanced suppressive
        effects of aversive events induced in rats by picrotoxine:
        possibility of a GABA control on behavioural inhibition,
        Pharmacol.Biochem.Behav., 10:463-469.
Stark, P., and Fuller, K. W., 1972, Behavioural and biochemical
        effects of PCPA; 3-chlorotyrosine and 3-chlorotyramine. A
        proposed mechanism for inhibition of self-stimulation, Neuro-
        pharmacology., 11:261-272.
Stein, L., 1980, Behavioural neurochemistry of benzodiazepines,
        Arzneim.-Forsch., 30:868-873.
Stein, L., Wise, C. D., and Berger, B. D., 1973, Anti-anxiety action
        of benzodiazepines: decrease in activity of serotonin neurons
        in the punishment system, in: "The Benzodiazepines,"
        S. Garattini, et al., eds., Raven Press, New York, pp.299-326.
Stein, L., Wise, C. D., and Belluzzi, D., 1975, Effects of benzo-
        diazepines on central serotonergic mechanisms, in: "Mechanism
        of Action of Benzodiazepines," E. Costa and P. Greengard,
        eds., Raven Press, New York, pp.29-44.
Stein, L., and Wise, C. D., 1970, Behavioural pharmacology of central
        stimulators, in: "Principles of Psychopharmacology, W. G.
        Clark and J. del Giudice, eds., Academic Press, New York,
        pp.313-325.
Steranka, L. R., Barrett, R. J., and Sanders-Bush, E., 1977, Facil-
        itation of Sidman avoidance performance by p-chloramphetamine:
        role of biogenic amines, Neuropharmacology., 16:751-759.
Sullivan, J., Sepinwall, J., and Cook, L., 1978, Anticonflict evalu-
        ation of muscimol, a GABA receptor agonist, alone and in
        combination with diazepam, Fed.Proc., 13:2143.
Thiébot, M., Jobert, A., and Soubrié, P, 1979, Effets comparés du
        muscimol et du diazepam sur les inhibitions du comportement
        induites chez le rat par la nouveauté, la punition et le
        non-renforcement, Psychopharmacologia (Berl)., 61:85-89.
Thiébot, M. H., Jobert, A., and Soubrié P., 1980, Chlordiazepoxide
        and GABA injected into raphe dorsalis release the conditioned
        suppression induced by rats by a conflict procedure without

nociceptive component, Neuropharmacology, 19:633-641.

Ticku, M. K., and Davis, W. S., 1981, Effect of valproic acid on ^{3}H-dihydropicrotoxin binding sites at the benzodiazepine-GABA-receptor complex, Brain Res., 223:218-222.

Tye, N. C., Everitt, B., and Iversen, S. D., 1977, Hydroxytryptamine and punishment, Nature (Lond.), 268:741-743.

Tye, N., Iversen, S., and Green, A., 1979, Effects of benzodiazepine's serotonergic manipulations on punished responding, Neuropharmacology., 18:689-695.

Valdman, A. V., 1980, Preclinic prediction of the spectrum of psychotropic activity of tranquilizers, Pharmacol.Res.Commun., 12:225-266.

Valdman, A. V., Zvartau, E. E., and Kozlovskaya, M. M., 1976, Psychopharmacology of Emotions, Meditsina, Moscow.

Valdman, A. V., Zvartau, E. E., and Kozlovskaya, M. M., 1975, Experimental study of the action of psychotropic drugs on emotions, motivations and social behaviour of animals, in: "CNS and Behaviour Pharmacology," Vol. 3, Proc. 6th int. congr.Pharmacol., M. Airaksinen, ed., pp.207-211.

Valdman, A. V., Kozlovskaya, M. M., and Medvedev, O. S., 1979, Pharmacological regulation of emotional stress, Moscow, Meditsina.

Valdman, A. V., and Marusov, I. V., 1979, Role of GABA-ergic systems in the activating effect of diazepam, Bull.Exp.Biol.Med., 6:551-553.

Van de Laan J. W., de Boer, J., and Bruinvels, J., 1979, Di-n-propylacetate and GABA degradation, referential inhibition of succinic semialdehyde dehydrogenase and indirect inhibition of GABA transaminase, J.Neurochem., 32:1769-1780.

Waldmeier, P. C., Baumann, P. A., and Maitre, L., 1979, CGP 6085A, a new, specific inhibitor of serotonin uptake: neurochemical characteristics and comparison with other serotonin uptake blockers, J.Pharmac.Exp.Therap., 211:42-49.

Wise, C. D., Berger, B. D., and Stein, L., 1973, Evidence of α-noradrenergic reward receptors and serotoninergic punishment receptors in the rat brain, Biol.Psychiat., 6:3-21.

Wong, B. T., Horng, J. S., and Bymaster, F. P., 1975, d,l-N-methyl-3(o-methoxyphenoxy)-3-phenylpropylamine hydrochloride, Lilly 94939, a potent inhibitor for uptake of norepinephrine into rat brain synaptosomes and heart, Life Sci., 17:755-770.

Wuttke, W., and Kelleher, R. T., 1970, J.Pharmacol.Exp.Therap., pp.397-405.

Zakusov, V., Ostrovskaya, R., Kozhechkin, S., Markovich, V., Molodavkin, G., and Voronina, T., 1977, Further evidence for GABAergic mechanisms in the action of benzodiazepines, Arch. Int.Pharmacodyn.Ther., 229:313-326.

Zwartau, E. E., 1977, Hypothalamic self-stimulation under the chronic morphine treatment, Res.Comm.Chem.Path.Pharmacol., 16:707-720.

Zwartau, E. E., and Patkina, N. A., 1974, Motivational components and self-stimulation in behavioral reaction, elicited by electrical stimulation of hypothalamus, J.Higher Nerv.Activ. (Russ.), 24:529-535.

Peptidergic and Neurochemical Aspects

Neurotransmitter and Peptidergic Mechanisms of Intraspecies Aggression and Sociability

V.P. Poshivalov

Department of Pharmacology, Pavlov Medical Institute, Leningrad

1. INTRODUCTION

One of the main problems now facing psychopharmacologists is how the function of motivational systems of intraspecies behavior is controlled, with special reference to the agonistic patterns of offense and defense (Silverman, 1978; Valdman and Poshivalov, 1980; Miczek and Krsiak, 1979; Poshivalov. 1978, 1980 and 1981b). According to Poshivalov (1981b, 1982), Valzelli (1981) and Mandel et al. (1981) there are grounds for supposing that the best means of controlling aggression, defense and sociability during agonistic interaction between animals may partially be revealed by:

  a) experimental intervention at different stages in the metabolism of GABA, catecholamines, serotonin and acetylcholine
  b) specific alterations in the activity of opioid and hypothalamo-pituitary peptide systems.

The anti-aggression effect of psychotropic (and neuropharmacological) drugs alone, once identified, appears much more meaningful when associated with the study of compensatory and behavioral patterns accompanying this effect and of changes from the usual patterns of intra-species behavior. The works of Silverman (1978), Poshivalov (1974, 1978 and 1981b), Brain (1979), Krsiak (1975) and Valdman (1980) describe ethological methods for investigating various substances, showing how promising these methods are for actually showing selective drug action on specific forms of intraspecies behavior.

In our work we set out to study the pharmaco-ethological range of activity of opioids, hypothalamo-pituitary peptides and different categories of neuropharmacological agents and their role in the pharmacological control of intra-species aggression and sociability.

2. METHODS

ANIMALS

Experiments were performed on 560 CC 57W strain mice separately housed under laboratory conditions. Only aggressive male animals, weighing between 18 and 20 g, were selected for experimentation with neuropharmacological substances and peptides. The mice were kept under standard conditions; T = 22-24°C, under a reversed day-night cycle, i.e. bright daylight from 18.00 to 9.00 and diffuse red light for the remainder of the day. Water and food pellets were provided ad libitum.

BEHAVIORAL STUDY

All mice were kept isolated in specially constructed single cages measuring 9x12x15 cm for 6-12 weeks. Each was allowed to interact with its fellows from the group once a week, for an average of 4 min - for 5 or 10 min in some experiments. Ethological methods were used for evaluating the ensuing behavior (Grant and Mackintosh, 1963; Silverman, 1978; Adams, 1976, 1980), using a special "ethograph computer" (Poshivalov, 1978; Poshivalov et al., 1979). Rating of the main behavioral features is shown in Table 1.

DRUGS

The following were used: cocaine (1-30 mg/kg), apomorphine (0.5-1 mg.kg), amphetamine (0.5-5 mg/kg), propranolol (1-2.5 mg/kg), phentolamine (2.5 mg/kg), haloperidol (1-5 mg/kg), disulfiram (100 mg/kg), FD-008 (100 mg/kg), nisoxetine (5-40 mg/kg), L-DOPA (10 and 200 mg/kg), 5-hydroxytryptophan (5-HTP, 50 mg/kg), L-tryptophan (50 mg/kg), fluoxetine (10 mg/kg), parachlorophenylalanine (PCPA, 300+100+100 mg/kg), scopolamine (0.25-1.0 mg/kg), methyl atropine bromide (1-5 mg/kg), picrotoxin (1 mg/kg), bicuculline (0.5-2 mg/kg), thiosemicarbazide (TSC, 1-2 mg/kg), muscimol (0.2-1 mg/kg), gamma-acetylenic-GABA (GA-GABA, 50-100 mg/kg), morphine hydrochloride (2.5 mg/kg), naloxone hydrochloride (0.25-0.5 mg/kg) and diazepam (2.5 mg/kg).

The following peptides and psychotropic compounds were used:

a) opioid peptides, synthesized in the Cardiological Research Center of the Acad. Med. Sci. of the USSR (Moscow) - ciotorphine (CT, 25 mg/kg), synthetic anolog of enkephalin (SA-ENK, 25 mg/kg), methionine enkephalin (MET-ENK, 25-50 µg into the lateral ventricle) and neo-endorphin (NEO-END, 25 mg/kg);

b) hypothalamo-pituitary peptides - $ACTH_{1-24}$ (2.5-25 µg; 25 µg/ day for 5 days, $ACTH_{4-10}$ (2.5-25 µg), α-MSH (25 µg), melanostatin (MIF, 1-25 mg/kg), thyroliberin (TRH, 25 mg/kg), luliberin (LH-RH, 2.5-25 mg/kg) and somatostatin (SST, 2.5-5 mg/kg).

Table 1. Pattern of Ethological Effects of Disulfiram, and How These Are Rated

Behavior	Disulfiram 100 mg/kg 1 h	Disulfiram 100 mg/kg vs. control on 0.9% phys. saline or Tween-80 1 h	Disulfiram 100 mg/kg 4 h	Disulfiram 100 mg/kg vs. control on 0.9% phys. saline or Tween-80 4 h
Animal group (intraspecies) behavior				
Aggression:				
attack	0.071	-0.196	0.030	-0.236* ▶ ▽
pursuit (and attack)	0.027	+0.008	0.015	-0.004 ▽
threat	0.008	-0.028	0.003	-0.033 ▽
Sociability:				
sniffing body	0.073	+0.052	0.058	+0.036 △
sniffing nose	0.030	+0.011	0.046	+0.026 △
sniffing tail	0.000	0.000	0.003	-0.007 ▽
grooming back of neck	0.011	-0.018	0.006	-0.023* ▶
grooming body	0.022	+0.017	0.003	-0.002 ▽
sniffing genitalia	0.063	-0.026	0.036	-0.052* ▶
Defense:				
sideways-on posture	0.000	0.000	0.000	0.000 —
retreat	0.024	-0.120	0.006	-0.030 ▽
Ambivalence:				
tail-rattling	0.046	-0.076	0.015	-0.102* ▶
circling	0.005	-0.019	0.003	-0.210* ▶
Individual behavior				
locomotion	0.133	+0.049	0.195	+0.111 △
rearing	0.095	+0.030	0.079	+0.014 △
self-grooming	0.068	+0.008	0.040	-0.020 ▽
sitting	0.304	+0.184*	0.419	+0.300* ◀

Note: Significance of differences marked with an asterisk; p<0.05 by Wilcoxon's rank sum test (marked with shaded triangles). Scores awarded for general responsiveness: △ a rise; ▽ a fall.

The peptides and other substances were given intraperitoneally
in a volume of 0.1 ml per 10 g of body weight. A few opioid peptides
were administered into the lateral ventricle in a volume of 2 μl.
All compounds were given to aggressive animals only. Their effects
on behavior were measured 5, 15 and 30 min after injection and at
other intervals, as indicated in the tables.

STATISTICAL WORK

A special "BEHAV" program was set up for statistical processing
of the results. The compounds' effects were assessed by comparing
the statistical probability of each alternative action occurring
post-administration against the control receiving an injection of
Tween-80 solution or 0.9% physiological saline (0.1 ml per 10 g body
weight). The significance of the changes produced was measured
according to Wilcoxon's rank sum test. Table 1 provides an example
of this rating item by item under several main categories or behavior, as produced on "Nairy-2" and "ES 1022" computers.

3. GABA-ergic MECHANISMS

Our ethological studies of interaction between previously isolated mice (Poshivalov, 1981) show that these animals display "disinhibition" or release of behavior - hyper-responsiveness, hyperactivity, changes in behavior pattern and unrestrainable intra-species
aggression. These features are most probably bound up with a deficit
in inhibitory systems, particularly the GABA system.

As illustrated in Table 2, our experiments show that raising the
GABA system activity by direct stimulation of the GABA receptor,
using muscimol, increases intra-species sociability amongst previously isolated mice and reduces the incidence of attack. However, the
0.2-0.5 mg/kg dose of muscimol indicated in the table does not completely inhibit aggression; threatening behavior is maintained and
ambivalent behavior increases. Raising the dose of muscimol suppresses aggressive behavior unselectively.

The high affinity of muscimol for GABA receptors in vitro does
not guarantee the success of using this compound on behavior to
compensate selectively and adequately for the GABA deficit. One
could alternatively postulate multiple GABA receptors, on each of
which muscimol might act differently.

More selective compensation for a GABA deficit may be achieved by
blocking GABA breakdown using an inhibitor of the GABA metabolizing enzyme GABA-T (see Table 2). For instance, when the GA-GABA dose is 75 mg/
kg, sociability and locomotion predominate. A dose range of 50-100 mg/
kg GA-GABA produces a fairly selective but short-lived reduction in ag-

Table 2. Pattern of Ethological Effects of GABA-positive Compounds on Previously-isolated Mice

Behavior	Control vs. muscimol	Muscimol (30 min) dose (mg/kg)			Control vs. GA-GABA	GA-GABA (2 h) dose (mg/kg)			GA-GABA 75 mg/kg + PT 1.0 mg/kg	GA-GABA 50 mg/kg + DZ 1.0 mg/kg
		0.2	0.5	1.0		50	75	100		
Animal group behavior										
Aggression:										
attack	0.210*	0.111*	0.019		0.322*	0.174*	0.042	0.016	0.238*	0.056
threat	0.066*	0.065*	0.051	0.023	0.116*	0.037	0.024	0.065*	0.067*	0.052
Sociability:										
sniffing body	0.009	0.051*	0.108*	0.062	0.010	0.103*	0.168*	0.016	0.091*	0.182
sniffing nose	0.004	0.001						0.012		
sniffing genitalia	0.036	0.048	0.043	0.003	0.032	0.088*	0.061		0.067	0.085*
Defense:										
sideways-on posture				0.060*				0.049		
upright posture				0.017				0.012		
Ambivalence:										
tail-rattling	0.103*	0.064	0.042	0.026	0.081	0.088	0.032	0.028	0.064	0.033
Individual behavior										
locomotion	0.175*	0.171*	0.137*	0.126	0.163*	0.182*	0.291*	0.196*	0.282*	0.208*
rearing	0.071	0.170*	0.120*	0.010	0.091	0.083	0.095	0.036	0.062	0.087
self-grooming	0.110*	0.037	0.006		0.075	0.074	0.038	0.008	0.045	0.066
sitting	0.192*	0.272*	0.488*	0.665*	0.102*	0.163*	0.241*	0.549*	0.069	0.223*

Note: Table shows statistical probability (frequency) of performance of these actions.
* - highest probability (incidence); PT - picrotoxin (30 min); DZ - diazepam (30 min).

gression lasting 2-3 h, and potentiates sociability. The anti-aggression effect has also been noted with other GABA-T inhibitors – aminohydroxyacetic acid (Da Vanzo and Sydow, 1979) and sodium valproate
(Puglisi-Allegra and Mandel, 1980). Doses of 125-250 mg/kg of the
inhibitor of GABA re-uptake nipecotic acid produced a similar effect
on aggressive behavior in isolated animals (Puglisi-Allegra and
Mandel, 1980). These publications did not describe changes in intraspecies sociability, however.

It should be noted that aggressive behavior is gradually reinstated by the time the high level of GABA in the CNS, produced by
GA-GABA is stabilized (i.e. in 4 h); this shows how rapidly the compensatory mechanisms, inhibiting GABA-T and integrating behavior
into changed circumstances, are brought into operation. It also
shows that a single boost of GABA in the CNS is insufficient to
suppress aggression successfully over prolonged periods.

The GABA analogs phenibut and phepiron suppress aggression
and sociability in previously isolated mice to varying degrees
(Poshivalov, 1981b; Valdman and Poshivalov, 1982); the same is
true both of sodium hydroxybutyrate and other psychotropic agents
with a GABA-positive component of action, such as diazepam
(Poshivalov, 1981b). The reason for the activity of GABA-positive
compounds in suppressing aggression requires special scrutiny.

In this connection the work of De Feudis (1979) should be mentioned, showing that GABA's capacity for binding to heavy synaptosomal fractions was lower in the brain of isolated mice than of
grouped mice. A lower level of GABA was found in the striatum,
hippocampus, amygdala and olfactory bulb of isolated animals (Earley
and Leonard, 1977) – this, too, distinguished them from grouped
animals. A reduction in glutamate decarboxylase was found in the
brain of aggressive isolated animals (Blindermann et al., 1979). All
these findings point to the involvement of GABA deficiency in the
pathogenesis of aggressive behavior and "behavioral disinhibition or
release" in isolated mice.

When GABA-negative compounds are used to exacerbate the GABA
deficit in isolated animals (see Table 3), aggressive behavior grows
more pronounced; picrotoxin and bicuculline both act in this way
within a certain dose range. It should be mentioned that isolation
also lowers the convulsion threshold in mice; this, too, may be
explained by a GABA deficit. Not all GABA antagonists acted identically; the glutamic acid decarboxylase (GDC) inhibitors, thiosemicarbazide, for instance, reinforced aggression and suppressed sociability (see Table 3), while also producing homosexual tendencies. It
is worth noting that the GDC inhibitors may not merely reinforce
aggression, as shown in the present work, but actually generate it in
previously non-aggressive mice. The same has been shown in the case
of DL-allylglycine (Puglisi-Allegra et al., 1981).

Table 3. Pattern of Ethological Effects of GABA-negative Compounds on Previously-Isolated Mice

Behavior	Control vs. TSC and PT	Thiosemicarbazide (TSC, 1 h) 1 mg/kg	3 mg/kg	Picrotoxin (PT, 15 min) 1 mg/kg	Control vs. BC	Bicuculline (BC, 15 min) 1 mg/kg
Animal group behavior						
Aggression:						
attack	0.162*	0.273*	0.269*	0.314*	0.288*	0.254*
threat	0.110*	0.048	0.049	0.130*	0.126*	0.128*
Sociability:						
sniffing body	0.007	0.007	0.002	0.007	0.012	0.011
sniffing nose	0.009	0.008			0.007	0.002
sniffing genitalia	0.009	0.010	0.013		0.022	
Sexual:						
attempted mount		0.021				
mount		0.003	0.003			
Ambivalence:						
tail-rattling	0.073	0.075	0.053	0.093	0.081	0.118*
Individual behavior						
locomotion	0.162*	0.168*	0.204*	0.152*	0.170*	0.124*
rearing	0.092	0.103+	0.114*	0.067	0.089	0.064
grooming	0.098	0.067	0.078	0.065	0.076	0.010
sitting	0.164*	0.141*	0.130*	0.134*	0.185*	0.278*

Note: Table showing statistical probability (frequency) of performance of these actions.
      * - highest probability (incidence).

Apart from a few minor discrepancies in effect on behavior, compensating for the GABA deficit in aggressive isolated animals tends to produce a reduction in aggression, whereas exacerbating the deficit (within limits) leads to a rise in intra-species aggression. As regards intraspecies sociability, it should be mentioned that this is potentiated more or less selectively by GABA-positive compounds, including GABA analogs. The potentiating effect is less marked than in the case of benzodiazepine derivatives. When certain GABA agonists (GA-GABA) are combined with small doses of benzodiazepines, substantial suppression of aggression together with ambivalence and reinforced sociability may be observed; this bears out the importance of activating the benzodiazepine receptor complex promoting intra-species sociability. It may be that we are dealing with a complex effect occurring at the GABA-benzodiazepine receptor complex. GABA-positive compounds, including GABA analogs, generally suppress aggression to a greater degree and potentiate sociability to a lesser extent following acute i.v. administration than, for instance, benzodiazepine-type substances, such as diazepam and medazepam.

GABA receptors were also found to be a component of the GABA-benzodiazepine complex (Costa et al., 1978). These substances, which react either on the latter complex or on benzodiazepine receptors unconnected with GABA, are effective as anxiolytics and anti-aggression agents. Perhaps the GABA-benzodiazepine receptor complex is bound up with antiaggressive and sedative action while the "pure" benzodiazepine receptor element is involved with potentiating intra-species sociability. Lack of sociability in isolated mice, irrespective of their tendency towards aggression, may be a consequence of a smaller number of benzodiazepine receptors in the brain (Robertson, 1980).

4. CATECHOLAMINE MECHANISMS

Potentiation of intraspecies aggression in isolated individuals doubtless depends on the level of NA and DA receptor system activation. Such animals are highly sensitive to the effects of central sympathomimetic and adrenergic drugs; small doses of amphetamine and cocaine, as shown in Table 4, potentiate aggression and reduce other types of intraspecies behavior. Prolonged isolation reverses the effects of large doses of amphetamine. Amphetamine induces fighting amongst grouped mice (Miczek and Krsiak, 1979) and suppresses aggression amongst isolated animals according to our findings given in Table 5. A low dose of 0.2 mg/kg apomorphine stimulated aggressive behavior in non-aggressive isolated mice (Krsiak et al., 1981), while a high dose of 10-20 mg/kg of this compound blocked aggression in previously isolated individuals (Hodge and Butcher, 1975).

The heightened sensitivity of NA and DA receptors in these animals may be bound up with their adaptation to lower catecholamine

Table 4. Pattern of Ethological Effects of Catecholaminergic Compounds

Compound	Dose mg/kg i.p.	Time (h)	Animal group behavior					Individual behavior		General responsiveness	Notes
			Aggression	Sociability	Defense	Sexual	Ambivalence	Dynamic	Static		
Amphetamine	0.5	0.5	◁	▶		▷	▷	◀	▷	◀	
	5	0.5	▶	▶		▷	▶	◀		◁	stereotypy
Cocaine	1	0.5	◀	▶			▷	◀		N	
	10	0.5	▶	▶			▶	◀		N	stereotypy
Nisoxetine	5	0.5	◀	▶			▶	◀	▷	◀	
L-DOPA	10	0.5	◀	▶	◀	▷	▷	◁	◀	◀	gnawing
	200	0.5	▶	▶	◁		▷	◁	◀	◁	rearing
L-DOPA + fluoxetine	10	0.5	▶	▷	◁	▷	▷	▷	◁	◁	
	20	1									
FD-008 + L-DOPA	100	0.5	▶	▶	▷		▷	◁	◁	◁	
	10	1	▶	◀	▷		▶	◁	◀		
FD-008	100	0.5	▶	◀	▷		▶	◁	◀		
	100	4	◀	◀▷			▷		◀	◁	
	100	24									
FD-008 + amphetamine	100	1	▶	▷	▷		◁	◀	◁	◁	
	3	0.5	▷	▷◁	▷		▷		◀		gnawing
Disulfiram	100	1	▶	▶	▷		▶	◁	◀	◁	stereotypy
	100	4	▶	▷◀	▷		▶	▷	◀		
Propranolol	1	0.5	▶	▶			▷	▷	◀	▷	
	2.5	0.5	▷	▶			▶	▷		▶	
Phentolamine	2.5	0.5	▶	◁			▶	▷		▷	
	1	0.5	▶	▷			▷				
Haloperidol	3	0.5					▷		◀	▶	

See note to Table 1.

(CA) concentration, produced by CA depletion under the stressful circumstances of prolonged isolated and intraspecies confrontation (Welch and Welch, 1969). A drop in the speed of CA metabolism - low rate of synthesis, CA release and recycling - in aggressive isolated animals causes a smaller reduction of CA than in grouped mice following administration of α-methylparatyrosine, a tyrosine hydroxylase inhibitor.

Long-term administration of haloperidol to isolated aggressive mice over 27 days, followed by abrupt withdrawal, maximized aggression and minimized sociability in our experiments (Valdman and Poshivalov, in press). Withdrawal of haloperidol and other DA receptor antagonists on behavior are brought about by a somewhat complex pattern of interaction (Cools and Van Rossum, 1980) with central DA systems, involving:

a) direct suppression of post-synaptic activity at the level of neostriatal DA_e receptors (insensitive to dihydroxyphenylamino-imidazoline - DPI) and

b) reinforcement of post-synaptic activity of mesolimbic DA_i receptors (sensitive to DPI) by means of direct blocking of their action on mesolimbic a-type NA receptors controlling DA_i receptor function.

Data is available pointing to major adaptational changes in mesolimbic a-type NA receptors, which regulate mesolimbic DA_i receptors during subacute injections of haloperidol or apomorphine (Cools and Van Rossum, 1980). Our findings show that haloperidol withdrawal, together with the hypersensitivity produced by its effects, plus a putative pharmacological factor, then gradually brings about a progressive increase in aggression to exceed control level for these animals.

It should nevertheless be mentioned that the isolated animals' response to amphetamine and apomorphine during haloperidol withdrawal tended towards stereotypy rather than an increase in aggression, as had been the case prior to injection with haloperidol. This effect may be due either to the relatively high sensitivity of neostriatal DA receptors or simply to their large number. It should, however, be pointed out that the isolated animals' behavior after amphetamine administration during haloperidol withdrawal was reminiscent of the action of amphetamine prior to injection of the DA receptor antagonist, but with larger doses.

Temporary depletion of NA systems produced by 6-OHDA (6-hydroxy-dopamine) also induces a hypersensitivity syndrome in rats, as shown by behavioral hyper-reactivity, hyperactivity and heightened sensitivity to amphetamine. Ellison and Bryan (1977) noted that the above-mentioned behavioral changes and response to amphetamine following 6-OHDA administration are characteristic of isolated rats

during recovery, but are not observed in grouped animals after
6-OHDA. This indicates potentiation of the effects of a putative
pharmacological isolation factor.

It would appear that typical adrenergic antagonists would help
to promote a "normalization" of an isolated animal's behavior;
phentolamine and propranolol, however, as shown in Table 4, do
suppress aggressive behavior, but without specifically restoring
intra-species sociability. The overall responsiveness of the
animals is reduced, while intraspecies contacts take on an almost
passive nature. The acute action of the DA receptor antagonist,
haloperidol, as shown in our ethological study, is non-selective,
being accompanied by a number of other effects which restrict both
group and individual behavior and reflect the blocking of multiple
functions of catecholamines in the CNS.

Paradoxically, inhibitory effects may be produced on intraspecies aggression by administering L-DOPA, the direct precursor of
CA (see Table 4). This inhibitory effect is only achieved, however,
in conjunction with display of abnormal eccentric defensive behavior.
Many pharmacologists have regarded these defensive attitudes and
states of hyper-responsiveness as signs of aggression, and this has
led to much confusion in the interpretation of experimental findings.
At the same time, such discrepancies have clearly pointed to the need
for ethologically-directed studies investigating the pharmacology of
aggression and defense (Miczek and Barry, 1976).

Small doses only of L-DOPA and nisoxetine are able to reinforce
species-specific aggression, manifesting as attacking and biting,
hitting, pursuit, threatening behavior, etc. (see Table 4).

As our knowledge now stands it is not easy to determine how far
NA and DA systems are essentially involved in the balance between
aggression and defense, in view of the complex relationship between
DAe, DA^1 and α-adrenergic receptors, and also DA 1 and DA 2 receptor
systems, whether linked with adenylcyclase or not.

In view of the fact that FD-008 selectively lowers brain NA
concentration (Ishii et al., 1975), we may suppose that reinforcing
the effect of this amine is largely responsible for maintaining
species-specific aggression. These inferences find partial support
from results obtained with other DBH inhibitors, such as FLA 57 and
FLA 63 (Ross and Ogren, 1976), which bring about a reduction of
aggression in isolated animals. No final conclusion may be drawn on
the basis of the latter work, however, concerning these compounds'
action on intraspecies behavior, since intraspecies sociability was
not a point specifically evaluated.

Catecholamine agonists in low doses thus potentiate the
"unsociable" components of isolation; this is manifested in the

reinforcement of aggressive behavior and the drop in sociability. Large doses of agonists induce displays of defensive behavior, often of an abnormal type and accompanied by motor stereotypy. Catechol-amine's antagonists decrease intraspecies aggression with varying degrees of selectivity; they have a generally restrictive action, reducing non-specific responsiveness. Here, however, high threshold external stimuli may evoke primitive forms of defensive behavior. Intraspecies sociability is severely impaired by the action of NA and DA antagonists.

5.   SEROTONINERGIC MECHANISMS

Table 5 gives results obtained during this work showing the pattern of ethological effects of serotoninergic compounds. Table 5 indicates that an increased concentration of brain serotonin reduces aggression in isolated animals, promotes defensive tendencies and heightens general responsiveness, whereas a drop in 5-HT can re-inforce aggression or cause other deviations in the animals' be-havior. In particular, it was found that 5-HTP and L-tryptophan loading blocks aggression and raises hyper-reactivity; this demon-strates how these two pathological phenomena are independently inte-grated into behavior. There is another method of raising 5-HT con-centration: inhibiting 5-HT reuptake by fluoxetine, which tempor-arily and non-selectively reduces aggression, while promoting defen-sive behavior.

Some similarity between the effects of 5-HT and CA compounds on behavior may be seen, in the same way as the effects of high doses of amphetamine resemble those of L-tryptophan; large doses of both substances reduce aggression and reinforce defensive types of behavior. These data are borne out by the results of Poshlova et al. (1977), which point out that the effects of 5,6-dihydroxytryptamine (5,6-DHT) on behavior and of amphetamine on aggression and defense resemble L-tryptophan action. The behavioral effects of 5,6-DHT and amphetamine may to some extent reflect their serotoninergic in-fluences (e.g. post-denervational hypersensitivity following destruc-tion of serotoninergic neurones by 5,6-DHT). Other research (Brase and Loh, 1976) showed that amphetamine reinforces the activity of serotoninergic neurones in response to 5-HT release from the raphe nucleus and increases tryptophan concentration in the brain.

Prolonged isolation in itself lowers the concentration of 5-HT in the brain (Valzelli, 1980, 1981). Exacerbating this deficit by administering comparatively low doses of PCPA (70 mg/kg), the effect of which is seen 24 hours later, increases the incidence of attack amongst the isolated animals; this corresponds with a drop in 5-HT concentration in the brains of these animals. Our experiments with PCPA (300 mg/kg, after a time lapse of 4 hours) reinforced hyper-reactivity in isolated mice and induced patho-logical display of aggression combined with homosexual behavior

(see Table 5); intraspecies sociability was reduced. An increase
in the dose of PCPA, administered gradually, completely blocks aggression and increases the amount of static behavior display by the
individual.

PCPA's inhibitory effect on aggressive behavior partially
resembles the effects of 5-HT receptor antagonists (Malick and
Barnett, 1973). Mianserin, methysergide, cyproheptadine, sinanserin
and other 5-HT blockers are capable of countering classic 5-HTP
effects, i.e. suppressing the "head shaking" phenomenon after
systemic administration, and reducing intraspecies aggression, when
used in this way. It should be borne in mind, however, that further
study is required in the specificity of these behavioral changes
produced by 5-HT receptor blockers. The work of Malick and Barnett
(1973) does not make it clear what was actually occurring in the
animals when aggression was totally blocked, nor how sociability
and the other biologically essential forms of behavior were affected, nor whether behavior returned to normal. At this stage,
we can only say that a) 5-HT antagonists are able both to inhibit and
provoke intraspecies aggression by different mechanisms for blocking
serotoninergic transmission, and b) this action is dependent on conditions difficult to control in the laboratory. We may thus conclude
that 5-HT plays a part in modulating how aggression is to be inte-
grated.

On the other hand, damage to the raphe nucleus causes both
a drop in brain 5-HT concentration and aggression in isolated aggressive animals (Kostowski and Valzelli, 1974). These animals
never actually attacked an adversary after destruction of the
raphe nucleus, but were capable of defending themselves from attack. A dietary deficiency of tryptophan led to a drop in brain
5-HT in isolated aggressive mice and reduced the number of "fighter"
mice, together with the duration of fighting (Kantak et al., 1980).
According to our experiments, only a very substantial 5-HT deficit
suppresses aggressive behavior (see Table 5).

The extent of the drop in brain 5-HT concentration could be
a factor in determining which behavioral changes occur. A dose-
response and time-related study of PCPA action on isolated animals
would reveal the following mounting effects: initial exacerbation
of aggression, followed by build-up of sexual (here, homosexual) behavior, reinforcement of hyper-reactivity and finally total blocking
and reversal of aggression into defensive behavior.

The need to maintain the 5-HT deficit at a certain level in
order for aggression to occur also emerged from our experiments
combining administration of PCPA and fluoxetine, the 5-HT re-uptake
inhibitor. This combination reduces the likelihood of aggression and
PCPA-produced homosexuality appearing, while maintaining high general
responsiveness (see Table 5).

Table 5. Pattern of Ethological Effects of Serotoninergic Compounds

Compound	Dose mg/kg i.p.	Time (h)	Aggression	Sociability	Defense	Sexual	Ambivalent	Dynamic	Static	General responsiveness	Notes
L-tryptophan	50	0.5	▼	▼	▲		▽	▼	▲	N	
5-HTP	50	1	▼	▽	△	▼	▽	▽	△	▲	head shaking
Fluoxetine	10	4	▼	▼	△	▽	▼	▽	▲	▲	
	10	4	▽△	▽		▽	▽		△	△	
PCPA	300	1	▽	▽		▲HS	▽		▲	△	HS
	300	4	▽△	△▼		▲HS	▽△		△	△	HS
	300	24		△		▲HS	△▽	▽▲	▲	△	HS
	300 + 100 + 100	78	▼	▼			▽				
PCPA + L-DOPA	300 + 100 + 100	78	▼	▽	△	▲HS	▽	▽	▲	▽	HS, burrowing
	+ 100 +1		▼	▽△	△	▲HS	▽	△	▲		HS
	+4										
PCPA + fluoxetine	300 + 100 + 100	78	▽	▽		▼	▽	▽	▲	△	HS
	+ 10 +4										

Note: HS - homosexual behavior.
See also note to Table 1.

Some increase of brain DA, occurring together with a 5-HT decrease, exacerbates the pathological behavior. This is confirmed by experiments involving administration of L-DOPA after pre-treatment with PCPA (see Table 5). The incidence of attempted "mountings", burrowing and adoption of static posture by the individual all increase.

We are thus led to suppose that the fall in brain 5-HT concentration accompanying activation of CA systems is an important factor, in inducing not only aggressive behavior but also abnormalities in sexual behavior, reduced sociability and heightened defensive behavior.

6. CHOLINERGIC MECHANISMS

Combined findings from ethological investigations using scopolamine, a classical cholinolytic, and the cholinomimetic, physostigmine, can be seen in Tables 6 and 7. These compounds had already been extensively studied previously, but from completely different angles.

Many authors (Allikmets, 1974; Romaniuk, 1974) have postulated the importance of the part played by the cholinergic system of the brain in reinforcing or inducing diffuse rage and defense, with affective overlay. The main evidence for cholinergic system involvement in integrating displays of aggression was built up from data obtained during intracerebral administration of cholinomimetics. These compounds, particularly carbachol, induced diffuse aggression in cats when administered into different regions of the limbic system, diencephalon and mesencephalon (Allikmets, 1974; Romaniuk, 1974). Carbachol-induced rage was averted by administering the cholinolytics, atropine and scopolamine, both intracerebrally and systematically. All results obtained, however, are for one individual interacting with the experimentalist (involving e.g. approach, provocation and handling) but not with animals of its own species. How an animal's behavior responds specifically to its fellows has to be taken into account in measuring and assessing the effects of psychotropic agents (Valdman and Kozlovskaya, 1976).

The fact that diffuse rage or affectively-charged defense, rather than species-specific attack, occurred in the animals may be partially explained as due to the experimental conditions. At the same time, the doses of carbachol and other cholinergic compounds used in this work exceeded the physiological production of acetylcholine in similar behavioral situations; this somewhat limits the significance of these effects as far as investigating the mechanisms and control of intraspecies aggression is concerned.

Only a limited amount of research has actually been devoted to cholinergic aspects of intraspecies aggression. The work of Silverman (1978), in fact, is alone in analyzing the effect of nicotine using an ethological approach. In rat experiments, he found that administering 25 µg/kg nicotine subcutaneously specifically reduced aggressive actions and postures, and produced insignificant changes in other forms of behavior, viz. - exploratory, sexual, submissive and grooming.

In our experiments, 0.25 mg/kg scopolamine was found to suppress intraspecies aggression fairly selectively (Table 6). It should be mentioned that this selectivity was tested for intraspecies as well as individual activity, as distinct from previous work on the subject. It was indeed found that scopolamine increased intraspecies activity (see Table 6). The actual incidence of actions performed increased, but somewhat abortively; contacts, for example, were short-lived. Perhaps the curtailment of each approach made in seeking out a fellow animal was connected with the impairment of other physical and mental mechanisms involved in intraspecies behavior, such as focusing of attention. Scopolamine impairs "sustained attention" (Cheal, 1981).

The cholinomimetic, physostigmine (Table 7), exerted a contrasting effect to scopolamine in intraspecies behavior in isolated mice; it intensified aggression and suppressed sociability, while forcefully activating ambivalent behavior patterns. Increased doses of physostigmine suppressed aggression non-specifically and potentiated both defense and hyper-reactivity.

The repression by scopolamine of intraspecies aggression, which is fairly selective compared with, say, impairment of motor activity, may be explained as follows: cholinergic mediation is involved at the earliest stages of the initiation of molecular events relating to aggressive behavior, especially perception and recognition of aggressive stimuli, as well as focusing attention. It has, at least, been clearly established that olfactory and visual information is transmitted via the primary pathways of the CNS, which possess cholinergic links.

Hence, aggressive behavior and sociability in isolated animals must be regulated by a variety of factors. Control should essentially embody a degree of one or more of the following: restoration of the sensitivity of dopaminergic systems, restoration of normal catecholamine metabolism, compensation for GABA deficit, intensified function of the GABA-benzodiazepine complex and/or a reduction in cholinergic system activity.

Table 6. Pattern of Ethological Effects of Scopolamine

Behavior	Scopolamine 0.25 mg/kg	Scopolamine 0.25 mg/kg vs. control on 0.9% phys. saline	Scopolamine 1 mg/kg	Scopolamine 1 mg/kg vs. control on 0.9% phys. saline
Animal group behavior				
Aggression:				
attack	0.000	-0.196*	0.000	-0.257*
threat	0.000	-0.096*	0.000	-0.115*
Sociability:				
sniffing body	0.330	+0.264*	0.315	+0.310*
sniffing nose	0.000	0.000	0.000	0.000
sniffing genitalia	0.098	+0.066*	0.155	+0.128*
Defense:				
sideways-on posture	0.000	0.000	0.000	0.000
upright posture	0.000	0.000	0.000	0.000
Ambivalence:				
tail-rattling	0.000	-0.050*	0.000	-0.052*
Individual behavior				
locomotion	0.295	+0.006	0.292	+0.014*
rearing	0.192	+0.096*	0.165	+0.057*
self-grooming	0.012	-0.012	0.027	-0.012
sitting	0.073	-0.077	0.056	-0.054

Note: Table gives a) the statistical probability of each item/action occurring after
administration of a given compound, and b) how this probability differs from the control.
* – significance of difference where $p < 0.05$ (Wilcoxon's rank sum test).

Table 7. Pattern of Ethological Effects of Physostigmine

Behavior	Physostigmine (0.125 mg/kg)	Physostigmine (0.125 mg/kg) vs. control on 0.9 % phys. solution	M-atropine (5 mg/kg) + PS (0.125 mg/kg)	M-atropine (5 mg/kg) + PS (0.125 mg/kg) vs. M-atropine (5 mg/kg)
Animal group behavior				
Aggression:				
attack	0.244	+0.070*	0.164	-0.093*
threat	0.059	-0.028	0.071	-0.044
Sociability:				
sniffing body	0.006	-0.043*	0.000	-0.005
sniffing genitalia	0.001	-0.019*	0.008	-0.019
Defense:				
side-ways-on posture	0.000	0.000	0.000	0.000
upright posture	0.000	0.000	0.000	0.000
Ambivalence:				
tail-rattling	0.202	+0.155*	0.105	+0.053*
Individual behavior				
locomotion	0.309	+0.006	0.100	-0.188
rearing	0.079	-0.032	0.023	-0.085
self-grooming	0.004	-0.038	0.007	-0.032
sitting	0.095	-0.059	0.522	-0.412*

Note: The significance of difference of the statistical probability of each item/action occurring after administration of a) 5 mg/kg of methyl atropine vs. b) the control (0.9% physiological saline) is indicated by the asterisk. $p < 0.05$ (Wilcoxon's rank sum test).
PS - physostigmine.

7. OPIOID MECHANISMS

The function of NA and DA neuromediator systems may be modulated by endogenous neuropeptides such as enkephalins MIF-1, TRH. There are reasons for believing that endogenous opioids are involved in the process of shaping intraspecies sociability and attachments. It has been shown that opiates are able to reduce the stress caused by isolating mothers from their young (Panksepp et al., 1978) and may mediate several different types of motivation, such as thirst, hunger and sex. Little information is available, however, concerning the ability of opiates to produce anti-aggressive effects.

It was found that 50 μg met-enkephalin administered into the lateral ventricle reduced the incidence of attack and threatening behavior towards fellow-animals (n = 12, T^Δ = 0, p<0.05) and activated some forms of intraspecies sociability, such as sexual exploration and grooming of a partner (see Table 8). Met-enkephalin increased individual motor activity in isolated mice, especially grooming, locomotion, etc. In a number of its effects, including aggression and sociability, this peptide influences intraspecies behavior in the same way as morphine.

A dose of 25 mg/kg of SA-ENK, a synthetic analog of enkephalin, reduced the frequency of attacks accompanied by biting and striking (n=12, T^Δ=0, p<0.05) and produced different changes in sociability patterns. SA-ENK decreased the incidence of ambivalent types of behavior and intensified static activity. The dipeptide ciotorphine (CT) intensified ambivalent behavior and static activity (n = 6, T^Δ = 0, p < 0.05). Under the influence of CT, only a tendency towards reduction in aggression and sociability was observed.

A dose of 25 mg/kg neo-endorphin reinforced the display of intraspecies sociability and defense (n=12, T^Δ=0, p<0.05). Neo-endorphin reduced aggressive behavior without completely suppressing it, and intensified self-grooming. It should be mentioned that diazepam, one of the benzodiazepine group of compounds, potentiated the anti-aggression effect and reinforced sociability (Table 8). Simultaneous activation of opioid and benzodiazepine receptors therefore helps to produce an anti-aggression effect and an increase of sociability.

Met-enkephalin's ability to counter aggression in isolated mice, which is shared by SA-ENK, morphine and other opiates, may partially be explained as compensating for the deficit in endogenous opioids occurring during prolonged isolation. The observed rise in the number of opiate receptors in isolated mice provides indirect evidence that such a deficit exists (Bonnet et al., 1976). The soundness of this supposition was confirmed by our findings (Poshivalov, 1982), which demonstrated that a relative enlargement of the opiate's deficit in isolated mice, produced by a 0.25 mg/kg dose of naloxone, intensifies aggression in these animals. One characteristic feature

Table 8. Pharmaco-ethological Range of Opioid Peptide Activity

Peptide / Dose Behavior	Met-enkephalin 50 µg	SA-enkephalin 25 mg/kg	Ciotorphine 25 mg/kg	Neo-endorphin 25 mg/kg	Neo-endorphin 25 mg/kg + diazepam 2.5 mg/kg
aggression	▶	▶	▷	▷	▶
sociability	▽◀	▽◁	▷	◀	◀
defense	▷	◁	◁	◀	—
ambivalence	▶	▷	◀	▷	▷
locomotion	◀	▷	—	▷	▶
grooming	◀	◁	◀	◀	—
sitting	◀	◀	◁	—	◀

See note to Table 1.

is that the changes in intraspecies behavior under these conditions are not merely one-sided. Similar effects are apparently observed in experiments showing how exogenous opioid deficiency in deprived morphine-dependent rats leads to a general increase in responsiveness and aggression. Under these conditions, naloxone serves to intensify aggression and morphine successfully suppresses it (Lal, 1973). It should be emphasized that potential natural antagonists of opioids, such as the peptides α-MSH and MIF-1, also intensify aggression in isolated animals, as well as small doses of naloxone.

The unusually high level of aggression observed in mice following prolonged isolation from fellow animals in single cages may be regarded as a reaction to the stress of deprivation (Valdman and Poshivalov, 1980; Valzelli, 1973). The effects on behavior of intraspecies fellowship deprivation, however, may be extremely varied; typical features of the initial response to the stress of isolation in primates and other gregarious animals could be, for instance, anxiety, depression, and vocalization of tearfulness. As with morphine deprivation, all these effects on behavior produced by deprivation of intraspecies fellowship are intensified when opiate receptors are blocked by naloxone (Panksepp et al., 1980). Thus some forms of aggressive behavior, such as those brought on by isolation or morphine withdrawal, clearly depend on the level of opioid activity in the CNS.

Exogenous opioids, including morphine, produce dose-dependent suppression of intraspecies sociability, as well as restraining aggression in previously isolated males (Poshivalov, 1947). Synthetic enkephalins and their analogs, as shown earlier in Table 8, vary in their capacity for controlling intraspecies sociability amongst isolated animals. This fact could be linked with the differing intensity of reinforcing properties of peptides (Ignatov et al., 1981) and varying affinity towards opiate receptors fulfilling a variety of functions.

8. EFFECTS OF HYPOTHALAMO-PITUITARY PEPTIDES

Hypothalamo-pituitary peptides (HPP) appear to be directly connected with the regulation and interconnecting of aggression and defense, since the "trigger-zones" shaping these actions is actually located in the hypothalamus. It is in these areas that the peptidergic neurones are to be found which produce substances fulfilling a combination of neuromediator, modulatory and hormonal functions. Peptides may affect aggressive behavior by changing a number of functions:

a) sensory systems
b) specific trigger zone aggression
c) memory
d) evaluation mechanisms
e) effector mechanisms (peripheral systems and organs).

HPP's acute effects on behavior are connected with their central (psychotropic) action; time-related effects are complex and include hormonal influences on peripheral systems and organs; their extent may directly indicate an animal's behavior and status within the animal community as well as capacity for intraspecies acoustic and pheromonal communication, i.e. ability to adapt in a competitive environment.

HPPs were used separately and combined with: haloperidol, a neuroleptic; seduxen, a tranquilizer; apomorphine, a dopamino-mimetic; muscimol, a GABA-mimetic; and morphine, an opiate agonist. Results of this work are shown in Table 9. It was found that the following actively potentiated aggression: MIF-1, while intensifying some forms of ambivalent behavior, and LH-RH, while increasing loco-motion, whereas other HPP's, such as ACTH, α-MSH and somatostatin, displayed an anti-aggression effect but without any increase in intraspecies sociability.

MIF-1 was the HPP which reinforced aggression most intensively. Its ability to intensify aggression and narrow down the behavioral spectrum is due in part to this compound's NA- and DA-positive characteristics. Judging by its effects on behavior, the action of MIF-1 resembled that of small doses of L-DOPA and apomorphine. In addition, MIF-1 has a pronounced antagonistic action compared with the anti-aggression effect of the DA receptor antagonist, haloperidol (see Figure 1). It should be noted that MIF-1 counteracted the effects of morphine, muscimol and diazepam (see Figure 2) on aggres-sive behavior in isolated mice. These findings show that MIF-1 also exerts an anti-opiate, GABA-negative and BDZ-negative action, which could independently bring about its potentiation of aggression.

Acute injection of $ACTH_{1-24}$ promoted displays of defensive behavior; its anxiolytic properties were diminished when used in combination with diazepam. On the other hand diazepam exerted less restraint on aggressive behavior after pre-treatment with $ACTH_{1-24}$. The acute and chronic effects of $ACTH_{1-24}$ differed substantially; the aggression- and sociability-suppressing effect appeared initially, followed by the onset of aggressive behavior, then being confined to a stable level by subchronic administration of $ACTH_{1-24}$. The acti-vation of aggressive behavior produced by $ACTH_{1-24}$ within 15-30 min after administration may be mediated by the secretion of glucocorti-coids. The work of Kostowsky et al. (1970) illustrated that gluco-corticoids can raise the incidence of attack in mice. In a number of behavioral tests $ACTH_{1-24}$ shows antagonism to opiates and is seen to be a synergist of naloxone; these two aspects of its action may enable the peptide to intensify aggressive behavior in isolated mice.

$ACTH_{1-24}$ and its analog $ACTH_{4-10}$ are capable of reducing intra-species sociability in rats and mice (File and Clarke, 1980), whereas compounds of the benzodiazepine group reinforce this. $ACTH_{1-24}$ can

Table 9. Pharmaco-ethological Range of Activity of Hypothalamo-pituitary Peptides

Peptide / Dose	MIF-1 25 mg/kg	LH-RH 25 mg/kg	TRH 25 mg/kg	ACTH$_{4-10}$ 25 µg	ACTH$_{1-24}$ 25 µg	Somatostatin 5 mg/kg
Aggression	◀	◀	◁	◁	▷	▶
Threatening behavior	△▽	▽△	▶	▶	▽△	◀
Sociability	▷	◁	▷	▶	▽△	▶
Ambivalence	◁	◁	▷	▷	◀	◁
Locomotion	◁	▽△	◀	◁	▷	▶
Grooming	△▽	◀	◁	◁	◀	▽△
Sitting	▶	▶	▶	▷	◁	◀

See note to Table 1.

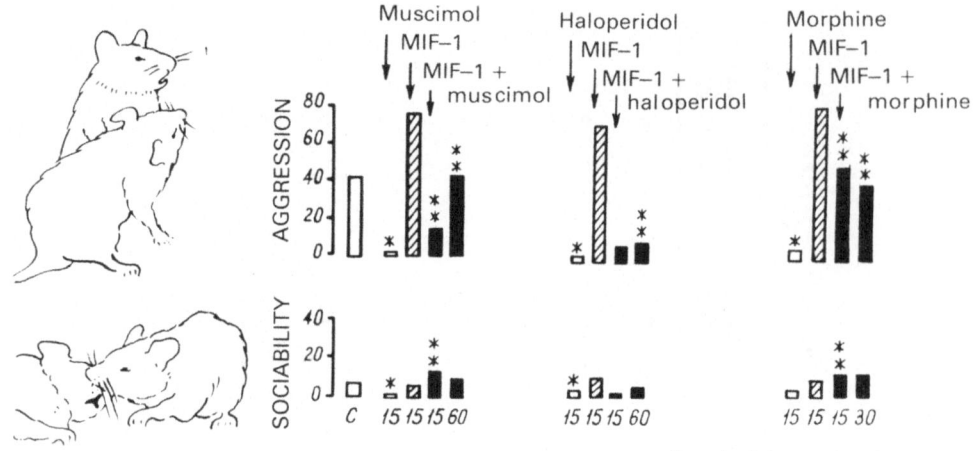

Fig. 1. Action of muscimol, haloperidol and morphine (after pre-
         treatment with MIF-1 25 mg/kg) on average frequency of
         attacks and sociability.   C - control on 0.9% physiological
         saline; muscimol (1 mg/kg, n = 12); haloperidol (1 mg/kg,
         n = 12) (unshaded columns); morphine (2.5 mg/kg, n = 12);
         Striped columns: MlF-1; Shaded columns: MIF-1 in combination
         with other compounds; * and ** - significance of differences
         (p<0.05 by Wilcoxon's rank sum test).

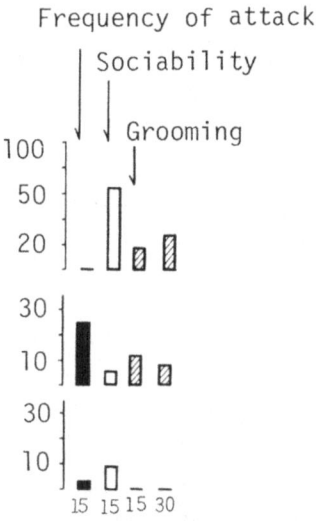

Fig. 2. The effect of MIF-1 combined with diazepam on aggressive
         behavior, sociability and grooming in isolated mice.
         Ordinate (in descending order): frequency of attack,
         sociability, grooming. Abscissa: time in min.

counteract some "opiate-like" effects of benzodiazepines; perhaps the antagonism of the peptide to the effects of diazepam in animal group behavior can be explained by these components of its action. As already mentioned, activating benzodiazepine receptors and compensating for the opiate deficiency in isolated animals constitute a substantial proportion of the mechanism reinforcing intraspecies sociability. This view is supported by findings from experiments using neo-endorphin and diazepam, as well as $ACTH_{1-24}$ (see Table 8). Neo-endorphin increases intraspecies sociability and this effect is potentiated by diazepam, as shown in Table 8. Under these circumstances aggressive behavior is completely blocked. These finding point to the significance of the opiate and benzodiazepine systems in interconnecting anti-aggression and sociability-promoting effects.

It should be pointed out that each of the opioid and hypo-thalamo-pituitary peptides investigated in this study has its own particular ethological range and characteristics of pro- or anti-

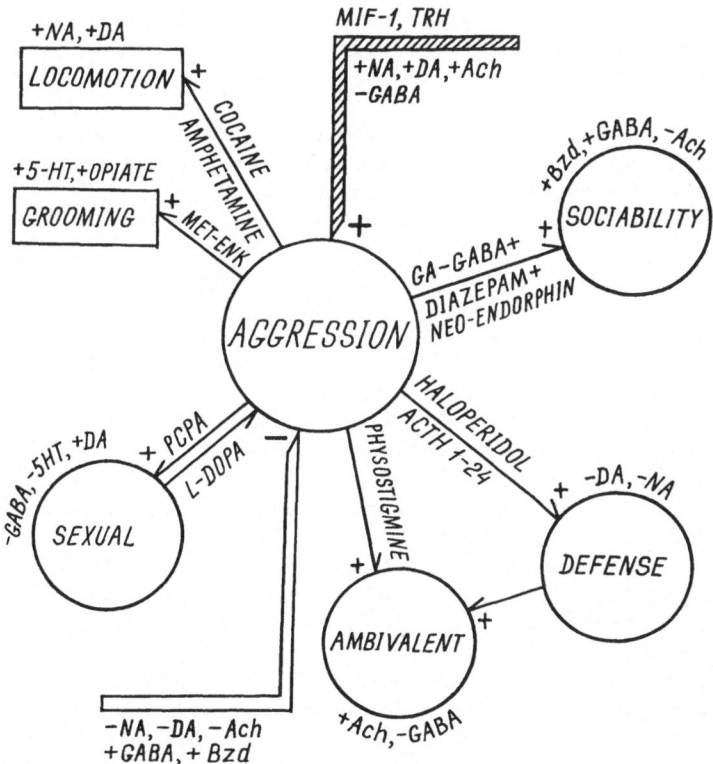

Fig. 3.   Main pathways of intraspecies behavior. Regulation and control by peptides and neuropharmacological substances. (+) - activation; (-) inhibition; Arrows show how different influences produce various changes in behavior.

aggression action, which may either be potentiated or suppressed by
certain psychotropic or neuropharmacological compounds.

9. CONCLUSION

A synthesis of the results of this work - the chief and most
controllable pathways for countering aggression pharmacologically -
are crystallized and expressed in diagrammatic form in Figure 3.
The information summarized in this diagram provides some answers to
the question of which peptides and neuropharmacological agents might
suppress aggressive behavior while increasing non-aggressive intraspecies sociability, without too high a cost in terms of maintaining
other biologically essential forms of behavior. Aggression may, for
example, be activated while potentiating locomotion, producing competitive steretoped motor activity by using cocaine or amphetamine,
promoting grooming and inducing a state of well-being by means of
morphine or met-enkephalin, reinforcing specific sexual responses
by administering LH-RH or thiosemicarbazide, by inducing fear using
$ACTH_{1-24}$ or defensiveness by means of haloperidol, or finally by
reinforcing sociability by administering diazepam, neo-endorphin or
GABA analogs.

Our findings show that reinforcement or restraint may focus on
selected patterns of species-specific behavior and hence that aggression may also be countered specifically.

REFERENCES

Adams, D. B., 1976, The relation of scent-marking, olfactory inves-
        tigation and specific postures in the isolation-induced fight-
        ing of rats, Behaviour, 56:286-297.
Adams, D. B. 1980, Motivational systems of agonistic behaviour in
        muroid rodents - a comparative review and neural model,
        Aggress.Behav., 6:295-346.
Allikmets, L. H., 1974, Cholinergic mechanisms in aggressive behav-
        iour, Med.Biol., 52:19-30.
Blindermann, J. M., DeFeudis, F. V., Maitre, M., Misslin, R., Wolff,
        P., and Mandel, P., 1979, A difference in glutamate-decar-
        boxylase activity between isolated and grouped mice,
        J.Neurochem., 32:1357-1359.
Bonnet, K. A., Miller, J. M., and Simon, E. J., 1976, The effect of
        chronic opiate treatment and social isolation on opiate recep-
        tors in the rodent brain, in: "Opiates and Endogenous Opioid
        Peptides," H. W. Kosterlitz, ed., Amsterdam, North-Holland
        Press, pp.335-343.
Brain, P., 1979, Effects of the hormones of the pituitary-gonadal
        azis on behaviour, in: "Chemical Influences on Behaviour,"
        k. Brown and S. Cooper (eds.), London, Academic Press,
        pp.255-328.

Brase, D., and Loh, H., 1976, The increase of brain tryptophan caused by amphetamine-like drugs - correlation with an increase in body temperature, Life Sci., 18:115-122.

Cheal, M. L., Scopolamine disrupts maintenance of attention rather than memory process, Behav.Neural.Biol., 33:163-187.

Cools, A. R., and Van Rossum, J. M., 1980, Minireview, Multiple receptors for brain dopamine in behavioural regulation: concept of dopamine-E and dopamine-I receptors, Life Sci., 27:1237-1253

Costa, E., Guidotti, A., and Toffano, G., 1978, Molecular mechanisms mediating the action of diazepam on GABA receptors, Br.J.Psychiat., pp.133-239.

Da Vanzo, J. P., and Sydow, M., 1979, Inhibition of isolation-induced aggressive behaviour with GABA-transaminase inhibitors, Psychopharmacology, 62:23-27.

De Feudis, F. V., 1979, Environment and central neurotransmitters in relation to learning, memory and behaviour, Gen.Pharmac., 10:281-286.

Earley, C. J., and Leonard, B. E., 1977, The effect of testosterone and cyproterone acetate in the concentration of γ-aminobutyric acid in brain areas of aggressive and non-aggressive mice. Pharmac.Biochem.Behav., 6:409-413.

File, S., and Clarke, A., 1980, Intraventricular ACTH reduces social interaction in male rats, Pharmacol.Biochem.Behav., 12:711-715.

Grant, E. C., and Mackintosh, J. H., 1963, A comparison of the social postures of some common laboratory rodents, Behaviour, 21:246-259.

Ellison, G. D., and Bryan, K. S., 1976, Evidence that supersensitivity after recovery from 6-hydroxydopamine is an artifact of isolation housing, Neurosc.Lett., 3:229-232.

Hodge, G. K., and Butcher, L. L., 1975, Catecholamine correlates of isolation induced aggression in mice, Eur.J.Pharmac., 31:81-93.

Ignatov Yu. D., Vasilev, Yu. N., Andreev, B. V., and Kovalenko, V. S. 1981, Comparative study of analgetic activity of neuropeptides and their influences on antinociceptive and reinforcement systems of the brain, in: "IV All Union Symposium on physiologically active substances research," Riga, p.8.

Ishii, Y., Natsugoe, K., and Umezawa, H., 1975, Pharmacological action of FD-008, a new dopamine-β-hydroxylase inhibitor. II. Effects on central nervous system in mice and rats, Arzneim Forsch., 2:213-215.

Kantak, K. M., Hegstrand, L. R., and Eichelman, B., 1980, Dietary tryptophan modulation and aggressive behavior in mice, Pharmac.Biochem.Behav., 12:675-679.

Kostowski, W., and Valzelli, L., 1974, Biochemical and behavioral effects of lesions of raphe nuclei in aggressive mice, Pharmac.Biochem.Behav., 2:277-280.

Kostowski W., Rewerski, W., and Piechocki, T., 1970, Effects of some

steroids on aggressive behaviour in mice and rats, Neuroendocrinology, 6:311-318.

Krsiak, M., 1975, Timid singly-housed mice - their value in prediction of psychotropic activity of drugs, Br.J.Pharmacol.,
55:141-150.

Krsiak, M., Sulcova, A., Tomasikova, Z., Dlohozkova, N., Kosar, E.,
and Masek, K., 1981, Drug effects on attack, defense and
escape in mice, Pharmacol.Biochem.Behav., 14:Suppl.I,
pp.47-52.

Lal, H., 1973, Morphine-withdrawal aggression - possible mechanism,
Psychopharmacol.Bull., 9:4:21.

Malic, J. B., and Barnett, A., 1976, The role of serotoninergic
pathways in isolation-induced aggression in mice,
Pharmac. Biochem.Behav., 5:55-61.

Mandel, P., Kempf, E., Mack, G., Haug, M., and Puglisi-Allegra, S.,
1981, Neurochemistry of experimental aggression, in: "Aggression and Violence: a Psychobiological and Clinical Approach,"
Milano, Saint Vincent, pp.61-71.

Miczek, K., and Barry, H., 1976, Pharmacology of sex and aggression,
in: "Behavioral Pharmacology," S. D. Glick and J. Goldfarb,
eds., C. V. Mosby, St. Louis, pp.176-257.

Miczek, K., and Krsiak, M., 1979, Drug effects on agonistic behaviour, in: "Advances in Behavioral Pharmacology," Vol. 2,
T. Thompson and P. Dews, eds., Academic Press, New York,
pp.87-162.

Panksepp, J., Herman, B., Conner, R., Bishop, P., and Scott, J. P.,
1978, The biology of social attachment: opiates alleviate
separation distress, Biol.Psychiat., 13:607-618.

Panksepp, J., Herman, B., Vilberg, T., Bishop, P., and DeEskinazi F.,
1980, Endogenous opioids and social behaviour,
Neurosc.Biobehav.Rev., 4:473-487.

Poshivalov, V. P., 1974, Pharmacological and psychophysiological
analysis of the aggressive behaviour with special reference to
zoosocial interactions, in: "Neuropharmacological Regulation
of the System Processes/Psychopharmacology of Behavior,"
Leningrad, Pavlov Medical Institute, pp.60-83.

Poshivalov, V. P., 1978, Ethological atlas for pharmacological research in laboratory rodents, Pavlov Medical Institute and
VINITI, Leningrad-Moscow, n. 3164-78, 3-42.

Poshivalov, V. P., 1980, The integrity of the social hierarchy in
mice following administration of psychotropic drugs,
Br.J.Pharmacol., 70:367-373.

Poshivalov, V. P., 1981a, Some characteristics of the aggressive
behaviour of mice after prolonged isolation: intraspecific and
interspecific aspects, Aggress.Behav., 7:195-204.

Poshivalov, V. P., 1981b, Pharmaco-ethological analysis of social
behaviour of isolated mice, Pharmacol.Biochem.Behav.,
14:Suppl.I, pp.53-59.

Poshivalov, V. P., 1982, Ethological analysis of neuropeptides and
psychotropic drugs: effects on intraspecies aggression and

sociability of isolated mice, Aggress.Behav., 8, 4.

Poshivalov, V. P., Hodko, S. T., and Besov, E. V., 1979, Ethograph-
    computer system for recording and analysis of zoosocial
    behaviour, J.Higher Nerv.Activ., 29:420-423.

Poschlova, N., Masek, K., and Krsiak, M., 1977, Amphetamine-like
    effects of 5,6-dihydroxytryptamine on social behaviour in the
    mouse, Neuropharmacology, 16:317-321.

Puglisi-Allegra, S., and Mandel, P., 1980, Effects of sodium n-dipro-
    pylacetate, muscimol hydrobromide and [R,S]-nipecotic acid
    amide on isolation induced aggressive behaviour in mice,
    Psychopharmacology, 70:287-290.

Puglisi-Allegra, S., Simler, S., Kempf, E., and Mandel, P.,1981,
    Involvement of the GABAergic system on shock-induced aggres-
    sive behaviour in two strains of mice, Pharmacol.Biochem.
    Behav., 14:Suppl, I, pp.13-18.

Robertson, H. A., 1980, Benzodiazepine and GABA receptors in animal
    models for aggression, in: "The Biology of Aggression, NATO
    Advanced Study Institute on the Biology of Aggression,"
    Chateau de Bonas, p.44.

Romaniuk, A., 1974, Neurochemical bases of defensive behaviour in
    animals, Acta Neurobiol.Exp., 34:205-214.

Ross, S. B., and Ogren, S. O., 1976, Anti-aggressive action of
    dopamine-β-hydroxylase inhibitors in mice, J.Pharm.Pharmac.,
    28:590-592.

Silverman, P., 1978, Animal Behaviour in the Laboratory, London,
    Chapman and Hall, pp.409.

Valdman, A. V., Theoretical Basis of Pathological States, Leningrad,
    Nauka, pp.195-206.

Valdman, A. V., and Kozlovskaya, M. M., Changes in animal group
    behavior relationships as an indicator of brain electrostimu-
    lation induced shifts in emotional state, in: "Neurophysio-
    logical Approach to the Analysis of Intraspecies Behaviour,"
    P. V. Simonov, ed., Moscow, Nauka, pp.74-110.

Valdman, A. V., and Poshivalov, V. P., 1980, Some aspects of a model
    of animal behaviour pathology, in: "Experimental and Clinical
    Psychopharmacology," G. C. Morozov, ed., Moscow, pp.3-18.

Valdman, A. V., and Poshivalov, V. P., 1982, Neuropharmacology of
    Intraspecies Aggression and Sociability in Isolated Mice, in
    press.

Valdman, A. V., and Poshivalov, V. P., Pharmacological Regulation of
    Intraspecies Behaviour, in press.

Valzelli, L., 1973, The "isolation syndrome" in mice, Psychopharm-
    acology, 31:305-320.

Valzelli, L., 1980, An Approach to Neuroanatomical and Neurochemical
    Psychophysiology, Edizioni Medico Scientifiche, Torino.

Valzelli, L., 1981 Psychopharmacology of aggression: an overview,
    Int.Pharmacopsychiat., 16:39-48.

Welch, B. L., and Welch, A. S., 1969, Aggression and the biogenic
    amine neurohumors, in: "Aggressive Behaviour," S. Garattini
    and E. B. Sigg, eds., Excerpta Medica, Amsterdam, pp.188-202.

Psychotropic Effect of Tuftsin, a Natural Phagocytosis-stimulating Peptide, and Some of its Analogs

A.V. Valdman, N.A. Bondarenko and M.M. Kozlovskaya

Institute of Pharmacology, Academy of Medical Sciences
of the USSR, Moscow

1. INTRODUCTION

Information has steadily built up over recent years on the influence exerted over the function of the central nervous system and emotional behavior by: a) neuropeptides released from brain tissue, b) the actual fragments formed during enzymatic cleavage of polypeptides, c) a number of peripheral peptides, and d) fragments of the latter.

The tetrapeptide tuftsin, or L-Thr-L-Lys-L-Pro-L-Arg, was first identified by Najjar and Nishioka in 1970 at Tufts University, Boston, USA. Its aminoacid sequence is included in the macromolecular (fragment 289-292) C_H2 heavy chain of human immunoglobulin. Tuftsin release is brought about by the action of specific enzymes. It is separated from the protein carrier, leukokinin, by the parallel action of tuftsin endocarboxypeptidase and a leukokinase. Tuftsin heightens the phagocytic activity of leucocytes both in vivo and in vitro, while increasing the migration of human polymorphonuclear leucocytes and immunological function. Radioimmunoassay reveals the presence of 255 µg/liter in the circulation of healthy subjects (Spirer et al., 1977). It is not yet known whether a similar tetrapeptide is present in or released by brain tissue.

We studied the central action of tuftsin on emotional behavior and response, taking the following considerations as our starting point:

1. The relationship between immunological system reactions and the central nervous system provides an interesting subject of study. Because some regulating mechanisms of the immune and nervous

processes have much in common, and since receptor ligands interact similarly with the membranes of nerve and immunocompetent
cells, it was considered important to evaluate the action of
this phagocytosis-stimulating peptide on CNS function.

2. The structure of tuftsin is similar in functional organization
    to that of a number of low molecular weight peptide receptor
    ligands (Chipens et al., 1973, 1979). These have fragments in
    common which contain identical or equi-functional aminoacids.
    They vary in their biological activity, are involved in ligand-
    receptor interaction and possibly in the generation of subse-
    quent intracellular information and transfer. We have previous-
    ly shown (Valdman et al., 1979; Kozlovskaya et al., 1981;
    Kozlovskaya et al., 1982) that a number of small peptides –
    C-terminal fragments of oxytocin 7–9 (melanostatin) and gastrin
    15–17 – potentiate emotional behavior when administered system-
    ically. These peptides influence the metabolism of biogenic
    amines (Ziles et al., 1979).

3. The aminoacid sequence Lys–Pro–Arg – a tripeptide incorporated
    in the structure of tuftsin – is contained in such highly active
    neuropeptides as substance P and neurotensin, both of which
    exert a profound influence in the regulation of central func-
    tion.

4. The peptide bonds, Lys–Pro and Arg–Pro, are fairly resistant
    to the action of enzymes of the trypsin group, indicating that
    the action of tuftsin may be prolonged when administered par-
    enterally, especially as tuftsin remains unchanged in the
    circulation.

The psychotropic action of tuftsin, a natural immunostimulant,
had not yet been investigated. The tuftsin we used in our experiments was synthesized, together with its analogues Thr–Lys–Pro–D–Arg,
Leu–Lys–Pro–Arg and Thr–Lys–Pro–Pro–Arg by V.N. Kalikhevich at
Leningrad State University.

In view of the fact that tuftsin activates immune processes and
phagocytosis, we first set out to discover whether this peptide has
any potentiating action on emotional/behavioral response.

2. THE POTENTIATING EFFECT OF TUFTSIN ON MODELS OF BEHAVIORAL DEPRESSION

2.1. EFFECT ON BEHAVIORAL DESPAIR

We used CBW strain male mice, weighing between 20 and 23 g,
which had been housed in a vivarium at a temperature of 21±1°C, with
free access to food and water. The mice were separated out into
single cells 2 hours before testing. The animals were placed in a
glass cylinder measuring 23 cm high and 10 cm across, containing 6 cm
of water (t = 21–23°C) for the test of Porsolt et al. (1978). Their

behavior was observed over a 30 min period and rated according to when they actually began to swim, when they first became immobilized, for how long and how frequently. The peptides were dissolved in distilled water and administered intraperitoneally. The control group was given an equivalent volume of vehicle.

Findings on the effect of varying doses of tuftsin and its analogs are given in Table 1.

Doses of 50–250 µg/kg tuftsin showed an activating effect. The total duration of immobility displayed was curtailed, while the active period was extended. Maximum effect was achieved by a 50 µg/kg dose. Increasing doses decreased the stimulating effect thereafter. Leu[1] resulted in an activating effect, unrelated to dose, in a 25 to 750 µg/kg dose range. Doses of 250–500 µg/kg D-Arg-tuftsin eliminated displays of immobility completely. No substantial effect was exerted on behavior by 100–400 µg/kg Thr-Lys-Pro-Pro-Arg, a pentapeptide known as an inhibitor of the stimulating influence of tuftsin on phagocytosis.

Psychostimulants as well as antidepressants are known to potentiate animal behavior in the Porsolt test (Porsolt et al., 1978; Maj. 1979; Gorca and Wojtasik, 1980). They work through the neuro-chemical mechanism of activating noradrenergic and dopaminergic processes (Zebrowska-Lupina, 1980), which leads us to believe that the potentiating effect of tuftsin is bound up with its action on CA mechanisms. Animal observations lasting 30 min showed tuftsin's activating effect to be of 8–12 minutes' duration. According to the findings of Kitada et al. (1981), the difference between psycho-stimulants and antidepressants, as revealed by using Porsolt's test, consists of the fact that psychostimulants precipitate 30 minutes' swimming time as against 5-6 min in the case of antidepressants. Here, tuftsin's stimulating effect on the behavioral despair model broadly resembles that of the antidepressants.

2.2.   TUFTSIN'S EFFECT ON THE RESERPINE MODEL OF DEPRESSIVE BEHAVIOR IN CATS

Depressive behavior was induced in cats by administering reserpine ("Rausedil", Gedeon Richter) at a dose of 0.1 mg/kg sub-cutaneously or 0.025 mg/kg intramuscularly over a 2-day period. The cats developed pronounced inhibition of emotional response and a reduction of motivated activity within 24 h following administration, together with hypodynamia. The animals adopted a stuporose posture. Adequacy of response was impaired. They adopted an emotionally negative attitude towards the experimental situation, and hence displays of positive emotion disappeared.

The animal's range of emotional/behavioral response, when the depressive effect had reached its peak, is shown in Figure 1(B).

Table 1. Effect of Tuftsin and its Analogs on the Duration of
          Immobility during a 6-min Test

Compound	Dose (μg/kg)	Duration of immobility in sec (M ± SEM)	% of control
Phys. saline		268.0± 8.8	
Tuftsin	750	233.5±11.4 n.s.	-13
	500	150.0±26.5**	-44
	250	137.5±55.0**	-49
	100	149.0±20.3*	-44
	50	95.8±18.5**	-64
Phys. saline		185.8±20.0	
Leu[1]-tuftsin	750	84.2±20.0*	-55
	500	78.0±19.6*	-58
	250	72.5±23.8*	-61
	100	65.8±23.9*	-65
	50	95.8±22.9*	-48
	10	72.5± 9.7*	-61
Phys. saline		147.5± 8.8	
D-Arg[4]-tuftsin	750	115.8±17.7 n.s.	-22
	500	96.2± 7.0*	-35
	250	91.7±19.4*	-38
	100	102.5± 3.5*	-31
	50	100.0± 6.2*	-32
	10	113.3±11.5 n.s.	-23

CBWA strain male mice, n = 6 per group.
* $p<0.05$.
**$p<0.01$ vs. control.
n.s. = non-significant.

Our method of section-by-section analysis is fully explained in
Chapter 1. A 500 μg/kg i.p. dose of tuftsin produced a short-lived
but clearcut anti-depressant effect on a model of this type (see
Figure 1C and D). Within 7 min of this treatment, the animal came
out of its state of torpor and apathy; its posture changed drasti-
cally. Purposeful activity resumed, such as hunting, food seeking
and exploratory activity; its responses recovered initiative and
adequacy. The effect began to fall off again 20 min later, however,
when features reminiscent of reserpine depression reappeared in the
animal's behavior.

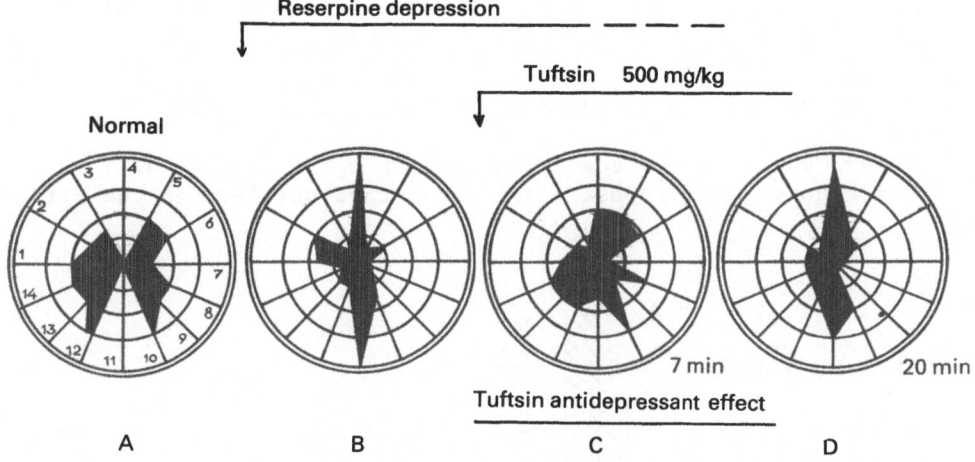

Fig. 1.   Tuftsin's effects on reserpine-induced depression in cats.
A - initial range of emotional/behavioral response; B - 24 h
after a 0.1 mg/kg subcutaneous dose of reserpine; C - 7 min
after a 500 µg/kg i.p. dose of tuftsin; D - 20 min after a
500 µg/kg i.p. dose of tuftsin. Labeling of segments of
circle as follows: 1 - rage or aggression; 2 - fear; 3 -
anxiety, 4 - negativity; 5 - curiosity; 6 - pleasure or
contentment; 7 - playfulness; 8 - hunting; 9 - friendliness;
10 - conflict; 11 - inadequacy of behavior; 12 - initiative;
13 - locomotor activity; 14 - alertness.

2.3.   TUFTSIN'S EFFECT ON MODEL BEHAVIORAL DEPRESSION PRODUCED BY
       STIMULATION OF THE SEPTAL REGION

Model behavioral depression was induced by electrical stimulation of cats' medial septal region, previously implanted with
electrodes, as part of investigations on individual cats. We had
shown earlier that stimulating this zone in rabbits produces depressive behavior characteristic of these animals (Valdman, Zwartau and
Kozlovskaya, 1976). It manifests an inhibition of motor activity
along the lines of a "stop reaction," with reduced respiration rate,
muscle tonus and responsiveness to external influences, together with
a soporific state, maintained long after stimulation has ended.

Similar effects develop in cats during a 180-300 sec period of
30 imp/sec stimulation of the medial septal region. Figure 2 shows
the response pattern associated with depressive behavior - i.e. the
latent period of a variety of behavioral signs.

A dose of 200 mg/kg tuftsin, administered 5-7 min before the
start of electrical stimulation of the brain, counters the development of depressive behavior a) during stimulation itself (extending

the latent period) and b) post-stimulation (reducing the soporific
state from 300-500 to 5-30 seconds' duration).

2.4.   EFFECT OF TUFTSIN ON EMOTIONAL/BEHAVIORAL RESPONSE

Table 2 shows the effect of tuftsin on a cat's range of
emotional/behavioral response - see our method outlined in the first
article.

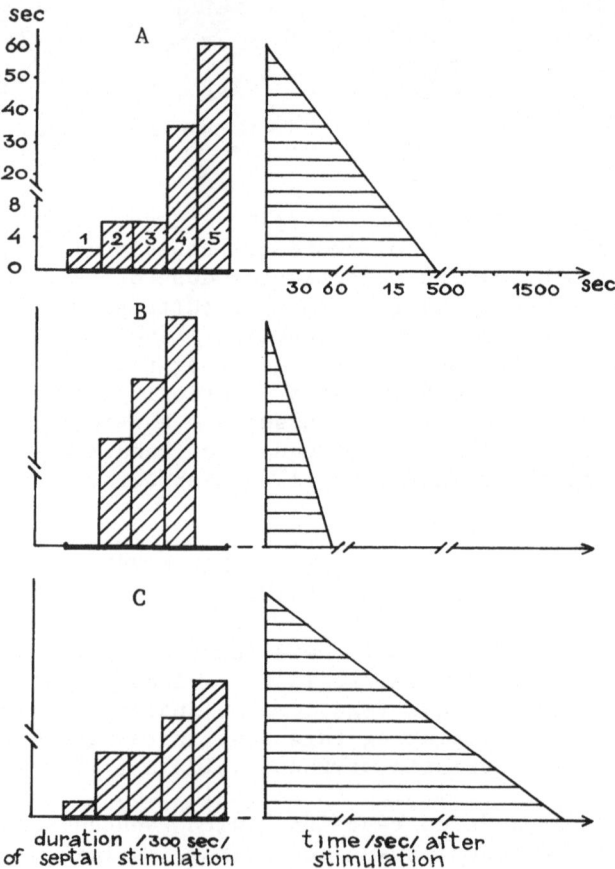

Fig. 2.   Tuftsin's modulating effect on cat behavior, altered by
          stimulating the medial septal region. Columns 1-5 represent
          latent periods (sec) for different features of the reaction
          to develop: 1 - "stop" reaction; 2 - accelerated breathing
          rate; 3 - drop in muscle tonus; 4 - drop in emotional
          responsivity; 5 - soporific state sets in. Horizontally
          shaded area: duration of soporific state post-stimulation
          (sec). A - before; B - 8 min after and C - 45 min after a
          200 µg/kg i.p. dose of tuftsin.

Table 2. Tuftsin's Effect on Cats' Emotional/behavioral Response
          Pattern

Range of emotional/ behavioral response	Normal range	Tuftsin (200-500 µg/mg)
Fear	1.8±0.3	0.6±0.4
Anxiety	2.0±0.3	0.6±0.4
Aggression	1.7±0.2	2.7±0.4
Conflict within the group	2.1±0.7	4.3±0.4
Display of positive emotions	2.7±0.5	1.5±0.6
Hunting	2.7±0.3	0
Friendly contacts	1.7±0.5	0
Exploration	2.5±0.2	2.3±0.4
Motor activity	2.3±0.1	2.7±0.5
Initiative	2.5±0.3	3.7±0.3
Alertness	2.0±0.1	2.7±0.3

Figures are the mean (M ± SEM) of points scored for displays of
emotional behavior (see the first article for details of method used).

A dose of 500 µg/kg of intraperitoneal tuftsin acted in two stages:
during the first 7-15 min it reduced display of panic-and-alarm,
increased alertness, exploration and awareness of surroundings,
including fellow-animals. The animal began to show increased initiative and competitiveness, as indicated by the incidence of approaches made to the feeding rack. It was able to raise its group
status. The cat's heightened level of activity and its attempts to
dominate led to increased conflict within the group and some further
aggression on the part of the test cat. Active forms of reciprocal
behavior, such as actively defensive behavior when provoked, also
increased. A mild sedative effect sets in within 29-35 min after
administering tuftsin; motor activity decreases and exploration
ceases. However, adequate displays of active defense do occur in
response to provocation, especially from fellow-animals. Status
within the animal group is well maintained, without any accompanying
increase in aggression.

Tuftsin likewise intensifies emotional response in rats, as
measured by the threshold of pain reaction (vocalization) to electric
shocks to the paws. The threshold of perceiving noxious influences
was reduced from 0.43±0.03 in the control to 0.13±0.02 mA by administering 500 µg/kg of tuftsin intraperitoneally (n = 13, p<0.01).
The threshold of producing aggression and fighting in two rats by
electrical stimulation (electrode floor) was lowered considerably -
from 2.3±0.25 in the control to 0.25±0.03 mA. Both the duration of
arousal during stalking, distinguished by maintaining a "boxer
stance" and rearing in response to a light tap were doubled by administration of tuftsin (Valdman et al., 1981).

3. **TUFTSIN'S MODULATING EFFECT ON BEHAVIOR UNDER STRESS OR IN A
    CONFLICT SITUATION (DEPENDENT ON INDIVIDUAL ANIMAL'S RESPONSE
    PATTERNS)**

Neuropeptides are widely thought to exercise a modulating or
regulating influence on both processes connected with neuromediator
function (Barker, 1977; Dismukes, 1979) and those involved with behavior, especially where they deviate from the physiological normal
state (De Wied, Gispen, 1977; Kovacs et al., 1979). The regulatory
effect of peptides on behavior reflects their action on neurome-
diators, and on catecholaminergic processes in particular (Fuxe et
al., 1980).

How animals respond in behavioral tests, especially when exposed
to stress or conflict, varies a great deal according to their original emotional response patterns, which in turn correspond to individual variations in the level of brain CA metabolism (Tizabi et al.,
1979; Bondarenko et al., 1981, a and b).

Bearing these factors in mind, we assessed the action of tuftsin
and its analogs on the behavior of animals with varying emotional
response levels which had previously been exposed to influences
affecting the brain's CA systems.

3.1. DIVISION OF ANIMALS INTO "EMOTIONAL" AND "NON-EMOTIONAL"
      SUB-GROUPS

So-called emotional and non-emotional classes of individuals
were selected from amongst a group of cross-bred male rats weighing
between 250 and 300 g, having been previously tested for this purpose
by sequential statistical analysis (Gubler, 1978).

The animals were kept in shared cages measuring 50 x 30 x 16 cm,
with 5 rats to a cage, under a natural day-night cycle and with free
access to food and water.

They were selected on the basis of their reactions to three
tests: open field, dark chamber with openings or holes, and how each
related to unfamiliar moving objects.

The open field used was daylight-lit, circular in shape,
measured 90 cm across and was divided into 20 squares along the
18 cm side. An opening was made in each corner. Concentric rings,
with a radius of a) 30 cm, and b) 15 cm divided the "field" into
3 zones: outer, middle and central. The outer ring measured 45 cm
in height. White noise level: 62 decibels. The animal was placed
in the outer section of the open field for 4 min. Records were kept
of the latent period (LP) of crossing the first square, total squares
crossed, the number of exits made into each section and how many

rearings occurred in each and finally the incidence of exploration of openings and of grooming.

The dark chamber with openings consisted of a rectangular container measuring 40 x 18 x 20 cm, in the roof of which 8 openings, each measuring 2.5 cm across, were made. One side of the chamber sloped downwards at an angle of 45°; this contained an opening 6 cm wide, in imitation of the entrance to a rat's nest. The animal was kept in the chamber for 4 min. A record was made of the latent period of the first peering out, how many times this occurred in all, defecation and the number of exits made from the chamber

Animal's reaction to unfamiliar moving objects were studied by placing them in a rectangular area measuring 60 x 40 x 20 cm, containing a mechanical toy, whose movements produced a 95 decibel noise level.

Our recordings lasted 30 sec, and featured the reaction of approaching and sniffing and that of active or passive avoidance, together with the degree of display of affect, rated according to a 3-point scale.

The following characteristics were noted in the animal group with low emotionality:

  a) an average level of motor activity with a high rate of exploration,
  b) short latent period for crossing the first square in the open field and first peering out (in the dark chamber with openings), and
  c) a relatively low incidence of grooming.

Highly emotional animals, however, exhibited either a high or low level of motor activity, low rate of exploration, a very extensive latent period for crossing the first square and for the first peering out, as well as increased incidence of grooming. They adopted an attitude of passive avoidance on coming into contact with a moving object, and cowered away from it.

Tables 3 and 4 give an itemized rating of the behavior of these two groups of animals.

The neurotoxin 6-hydroxydopamine (6-OHDA), according to Jonsson et al. (1974), was used to produce indicative changes in the functional activity of catecholaminergic systems. A 0.1 ml dose of 100 mg/kg 6-OHDA (Sigma, USA) was given subcutaneously to 3-day old Wistar rats. Animals weighing 220-270 g were then removed for experimental use 14-16 weeks later. Their behavior was rated by the same tests as those used for untreated rats. Statistical evaluation was carried out by the non-parametric Mann-Whitney U-test.

The emotional/behavior response of those animals treated with
6-OHDA during the post-natal period was higher overall than in untreated mice. This had already been noted by Fukuda et al. (1977)
and Miller et al. (1981). These authors, however, had calculated the
statistical mean of all data obtained for the whole animal group,
whereas our findings established that the group of rats pre-treated
with 6-OHDA was dissimilar (Bondarenko et al., 1981). Rats pretreated with 6-OHDA should also be divided into "emotional" or
"non-emotional" (or rather, less emotional) sub-groups, depending on
individual emotional/behavioral response patterns. These groups vary
significantly for all the values under consideration (see Tables 3
and 4). The difference in their response patterns is due to the fact
that the degree of CA terminal destruction and the relationship
between dopaminergic and noradrenergic terminal breakdown varies from
one individual to another (Nomura et al., 1976).

These different sub-groups of animals could therefore be
arranged in ascending order of emotionality in this way: non-
emotional and untreated → non-emotional and pre-treated with
6-OHDA → emotional, untreated → emotional and pre-treated with
6-OHDA.

3.2. THE EFFECT OF TUFTSIN AND ITS ANALOGS ON OPEN FIELD BEHAVIOR,
      DEPENDING ON AN ANIMAL'S RESPONSE PATTERN

Open field behavior is known to be motivated by two opposing
impulses: fear and exploration (Levin, 1962). Where neither of the
two competing forces prevails, the mixed activity of self-grooming
occurs. Should either of the two begin to fall off, allowing the
other to predominate, grooming decreases (Bolles, 1960). The data
provided in Table 3 show how the different tetrapeptides under study
act differently on behavior, depending on the animal's emotional/
behavioral response pattern.

Tuftsin increased exploratory activity, i.e., the number of holes
investigated, in all groups of animals, especially in the "non-
emotional" class, while raising the incidence of rearing in the
"emotional" category. At the same time, fewer squares were crossed.
These changes indicate how tuftsin's action potentiates and motivates
exploration, while reducing the opposing fear motivation.

Leu[1]-tuftsin, on the other hand, reduces exploratory activity,
rearing and ambulation while raising the incidence of grooming; this
could be interpreted as an indication of frustration, replacing activity, and reflects the generally increased level of emotionally
negative response. D-Arg[4]-tuftsin fails to produce any major changes
of behavior in "non-emotional" animals, while exercizing a regulating
(both potentiating and sedative) effect on "emotional" animals, producing a rise in exploration and a fall in the incidence of grooming.

Table 3. Levels of Effect of Tuftsin and its Analogs on Open Field Behavior in 4 Groups of Rats, Types I and II, both Treated and Untreated (M ± SEM)

Compound	Dose (µg/kg)	Animal group	Sub-group	Ambulation (No. squares crossed)	Rearings (count)	Grooming (counts)	Exploratory activity (No. openings sniffed at)
Physiological saline		untreated	1	86.2±10.3	1.6±0.4	8.3±2.7	1.9±0.3
			2	64.0± 5.9	12.7±1.8	1.8±0.6	5.7±2.0
		6-OHDA	1	96.5±11.7*	3.1±0.7*	9.7±1.8	0.3±0.0**
			2	69.0± 5.0	7.4±1.3*	2.1±0.3	3.1±1.5
Tuftsin	200	untreated	1	75.9± 5.1	7.8±3.1**	6.2±2.3	3.8±1.0
			2	48.4± 6.0**	14.3±5.1	1.9±0.4	9.0±2.6**
		6-OHDA	1	97.3± 7.6	6.4±2.3**	7.3±1.6	5.1±1.7*
			2	50.0± 6.4	8.1±2.6	3.9±1.1	7.3±2.0
Leu^1-tuftsin	200	untreated	1	64.5± 7.0	2.6±0.8	6.9±2.2**	1.3±0.6*
			2	55.2± 8.6	6.2±2.1**	4.4±1.7**	2.7±0.8**
		6-OHDA	1	74.6± 8.1*	1.3±0.3*	8.3±2.0	0.9±0.2*
			2	65.8± 6.2	5.3±1.6*	3.6±1.4	1.4±0.7*
D-Arg^4-tuftsin	200	untreated	1	63.6± 6.8*	4.2±0.9*	2.0±0.8	5.6±1.1**
			2	54.0± 4.6	7.0±1.6	1.1±0.7	6.3±2.0
		6-OHDA	1	107.2±12.3	2.7±1.6	3.2±1.5*	6.4±2.3**
			2	75.6± 7.1	6.1±3.0	1.9±1.0	4.9±1.6

Differences significant between a) sub-divided groups in tests with control, and b) between tests with control, and after treatment.
* p<0.05.
** p<0.01.

Table 4. The Effect of Tuftsin and its Analogs on Avoidance Behavior

Compound and dose (µg/kg)	Animal group and type of response		No. failed escapes	Level of affect displayed (scores in fixed points)	Latency of escape (sec)
Physiological saline	untreated	1	28.3±4.2	9.4±3.7	∞
		2	none	none	26.5±8.4
	6-OHDA	1	42.3±8.3	18.2±4.8	∞
		2	8.7±1.8	7.6±2.0	43.6±5.9
Tuftsin 200	untreated	1	9.7±2.6**	3.8±1.4**	31.6±5.9
		2	none	none	13.2±3.5**
	6-OHDA	1	12.4±5.2**	7.5±3.3**	41.4±7.1**
		2	none**	2.9±1.8*	18.2±4.0**
Leu^1-tuftsin 200	untreated	1	28.8±7.6	16.4±5.1*	∞
		2	13.1±3.8**	3.2±0.9**	24.3±4.5
	6-OHDA	1	44.7±9.2	17.8±5.7	∞
		2	9.3±3.1	7.2±4.2	47.8±7.5
D-Arg^4-tuftsin 200	untreated	1	5.9±2.0*	none**	18.5±4.3**
		2	none	none	6.7±1.6*
	6-OHDA	1	9.4±4.6**	3.8±2.0**	28.1±4.4**
		2	none**	none**	22.7±5.5**

Note: Rats: 1 - emotional; 2 - non-emotional; 6-OHDA - pretreated group.
* $p < 0.05$.
**$p < 0.01$ vs. control testing (saline).

3.3. THE EFFECT OF TUFTSIN AND ITS ANALOGS ON GOAL-DIRECTED
      BEHAVIOR IN A STRESSFUL SITUATION, DEPENDING ON RESPONSE
      PATTERN AND STATE OF BRAIN CA SYSTEMS

The action of tuftsin and its analogs on goal-directed avoidance
patterns under conditions of emotional tension was rated according to
Henderson's (1970) species adaptable methods.

A glass cylinder measuring 12 cm across and 22 cm high is hung
centrally, 17.5 cm from the base, within a 40 cm high cylindrical
vessel, 35 cm in diameter, which is filled with water at a temperature of 22°C. The bottom end of the cylinder is sunk 2.5 cm below
the water's surface. Visible wire netting covers the inner walls of
the vessel.

Animals were placed in the cylinder, immersed in the water, for
2 min. In our version of this test, the animal can only manage to
escape by diving under water. The walls of the cylinder do not allow
it to get out by jumping upwards. The animal sees the wire netting,
providing a means of escape from the vessel, through the glass. But
escape would require the unusual avoidance procedure of diving under-
neath the walls of the cylinder, thereby overcoming the competing
biological fear response before plunging its head under water. The
stress factors involved in this test are as follows: actually being
in the water, partial immobilization (being unable to adopt the
horizontal swimming position) and a combination of new and unac-
customed stressors. A recording was made of: latency of initial
motor reactions, the number of unsuccessful attempts made to escape
by jumping upwards and the degree of affect displayed, for which
fixed points were awarded according to a rating scale, plus the
latency of avoidance reaction. The significance of differences was
determined by the Mann-Whitney non-parametric U-test.

The success of the animals in resolving the avoidance problem
was inversely proportional to the degree of affective response shown
when placed in the cylinder. An increase in initial emotional
response level in those animals which had been pre-treated with
6-OHDA, caused by upsetting the balance of brain CA system activity,
disrupts avoidance behavior, as seen by the increased number of
failed escapes and extended latency of escape. The more highly
emotional groups of animals - the emotional, untreated and the
emotional, 6-OHDA-pre-treated categories - failed to show any avoidance behavior during testing - latent period of escape = ∞.

Tuftsin improved avoidance behavior, enabling the animals to
solve this predictive problem, in the case of rats with damaged CA
terminals, as well as "emotional" and "non-emotional" control
animals. This clearly distinguished tuftsin from other compounds
achieving their effects via CA mechanisms, such as amphetamine,
haloperidol and clonidine, which exert a varying and occasionally

opposing influence on animals displaying differing degrees of
emotional responsiveness, especially where the CA terminals have
been destroyed (Bondarenko et al., 1981).

Leu[1]-tuftsin increased display of affect and number of failed
escapes; it also severely impaired avoidance reaction. This peptide
reduced exploratory activity in an open field and appeared to reinforce the state of frustration.

D-Art[4]-tuftsin reduced affective responses thoughout all the
animal groups and speeded up avoidance behavior by greatly reducing
the number of failed escapes. These changes correlate with behavior
displayed by animals under the influence of the peptides in an open
field.

The above findings show in particular the need to optimize the
modulating influence on emotional responsiveness in order to make
goal-directed behavior (escape) operative in a stress situation, as
well as predictive activity.

4. NEUROCHEMICAL MECHANISMS OF TUFTSIN'S CENTRAL EFFECTS

The results of behavioral investigations into tuftsin's action
indicate the connection between its psychotropic effects and the
state of brain CA systems - hence, the further experiments intended
to show how far CA neuromediator mechanisms are involved with potentiating the central action of tuftsin and its analogs.

4.1. TUFTSIN'S EFFECT ON BRAIN MONOAMINE CONCENTRATIONS

Klusha investigated tuftsin's effect on monoamine concentration
(Kozlovskaya, Klusha and Bondarenko, 1982), using spectrofluori-
metry (Hitachi MPH - 2A). Concentration of NA, DA, 5-HT and their
metabolites were measured in the brain of Wistar rats weighing
200-250 g, by the methods of Curzon and Green (1970), Schellenberger
and Gordon (1971) and Spano and Neff (1971). Tuftsin was dissolved
in physiological saline and administered in a volume of 5 µl into
the lateral ventricle through an implanted cannula (coordinates -
A = -1, L = 1.5, v = 4). The animals were decapitated 5 or 20 min
later. The mean level (M±SEM) of monoamines and their metabolites,
in µg per g of brain tissue, was calculated as a percentage of the
control.

The results of these investigations are given in Table 5.
Tuftsin showed a tendency to raise DA and HVA concentrations within
5 min. This correlated with the rise in motor activity and grooming
following intraventricular administration of the drug.

Table 5. Tuftsin's Effect on Brain Monoamine and Metabolite Concentrations in Rats Following Intraventricular Administration

Compound	Dose (µg/rat)	Decapitation (min after-wards)	"Monoamine" concentration as % (M ± SEM)					
			NA	DA	HVA	5-HT	5-HIAA
Phys.saline (control)		5	100± 7	100±12	100±14	100±12	100±15
Tuftsin	10	5	93± 5	125±20	124±17	81±11	83±12
	20	5	111± 6	122±13	110±16	116±12	88±17
	100	5	90± 4	91±16	96±15	118±18	104± 8
Phys.saline (control)		20	100±13	100±14	100±11	100±12	100±13
Tuftsin	10	20	109± 6	98±12	91± 8	105± 7	97± 8
	20	20	96±19	87±20	96± 6	86±16	115± 8
	100	20	110±12	94±14	96±12	88±20	83±11

These changes were not statistically significant, however. It may be that estimating monoamine concentrations in the brain as a whole is concealing shifts in turnover in separate brain regions.

4.2. ACTION OF TUFTSIN AND ITS ANALOGS ON DOPAMINERGIC MECHANISMS

We used the rotational model produced by unilateral destruction of the DA terminals of the striatum to study peptide action on dopaminergic mechanisms (after Ungerstedt, 1971, as modified by Pycock et al., 1975).

A group of 30 Wistar male rats weighing 220-250 g were anesthetized using ether. A dose of 16 µg 6-OHDA HBr (Sigma) was administered into the right, rostro-medial head of the caudate in a volume of 4 µg saline, containing 0.2 mg/ml ascorbic acid, added at the rate of 1 µl per 15 sec (AP + 0.5; L - 2.8; H - 4.8; Fifkova and Marsala's (1960) atlas). Ten Wistar rats which had received 4 µl of saline served as controls. Adoption of an asymmetrical posture was noted within 24 h of operating, with the animal inclining towards the side of the damaged DA terminals with its head, tail and body but with the forelimbs turned outwards, away from the rest of the body. The control number of spontaneous ipsilateral rotations made per 15 min (diameter 35 cm) was measured. Each animal was tested for amphetamine's and apomorphine's effect on circling behavior - producing reinforcement of ipsilateral and contralateral rotation respectively. Haloperidol and the DBH (dopamine β-hydroxylase) inhibitor, DTC (diethyldithiocarbamate), were used in this study. All compounds were administered intraperitoneally.

Side by side with the rotometer investigations, stereotyped behavior was rated, awarding fixed points according to Lat et al.'s (1981) system. The aggression displayed as each animal approached a metal rod was also measured in some experiments (Yamamoto and Uiki, 1978), applying the same dosage schedule of peptide and test compound. The statistical assessment was carried out using the Mann-Whitney U-test. Table 6 shows the effect of tuftsin and its analogs on circling behavior and its interaction with psychotropic compounds which act on DA brain systems.

Unilateral damage to the DA terminals of the nigro-striatal system by pre-treatment with the neurotoxin 6-OHDA creates DA activity asymmetry, and this in turn produces circling towards the side of the lesion. The action of direct (apomorphine-like) and indirect (amphetamine-like) DA agonists is illustrated by means of this model. It is chiefly DA, released from the mesolimbic DA neurones in the nucleus accumbens septi, which causes ispsilateral circling, while stereotyped behavior is mainly due to the action of nigro-striatal DA neurones. The destruction of striatal DA terminals by administering 6-OHDA disrupts the stereotypy produced by amphetamine, but does not activate motor activity (Moore and Kelly, 1978).

Of the three compounds studied, tuftsin substantially increases ipsilateral circling compared with the control, Leu[1]-tuftsin exerted no significant effect and D-Arg[4]-tuftsin reduced ipsilateral and promoted contralateral circling.

Amphetamine being a presynaptic releaser which reinforces ipsilateral circling, both tuftsin and its Leu[1] analog reduced circling after pre-treatment with this drug, whereas D-Arg[4]-tuftsin greatly inhibited this phenomenon, while also inducing contralateral circling.

Following unilateral destruction of the DA terminals, apomorphine induces contralateral circling due to a stronger action on the supersensitive striatal DA receptors on the lesioned side. D-Arg[4]-tuftsin greatly reduced this effect - the Leu[1]-tuftsin analog to a lesser extent. The latter simultaneously induced ipsilateral circling. Tuftsin itself substantially reinforced the effect of apomorphine.

After pretreatment with haloperidol, which blocks DA receptors, all the peptides produced circling behavior. This demonstrated their DA-mimetic effect. The direction of these movements varied, however; ipsilateral in the case of tuftsin and its Leu[1] analog, and contralateral with D-Arg tuftsin. Following prior inhibition of DBH by DTC, the effect of tuftsin and its D-Arg[4] analog remained unchanged from its level before DTC pre-treatment; Leu[1] tuftsin, however, substantially restored the number of ipsilateral rotations suppressed by DTC.

It may be concluded from these findings that neither tuftsin nor its Leu[1] analog exert any direct influence on post-synaptic DA structures, since they themselves do not actually bring about any contralateral circling, even where the destruction of terminals has produced supersensitivity. DA releasers, such as amphetamine, acting on the nucleus accumbens, are known to reinforce a) motor activity, especially when previously inhibited, and b) circling in cases of unilateral lesion of striatal DA terminals (Moor and Kelly, 1978). Since tuftsin and its Leu[1] analog promote both circling and ambulation in an open field, it could be postulated that these peptides have the ability to enhance DA release at the mesolimbic level. This hypothesis is contradicted, however, by the fact that they do not reinforce amphetamine-produced circling - quite the reverse. Tuftsin and its Leu[1] analog do, however, promote circling behavior following suppression of ipsilateral circling caused by inhibition of DA-synthesis by DBH inhibitors. This all goes to show, rather, that tuftsin exercises a modulating influence on DA process of the mesolimbic system.

A study of the influence of peptides on displays of such behavior as stereotypy and aggression provides further information on how peptides act on brain DA mechanisms (see Table 6).

Table 6. The Effect of Tuftsin and its Analogs on Circling and Drug-induced Behavior

Compound	Dose (mg/kg)	How long after pre-treatment (min)	Circling behavior (15 min)		Intensity of stereotyped behavior (in fixed points)	Level of aggression (in fixed points)
			ipsilateral	contra-lateral		
Phys. saline (control)			14.6±5.1	none	none	none
Tuftsin	0.2		23.7±4.5*	none	none	0.2±0.04
Leu^1-tuftsin	0.2		19.0±3.4	none	none	1.3±0.4**
D-Arg^4-tuftsin	0.2		7.7±3.0*	4.2±1.8**	none	none
Amphetamine (Am)	1.0		44.2±8.6*	none	4.1±1.2*	1.5±0.6*
Am + tuftsin	1.0±0.2	15	27.4±5.3**	none	7.7±2.8*	0.4±0.06
Am + Leu^1-tuftsin	1.0±0.2	15	24.5±4.1**	none	6.9±2.0*	1.3±0.5
Am + D-Arg^4- tuftsin	1.0±0.2	15	6.1±2.5	7.4±3.0*	2.8±0.9*	none
Apomorphine (Ap)	0.5		none	48.4±0.1*	7.1±0.8*	4.0±1.6*
Ap + tuftsin	0.5±0.2	20	none	55.3±9.4	6.0±3.1	2.2±0.9*
Ap + Leu^1-tuftsin	0.5±0.2	20	12.0±4.4**	38.1±7.5*	6.6±3.5	1.4±0.6**
Ap + D-Arg^4-tuftsin	0.5±0.2	20	none	17.6±5.0**	3.6±0.7**	none
Haloperidol (H)	1.0		none	none	none	none
H + tuftsin	1.0±0.2	45	3.0±0.9**	none	none	1.1±0.3**
H + Leu^1-tuftsin	1.0±0.2	45	9.2±2.6**	none	none	0.6±0.02*
H + D-Arg^4-tuftsin	1.0±0.2	45	none	4.0±1.3**	none	none
Diethyldithio-carbamate (DTC)	35.0		8.3±2.4$^+$	none	none	none
DTC + tuftsin	35.0±0.2	30	27.1±5.4**	none	none	2.3±0.7**
DTC + Leu^1-tuftsin	35.0±0.2	30	24.5±4.1**	none	none	3.3±0.9**
DTC + D-Arg^4-tuftsin	35.0±0.2	30	6.1±2.5	7.4±3.0**	none	none

Male Wistar rats. Pretreatment (24 h) – 6-OHDA, micro-injected into the head of the caudate.

+ $p < 0.05$ – effect of the test substance vs. control (saline).

* $p < 0.05$.

**$p < 0.01$ – effect of peptides vs. test substance.

Stereotyped behavior is known to be particularly closely connected with activation of nigrostriatal DA neurones (Pejnenburg et al., 1975). Tuftsin and its Leu[1] analog potentiate amphetamine stereotypy (sniffing, rearing, etc.) and extend its duration. The effect produced is similar to that of increasing the dose of amphetamine. As with a large dose of amphetamine, the increase in stereotypy does not occur together with intensified motor activity - on the contrary, the number of ipsilateral rotations decreases. This provides evidence that peptide action is focused on the DA system of the striatum, but not on the nucleus accumbens, with which locomotor activity is associated.

D-Arg[4]-tuftsin reduced both amphetamine- and apomorphine-induced stereotypy. Licking and gnawing are chiefly manifested in apomorphine stereotypy (Fray et al., 1980). D-Arg[4]-tuftsin made this type of behavior lapse back into amphetamine-type stereotypy. Tuftsin exerted no substantial influence on the effects of apomorphine. Tuftsin and Leu[1] analog, but not D-Arg[4]-tuftsin, slightly intensified aggression, both in the control and after pretreatment with haloperidol and DTC: degree of aggression was connected with the current state of the NA systems of the hypothalamus. They did not, however, bring about any displays of DA-dependent stereotypy, whereas apomorphine-induced aggression was reversed by all the peptides. (Stereotypy was only eliminated by D-Arg[4]-tuftsin).

Hanson et al. (1981) have shown that substance P exercises a modulating influence over the activity of DA receptors. It is present in high concentration in the substantia nigra (Brownstein et al., 1976) and in the axon terminals of the striato-nigral tract - a feedback loop involved in the regulation of nigro-striatal function. Substance P acts as a modulator of DA neuronal transmission, enhances the metabolism of DA in the striatum (Waldmeir et al., 1978) and increases the incidence of rearing and sniffing - the amphetamine-type features - displayed in stereotyped behavior in rats. Both disappear after the destruction of caudate DA terminals by administering 6-OHDA (Kelly and Iversen, 1979). In this connection it is important to note that the aminoacid sequence Lys-Pro-Arg in the tuftsin molecule is a reversal of the N - terminal 1-3 fragment (Arg-Pro-Lys) of the substance P molecule, as well as of the bradykinin-potentiating peptide (segment 6-8) and fibrinopeptide II (segment 13-15). Hence the structure of tuftsin has something in common with the "common" fragments of a number of low-molecular peptides (Chipens et al., 1979).

Substituting D-Arg for Arg in the tuftsin molecule substantially changes the spectrum of the peptide's range of action. D-Arg[4]-tuftsin reduces motor and emotional response and displays of aggression, as distinct from tuftsin, which increases the first two, especially after previous sedation. Tuftsin would not appear to exert potentiating effect on post-synaptic DA receptors, judging from the fact that even where super-sensitivity has been produced by degeneration of DA terminals, this peptide does not give rise to contralateral circling. D-Arg[4]-tuftsin's action contains a post-synaptic DA-mimetic component, since it produces contralateral circling on its own as well as following pre-treatment with haloperidol and DTC. But further research is needed to decide whether D-Arg[4]-tuftsin's ability to produce contralateral circling is brought about by its direct action on DA receptors. At the same time it acts as an antagonist by eliminating or attenuating the effects of apomorphine – contralateral circling, stereotypy and aggression. This peptide does not antagonize the effects of haloperidol; it attenuates ipsilateral circling induced by amphetamine, the DA releaser, while considerably facilitating tyrosine hydroxylase (TH), as does haloperidol. D-Arg[4]-tuftsin may be displaying DA inhibitory action on the nucleus accumbens alone; this brings about a drop in motor activity and in the number of amphetamine- and apomorphine-induced rotations. Tuftsin's action may apparently be explained by its pre-synaptic effects, primarily on the DA elements of the mesolimbic system, but not the nigro-striatal system, since it fails to induce stereotypy.

4.3. TUFTSIN'S INFLUENCE ON BRAIN TYROSINE HYDROXYLASE ACTIVITY

Tyrosine hydroxylase activity was measured after administering the peptides intraperitoneally in order to obtain more direct evidence of tuftsin's action on central CA systems and its penetration through the blood-brain barrier. It was thought that changes in the rate-limiting enzyme of CA synthesis could be a better indicator of the relationship between the central effects of tuftsin and the state of NA and DA systems of the brain.

The animals were decapitated in order to ascertain TH activity. Structures of the striatum and hypothalamus were separated and put on ice. A 10% homogenate of the brain structures was prepared in 0.05 M tris-maleate buffer (pH 6.1), containing 0.2 Triton X 100. The homogenate was centrifuged at 12,000 g for 15 min, and the supernatant used as a source of enzyme. The dopa fluorimetric method was used to determine the level of TH activity – indicated by the rate of formation of the reaction product (Yamauchi and Fujisava, 1978). Statistical analysis of the results was made using Student's t-test.

Figure 3 shows the pattern of TH activity following administration of 200 µg/kg of the peptides i.p. Changes were measured 7, 15 and 30 min after administration. Under these conditions both tuftsin and its Leu[1] analog produced a time-related decrease in striatal TH activity, with a somewhat smaller reduction in the hypothalamus, after an initial phase of slight activation. D-Arg[4]-tuftsin has a pronounced activating effect on TH.

Tuftsin exercises a dual influence on behavior, which it can either potentiate or inhibit. Its effects are dose- and time-related (Valdman et al., 1981, 1982).

TH activity

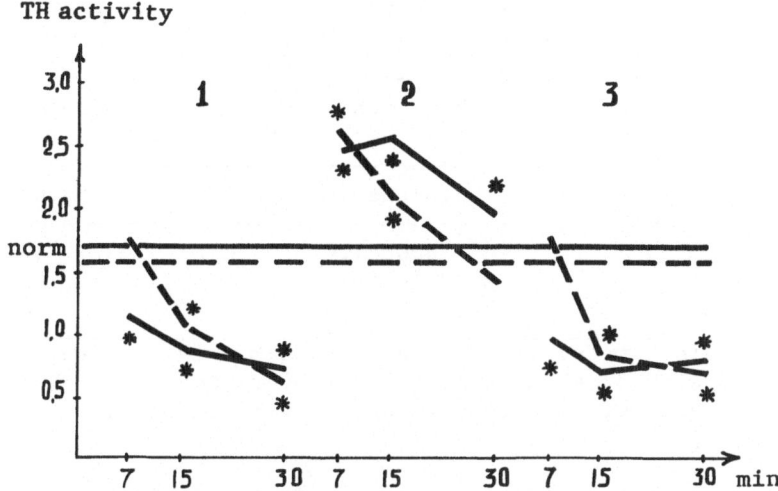

Fig. 3. Changes (nmol per 1 mg/protein per 15 min) in TH activity
in   rat hypothalamic and striatal structures at varying
intervals after administering 200 µg/kg tuftsin and its
analogs. Unbroken line – TH activity in the striatum;
Broken line – TH activity in the hypothalamus. 1 – change
produced in TH activity by treatment with tuftsin; 2 –
change produced in TH activity by treatment with D-Arg[4]-
tuftsin; 3 – change produced in TH activity by treatment
with Leu[1]-tuftsin. *p<0.01 – significance of differences
in comparison with the normal.

        Table 7 shows the effects of the largest dose, of 500 µg/kg i.p.
TH activity was estimated by direct spectrophotometry (Mineyeva-
Vialych, 1976) in solubilized homogenates of the hypothalamus and
striatum. TH activity increased both in the hypothalamus and even
more so in the striatum after 10 min. Activation of the enzyme was
still increasing in the former after a further 10 min, whereas it had
begun to drop in the striatum, though still exceeding control value.

        Changes in TH activity produced by tuftsin in vivo reflect the
fast-developing adaptive (kinetic) readjustments in the rate-limiting
enzyme of CA biosynthesis. They can be attributed to a rise in CA
metabolism or release; this may be considered evidence of the central
activity of tuftsin. Differences have emerged between the level of
activation of a) hypothalamic, and b) striatal CA processes
(respectively NA-dominated and DA-dominated).

        Interpreting these findings is complicated by the fact that
altered TH activity in the striatum may be produced by pre-synaptic
activation of the DA receptors, either directly or as a function of
altered DA release or reuptake. It could equally be caused by re-

Table 7. Tyrosine Hydroxylase Reaction Rate in Rat Brain after
          500 μg/kg Tuftsin

Experimental variables	Rate of reaction, μmol/mg protein per min (M ± SEM)	
	hypothalamus	striatum
Control	16.0± 1.0	33.3± 2.8
Tuftsin (10 min)	25.0± 6.5	87.1± 6.2
Tuftsin (20 min)	44.3±14.4	55.7±11.1

Solubilized brain homogenate. Buffer - 0.05 M tris maleate, pH 6.1;
0.11 mM tyrosine; 0.17 mM 6,7-diethyltetrahydropterin; T - 30°C.
Direct spectrophotometric method (Mineyeva-Vialych, 1976).

inforcing the inhibitory nigro-striatal influences, directed onto the
striatum, either directly or via serotoninergic mechanisms (Bunney
and Aghajanian, 1978).

When tuftsin was brought into direct interaction with purified
TH in vitro, using the method of Mineyeva et al. (1979), a 10^{-5} M
concentration and particularly a 10^{-4} M concentration was seen to
inhibit TH activity. Tyrosine, the substrate, protects TH from the
inhibitory effect of tuftsin. Fluphenazine, which acts as an allosteric regulator of TH activity (Rayevsky et al., 1979), almost
completely eliminated the inhibitory effect of tuftsin (by 80%) -
see Table 8.

These data show that the interaction between tuftsin and TH
does not develop via the active center of the enzyme. Tuftsin's
modulating effect in vivo may occur as a result of conformational
changes either in the enzyme itself or in the lipid components of
the membranes to which it is bound. (Interaction between small
peptides and lipid membranes are dealt with in the eighth article).

It is highly significant that in spite of the short duration
(ca. 20 min) of tuftsin's effect on behavior, the interaction between
tuftsin and TH over a fairly long period changes the initial DA
turnover rate. If α-MT is administered after pre-treatment with
300 μg/kg tuftsin, it fails to produce the usual decreases of DA
concentration in rat striatum (Zhelyazkov, 1982 - personal communication). DA concentration in the control (n = 8) was 6.58 μg/g
fresh tissue, 120 min after pretreatment with 200 mg/kg α-MT i.p.
(n = 8); it reached 1.74 and 7.35 when 300 μg/kg tuftsin had been
given 2 h earlier (n = 4). When tuftsin was given 30 min before
α-MT, the DA concentration in the striatum was 5.67 μg/g (n = 6) -
hardly distinguishable from the control. The mechanism of this
effect has yet to be fully investigated; is tuftsin protecting TH

Table 8. Tuftsin's Effect on Tyrosine Hydroxylase of Rat
          Hypothalamus in Vitro

Experimental variables	Reaction rate, μmol/mg protein per min (M $\pm$ SEM)
Control	355$\pm$10.0
Tuftsin 10^{-4}M	90$\pm$ 4.5
Tuftsin 10^{-5}M	255$\pm$15.3
Tuftsin 10^{-4}M + fluphenazine	278$\pm$13.9

Direct spectrophotometric method. Buffer - 0.5 M tris-maleate,
pH 6.1; 0.11 mM tyrosine; 0.17 mM 6,7-dimethyltetrahydropterin;
30 mg/protein per ml assay; T - 30°C.

from the inhibitory effect of α-MT, or are the conditions of DA
metabolism actually changing?

Tuftsin at concentrations of 5 x 10^{-5} and 5 x 10^{-4} inhibits
reuptake of NA and 5-HT by the crude synaptosomal fraction of rat
brain while increasing the accumulation of DA (see the eighth
article). These findings point to an interaction between tuftsin
and neuromediator transport mechanisms. According to our observations, tuftsin binds neither to benzodiazepine nor imipramine
receptors (through binding of ^{3}H-diazepam or ^{3}H-imipramine).
Naloxone fails to reverse the analgesic effect of tuftsin and some
of its analogs, observed when doses of 200 μg are injected into the
lateral ventricle, which shows tuftsin's effect to be unconnected
with opiate receptors (Herman and Stachura, 1980).

5. CONCLUSION

The above findings show that tuftsin - a naturally occurring
peptide - which has an activating effect on phagocytosis and immunogenesis also exerts a central stimulating action mediated by the
CA brain systems. We began by demonstrating the central psychotropic
effect of tuftsin and some of its analogs. Tuftsin's potentiation of
emotional/behavioral response closely resembles the action of other
small peptides, such as TRH and MIF, which have some structural
fragments in common. The means of action of these small peptides
on brain CA systems could well be virtually identical to that of
the regulatory peptides, which are formed and released together with
catecholamines from the terminals of cell structures belonging to
the APUD system. The modulating effect, peculiar to "modulating
peptides", consists of a varying effect, depending on the state of
the biological entity at the outset. The mechanism of this effect
has not yet been investigated, but one interesting hypothesis has
been put forward by Chipens et al. (1979) on the interaction between

the peptide regulator and the receptor, in the non-polar shape of a
quasi-cyclic structure, which is formed in the non-polar biophase.
In the aqueous phase the tuftsin molecule assumes a unfolded conformation, whereas a quasi-cyclic structure forms in the biophase
(Chipens et al., 1979). The aminoacid proline is responsible for
the bend of the peptide chain. The lysine amino group and the car-
boxyl group of the C-terminal arginine take part in the formation of
the quasi-cyclic structure. It was postulated that the formation of
compact quasi-cyclic structures in the course of binding with lipophilic receptors is one of the universal principles governing the
action of natural peptide bioregulators.

No findings have directly explained the link between tuftsin's
central action and its phagocytosis-stimulating effects to date.
Phagocytosis is known to be stimulated initially as a result of
interaction between tuftsin's positively charged molecular groups
and the negatively-charged lipid components of leucocyte membrane
(Constantopoulos and Najjar, 1973). The interaction between tuftsin
and TH does not develop via the active center of the enzyme, but
could result from conformational changes in the surrounding membrane
lipids. This, in turn, might affect neurotransmitter transport (see
the eighth article).

Changes in tuftsin's aminoacid molecular sequence produces both
changes in its phagocytosis-stimulating ability and the emergence of
certain psychotropic properties.

The following observations were made by comparing and contrasting the psychotropic properties of the peptides under investigation
with their effect on phagocytosis and immunogenesis: $D-Arg^4$ does not
have tuftsin's capacity for reinforcing the leucocyte's phagocytosis
activity, but does match its capacity in stimulating immunogenesis.
Leu^1-tuftsin stimulates phagocytosis even more powerfully than
tuftsin. A dose of $D-Arg^4$ analog, given intraventricularly, produces
a stronger and longer-lasting analgesic effect than tuftsin, while
$D-Leu^1$ analog's action resembles that of tuftsin (Herman and
Stachura, 1980). The various tuftsin derivatives do not produce
identical activation of phagocytosis and immunogenesis. Sometimes
they exert opposing effects, which shows that they derive from a
variety of mechanisms. The transformation of C-terminal arginine
into the D-stereoisomer changes the conditions of interaction between
the peptide and receptor or the latter's "surroundings", and it is
this which mainly distinguishes $D-Arg^4$-tuftsin's pharmacological
effects from those of the naturally occurring tetrapeptide.

REFERENCES

Barker, J. L., 1977, Physiological role of peptides in the nervous
        system, in: "Peptides in Neurobiology," H. Gainer, ed.,
        Plenum, New York, pp.295-343.

Bolles, R. C., 1960, Grooming behaviour in the rat, J.Comp.Physiol. Psychol., 53:306-309.

Bondarenko, N. A., Valdman, A. V., and Kamysheva, V. A., 1981, Change in the psychotropic effect on emotional reactivity and behaviour under stress depending on the status of the brain catecholaminergic system, Bull.Exp.Biol.Med., 7:35-38.

Bondarenko, N. A., Kamysheva, V. A., Mineeva, M. F., and Valdman, A. V., 1981, Effect of chronic stress on behaviour, somatic state and brain tyrosine hydroxylase activity in "emotional" and "non-emotional" rats, Bull.Exp.Biol.Med., 1:20-22.

Brownstein, M. J., Mroz, E. A., Kizer, J. S., 1976, Palkovits, M., ans Leeman, S. E., Regional distribution of substance P in the rat brain, Brain Res., 116:299-305.

Bunney, B. C., and Aghajanian, G. K., 1978, D-amphetamine-induced depression of central dopamine neurons: Evidence for medication by both autoreceptors and a striato-nigral feedback pathway. Naunyn-Schmiedeberg's Arch.Pharmacol., 304:255-261.

Chipens, G. L., Aunaz, P., Klusha, V. E., et al, 1973, Some concepts on the molecular mechanism of peptide hormone action at receptor level, in: "Peptides," H. Henson, ed., North Holland Publ., Amsterdam, pp.437-449.

Chipens, G., Nikiforovich, G., Mutulis, F., Veretennikova, N., et al., 1979, Cyclic analogs of linear peptides, in: "Peptides, Structure and Biological Function," E. Gross and J. Meienhofer, eds., Piers Chem. Co., Rockford, Illinois, pp.567-570.

Constantopoulos, A., and Najjar, V. A., 1973, The requirement for membrane sialic acid in the stimulation of phagocytosis by the natural tetrapeptide tuftsin, J.Biol.Chem., 248:3819-3822.

Curzon, G., and Green, A. R., 1970, Rapid method for the determination of 5-hydroxytryptamine and 5-hydroxyindoleacetic acid in small regions of rat's brain. Brit.J.Pharmacol., 39:653-655.

De Wied, D., and Gipsen, W. H., 1977, Behavioural effects of peptides, in: "Peptides in Neurobiology," H. Gainer, ed., Plenum, New York, pp.397-448.

Dismukes, R. K., 1979, New concept of molecular communicationary neurons, Behavioral Brain Sci., 2:409-448.

Fifkova, E., and Marsala, J., 1960, Stereotaxic atlases for the cat, rabbit and rat, in: "Electrophysiological Methods in Biological Research," C. Scient, I. Holubar and J. Ipser, eds., Prague.

Fray, P. J., Sahakian, B. J., Robbins, T. W., Koob, G. F., and Iversen, S. D., 1980, An observational method for quantifying the behavioural effects of dopamine agonists: Contrasting effects of d-amphetamine and apomorphine, Psychopharmacology (Berlin), 69:(3), 253-259.

Fukuda, T., Yamada, K., Suenaga, N., and Takishita, S., 1977, Changes in spontaneous activity and emotional responses of rats treated with 6-hydroxydopamine at the suckling age, Arch. Int.Pharmacodyn., 230:100-111.

Fuxe, K., Agnati, L., Andersson, K., Locatelli, V., et al., 1980,
     Concepts in neuroendocrinology with emphasis on neuropeptide-
     monoamine interaction in neuroendocrine regulation, in: "Pro-
     gress in Psychoneuroendocrinology," F. Brambilla, G. Racagni,
     and D. de Wied, eds., Elsevier, Amsterdam, pp.47-61.
Gubler, E. V., 1978, Calculating methods of analysis and diagnosing
     of pathological processes, Meditsina, Moscow.
Hanson, G., Alphs, L., Pradhan, S., and Lovenberg, W., 1981, Response
     of striatonigral substance P systems to a dopamine receptor
     agonist and antagonist, Neuropharmacology, 20:541-548.
Henderson, N., 1970, Behavioural reactions of Wistar rats to con-
     ditioned fear stimuli, novelty and noxious stimulation,
     J.Psychol., 75:19-34.
Herman, Z. S., and Stachura, S., 1980, Analgesic activity of some
     tuftsin analogs, Naturwissenschaften, 67:613-614.
Jonsson, G., Pycock, G., Fuxe, K., and Sachs, C., 1974, Changes in
     the development of central noradrenalin neurons following
     neonatal administration of 6-hydroxydopamine, J.Neurochem.,
     22:419-426.
Kelly, P. H., Seviour, P. W., and Iversen, S. D., 1975, Amphetamine
     and apomorphine responses in the rat following 6-OHDA lesions
     of the nucleus accumbens septi and corpus striatum, Brain
     Res., 94:507-522.
Kelly, A., and Iversen, S. D., 1979, Substance P infusion into
     substantia nigra of the rat: Behavioural analysis and involve-
     ment of striatal dopamine, Europ.J.Pharmacol., 60:171-179.
Kitada, V., Miyauchi, T., Satoh, A., and Satoh, S., 1981, Effects of
     antidepressants in the rat forced swimming test, Europ.J.
     Pharmacol., 72:145-152.
Kovacs, G. L., Bohus, B., and Versteeg, D. H., 1979, The interaction
     of neuropeptides with monoaminergic neurotransmission, Neuro-
     science, 4:1529-1537.
Kozlovskaya, M. M., Klusha, V. E., and Bondarenko, N. A., 1982,
     Comparison of psychotropic and neurochemical activity of short
     peptides, in: "Neurochemical Basis of Psychotropic Effects,"
     A. V. Valdman, ed., Inst. Pharmacol. Acad. Med. Sci., Moscow,
     pp.95-105.
Gorka, Z., and Wojtasik, E., 1980, The effect of antidepressants on
     behavioural despair in rats, Pol.J.Pharmacol.Pharm.,
     32:463-468.
Lat, H., Carino, M. A., Sperry, R., and Horita, A., 1981, Effects
     of micro-injection of 2-chloro-2-(3-dimethylaminoethoxy)-
     dibenzo(Br)-thiepine (zotepine) and thioridazine on stereo-
     typic behaviour and motor activity, J.Pharm.Pharmacol.,
     33:252-254.
Levin, S., 1962, The effects of infantile experience on adult behav-
     iour, in: "Experimental Foundations of Clinical Psychology,"
     A. J. Becrach, ed., New York, 4, 126.
Maj, J., 1980, Studies on the action of antidepressant drugs of
     second generation, Pol.J.Pharmacol., 32:437-449.

Miller, F. E., Heffner, T. G., Kotake, G., and Seiden, L. S., 1981,
     Magnitude and duration of hyperactivity following neonatal
     6-hydroxydopamine is related to the extent of brain dopamine
     depletion, Brain Res., 229:123-132.
Mineyeva-Vialych, M. F., 1976, Direct spectrophotometric techniques
     for the assessment of tyrosine hydroxylase, Vop.Med.Khim.,
     22:374-279.
Mineyeva, M. F., Kudrin, V. S., and Rayevsky, K. S., 1978, Brain
     tyrosine hydroxylase: kinetic properties and regulation of the
     activity, Annali dell'Instituto Superiore di Sanità, 14:83-88.
     Proceedings of the Second Soviet-Italian Symposium on
     Transmitters in the Action of Psychotropic Drugs, V. G. Longo
     and V. V. Zakusor, eds., Instituto Superiore di Sanita, V le
     Regina Elena, 299-Roma.
Moore, K. E., and Kelly, P. H., 1978, Biochemical pharmacology of
     mesolimbic and mesocortical dopaminergic neurons, in: "Psy-
     chopharmacology: a Generation of Progress," M. A. Lipton et
     al., eds., Raven Press, New York, pp.221-234.
Morrison, B. I. and Hill, W. F., 1967, Socially facilitated reduction
     of the fear response in rat raised in groups or in isolation,
     J.Comp.Physiol.Psychol., 63:6376-6381.
Najjar, V. A., and Nishioka, K., 1970, Tuftsin, a natural phago-
     cytosis stimulating peptide, Nature, 228:672-673.
Nomura, Y., Naiton, F., and Segava, T., 1976, Regional changes in
     monoamine content and uptake of the rat brain during postnatal
     development, Brain Res., 305-315.
Pijnenburg, A. J., Honig, W. M., and Van Rossum, J. M., 1975, Antag-
     onism of apomorphine and d-amphetamine-induced stereotyped
     behaviour by injection of low doses of haloperidol into the
     caudate nucleus and the nucleus accumbens, Psychopharmacologia
     (Berlin), 45:65-71.
Porsolt, R. D., Bertin, A., and Jalfre, M., 1977, Behavioural despair
     in mice: a primary screening test for antidepressants, Arch.
     Intern.Pharmacodyn.Therap., pp.327-336.
Pycock, C. D., Tarsy, D., and Marsenn, C. D., 1975, Inhibition of
     circling behaviour by neuroleptic drugs in mice with uni-
     lateral 6-hydroxydopamine lesions of the striatum, Psycho-
     pharmacology, 45:211-219.
Rayevsky, K. S., Mineyeva, M. F., and Kudrin, V. S., 1978, The effect
     of neuroleptics on brain tyrosine hydroxylase, Annali
     dell'Instituto Superiore di Sanità, 14:89-96.
Schellenberger, M. K., and Gordon, J. H., 1971, A rapid simplified
     procedure for simultaneous assay of norepinephrine, dopamine
     and 5-hydroxytryptamine from discrete brain areas, Analyt.
     Biochem., 39:355-372.
Spano, P. F., and Neff, N. H., 1971, Procedure for simultaneous
     determination of dopamine, 3-methoxy-4-hydroxyphenylacetic and
     3,4-dihydroxyphenylacetic acid in brain, Analyt.Biochem.,
     41:113-118.

Spiere, Z., Zakuth, V., Bogair, N., and Fridkin, M., 1977, Radio-immunoassay of the phagocytosis-stimulating peptide tuftsin in normal and splenectomized subjects, Europ.J.Immunol., 7:69-74.

Tizabi, Y., Thoa, N., et al., 1979, Behavioural correlation of catecholamine concentration and turnover in discrete brain areas of three strains of mice, Brain.Res., 166:199-205.

Ungerstedt, U., 1971, Postsynaptic supersensitivity after 6-OHDA induced degeneration of the nigro-striatal dopamine system, Acta.Physiol.Scand., Suppl. 367:69-93.

Valdman, A. V., Kozlovskaya, M. M., Ashmarin, M. P., Mineeva, M. F., and Anokhin, K. V., 1981, Central effects of the tetrapeptide tuftsin, Bull.Exp.Biol.Med., 62, 7:31-33.

Valdman, A. V., Bondarenko, N. A., Kamsheva, M. A., Kozlovskaya, M. M., Kalikhevich, V. N., and Ardemasova, Z. A., 1982, Analysis of the neurochemical mechanisms of psychotropic effects of tuftsin and its analogs, Bull.Exp.Biol.Med., 93, 4:57-60.

Valdman, A. V., Bondarenko, N. A., Kozlovskaya, M. M., Rusakov, D. Yu., Kalikhevich, V. N., and Ardemasova, Z. A., 1982, Comparative study of psychotropic activity of tuftsin and its analogs, Bull.Exp.Biol.Med., 93, 4:49-52.

Valdman, A. V., Zvartau, E. E., and Kozlovskaya, M. M., 1976, Psychopharmacology of Emotions, Meditsina, Moscow.

Valdman, A. V., Kozlovskaya, M. M., Klucha, V. E., and Svirskis, Sh. V., 1980, Study of the psychopharmacological spectrum of melanostatin, Bull.Exp.Biol.Med., 89:693-696.

Waldmeier, P. C., Kam, P., and Stocklin, K., 1978, Increased dopamine metabolism in rat striatum after infusions of substance P in the substantia nigra, Brain Res., 159:223-227.

Yamamoto, T., and Ueki, S., 1978, Effects of drugs on hyperactivity and agression induced by raphe lesions in rats, Pharmacol. Biochem.Behav., 9, 6:821-826.

Yamauchi, T., and Fujisava, H., 1978, A Simple and sensitive fluorometric assay for tyrosine hydroxylase, Analyt.Biochem., 89:143-150.

Zile, R. K., Odynets, T. S. and Klusha, V. K., 1979, Effect of some fragments of peptide hormones on the contents of biogenic monoamine from mouse brain, Biochemia., 44:93-96.

Zebrowska-Lupina, I., 1980, Presynaptic α-adrenoceptors and the action of tricyclic antidepressant drugs in behavioral despair in rats, Psychopharmacology, 71:169-172.

Monoaminergic Mechanisms of Psychotropic Action of Short Peptides

A.V. Valdman[1], M.M. Kozlovskaya[1], N.A. Bondarenko[1],
V.V. Rojanetz[1], N.A. Avdulov[1], V.E. Klusha[2]
and R.K. Mucinieze[2]

[1] Institute of Pharmacology, Academy of Medical Sciences
of the USSR, Moscow
[2] Institute of Organic Synthesis, Latvian Academy of Sciences, Riga

Conceptions of the physiological role of peptide hormones have changed profoundly in recent years. Not only were the "classical" neurohormones, such as vasopressin and oxytocin, found to have a direct action on mental processes (De Wied, 1980), but it was also established that hypothalamic "regulatory" peptide hormones, known to be widespread in the extrahypothalamic areas of the brain, produce a complete range of psychotropic effects unconnected with hormonal function (De Wied and Gispen, 1977; Barker, 1977; Prange et al., 1978). Many so-called gut peptides, such as cholecystokinin, gastrin and vasoactive intestinal polypeptide, amongst others, have been discovered in brain structures, especially in the cerebral cortex. Their neurophysiological effects have been observed, although their functional involvement has yet to be fully explored.

Furthermore, any function peptide hormone fragments may possess has received insufficient attention to date. These fragments may form endogenously, as products of the enzymatic cleavage of the parent hormones. One of the principles of peptidergic regulations is the release of an active peptide from a larger macromolecular precursor by the action of proteolytic enzymes, present in neuronal membranes. Membrane-bound enzymes release certain aminoacid sequences, which may perform a wide variety of biological activities. Burbach and De Wied (1980) showed that endopeptidase and aminopeptidase, associated with brain synaptosomal membranes, form a) α-endorphin, or des-Tyr-α-endorphin, at pH = 5.9 or b) α-endorphin, or des-Tyr-α-endorphin, at pH = 6.7. from the β-endorphin molecule. The selective formation of these two types of endorphin, as a function of the pH of the brain environment, may prove to be one of the mechanisms of peptidergic regulation of adaptive behavior, since these two types of endorphin have a different physiological action

203

(De Wied, 1981). It has also been thought that the active peptide
may be released at dendrite membranes (Guillemin, 1978).

Thus partial enzymatic degradation of the peptide does not
eliminate its biological action; in fact, proteolysis of the peptide
may in itself constitute one of the bases of peptide behavioral
regulation. Hence, one important point: the individual components
which go to make up the structure of a peptide hormone have different
physiological effects (De Wied, 1980). According to Chipens et al.
(1973), the fragments may produce similar ligand-receptor interactions.

The half-life of regulatory peptides lasts a matter of minutes,
while their effect on physiological processes may extend for much
longer – perhaps hours, or even days. Potentiation of their physiological effects by the mediation of brain monoaminergic mechanisms is
one possible explanation of this discrepancy. The nature and possible mechanisms of such a modulating effect has yet to be properly
investigated. Their effects are most likely to involve catabolism or
monoamine (MA) biosynthesis, MA release and reuptake, changes in the
functional state of specific receptors or in the functional characteristics of biological membranes in general and their lipid matrix
in particular.

2.   COMPARISON BETWEEN THE EFFECT OF SHORT PEPTIDES ON BRAIN MONOAMINES (BM) AND ON BEHAVIOR

2.1.1. Selecting Peptides for Investigation

The following C-terminal tripeptides, deriving from peptide
hormones and differing in structure and function, were investigated:
luliberin 8-10 (APG), gastrin 13-15 (MAP), oxytocin 7-9 (MIF), substance P 9-11 (GLM), and the tripeptide thyroliberin (TRH). The
peptides' primary structure is shown in Table 1. The choice arose
out of a number of considerations. In our investigations of the
possible influence of N- and C-terminal tripeptides on the concentration of BM and their metabolites in the brains of mice pretreated
with haloperidol, we discovered (Zile et al., 1979) that it is only
C-terminal tripeptides having a terminal amide group that are capable
of modulating the DA processes inhibited by this neuroleptic. One
suggestion made was that the amine group brings the properties of the
above-mentioned fragments somewhat closer to those of MA, while the
C-terminal portion of the peptide molecule is more important for
maintaining its biological activity. The aminoacid sequence of the
tripeptides under study is seen in a number of biologically active
peptides. Hence, the structure of GLM is common to tachykinins,
physalaemin and eledoisin amongst others; that of MAP is shared by
caerulein-like peptides, caerulein itself, cholecystokinin, gastrin,
etc. while APG shows similarities to the C-terminal tripeptide of
vasopressin.

Table 1. Primary Structure of the Peptides Under Study

Peptide	Aminoacid sequence of the fragment	Abbreviation
Thyroliberin (thyrotropin releasing hormone)	p Glu^1-His^2-Pro^3-NH_2	TRH
Luliberin (luteinizing hormone releasing hormone)	Arg^8-Pro^9-Gly^{10}-NH_2	APG
Oxytocin	Pro^7-Leu^8-Gly^9-NH_2	PLG or MIF
Gastrin	Met^{15}-Asp^{16}-Phe^{17}-NH_2	MAP
Substance P	Gly^9-Leu^{10}-Met^{11}-NH_2	GLM

All peptides were synthesized at the Institute of Organic Synthesis, Latvian Acad.Sci., Riga.

2.1.2. BM Concentration Following Intraperitoneal Administration

BM concentration was measured in homogenates of whole mouse brain, excluding brain stem and cerebellum. It was assessed spectrofluorimetrically, using the methods of Shellenberger et al. (1971), Spano et al. (1971) and Curzon and Green (1970). The brain tissue was divided into two: in one portion NA and DA were estimated according to the amount of fluorescent product generated. HVA concentration was determined after adsorption onto alumina. 5-TH and 5-HIAA levels were measured in the other portion of the brain. BM concentration was calculated by comparison with the fluorescence of standard solutions. The average BM concentration in µg/g of brain tissue was determined in a group of 10-12 animals. The significance of differences between the experimental and the control group, which had received physiological saline i.p., was calculated using Student's t-test ($p < 0.05$).

No substantial change in BM concentration occurred when a single dose of 5 mg/kg of the peptides under study was administered i.p. The animals were decapitated and the brain removed 30 min after administering the peptide. No noticeable effect was produced on the animals' behavior during this interval.

The lack of significant change in BM may be due to a number of reasons: insufficient penetration of peptides into the brain across the bloodbrain barrier, rapid peptide inactivation within the organism, and so forth. It is equally possible that an assessment of MA in the brain as a whole is masking changes taking place in separate brain structures. Telegdy et al. (1980), Fekete et al. (1980) and others showed that the effect of peptides on BM concentration differs from one morphological region to another.

2.1.3. Behavioral Reaction and BM Concentration After Intraventricular Administration

Wistar rats weighing 200-250 g were used. A stainless steel cannula, 0.45 mm in diameter, was implanted in the lateral ventricle; stereotaxic coordinates: A: -1, L: 1.5, V: 3.5. The peptides, in solution, were administered in a volume of not more than 10 μl, at a rate of 10 μl/min. An equivalent volume of sterile physiological saline was used as control. Each animal was housed in a plastic cage measuring 34 x 54 cm, and its behavior observed for 20 min. A recording was made of its general locomotor activity, amount of rearing, circling, grooming, gnawing, head twitches, wet-dog shakes, ptosis, catalepsy, etc. The animals were decapitated 20 min after administering the peptides. BM level was ascertained by the above method (see Section 2.1.2).

Substantial changes in BM level were noted when the same peptides were administered intraventricularly (see Table 2). TRH considerably raised the concentration on NA, DA and 5-HT, as well as that of their metabolites (by 30-40%). MIF activates the DA system only, increasing DA and HVA concentration. APG would appear only to reinforce 5-HT metabolism, judging by the rise in 5-HIAA concentration. MAP produces a dose-dependent reduction in DA and HVA level, and increases the content of 5-HT and 5-HIAA, and GLM decreases the DA concentration but increases HVA, with an accompanying drop in 5-HT concentration and a rise in 5-HIAA level.

Table 2.   The Influence of Intracerebroventricular Tripeptides on Brain Monoamine (BM) Concentration

Compound	Dose (μg)	BM concentration (M ± SEM, % of control; n = 12)				
		NA	DA	HVA	5-HT	5-HIAA
Phys. saline (control)		100± 9	100± 8	100± 7	100±14	100±13
TRH	20	124±10*	144±12*	146±12*	143±10*	135±13*
MIF	50	112± 7	124±4*	119± 8*	95± 9	105± 7
	100	110±13	143±13*	135± 8*	90± 7	95±12
APG	50	120±13	100± 5	93±12	100±13	127± 9*
	100	115±10	114± 6	110± 9	121±15	133±10*
MAP	50	90± 7	72± 4*	85± 3*	120±11	105±11
	100	110± 6	63± 7*	78± 4*	140±10*	128± 8*
	200	105±10	59± 6*	69±12*	157±17*	135±15*
GLM	50	103± 7	85±16	132± 7*	96±12	103± 7
	100	105±10	87± 5*	140±10*	90±10	124±10*
	200	112±13	80± 8*	148±12*	83± 9*	135± 9*
	360	95±18	75±12*	158± 8*	78±13*	146±14*

* $p < 0.05$ vs. control.

A change in brain MA activity after intraventricular peptide administration is obviously another reason for the effects produced on behavior. Thus, TRH induces motor activity, head twitches, wet-dog shakes and head weaving; these signs may easily be explained by TRH intensifying metabolism of brain MA. MIF acts similarly to TRH, but the effects displayed are less pronounced. A rise in motor activity, head twitches, sneezing and grooming are all observed. These features are evidently connected with accelerated DA turnover.

The rise in behavioral effects is dose-related. Dose-dependent. tremorgenic activity typical of APG must clearly be produced by raised 5-HT metabolism. MAP exerts a depressing effect. Locomotor activity drops and a cataleptic-type state develops, with brief circling motions (Straub's symptom). These reactions may be due to a drop in DA and HVA level and a rise in 5-HT and 5-HIAA concentration. Low doses of GLM stimulate grooming, whereas higher doses, which raise DA and 5-HT turnover (causing a drop in DA and 5-HT and a rise in HVA and 5-HIAA) produces shaking and a rise in motor activity.

When the peptide is administered exogenously into the cerebral ventricle, its concentration far exceeds the likely natural level. This is even more true of the fragments of peptide hormone, which are formed and released in localized brain regions. It is thus rather difficult to assess the physiological significance of the peptides' effects when administered intracerebrally. It should also be borne in mind that when the peptide is administered into the cerebral ventricles or cisterns it is very rapidly transported into the peripheral blood stream. The passage of the peptide from the cerebrospinal fluid into the blood is dynamically similar to the movements of inert material, such as microspheres (Passaro et al., 1982).

Thus, two sub-groups of tripeptides may be distinguished when they are administered intracerebroventricularly, exerting opposing influences on the DA system. TRH and MIF both raise DA level; it is reduced by MAP and GLM.

2.2. THE EFFECT PRODUCED ON CATS' EMOTIONAL/BEHAVIORAL RESPONSE
      BY INTRAPERITONEAL ADMINISTRATION OF MIF AND MAP

The fact that many naturally occurring and synthetic peptides have a regulatory effect on behavior and mental function is as yet little understood, however. If peptides are to be used as pharmaceuticals regulating central functions, the standard methods of administering drugs will need to be investigated rather than intracerebroventricular administration.

The effect of peptides on the range of cats' emotional behavior when this behavior relates to an animal group was assessed by the method established in the first article. Doses of 20-40 mg/kg were administered intraperitoneally. Table 3 gives combined values, expressing the mean (M ± SEM) of fixed points scored for degree of peptide effect on a set of emotional/behavioral responses.

MIF produces substantial changes in the range of emotional/ behavioral response 30 min after intraperitoneal administration. Effects or display related to positive emotional state were reduced; no pleasurable or playful responses occurred, while exploratory activity (or curiosity) declined. Actively defensive reactions in response to aversive test stimuli increased. Reciprocal behavior between the animal and its fellows became increasingly conflicting, and the number of fights grew in length and intensity. Spontaneous motor activity and behavioral initiative both declined. The changes in animals' behavioral response patterns produced by MIF are indistinguishable as far as behavioral display is concerned from those produced by administering L-DOPA and amantadine. A certain jumpiness of behavior is typical of all three compounds, with a rise in affect and conflict accompanied by a measure of motor inhibition, whereby negative emotions are increasingly displayed and positive ones decline (Valdman et al., 1979). According to the findings of North et al. (1973), doses of 10-15 mg/kg have no effect on stereotyped behavior produced in cats by L-DOPA, while suppressing an L-DOPA-induced rise in motor activity. Small doses of 0.1-0.5 mg/kg MIF produced no effect on motor activity, but 10-50 mg/kg caused a reduction in locomotion and the adoption of a sphinx-like pose, although the animal was not sedated and appeared fairly alert. These authors also noted similarities between the effects of MIF on one hand and L-DOPA and amphetamine on the other.

Table 3. Effect of MIF and MAP on Cats' Range of Emotional Behavior

Range of emotional/ behavioral response	Score for intensity of display (M ± SEM)		
	Control	MIF (20–40 mg/kg)	MAP (20–40 mg/kg)
Fear	1.8±0.3	1.8±0.3	0.8±0.3*
Anxiety	2.0±0.3	1.7±0.5	1.0±0.1*
Negativity	0	1.0±0*	1.0±0*
Aggression	1.7±0.2*	3.0±0.5*	2.6±0.5
Conflict	2.1±0.7	3.6±0.6*	2.6±0.3
Signs of pleasure	2.7±0.5	0.8±0.1*	1.3±0.1*
Hunting	2.7±0.3	1.0±0.3	0.8±0.2*
Friendliness	1.7±0.5	0.8±0.5	0.4±0.1*
Exploration	2.5±0.2	1.5±0.5	1.0±0
Motor activity	2.3±0.1	1.0±0.1	2.0±0
Initiative	2.5±0.3	1.6±0.6	2.0±0.2
Alertness	2.0±0.1	1.0±0.3	1.0±0

* $p < 0.05$.

MAP given intraperitoneally also suppressed effects related to
positive or pleasurable emotional state. It did not, however,
reinforce affective response to aversive testing devices. The
animals' behavior took on a passively defensive aspect, and became
less alert. A connection may be made between these behavioral effects and the changed BM concentration produced by MAP - reduced DA
and HVA, and increase in 5-HT turnover.

3. THE MODULATING EFFECT OF SHORT PEPTIDES AS A FACTOR IN INITIAL EMOTIONAL/BEHAVIORAL RESPONSE PATTERN AND THE STATE OF MONOAMINERGIC BRAIN SYSTEMS

3.1. THE "MODULATING" EFFECT OF PEPTIDES

Neuropeptides are widely thought to exert a modulating effect on
processes related to "classical" neurotransmitters. The expressions
"neuromodulator" and "modulating effect" have been much used in
recent years, but rather loosely; the phenomena which they are
intended to describe are not in fact the same. "Modulating", as
most widely employed, refers to an effect which, on account of its
electrophysiological characteristics, cannot be described by means
of the terms "stimulation" (EPSP) or "inhibition" (IPSP) and differs
from accepted ideas of the action of synaptic transmitters. Barker
(1977) suggested that neuropeptide action on neuronal processes could
occur extra-synaptically, resulting from the regulation of changes in
the way peak potentials or pacemaker activity are produced.

The contrast between the peptide modulating effect on synaptic
transmission and the action of neurotransmitters themselves is mainly
a matter of degree, however. Accepted ideas on neurochemical transmitters, as mediators of synaptic conduction into the monoaminergic
systems of the CNS, have also evolved substantially in recent years
(Dismukes, 1979). Most data point to the extra-synaptic release of
"classical" neurotransmitters such as NA and 5-HT (Vizi, 1980).
Electronmicroscopy has shown that the majority of axonal terminals or
varicosities of monoaminergic neurones do not form true synaptic
contact (Descarries et al., 1975, 1977). Thus, the transmitter
released from these extrasynaptic formations is spread diffusely, and
may react on the surrounding cell populations, which are involved in
different types of function. Nor does the biologically active amine
fulfill the function of a transmitter of specific information to a
particular "recipient". It does exert some modulating effect, changing the functional state of the nerve elements. It should also be
remembered that the terminals of NA and 5-HT neurones, which are
widespread in brain structures, derive from a comparatively small
number of cell bodies, situated in brain stem structures, such as the
raphe nucleus and nucleus locus caeruleus (LC). Axons from these
neurones branch extensively. A single neurone supplies large masses

of the brain; this in itself would preclude transmission of any type
of localized signal, apart from the fact that these monoaminergic
neurones receive no specialized afferent input. Hence, their role
does not consist of transmitting discrete information concerning
fast-moving processes. Low-frequency activity, varying over a
limited frequency range, is typical of these LC and raphe nucleus-
neurones. These monoaminergic systems may be acting as modulators
of target cells and exerting a tonic (trophic) effect on groups of
cells. Even post-synaptic processes, produced by specific neurotransmitters, can exercise a modulating influence on NA. It was
shown iontophoretically that application of NA, DA and 5-HT can
produce changes lasting from between a few seconds and a minute (as
distinct from synaptic potentials, which take place within a span of
msec). Presumably monoaminergic pathways from the LC, the substantia
nigra and the raphe nucleus generate long lasting, slow, postsynaptic
responses, their function being unconnected with the transmission of
rapid information to discrete neurones. Hence, this type of neuronal
communication does not require circumscribed synaptic contact. This
same basic type of interneuronal regulation also characterizes neuropeptides which may be released extra-synaptically and may modulate
synaptic function, producing a long-term change in the excitation of
certain neuronal populations.

    Thus, the so-called neurotransmitters (NA, DA, 5-HT) do not
differ so fundamentally from neuropeptides. Both may act as true
mediators in synaptic contacts in certain localized regions of the
brain, transmitting discrete signals. When released extra-synaptically, however, where they exert a little-understood influence on
either pre/post-synaptic elements of more complex processes related
to the membrane status or the cell genome, they are instrumental
either in changing the efficacy of input or switching from one
sensory pathway to another. Where this is the case, they are acting
as "modulators", signifying alterations in the functional state of
large cell populations.

3.2. THE INFLUENCE OF SHORT PEPTIDES ON RAT BEHAVIOR DURING
      AN EMOTIONALLY STRESSFUL SITUATION, TAKING INDIVIDUAL
      RESPONSE PATTERN INTO ACCOUNT

3.2.1. Open Field (OF) Behavior of "Emotional" and
        "Non-emotional" Rats

    Cross-bred male rats weighing 250-300 g, which had previously
been split up into an "emotional" and a "non-emotional" group, were
used in the experiments. These groups were further subdivided into
untreated rats and those given 6-OHDA during the neonatal period.
(See the seventh article, Section 3.1 for details of the method
used).

Response level differed substantially amongst these four animal sub-groups. They could be arranged in ascending order of emotionality as follows: non-emotional and untreated → non-emotional but pretreated with 6-OHDA → emotional but untreated → emotional and pretreated with 6-OHDA. The non-emotional animals' general level of motor activity in an open field was found to be lower, judging by the incidence of rearing and the greater number of holes (openings) explored and the more moderate pace of movement in general. The incidence of grooming was lower and the amount of defecation less than for the emotional sub-groups (Bondarenko et al., 1981). Although animals belonging to the "emotional" category did cross a large number of squares, they showed a low level of exploratory activity, judging by the incidence of rearing and the number of holes investigated. They ventured less into the central area of the OF. It was concluded that the degree and duration of locomotor hyperactivity shown after neonatal administration of 6-OHDA correlates with the extent of damage to DA neurones (Miller et al., 1981).

Investigations using Denenberg's (1969) OF behavior method traditionally apply the criteria of reduced exploratory activity, judged by the number of squares crossed, and increased defecation, as indicators representing emotionality in rats. Our findings, however, would suggest that it is not enough to take these factors alone into account, especially for the purposes of pharmacological research. The process of handling animals and injecting them suffices to induce defecation, as was illustrated in subsequent results. Our findings on peptide action on the parameters of OF behavior in rats with varying response levels are given in Table 4.

MAP reduced ambulation in all groups of animals. The emotional and the non-emotional categories, however, showed this effect differently. The ratio between motor activity levels, emotional/non-emotional, was 1.35 for untreated controls rats and 1.4 for those pretreated with 6-OHDA; this illustrates the emotional animal type's high level of motor activity. After MAP, the ratio was 0.7 for the untreated group and 0.5 for those given 6-OHDA. The values had been reversed, the general level of motor activity becoming relatively lower in emotional animals. The level of motor activity of the non-emotional group, expressed as a percentage of the control, was 42% for untreated animals and 66% for those given 6-OHDA following administration of MAP, against 21% and 26% respectively for the "emotional" sub-group.

A comparison between untreated and 6-OHDA pretreated animals' level produced the following results. The ratio of motor activity in untreated/6-OHDA pretreated groups against control is close to one for emotional and non-emotional sub-groups (0.8 and 0.98 respectively). Administering MAP brings these values to 0.7 for the emotional and 0.5 for the non-emotional sub-groups, which amounts to relative motor response inhibition within the untreated group.

Table 4. The Effect of Short Peptides on Open Field Behavior in Rats with Different Emotional Response Patterns

Substance (mg/kg)	Animal group (type of emotional response pattern)		Values recorded (M ± SEM) over 4 min			
			Perambulation (No. squares crossed)	Rearing (counts)	Incidence of grooming	Exploration (No. openings investigated)
Physiological saline	non-emotional	1	64.0±5.4	12.7±1.8	1.8±0.6	5.7±2.0
		2	69.0±5.0	7.4±1.3	2.1±0.3	3.1±1.5
	emotional	1	86.2±10.3	1.6±0.4	8.3±2.7	1.9±0.3
		2	96.5±11.7	3.1±0.7	9.7±1.8	0.3±0.0
MAP 5.0	non-emotional	1	27.6±3.3**	7.7±2.1*	2.1±1.0	3.9±1.7*
		2	46.6±5.9**	8.2±2.2	2.1±0.7	4.3±1.5
	emotional	1	18.7±3.9**	1.6±0.3	5.0±1.7*	2.0±0.9
		2	25.9±6.0**	1.6±1.2	7.2±0.9*	2.2±0.4*
MIF 5.0	non-emotional	1	25.3±4.6**	6.0±2.1**	2.4±0.8	6.0±2.4
		2	53.2±6.4	6.3±2.1	4.3±1.5*	1.7±0.5
	emotional	1	43.4±3.1**	1.2±0.9	6.2±3.9	1.1±0.4
		2	128.7±18.2**	2.2±1.0	9.7±3.2	0.4±0.0

1 = untreated.
2 = pre-treated with 6-OHDA.
Differences significant between test and control animals. Sub-groups: n = 8.
* $p < 0.05$; ** $p < 0.01$.
Compounds administered 10 min prior to placing in OF; repeated after 168 h.

MIF likewise reduced motor activity. It increased the motor
activity ratio emotional animals/non-emotional animals as against
control to 1.7 for the untreated sub-group (vs. 1.35 for the
control), and 2.4 for rats pretreated with 6-OHDA (vs. 1.4 for the
control). Hence the shift occurred in the opposite direction from
that caused by MAP. Motor activity was thus maintained at higher
levels in the emotional type of animal. The ratio between motor
response for untreated animals/6-OHDA pretreated animals was 0.5 (vs.
control - 1) for non-emotional types and 0.3 (vs. control - 0.9) for
emotional animals after administration of MIF; this again points to
the relatively lower inhibition of motor activity in animals
pretreated with 6-OHDA.

The results obtained could be explained as follows: the mesolimbic cortical systems of DA terminals is known to be instrumental
in the development of locomotor and exploratory activity in rats
(Fink and Smith, 1980), especially in new and unfamiliar surroundings, as with the OF. Small doses of DA, L-NA or apomorphine applied
locally to the region of the nucleus accumbens during open field
exposure causes inhibition of motor activity and rearing in untreated
animals (Van Ree and Wolternik, 1981; Svensson and Ahlenius 1982).
High doses of these compounds produce hypermotility. Inhibition of
motility is considered a presynaptic receptor phenomenon. The
nucleus accumbens systems appears to form part of the inhibitory
polysynaptic pathways involved in NA and DA neurotransmission. This
system may be traced from the ventral tegmental area over the nucleus
accumbens to the ventral pallidum.

MIF activates DA turnover (see Table 2), apparently because of
increased release of mediator from the presynaptic terminals. MIF's
modulating influence on motor activity may result from the presynaptic action of the catecholamines released onto the inhibitory
receptors and the nucleus accumbens. Rearing is inhibited as well as
motor signs in untreated non-emotional animals. Low doses of apomorphine, microinjected into the nucleus accumbens, produced the same
effect. MIF's inhibiting action on motor signs is less pronounced in
6-OHDA pretreated animals, possibly because of degeneration of presynaptic terminals; this is also the reason why the presynaptic
inhibitory action of the mesolimbic system on motor signs and exploration fails to occur. Motor activity is actually halved after
administering MIF to the group not treated with 6-OHDA. The fact
that MIF causes a rise in motor signs in the "emotional" rat-group
pre-treated with 6-OHDA might be due to the post-synaptic effect of
the DA released on DA receptors rendered supersensitive by the
denervation of presynaptic terminals. The onset of some features of
stereotyped DA-mediated behavior, such as grooming, gnawing and
scratching provides evidence of this.

A connection may be made between MAP's inhibitory action on
motor signs plus exploratory activity and its potentiating effect on

5-HT turnover (see Section 2.1.3), raising 5-HT and 5-HIAA concentrations.   Inhibition of motor activity in rats constitutes one of
5-HT's most clear-cut central effects (Jacobs et al., 1975).   The
appearance of shaking behavior, especially in 6-OHDA pretreated rats,
bears out the serotoninergic origin of the MAP effect.   The shaking
also reflects stimulation of central 5-HT receptors (Vetulani et al.,
1979).

3.2.2. Goal-directed Behavior Under Stress in Relation to Rats' Response Pattern

The influence of small peptides was assessed on the goal-
directed escape reaction under conditions of emotional tension on
untreated and 6-OHDA pretreated sub-groups, using Henderson's (1970)
species-adaptable test.   Details of this method are given under
Section 3.3 of the seventh article.

Success in solving the predictive problem of escaping from a
stressful situation by diving beneath the walls of a cylinder is
inversely proportional to the degree of affect displayed by the
animal.   Animals in an emotionally excited state make many attempts
at escape which fail to achieve the desired goal, such as jumping
upwards, attempting to climb up the walls of the cylinder, etc.
This means that it takes the animal longer to find (or not to find)
the right solution during the 120 sec duration of the test.   This
relationship is depicted in Table 5, where the animal sub-groups
are arranged in ascending order of emotional responsiveness.

Owing to its sedative effect which, in turn appears to be caused
by activation of 5-HT processes, a dose of 5 mg/kg MAP modulates the
level of affect displayed and reduces the number of failed escapes
from the stressful situation.   Except for the most reactive sub-group
of 6-OHDA pretreated rats, it potentiates goal-directed avoidance
behavior, requiring correctly predictive action in a stressful situation.

An i.p. dose of 5 mg/kg MIF did not exert a favorable influence
on the behavior of animals belonging to the emotional sub-group in
the above situation.   Latent avoidance period was reduced in
non-emotional animals, however.   A dose of 0.5 mg/kg TRH greatly
increased the display of affect under stress to the point of complete
behavioral inadequacy.

A clearer picture of the differences between MIF's and MAP's
effects on avoidance behavior, as dictated by an animal's response
patterns and the state of the CA brain system, may be gathered from
the dose-related curves presented in Figure 1.

When low doses of 0.1-1.0 mg/kg MIF are administered i.p. an
improvement is noted in the behavior of the most responsive group -

Table 5. Short Peptide Effects on Avoidance Behavior

Compound administered (mg/kg)	Animal group and response pattern		No. of failed escapes	Latent period of escape (sec) over 120 sec test period	Score for degree of affect displayed
Physiological saline n = 16	non-emotional	1	0	26.5±8.4	0
		2	8.7±1.8	43.6±5.9	7.6±2.0
	emotional	1	28.3±4.2	∞	9.4±3.7
		2	42.3±8.4	∞	18.2±4.8
MAP 5.0 n = 22	non-emotional	1	3.8±1.2**	11.4±3.8**	0
		2	5.0±1.4	16.3±4.4**	2.9±0.9*
	emotional	1	17.2±3.1*	66.2±5.2**	5.7±2.2*
		2	30.3±7.3	∞	15.5±4.8
MIF 5.0 n = 18	non-emotional	1	0	6.7±2.9**	2.1±0.4**
		2	8.6±3.2	22.7±4.9**	5.3±2.4**
	emotional	1	21.2±3.7	∞	12.4±3.8
		2	43.7±5.6	∞	13.1±7.2
TRH 0.5 n = 16	non-emotional	1	70.0±13.1**	∞	14.5±4.0**
	emotional	1	56.1±8.7**	∞	15.1±5.8

1 = untreated; 2 = pre-treated with 6-OHDA.
Differences significant between control and test animals.
Sub-groups: n = 8.
* p < 0.05; ** p < 0.01.
Compounds administered 10 min prior to placing in OF; no repeat tests.

No. of failed escapes   Latency of escape (sec) (100 = no escape)

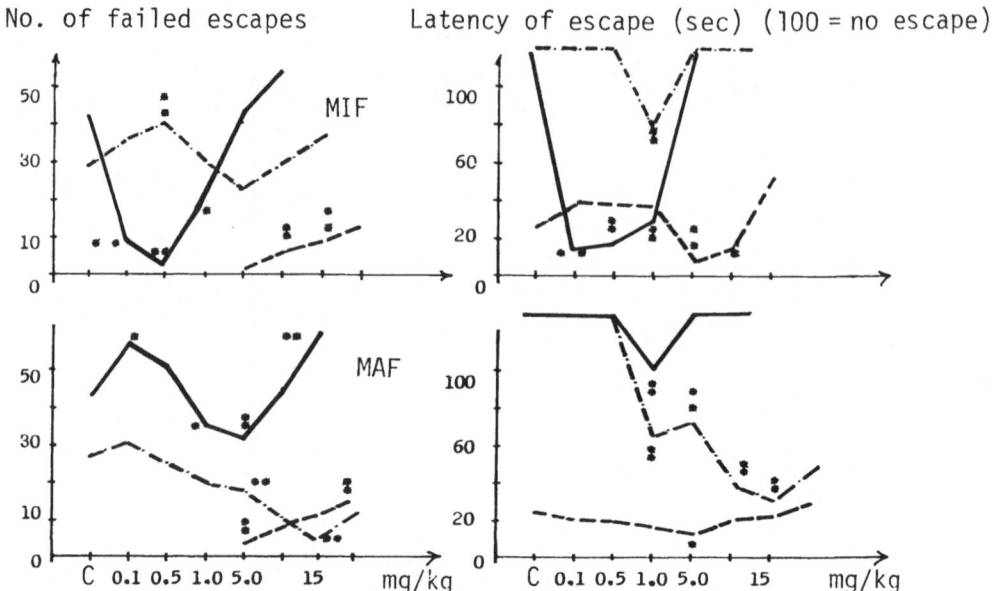

Fig. 1. The effect of increasing doses of short-chain peptides on
         avoidance behavior. (——) emotional rats (pretreated with
         6-OHDA); (-·-·) emotional intact; (---) non-emotional
         intact; C - saline; n = 6 animals per dose for each group.
         Significance of differences compared with control (saline).
         * p < 0.05; ** p < 0.01 (Mann-Whitney U-test). Compounds
         administered 10 min before testing procedure. No
         replication tests.

the emotional category of which had been pretreated with 6-OHDA
during the neonatal period. An improvement in behavior of animals
belonging to the emotional untreated sub-group was produced by a
narrow dose range of 0.5-1 mg/kg. On raising the dose to 1-5 mg/kg
emotional animals' avoidance behavior broke down again. A dose of
5-10 mg/kg MIF heightened affect and hampered achievement of avoidance in non-emotional untreated animals.

     Administering MAP produced a two-stage effect in the emotional,
6-OHDA treated group only. When given in a low dose range of 0.1-1
mg/kg, MAP reduces the number of failed attempts, and the animals
manage to escape; the degree of affect displayed by the animals under
stress is also decreased. As the dose is raised, however, responsiveness begins to increase again, and the animals no longer manage
to escape, while the emotional but untreated group show improved
behavior.

4. THE INFLUENCE OF SHORT PEPTIDES ON BEHAVIOR AND BM LEVEL DURING CHANGES IN FUNCTIONAL STATE OF MONOAMINERGIC SYSTEMS

4.1. THE EFFECT OF SHORT PEPTIDES ON BEHAVIOR AND BM LEVEL IN RESERPINE-PRETREATED OR HALOPERIDOL-TREATED MICE

A dose of 5 mg/kg reserpine (Rausedil, from Gedeon Richter) was given intraperitoneally to mice weighing 18-22 g, 24 hours prior to the experiment. A dose of 5 mg/kg of peptide hormone fragments was given intraperitoneally. The rectal temperature of the reserpine pretreated mice was taken again 30 min after administering the peptide. A dose of haloperidol 5 mg/kg (Gedeon Richter) was adminis- tered intraperitoneally, and 15 min later the percentage of animals within the group showing a cataleptic state was noted. The peptides were given 15 min before haloperidol. The percentage of animals showing signs of catalepsy was ascertained once every 20 min there- after. The animals were decapitated 30 min after peptide adminis- tration and the concentration of monoamines and their metabolites were measured spectrofluorimetrically by the method described in Section 2.1.2.

As shown in Table 6, reserpine reduces the concentration of brain monoamines and their metabolites, exhausting supplies of BM while also causing hypothermia. All the tripeptides under study had a dose-related influence on reserpine-induced hypothermia when administered intraperitoneally. The TRH, MIF, APG group of peptides, however, counters hypothermia; this effect lasts for 1-2 h in the case of TRH and APG and up to 4 h with MIF. The MAP, GLM group of peptides intensifies reserpine-induced hypothermia.

When BM concentrations were measured, TRH, MIF and APG were found to raise HVA concentration, thereby reversing reserpine's effect on this measurement. Similarly, TRH substantially increases DA concentration which had been reduced by reserpine. APG tends to raise DA concentration, a possible indication of these peptides' ability to release DA and reverse reserpine's inhibition of the DA system. Neither MAP nor GLM combat reserpine's effects where DA and HVA are concerned. None of the tripeptides studied reversed reser- pine's effect on 5-HT and 5-HIAA.

Haloperidol reduces DA concentration and raises HVA level, thereby raising DA turnover (see Table 6). It also produces cata- lepsy. Peptides exerted diverse effects on animals treated with haloperidol. The three tripeptides, TRH, MIF and APG, attenuated haloperidol's cataleptic action for about 2 h, whereas MAP and GLM produce the reverse effect, powerfully potentiating the action of haloperidol, especially during the first 30 min after administration (see Figure 2).

Following pretreatment with haloperidol, TRH, MIF and APG sub- stantially inhibit the rise in HVA concentration produced by this

Table 6. The Effect of Tripeptides on the Concentration of BM and
          Their Metabolites in the Brain of "Model" Mice

Animal group	Compound	BM concentration (M ± SEM) %				
		NA	DA	HVA	5-HT	5-HIAA
Untreated	Phys.saline	100±12	100±12	100±15	100±17	100±20
Reserpine pre-treated	Reserpine (5 mg/kg)	60±10*	40±4*	60±7*	50±5*	60±10*
	TRG	70±10*	60±5***	90±12**	55±10*	62±10*
	MIF	67±5*	43±7*	83±9**	55±15*	61±10*
	APG	60±13*	55±8*	88±10**	50±9*	60±12*
	MAF	60±7*	38±5*	55±5*	48±5*	58±10*
	GLM	63±7*	42±9*	58±6*	56±8*	68±9*
Haloperidol pre-treated	Haloperidol (5 mg/kg)	90±15	50±5*	500±10*	95±13	99±10
	TRG	95±10	55±6*	350±10***	95±10	104±12
	MIF	95±8	50±7*	400±5***	95±12	100±13
	APG	90±12	55±4*	450±5***	90±9	101±10
	MAF	101±5	53±7*	530±10***	90±10	97±9
	GLM	92±4	45±8*	535±8*	92±10	98±9

* $p \leqslant 0.05$ vs. control (phys. saline); ** $p \leqslant 0.05$ compared to the
effect of peptides administered intraperitoneally at a dose of
5 mg/kg.

drug, while DA and NA concentration remains unchanged. The two
tripeptides MAP and GLM, however, further raise HVA concentration,
thereby potentiating haloperidol's effects.

    It is therefore safe to say that whereas intraperitoneal
administration of the above peptides fails to produce noticeable
modifications in the concentration of brain monoamines and their
metabolites, as shown in Section 2.1.2, they do, nonetheless, exert
a clear-cut effect when BM metabolism has been altered by e.g.,
reserpine or haloperidol. Furthermore, the action of different
peptides follows different trends, depending on what influence they
may exert when administered intraventricularly (see Section 2.1.3).

4.2. MIF'S EFFECTS ON DEPRESSIVE BEHAVIOR INDUCED IN CATS BY
      RESERPINE OR HALOPERIDOL

    MIF counters the action of reserpine and haloperidol on altered
brain DA metabolism. It also combats hypothermia and signs of catalepsy. We assessed the emotional behavior displayed by cats using
the method explained in the first article, to gain a fuller picture

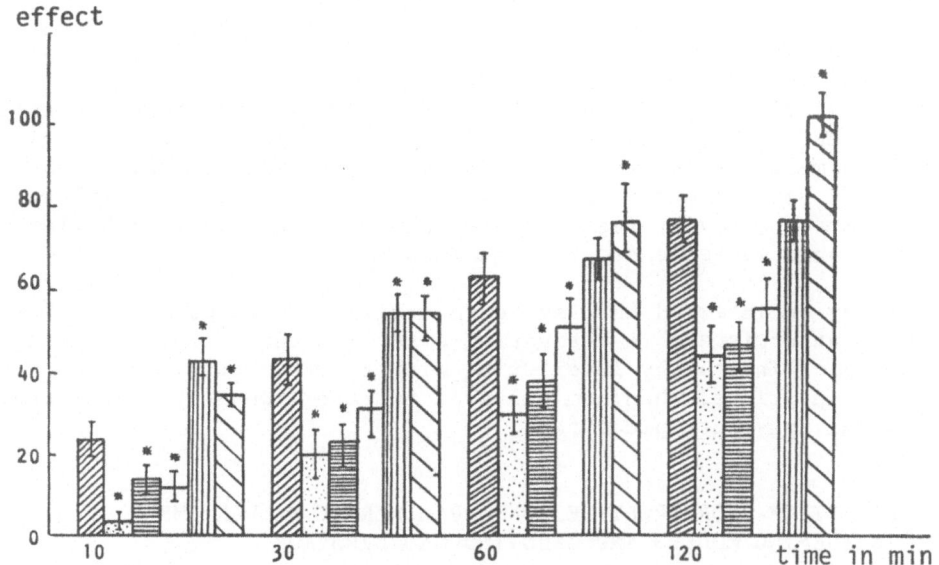

Fig. 2.   Effects of various peptides (5 mg/kg i.p.) administered to
mice, on the cataleptic effect of 5 mg/kg i.p. haloperidol.
Key to columns: ▨ physiological saline (control); ☐ APG;
▩ TRH; ▥ MHF; ▤ MIF; ▨ GLM.   * p ⩽ 0.05.

of this peptide's effects on behavior which has previously been
altered by administering reserpine and haloperidol. Depressed
behavior was reproduced by treatment with 0.25 mg/kg i.p. reserpine
(Rausedil, from Gedeon-Richter) or 1 mg/kg of haloperidol administered intramuscularly twice a day for 2 days.

The following signs indicated how far the reserpine-induced
depression caused the animal's emotional response range to contract:
positive emotions were suppressed, affective displays of aggression
and fear were reduced, together with purposeful activity such as
exploration, food-seeking and hunting; hypothermia set in,
accompanied by the adoption of a "torpid" posture, with the paws
tucked under, the head held very low, the tail flicked to one side
and the ears laid back and kept still. Rectal temperature also fell,
but not so reliably as in the case of rodents.

A single 20 mg/kg i.p. dose of MIF failed to influence the state
of depression, while substantially restoring rectal temperature,
which had been lowered by giving reserpine. When 20 mg/kg i.p. doses
of MIF were repeated twice a day for 3-4 days, actively defensive
behavior was stimulated in response to provocation. Animals
occasionally emerged from their torpor. Goal-directed activity and
reactions prompted by positive emotions was not reinstated, however.

The following behavior was typical of depression induced by
haloperidol: complete suppression of positive emotions, uninterrupted
vocalization, a reduction in motor activity and signs of catalepsy,
whereby an awkward posture may be maintained for up to
5 min. A single i.p. dose of MIF promoted the latter effect on this
model, as shown in Figure 3. A 20 mg/kg dose repeated twice daily
produced a more clear-cut effect. Catalepsy was either reduced or
eliminated, goal-directed exploratory behavior was restored, alertness and motor activity were heightened and the emotional/behavioral
response range was extended in some measure.

It may be inferred that systemic administration of MIF produces
a slight but unmistakable activating effect on the model of depressive behavior induced by interfering with BM metabolism, and hence
that short peptides, when administered intraperitoneally, exert their
own central, psychotropic effect.

4.3. EFFECT OF SHORT PEPTIDES ON DOPAMINERGIC MECHANISMS
      IN TURNING MODELS AND DRUG-INDUCED BEHAVIOR

Short peptides' action on pre- and post-synaptic DA receptors
of the mesolimbic and nigro-striatal systems was studied using
Ungerstedt's (1971) turning model, which is prepared by unilateral
destruction of striatal DA terminals by 6-OHDA. The eighth article,
Section 4.2 contains a detailed description of the method used. The
turning model provides a behavioral yardstick for measuring hypersensitivity of the nigro-striatal DA systems. Peptides' effects can
easily be tested by the degree and direction of circling when alterations have been produced in the functional state of the presynaptic
or post-synaptic DA receptors, by either 6-OHDA-induced damage or
terminal breakdown-induced supersensitivity, respectively. The peptides' effects on the pre- and post-synaptic components of the mesolimbic and nigro-striatal DA systems may be ascertained by the change
in drug induced behavior, taking stereotypy as well as circling into
account, as produced by amphetamine or apomorphine. There is much
evidence to show the DA-ergic inputs into the n. accumbens are
involved in exploratory activity, locomotor activity and sniffing.
DA-ergic inputs into the caudate-putamen are related to stereotyped
behavior (Kelly et al., 1975; Makanjuola and Ashcroft, 1982).

Ipsilateral circling is somewhat reinforced in comparison with
control on the turning model by administering 10 mg/kg MIF i.p. (see
Table 7). These changes are not statistically significant, however.
A dose-related increase in ipsilateral circling etc. is produced by
1 mg/kg of MIF (or TRH) following pre-treatment with amphetamine.
This, too, is illustrated in Table 7.

Amphetamine's potentiating action on circling is mediated by
presynaptic DA release. The fact that the two tripeptides, MIF and

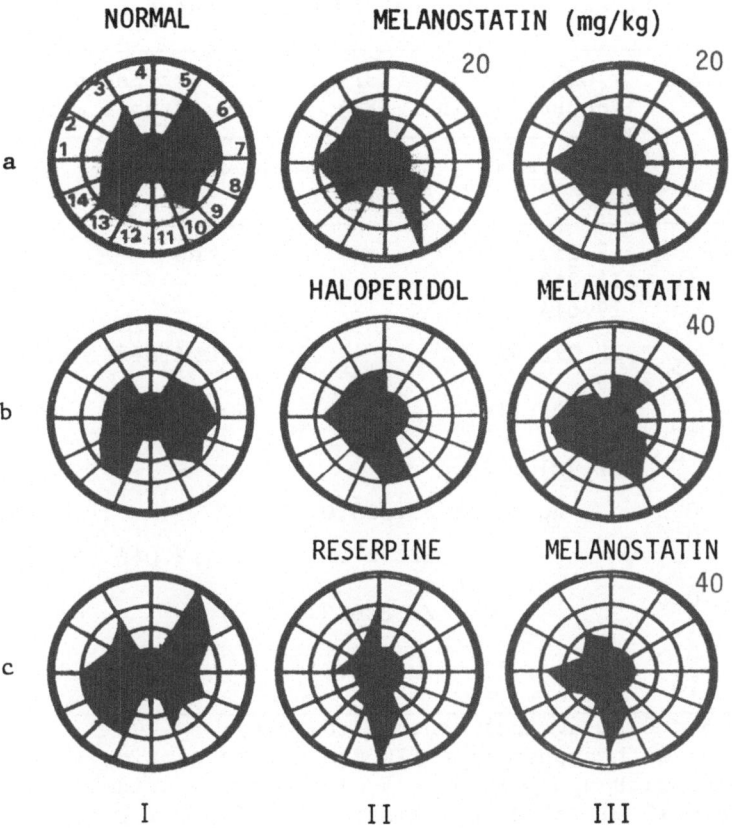

Fig. 3.   Changes produced in cats' emotional response range by
          melanostatin (MIF). (a), (b), (c) animals' emotional
          response range. I: before exposure; A: (II,III) following
          pre-treatment with 20 mg/kg – single dose and repeated doses
          respectively. (b) and (c): (II) following pre-treatment
          with haloperidol and reserpine respectively; III: following
          a once-repeated dose of MIF at a total dose of 40 mg/kg.
          1-14 – constituents of the range of emotional response: (1)
          rage or aggression; (2) fear; (3) anxiety; (4) negativity;
          (5) curiosity (exploration); (6) pleasure or contentment;
          (7) playfulness; (8) hunting; (9) friendliness; (10)
          conflict; (13) motor activity; (14) alertness.

TRH, in turn potentiate this amphetamine effect, indicates heightened
mesolimbic DA release. The number of rotations per minute depends on
the activity of the nucleus accumbens, according to the model
explaining the interaction between the nucleus accumbens and the
caudate nuclei involved in circling behavior (Moore and Kelly, 1978).
TRH was shown to release ^{3}H-DA from sections of the nucleus accumbens

Table 7. MIF's Effect on Circling and Drug-induced Behavior

Compound and dose (mg/kg)		Circling behavior (counts/5 min ± SEM)		Stereotyped behavior (how often/5 min ± SEM)	
		Ipsilateral	Contra-lateral	Head weaving	Rearing
Control (saline)		7.2±3.1	no	no	no
MIF	10	11.3±2.2	–	–	2.3±0.8
Amphetamine	1	19.6±4.8	–	–	6.7±1.9
Amphetamine + MIF	1 5	28.3±3.5*	–	12.0±4.8**	26.6±7.9**
Amphetamine + MIF	1 10	42.4±6.1**	–	7.4±2.0**	15.3±5.0**
Amphetamine + MIF	1 15	71.4±10.3**	–	–	6.2±2.8
Apomorphine	0.5	no	29.3±5.0	–	38.2±10.1
Apomorphine + MIF	0.5 5	–	22.9±7.1	2.4±2.0**	44.8±11.6
Apomorphine + MIF	0.5 10	–	23.8±6.5	–	62.5±11.9*
Apomorphine + MIF	0.5 15	2.6±0.9**	19.1±2.8**	–	88.7±16.4**

* p < 0.05, ** p < 0.01 against control (U-test).
i.p. pretreatment with amphetamine (20 min before peptide).
i.p. pretreatment with apomorphine (15 min before peptide).
Behavior assessed 10 min after administration of MIF i.p.

in vitro, but not of the caudate (Kerwin and Pycock, 1979). There is
no such direct evidence in the case of MIF. The effect produced on
behavior by microinjecting TRH into the region of the nucleus
accumbens resembles that of microinjected DA; it disappears once the
presynaptic DA terminals in this region have been destroyed. This is
an additional indication of release of DA from nucleus accumbens
terminals in vivo (Heal and Green, 1979).

    The above doses of MIF and TRH do not in themselves produce
amphetamine type stereotypy. Cox et al. (1976) obtained similar
findings from their investigations of MIF. MIF does possess a
diphasic action on amphetamine-induced stereotypy, reinforcing this
effect at a 5-10 mg/kg dose range, while reducing rearing and sniffing at higher than 15 mg/kg, accompanied by an increase in the
number of rotations made. Doses of 5-10 mg/kg MIF have no effect on
apomorphine-induced contralateral circling; this treatment does, however, reinforce dopamine-dependent stereotyped behavior, which is
mainly connected with the nigro-striatal system. A larger dose of
15 mg/kg MIF partially reverses apomorphine-induced circling; ipsi-

lateral circling gradually begins to appear as the overall number of
contralateral rotations decreases.

While an i.p. dose of 0.5 mg/kg apomorphine produced contralateral circling at the rate of 48.4 ± 10 rotations per 15 min,
0.5 mg/kg TRH i.p. almost completely reversed the direction of the
circling. Animals made 28.8 ± 4.8 ipsilateral as against 6.3 ± 3.5
contralateral circles in the course of 15 min (p < 0.01).

Since stereotypy is mainly connected with the striatal DA
system, while locomotor activity (in this case, circling) is related
to the mesolimbic DA system of the nucleus accumbens septi, it may be
inferred that both MIF and TRH operate on the latter. TRH microinjected into the striatum does not reproduce DA-induced effects, and
fails to produce circling behavior. This conforms with immunohisto-
fluorescent data on the presence of numerous TRH-containing fibers in
the central segment of the nucleus accumbens but not in the striatum
(Hökfelt et al., 1975). It is in this area of the nucleus accumbens
that maximal binding of TRH occurs. MIF's effect may well be
achieved in a similar fashion.

An i.p. dose of 10 mg/kg MAP produces neither circling nor
stereotyped behavior on the circling model (see Table 8). Following
pretreatment with amphetamine, however, doses of 1-15 mg/kg of this
peptide produce a dose-related increase in ipsilateral circling and
DA-induced stereotypy, complete with rearing, sniffing, etc. MAP
reduced contralateral circling and stereotypy produced by apomorphine. MAP, however, brought on serotoninergic-like stereotypy,
with head-weaving, etc. These findings illustrate MAP's ability to
potentiate amphetamine-induced release of DA from the terminals of
the mesolimbic DA system. Telegdy (1980) found that the C-terminal
tripeptide, gastrin, raises DA concentration in the septum but lowers
it in the striatum. The concentration of 5-HT was observed to rise
in the hypothalamus but decline in the striatum.

The effect of tripeptides on the turning model leads us to
believe that MIF, TRH and MAP do not activate post-synaptic DA receptors directly, since they do not induce contralateral circling in
themselves, even when 6-OHDA-induced damage to the striatal presynaptic terminal has made these receptors supersensitive. The
modulating effect of peptides is shown in their facilitation of DA
release from the mesolimbic DA-ergic system, once this process has
been initiated - by amphetamine, in this case. The peptides did not
facilitate the interaction between DA and the post-synaptic receptors
of the striatum, for apomorphine-induced circling was not reinforced,
but actually inhibited. MIF's reinforcement of DA-related stereotypy
and initiation of 5-HT-related stereotypy follow the same lines as
the effect produced by these peptides on brain monoamine metabolism
(see Section 2).

Table 8. MAP's Effect on Circling and Drug-induced Behavior

Compound and dose (mg/kg)		Circling behavior (counts/5 min ± SEM)		Stereotyped behavior (how often/5 min ± SEM)	
		Ipsilateral	Contra-lateral	Head weaving	Rearing
Saline solution		7.2±3.1	no	no	no
MAP	10	8.7±3.4	–	–	–
Amphetamine	1	19.6±4.8	–	–	6.7±1.9
Amphetamine + MAP	1 5	26.4±3.3*	–	16.7±5.4**	2.1±0.6**
Amphetamine + MAP	1 10	34.0±5.0**	–	6.2±3.1**	17.7±6.0*
Amphetamine + MAP	1 15	54.3±8.2**	–	–	28.1±7.9**
Apomorphine	0.5	–	29.3±5.0	–	38.2±10.1
Apomorphine + MAP	0.5 5	–	18.5±4.2*	19.2±7.0**	4.2±1.7**
Apomorphine + MAP	0.5 10	–	21.3±3.7*	32.6±5.8**	–
Apomorphine + MAP	0.5 15	–	13.2±4.8**	11.0±3.3**	17.5±6.1**

* $p < 0.05$, ** $p < 0.01$ against control (U-test).
i.p. treatment with amphetamine (20 min before peptide).
i.p. treatment with apomorphine (15 min before peptide).
Behavior assessed 10 min after administration of MAP (i.p.)

4.4.   THE EFFECT OF SHORT PEPTIDES ON BM FOLLOWING INHIBITION OF MA-SYNTHESIZING ENZYMES

Pharmacological inhibition of enzymes involved in MA bio-synthesis leads to functional changes in BM systems. It might be expected that our conception of the primarily modulating action of peptide regulators might emerge more clearly or some new properties of peptide action on BM turnover might become apparent under these circumstances. Results are given in Table 9.

Pretreatment with α-methyl-para-tyrosine (α-MT) 3 h beforehand reduced the concentration of NA and DA in rat brain. Following administration of α-MT, which inhibits TH and blocks formation of DA from tyrosine, TRH and MIF appear to accelerate DA turnover, judging from the increase in HVA concentration. NA and DA levels fall even more than when α-MT is given alone, an indication that TRH and MIF may be capable of promoting the release of stored catecholamines, even though activation of turnover because of de novo synthesis cannot be excluded. Komissarof et al. (1981) had shown that TRH

Table 9. The Effects of TRH, MIF and MAP, Administered After Pre-treatment with α-MT, DDC and PCPA i.v., on BM Concentrations in Rats

Drugs	Dose	BM concentration (%) (M ± SEM)				
		NA	DA	HVA	5-HT	5-HIAA
Saline (control)		100±9	100±8	100±11	100±10	100±13
α-MT-saline	50mg/kg	68±8*	75±5*	90±8	105±6	111±12
DDC + saline	400mg/kg	60±5*	120±8*	120±7*	113±15	92±17
PCPA + saline	300mg/kg	100±5*	80±10	113±7	52±4*	87±5*
α-MT+TRH	50mg/kg + 20μg	58±4*/**	56±6*/**	130±3*/**	122±9*/**	130±5*/**
α-MT+MIF	50mg/kg + 100μg	58±3*/**	60±4*/**	144±12*/**	89±7	115±10
α-MT+MAP	50mg/kg + 200μg	66±5*	70±10*	80±6*	131±7*/**	136±10*/**
DDC+TRH	400μg/kg + 20μg	73±10*/**	132±13*	136±6*/**	139±10*/**	125±5*/**
DDC+MIF	400μg/kg + 100μg	46±9*	140±10*	131±7*/**	95±6	110±11
DDC+MAP	400μg/kg + 200μg	69±4*/**	91±7*/**	87±6*/**	148±17*/**	128±8*/**
PCPA+TRH	300mg/kg + 20μg	90±5	93±12	140±9*/**	56±7*	126±4*/**
PCPA+MIF	300mg/kg + 100μg	86±7	89±4	127±6*	54±3*	86±6*
PCPA+MAP	300mg/kg + 200μg	93±10	111±6	89±8	66±6*/**	105±6**

* $p < 0.05$ vs. control; ** $p < 0.05$ vs. effects of α-MT, DDC and PCPA.
α-MT administered intraperitoneally 3 h before peptides.
DDC administered intraventricularly 30 min before peptides.
PCPA administered intraperitoneally 24 h before peptides.

potentiates ^{14}C-NA release from electrolytically destroyed slices of
rat brain hemispheres. TRH increased 5-HT and 5-HIAA concentrations
following pre-treatment with α-MT, as in the control. MIF does not
have any noticeable effect on 5-HT turnover.

MAP likewise produced no noticeable changes in catecholamine
concentration following pretreatment with α-MT compared with the
effect produced by α-MT given alone, although it did reduce DA and
HVA concentration in the untreated control (see Table 2). It appears
that MAP is less able to potentiate CA release than either TRH or
MIF. MAP's characteristic effect on 5-HT turnover following pre-
treatment with α-MT remains unchanged.

Diethyldithiocarbamate (DDC), a dopamine β-hydroxylase (DBH)
inhibitor, reduces the concentration of brain NA. DA and HVA concentrations rise, resulting from increased DA turnover when NA
synthesis is decreased. When followed by TRH, both NA and DA concentrations increase, indicating that this peptide has the capacity to
promote catecholamine synthesis at various stages of CA biosynthesis.
Transformation of NA into DA would also appear to be enhanced, since
HVA concentration rises considerably. But the question of whether
TRH is able to counter DDC's effect on TH remains unanswered. TRH's
potentiating effect on 5-HT turnover does not alter under these
circumstances. When DDC and TRH are used in combination, TRH's
potentiating effect on motor activity is enhanced; head-twitching
and wet-dog shakes both increase. This observation correlates with
changes in BM.

After pretreatment with DDC, MIF also raises DA and HVA concentration, compared with the action of DDC alone. MIF only produces a
slight effect on behavior, however, when administered after DDC.

MAP cancels the increase in DA and HVA produced by pretreatment
with DDC, while raising NA concentration. The behavioral responses
of MAP-treated animals are not changed by pretreatment with DDC.
There is evidence to show that MAP increases NA concentration in
different regions of the brain in untreated rats (Telegdy, 1980).
NA uptake is also potentiated by the peptide (see Section 2).
Further study will show whether MAP potentiates NA synthesis when
DBH has been inhibited by pretreatment with DDC.

Parachlorophenylalanine (PCPA) – a non-specific tryptophan
hydroxylase (TRH) inhibitor, reduces concentrations of 5-HT and
5-HIAA within 24 h.

Following pretreatment with this compound TRH raises HVA and
5-HIAA concentration, but without reducing 5-HT level. TRH apparently potentiates 5-HT synthesis, as well as accelerating turnover.
TRH's behavioral effects change after pretreatment with PCPA: motor
activity increases, while head twitching and grooming are reduced –

proof that the functional activity of the serotoninergic system must be maintained in order for TRH to produce its full range of behavioral effects.

PCPA's inhibitory effect on 5-HT and 5-HIAA production was not antagonized by MIF. HVA concentration rose, but without an accompanying increase in DA level. Pretreatment with PCPA produced some changes in MIF's effects on behavior - a rise in motor activity and a decrease in grooming.

MAP increased 5-HT and 5-HIAA concentrations following pretreatment with PCPA, indicating that this peptide may activate 5-HT synthesis and release. This was accompanied by a drop in HVA concentration.

TRH and MIF would thus appear to affect the release and uptake of brain monoamines rather than BM synthesis, whereas MAP influences NA and 5-HT synthesis.

4.5. THE EFFECTS OF TRH AND MAP ON BM LEVEL FOLLOWING DA
      AND 5-HT RECEPTOR BLOCKADE

The effects of administering peptides intracerebroventricularly (i.c.v.) were measured after changes in BM concentration following injection by this route of the DA receptor blocker, haloperidol and of methysergide - a 5-HT receptor blocker. This investigation was performed on rats, using the methods described in Section 2. Results are presented in Table 10.

Table 10. The Effect of TRH and MAP on BM Concentration Following
           Pre-treatment with Haloperidol or Methysergide

Drugs	BA concentration (%) (M ± SEM)				
	NA	DA	HVA	5-HT	5-HIAA
Saline (control)	100±8	100±8	100±11	100±10	100±8
Haloperidol + saline	100±6	68±7*	163±10*	100±4	96±3
Methysergide + saline	100±3	96±11	96±5	93±10	180±12*
Haloperidol + TRH	120±4*,**	80±7*,**	106±8,**	114±10	128±10*,**
Haloperidol + MAP	94±6	59±6*	130±5*,**	92±6	117±3*,**
Methysergide + TRH	102±3	112±5	125±4*,**	128±6*,**	130±15*,**
Methysergide + MAP	90±7	93±6	70±6*	92±5	128±6*,**

* p < 0.05 vs. control, ** p < 0.05 compared with effect of haloperidol or methysergide.
Haloperidol: 10 μg i.c.v. 30 min before TRH (20 μg i.c.v.).
Methysergide: 50 μg i.c.v. 30 min before MAP (200 μg i.c.v.).

Haloperidol showed the same effects on BM whether administered
i.c.v. or i.p., i.e. DA level was reduced while that of HVA rose.
When administered i.c.v. under the same circumstances, TRH exerted a
modulating influence: HVA concentration dropped to control level,
while DA level rose, in contrast to haloperidol's effect; TRH also
promoted an increase in NA level to above that of the control. MAP,
when given i.c.v. following treatment with haloperidol, somewhat
reduced HVA concentration, while DA level did not merely fail to
increase, but showed a tendency to fall. The potentiation of 5-HT
metabolism produced by both peptides after prior treatment with
haloperidol remained unchanged, judging by the rise in 5-HIAA concentration.

Administration of i.c.v. methysergide potentiated 5-HT metabolism, thereby producing a rise in 5-HIAA. TRH administered in this
way exerted its characteristic effect on DA metabolism of increasing
DA and HVA, while modulating methysergide's effect on 5-HT metabolism: the rise in 5-HIAA level is checked, but with an accompanying
rise in 5-HT concentration. MAP administered i.c.v., following
pre-treatment with methysergide, showed its usual effect of increasing HVA concentration, while reducing 5-HIAA level, which had been
raised by methysergide, without any accompanying rise in 5-HT concentration.

The above findings demonstrate that peptides given following
pretreatment with haloperidol and methysergide tend to counter
changes in DA and 5-HT metabolism produced by treatment with blockers
of their corresponding receptors. Further study is required as to
how peptides act in these circumstances. It may nevertheless be
concluded, by surveying the above effects of tripeptides as a whole,
that such C-terminal fragments' action on monoamine metabolism is
mainly at the presynaptic level.

5. INVESTIGATION OF THE EFFECTS OF SHORT-CHAIN PEPTIDES ON MA TRANSPORT, MEMBRANE RECEPTORS AND THE LIPID PHASE OF BIOLOGICAL MEMBRANES

At present, opinions vary on the biological targets of each
different short peptide at cellular level. Although specific receptors have been identified for some naturally occurring peptides such
as enkephalins and endorphins, targets for other biologically active
substances belonging to this group have not yet been identified.
There is no evidence that specific receptors exist for each of the
numerous fragments of peptide bioregulators which are able to perform
a given pharmacological function. Interaction between short peptides
and monoaminergic mechanisms need not necessarily be bound up with
the peptides' effects on pre- or post-synaptic receptors. Peptides
may, however, be acting as modulators of transport and synthesis of
transmitters, as a result of their effect on biological membranes.

The neurochemical mechanisms of neuropeptide action should not merely
be assessed by their effect on neurotransmitter metabolism without
taking into consideration changes which may be produced by peptides
on the state of biological membranes. More attention should be paid
than formerly to studying changes in membrane lipids and the part
they play in the dynamics of ligand-receptor reaction. We have
carried out research on some aspects of this question.

5.1. THE INFLUENCE OF SHORT PEPTIDES ON IMIPRAMINE AND
BENZODIAZEPINE RECEPTORS, AND β-ADRENORECEPTORS OF
MOUSE BRAIN

The psychotropic action of a number of the short peptides under
study, including TRH, MIF, MAP and tuftsin may be classified as
tranquilizing, antidepressant or stimulant. Research into membrane
benzodiazepine, imipramine, noradrenaline and other receptors has
scored many major successes in recent years. It has proved possible
to demonstrate the affinity of certain compounds under study for
these membrane receptors in vitro by means of radioligand receptor
analysis. No equally clear data of this nature have been obtained
for short peptides. We set out to ascertain whether a number of
short peptides which produce a clearly-defined psychotropic effect on
emotional/behavioral response in such animals as cats and rats have
the capacity for binding to benzodiazepine, imipramine and β-adrenoreceptors of the brain membranes.

Male CBWA tetrahybrid mice were used in the work. Straight
after decapitation the mouse brain was homogenized in a mixture of
0.32 M sucrose 50 Mm tris-HCl and 1 mM EDTA at a temperature of
0°-4°C, using a Teflon homogenizer. An unpurified synaptosomal
fraction (P_2) was separated out using differential centrifugation,
and the material re-suspended in 50 mM tris - HCl (10 ml per 1 g of
original tissue), frozen at a temperature of $-20°C$ and kept for no
longer than one week. The specific binding of ^3H-dihydroalprenolol,
^3H-imipramine and ^3H-diazepam were measured by the method of Raisman
et al. (1980) and Ticku et al. (1981). The volume of the incubation
mixture was 0.5 ml; 0.25-0.4 mg membrane protein was added for each
test. Protein concentration was measured by the Lowry et al. (1951)
method. The radioactivity of the filters was measured using an
SL-4000 liquid scintillation counter with Bray's solution. The
peptide solutions were prepared 2-3 hours before the experiments.
They were not reused. Results of this work are given in Table 11.

As will be gathered from the findings shown in the above table,
none of the peptides studied affects specific binding of the ligands
when used over a wide range of concentrations. These peptides thus
resemble neither imipramine, benzodiazepine nor β-adrenoreceptors in
this respect. It should be noted, however, that this inference only
holds for a given set of experimental conditions. Binding of ^3H-

Table 11. The Effect of Short Peptides on Some Mouse Brain
           Receptors

Peptide	Concentration (M)	B_i/B_{max} (%) (M ± SEM)		
		^{3}H–DHA	^{3}H–IMI	^{3}H–DIAZ
Tuftsin	10^{-9}		105±9	100±5
	10^{-6}		100±10	88±10
	10^{-3}	95±9	98±6	103±8
MIF	10^{-9}		100±11	98±7
	10^{-6}		90±10	100±11
	10^{-3}	104±8	106±5	110±10
MAP	10^{-9}		102±9	106±5
	10^{-6}		103±7	100±9
	10^{-3}	100±2	90±9	94±8
TRH	10^{-9}		100±10	98±8
	10^{-6}		97±11	100±9
	10^{-3}	90±12	100±5	110±10

M ± m – values represent mean (± SD) of 2–3 independent experiments.

diazepam and ^{3}H–imipramine are thus usually investigated at a temperature of 0°C and of ^{3}H–dihydroalprenolol at 25°C in accordance with accepted practice. It may be that at physiological temperatures, when the phase state of membrane components is altered, the effect of peptides on these receptors will be different.

Thus it may be seen that the mechanisms of peptides' tranquilizing or antidepressant effect on behavior differ from those of benzodiazepine tranquilizers and tricyclic antidepressants. As to their effects on interaction with monoamine, these were not potentiated by acting directly on β-adreno receptors – at least as far as NA is concerned.

5.2. INTERACTION BETWEEN SHORT PEPTIDES AND MEMBRANE LIPIDS

The effect of peptides on ligand–receptor interaction may result from their potentiating action on the biological membrane. In their role as membranotropic compounds, short peptides may influence either protein or lipid components of biological membranes. In recent years a great deal of data have emerged on lipid involvement in membrane processes (Loh and Law, 1980). Short peptides, being fragments of protein molecules, could well have some affinity for the lipid phase of a membrane, but since the total charge of most of them at physiological pH values does not equal 0, it is unlikely that they would penetrate into the hydrophobic area of the bilayer. The polar heads of lipids could well be the site where they are positioned on the membrane, if this were the case.

A close relationship between the functioning of the glycoprotein receptor for TSH, LH and the state of membrane phospholipids has been shown (Aloj et al., 1977). Phospholipase C, splitting off the polar head of membrane lipids, reduces ^{3}H-TSH and ^{3}H-TRH binding, due to a reduction in the affinity of binding sites, without any decrease in their number (Hayes et al., 1971; Barden et al., 1973). Stereo-specific binding between opiate ligands and brain membranes is reduced by prior treatment of membranes with phospholipase A_2 (Abood et al., 1976). The phagocytosis-stimulating peptide, tuftsin, inter-acts initially with the negatively-charged sialic acid groupings at the leucocyte membrane (Constantopoulos and Najjar, 1973).

Lipids' contribution towards receptor mechanisms may well lie in the provision of recognition sites for ligands, acting either alone or in conjunction with membrane protein. The lipids surrounding the receptor molecule and dispersed throughout the membrane determine the receptor's three-dimensional structure. Any change in the state of the membrane lipids could alter the conformation of the receptor protein, thereby changing the affinity of ligand binding. The physical state of the lipid sphere or domain surrounding the receptor determines the receptor's mobility along the membrane; this is reflected in the integration (interaction) between ligand-receptor complex and effector system (the adenylate cyclase system or ionic channel).

The state of the membrane's lipid components is also crucial to the ligand/receptor interaction at CA-ergic synaptic processes (Loh and Law, 1980). Both the number of specific β-adrenergic binding sites demonstrated by the radioligand (^{3}H-DHA) and the coupling of the agonist-adrenergic receptor complex to adenylcyclase are substantially modified in accordance with the state of the mem-brane lipids (Limbid and Lefkowitz, 1976; Hirata et al., 1979; Berrige, 1980). A reduction in the number of ^{3}H-DHA binding sites is observed after treatment of the erythrocyte membrane with phos-pholipase C, thus splitting off the polar heads of the lipids (Limbid and Lefkowitz, 1976). This indicates possible involvement of phos-pholipid groupings for the recognition of β-adrenergic ligands.

These factors led to the assumption that interaction with mem-brane lipids could be one of the factors modulating short peptides' action on MA brain processes (Waldman, 1982). Peptide binding con-stants and their influence on the stability of the surface charge of the membranes were also ascertained in order to test this hypothesis on phospholipid-cholesterol membrane models or liposomes (see Table 12).

Model phospholipid membrane vesicles (liposomes) were given a rapid injection of an ethanol solution of total hen's egg phospho-lipid fraction in a buffer solution of tris-HCl, pH 7.4 (Batzri et al., 1973). The liposomes obtained had a single bilayer membrane.

Table 12. Outline of Interaction Between Small Peptides and Model
           Phospholipid Membranes

Peptide	K_b	N	$K_b N$	f_n, %
MIF	0.08	34.3	2.7	0.3
MAP	0.23	33.9	7.8	0.3
Tuftsin	0.05	28.7	1.4	0.3

K_b – binding constant, μM^{-1}; N – number of binding sites;
f_n(%) – change of the electric charge of membrane surface.

The compound 1-anilinonaphthalene-8-sulfonate (1,8-ANS) served as
probe for the polar heads of the lipids. Its fluorescence was
excited at 360 nm with emission at 480 nm. Under these circumstances
virtually all fluorescence recorded was produced by luminescence
connected with the membrane probe. The experiments were performed
using standard concentrations of membranes (0.2 mg/kg) and probe
(10 μm). Aqueous solutions of peptides were added to the suspension
by microsyringe. The Opton spectrofluorometer was used to record
fluorescence. Results are given in Table 12.

All the peptides studied were observed to interact with the
lipid component of the membrane. MAP showed the greatest and tuftsin
the least affinity. The peptides also acted identically on the
stability of the surface of model membranes, without changing
1,8-ANS's affinity for the membrane. It may thus be inferred that on
interacting with model membrane, short peptides are altering the
structure of the polar heads of the phospholipid bilayer.

5.3. THE EFFECT OF SHORT PEPTIDES ON ACCUMULATION OF MONOAMINES BY RAT BRAIN SYNAPTOSOMES

The action of Na/K-ATP-ase, which is involved with the process
of neurotransmitter transport, is known to be modulated by changes in
the state of membrane lipids (Barnett and Palozzotto, 1975). The
effect of various concentrations of peptides on synaptosomal uptake
of ^{3}H-NA, ^{3}H-DA and ^{3}H-5-HT was assessed in this connection – see
Table 13.

We performed the experiments on crude synaptosomal fraction of
the brain of white cross-bred male rats weighing 180–200 g. In
standard experiments 50 μl synaptosomal suspension (an average of 1
mg protein per 1 ml) was incubated in a mixture containing 100 mM
NaCl, 6 mM KCl, 2 mM $CaCl_2$, 1.14 mM $MgCl_2$, 5 mM Na_2HPO_4, 10 mM
glucose, 100 mM sucrose, 0.125 mM pargyline, 0.54 mM ethylenediamine-
tetraacetate, 1.14 mM ascorbic acid, 30 mM tris-HCl-buffer, pH 7.4

Table 13.   The Effect of Peptides on Neurotransmitter Accumulation
            (in crude synaptosomal fraction of rat brain).

Peptide	Concentration (M)	Accumulation of neurotransmitters (% vs. control)		
		^{3}H–NA	^{3}H–DA	^{3}H–5–HT
Control		100±10	100±10	100±9
	$5 \cdot 10^{-7}$	96±10	94±10	57±6
MIF	$5 \cdot 10^{-6}$	105±11*	96±10	54±6*
	$5 \cdot 10^{-5}$	101±11*	100±10*	27±3*
	$5 \cdot 10^{-7}$	103±11	97±10	109±11
MAP	$5 \cdot 10^{-6}$	140±14*	95±10	118±12
	$5 \cdot 10^{-5}$	147±14	70±8*	60±7*
	$5 \cdot 10^{-6}$	110±11	110±11	116±12
Tuftsin	$5 \cdot 10^{-5}$	63±7*	117±12	74±8*
	$5 \cdot 10^{-4}$	59±6*	131±13*	

with corresponding concentrations of labelled transmitters and peptides. Incubation took place for 20 min at a temperature of 37°C
with a constant mixing.   Snyder and Coyle's (1969) method of separating the synaptosomes from the surrounding incubation medium and
recording the amount of transmitter which has been bound was adopted.
Protein was measured by the method of Lowry et al. (1951). The
results were evaluated statistically and means obtained ($p = 0.05$).
Results are shown in Table 13.

     MIF failed to exert any influence on NA and DA uptake in crude
synaptosomal fraction at concentrations of 5×10^{-7} and 5×10^{-5} but
did not substantially reduce 5-HT accumulation. This correlates in
some measure with MIF's antireserpine action in mice, as well as
the rise in tremor produced by intraventricular administration,
associated with activation of serotoninergic receptors. MAP, the
C-terminal fragment of gastrin, only reduced DA and 5-HT uptake at
a concentration of 5×10^{-5}M. NA accumulation, however, increased
substantially.

     Tuftsin at a concentration of 5×10^{-5} – 5×10^{-4} inhibits NA and
5-HT uptake while raising DA accumulation. These findings show the
complex interaction pattern between a number of short peptides and
transmitter transport mechanisms.

     It should be emphasized that the method used did not allow for
strict differentiation between the peptides' influence on reuptake
and accumulation of neurotransmitter; it only illustrates an overall
process. The fact that the different peptides studied affect the way
in which different changes in monoamine accumulation occur could well

be bound up with their varying effects on the secretion and reuptake
of transmitters.

The effect produced by these peptides on synaptosomal neurotransmitter transport may be partially caused by their interaction
with the lipid phase of membranes. This would indicate the underlying possibility of short peptides achieving their psychotropic
effect on MA metabolism through their influence on the lipid phase
of biological membranes, but does not preclude the existence of
other, more specific "targets" for these compounds in vivo. A close
correlation was noted between the specific number of peptide binding
sites on membranes and their influence (5×10^{-5}) on the accumulation
of serotonin by synaptosomes (r = 0.8).

The effect of the short peptides studied on neurotransmitter
accumulation is obviously brought about mainly by the total number of
molecules bound to the membrane, which is proportional to the constant K_bN, and the changes which result from this in the stability of
the membrane's surface charge, which is proportional to the constant
f_n. One property the peptides have in common is their identical
effect on the stability of the surface charge of the lipid portion of
the membrane. Thus, the more "frequently" these peptides occur
within the membrane (proportional to N) the more they change its
state. In view of the important part played by membrane lipids in
ionic transport, it might also be assumed that peptides, binding to
the lipid phase of membrane, alter membrane potential and ionic
gradient. This, in its turn, could also change the binding and
neuronal transport of MA.

REFERENCES

Abood, L. G., and Takeda, F., 1976, Enhancement of stereospecific
        opiate binding to neural membrane by phosphatidylserine,
        Eur.J.Pharmacol., 39:71-77.
Aloy, S. M., Kohn, L. D., Lee, G., and Meldolesi, M. F., 1977, The
        binding of thyrotropin to liposomes containing gangliosides,
        Biochem.Biophys.Res.Commun., 74:1053-1059.
Barden, N., and Labrie, F., 1973, Receptor for thyrotropin releasing
        hormone in plasma membranes of bovine anterior pituitary
        gland. Role of lipids, J.Biol.Chem., 248:7601-7606.
Barker, J. L., 1977, Physiological roles of peptides in the nervous
        system, in: "Peptides in Neurobiology," H. Gainer, ed.,
        Plenum, New York, pp.295-343.
Barnett, R. E., and Palazzotto, J., 1975, Mechanism of the effects of
        lipid phase transitions on the Na^+, K^+-ATPase and the role of
        protein conformational changes, Ann.N.Y.Acad.Sci., 242:69-76.
Berridge, M. Y., 1980, Receptors and calcium signalling, TIPS,
        1:419-424.

Batzri, S., and Korn, E. D., 1973, Single bilayer liposomes prepared
    without sonification, Biochim.Biophys.Acta, 298:1015-1019.
Bondarenko, N. A., Valdman, A. V., and Kamysheva, V. A., 1981, Change
    in the psychotropic effect on emotional reactivity and be-
    haviour under stress depending on the status of brain cat-
    echolaminergic system, Bull.Exp.Biol.Med., 60:35-38.
Burbach, P., and De Wied, D., 1980, Adaptive behaviour and endorphin
    biotransformation, in: "Enzymes and Neurotransmitters in
    Mental Disease," E. Usdin et al., eds., J. Wiley & Sons Ltd.,
    pp.103-114.
Chipens, G., Auna, Z., Klusha, V., et al., 1973, Some concepts on the
    molecular mechanisms of peptide hormone action at receptor
    level, in: "Peptides," H. Hanson, ed., North Holland Publ.,
    Amsterdam, pp.437-449.
Constantopoulos, A., and Najjar, V. A., 1973, The requirement for
    membrane sialic acid in the stimulation of phagocytosis by the
    neutral tetrapeptide tuftsin, J.Biol.Chem., 248:3819-3822.
Cox, B., Kastin, A., and Schmieden, H., 1976, A comparison between a
    melanocyte-stimulating hormone inhibitory factor (MIF) and
    substances known to activate central dopamine receptors,
    Eur.J.Pharmacol., 36:141-147.
Curzon, G., and Green, A. R., 1970, Rapid method for the determi-
    nation of 5-hydroxytryptamine and 5-hydroxyindoleacetic acid
    in small regions of rats brain, Brit.J.Pharmacol., 39:653-655.
Denenberg, V. H., 1979, Open-field behaviour in the rat: what does it
    mean? Ann.N.Y.Acad.Sci., 159:852-859.
De Wied, D., 1980, Behavioral action of neurohypophysial peptides,
    Proc.Roy.Soc.Lond.B., 210:183-195.
De Wied, D., and Gispen, W. H., 1977, Behavioural effects of pep-
    tides, in: "Peptides in Neurobiology," H. Gainer, ed., Plenum,
    New York, pp.397-448.
De Wied, D, 1981, Neuropeptides in normal and abnormal behaviour, in:
    "Endocrinology, Neuropeptides," E. Stark, J. Makara, and E.
    Ednroczi, eds., Adv.Physiol.Sci., 13:23-28.
Descarries, L., Beaudet, A., and Watkins, K., 1975, Serotonin nerve
    terminals in adult rat neocortex, Brain Res., 100:536-588.
Descarries, L., Watkins, K., and Lappierre, Y., 1977, NA axon
    terminals in the cerebral cortex of rat, Brain Res.,
    133:197-222.
Dismukes, R. K., 1979, New concepts of molecular communications among
    neurons, Behav.Brain Sci., 2:409-448.
Fekete, M., Bokor, M., Penke, B., Kovacs, K., and Telegdy, G., 1981,
    Effect of CCK-8 on brain MA, Neurochem.Internat., 3:165-169.
Fink, J. S., and Smith, G. P., 1979, Decreased locomotion and invest-
    igatory exploration after denervation of catecholamine term-
    inal fields in the forebrain of rats, J.Comp.Physiol.Psychol.,
    93:34-65.
Guillemin, R., 1978, Biochemical and physiological correlates of
    hypothalamic peptides, in: "Hypothalamus," S. Reichlin and R.
    Baldessarini, eds., Raven Press, New York, pp.155-194.

Hayes, B., and Jacquemin, C., 1971, Interaction de la thyréostimuline avec ses récepteurs cellulaires. Effet de la phospholipase C sur la fixation et l'activité biologique, FEBS Lett., 18:47-52.

Heal, D. Y., and Green, A. R., 1979, Administration of thyrotropin releasing hormone (TRH) to rats releases dopamine in n. accumbens but not n. caudatus, Neuropharmacology, 18:23-31.

Henderson, N., 1970, Behavioural reactions of Wistar rats to conditioned fear stimuli, novelty and noxious stimulation, J.Psychol., 75:19-34.

Hirata, F., Strittmatter, W. Y., and Axelrod, Y., 1979, β-Adrenergic receptor agonists increase phospholipid methylation, membrane fluidity and β-adrenergic receptor-adenylate cyclase coupling, Proc.Nat.Acad.Sci.,USA, 76:368-372.

Hökfelt, T., Fuxe, K., Johansson, O., Jeffcoate, S., and White, N., 1975, Thyrotropin releasing hormone (TRH)-containing nerve terminals in certain brain stem nuclei and the spinal cord, Neuro.sci.Lett., 1:133-139.

Jacobs, B. L., Wise, W. D., and Taylor, K. M., 1975, Is there a catecholamine serotonin interaction in the control of locomotor activity? Neuropharmacology, 14:501-506.

Kelly, P. H., Seviour, P. M., and Iversen, S. D., 1975, Amphetamine and apomorphine responses in the rat following 6-OHDA lesions of the nucleus accumbens septi and corpus striatum, Brain Res., 94:507-522.

Kerwin, R. W., and Pycock, C. Y., 1979, Thyrotropin releasing hormone stimulates release of ^{3}H-dopamine from slices of rat nucleus accumbens in vitro, Br.J.Pharmacol., 67:323-325.

Kozlovskaya, M. M., Klusha, V. E., and Bondarenko, N. A., 1982, Comparison of psychotropic and neurochemical activity of short peptides, in: "Neurochemical Basis of Psychotropic Effects," A. V. Valdman, ed., Inst. Pharm. Acad. Med. Sci., Moscow, pp.95-105.

Limbid, L. E., and Lefkowitz, R. Y., 1976, Adenylate cyclase-coupled beta-adrenergic receptors: effect of membrane lipid-perturbing agents on receptor binding and enzyme stimulation by catecholamines, Mol.Pharmacol., 12:559-567.

Loh, H., and Law, P., 1980, The role of membrane lipids in receptor mechanisms, Ann.Rev.Pharmacol., 20:201-234.

Lowry, O. H., Rosebrough, N. J., and Randall, R. J., 1951, Protein measurement with the Folin phenol reagent, J.Biol.Chem., 193:265-275.

Makanjuola, R. O., and Ashcroft, G. W., 1982, Behavioural effects of electrolytic and 6-hydroxydopamine lesions of the accumbens and caudate-putamen nuclei, Psychopharmacology, 76:333-340.

Miller, F. E., Heffner, T. G., Kotake, C., and Seiden, L. S., 1981, Magnitude and duration of hyperactivity following neonatal 6-hydroxydopamine is related to the extent of brain dopamine depletion, Brain Res., 229:123-132.

Moore, K. E., and Kelly, P. H., 1978, Biochemical pharmacology of mesolimbic and mesocortical dopaminergic neurons, in: "Psychopharmacology: a Generation of Progress," M. A. Lipton, A. Di Mascio, and K. F. Killam, eds., Raven Press, New York, pp.221-234.

North, R. B., Harik, S. I., and Snyder, S. H., 1973, L-prolyl-L-leucyl-glycine-amide (PLG): influences on locomotor and stereotyped behaviour of cats, Brain Res., 63:435-439.

Passaro, E., Dbas, H., Oldendorf, W., and Yamada, T., 1982, Rapid appearance of intraventricularly administered neuropeptides in the peripheral circulation, Brain Res., 241:335-340.

Raisman, R., Briley, and M. S., Langer, S. Z., 1980, Specific anti-depressant binding sites in rat brain, characterized by high-affinity ^{3}H-imipramine binding, Eur.J.Pharmacol., 61:373-380.

Schellenberger, M. K., and Gordon, J. H., 1971, A rapid simplified procedure for simultaneous assay of norepinephrine, dopamine and 5-hydroxytryptamine from discrete brain areas, Analyt.Biochem., 39:355-372.

Snyder, S. H., and Coyle, J. T., 1969, Regional differences in ^{3}H-norepinephrine and ^{3}H-dopamine uptake into rat brain homogenates, J.Pharmacol.Exp.Ther., 165:78-86.

Spano, P. F., and Neff, N. H., 1971, Procedure for simultaneous determination of dopamine, 3-methoxy-4-hydrophenylacetic and 3,4-dihydroxyphenylacetic acid in brain, Analyt.Biochem., 41:113-118.

Svensson, L., and Ahlenius, S., 1982, Functional importance of nucleus accumbens noradrenaline in the rat, Acta Pharmacol. Toxicol., 50:22-24.

Telegdy, G., 1980, The effect of neurohormones on the brain and endocrine system, Acta Physiol.Acad.Sci.Hungaricae, 55:273-281.

Telegdy, G., Fekete, M., Varszegi, M., and Kádár, T, 1980, Effect of peptide hormones on biogenic amines of the central nervous system, in: "Advances in Pharmacological Research and Practice," J. Knoll, ed., Vol. VII, "Aminergic and Peptidergic Receptors," E. Vizi and M. Wolleman, eds., Pergamon Press, Akademia Kiado, pp.169-185.

Ticku, M. K., and Olsen, R. W., 1978, Interaction of barbiturates with dihydropicrotoxin binding sites related to the GABA receptor-ionophore system, Life Sci., 22:1643.

Ungerstedt, U., 1971, Postsynaptic supersensitivity after 6-OHDA induced degeneration of the nigro-striatal dopamine system, Acta Physiol.Scand., Suppl., 367, pp. 69-93.

Valdman, A. V., 1982, Neuropeptides and emotional behaviour, in: "Advances in Pharmacology and Therapeutics," Vol. I, "CNS Pharmacology, Neuropeptides," H. Yoshida, Y. Hagihara, and S. Ebashi, eds., Pergamon Press, Oxford, pp.165-173.

Valdman, A. V., Kozlovskaya, M. M., Klusha, V. E., and Svirskis, S. V, 1980, Study of the psychopharmacological spectrum of melanostatin, Bull.Exp.Biol.Med., 89:693-696.

Van Ree, J. M., and Wolternic, G., 1981, Injection of low doses of
    apomophine into the nucleus accumbens of rats reduces
    locomotor activity, Eur.J.Pharmacol., 72:107-111.
Vetulani, J., Byrska, B., and Reichenberg, K., 1979, Head twitches
    produced by serotonergic drugs and opiates after lesion of the
    mesostriatal serotonergic system of the rat, Pol.J.Pharm.
    Pharmacol., 31:413-423.
Vizi, E., 1980, Non-synaptic modulation of transmitter release:
    pharmacological implication, Trends Pharmacol.Sci., 7:172-175.
Zile, R., Odynets, T., and Klusha, V., 1979, Effect of some fragments
    of peptide hormones on the content of biogenic monoamines from
    mouse brain, Biochemia, 44:93-96.

The GABA System as a Factor in Adaptation and Pharmacological Treatment of Stress

Yu.D. Ignatov, B.V. Andreev and G.E. Galustyan

Department of Pharmacology, Pavlov Medical Institute, Leningrad

The part played by the catecholamine, cholinergic and sero-toninergic systems in generating stress and the countering of stress by pharmacological agents has recently been fairly fully investigated (Maynert and Levy, 1964; Broverman et al., 1974; Anichkov, 1974; Anokhina, 1975; Matlina et al., 1975; Palkovits et al., 1976; Valdman et al., 1979; Andreev, 1978). Yet current ideas on the neurochemical mechanisms of stress do not preclude the existence of other neuro-transmitter systems. Pharmacological correction of stress may be bound up with an agent's action on the γ-aminobutyric acid (GABA) system. This system is involved in regulating neuronal activity (Curtis and Johnston, 1974; Roberts, 1976; Sytinsky et al., 1978) and in modulating monoaminergic (Biswas and Carlsson, 1977; Pycock and Horton, 1978) and neuroendocrine changes (Tapia, 1980), both of which are instrumental in generating and reinforcing pathological changes associated with stress (Matlina et al., 1975; Broverman et al., 1974; Anokhina, 1975; Valdman et al., 1979; Lapin, 1979). This involvement highlights the part played by the neurochemical system in activating the naturally occurring inhibitory mechanisms which play a part in adaptation. Unlike other neurotransmitters, GABA is not only import-ant for synaptic transmission, but also in energy balance in the nervous system (Baxter, 1970; Bunyatyan, 1976; Sytinsky et al., 1978); adjustment of the latter might provide a radical new approach to the pharmacological treatment of stress.

Only a limited number of publications deal with GABA levels during stress, however, and even these offer contradictory results (Nilova,1963; Hahn et al., 1975; Singh et al., 1979; Earley and Leonard, 1979; Meerson, 1981). Furthermore, such an approach is not without its shortcomings, in that the majority of these authors highlight the studies made by themselves on one of the three major components of the GABA system: GABA itself, GABA-transaminase (GABA-T) or glutamate decarboxylase (GAD). They have not usually followed up how these variables change in particular brain structures at different intervals during exposure to stress. Little attention has been paid to the relationship between changes in the GABA system and emotional/behavioral and somatic responses to stress of varying intensity. Our knowledge concerning the connection between GABA metabolism and changes in the brain's energy metabolism under stress is only very slight.

We therefore proceeded to investigate these topics in our laboratories.

1. METHODS

Experiments were performed on albino male rats weighing 150-200 g. The animals were normally provided with food and drink ad libitum, but were deprived of food for the 24 h preceding the experiments. They were subjected to stress on a regular basis from 10.00-10.30 to 13.00-13.30 h. The stress was of two different types, a) immobilization stress or IS of 3 h duration, and b) sporadic electrical stimulation of the tail following the 3-hour immobilization period, referred to as painful/emotional stress or PES (Andreev et al., 1981). The electrical stimulation was applied once per 3 min as 30 sec square impulses with a frequency of 20 imp/s, lasting 10 ms; amplitude was determined on each occasion by the vocalization arousal threshold. After exposure to stress, the animals' behavior was studied in an open field with openings, consisting of a rectangular arena measuring 100 x 100 x 30 cm; each square measured 20 cm, and the holes - 2.8 cm in diameter, lit by 50 lux. Latency of escape from the center of the arena, the number of squares crossed, the incidence of rearing and the number of holes explored were all measured over an interval of 2 min. Emotional response was rated according to a 3-point scale, as follows:

    0 - does not react to grasping hand
    1 - either resists or squeaks faintly
    2 - both resists and squeaks
    3 - resists and attempts to bite the hand.

Aggression is rated according to a 5-point scale, viz:

    0 - the animal fails to respond to an approaching metal rod
    1 - turns round
    2 - sniffs the rod
    3 - adopts a mock-fighting posture
    4 - bites occasionally
    5 - attacks, multiple bites.

Material for biochemical and histochemical investigations was processed 35 min after exposure to stress. An area of the fore-brain was separated - rostral septum, amygdala, a portion of the anterior commissure, corpus callosum and rostral portion of the striatum - together with brain stem - mamillary bodies, structures of the mid-brain and of the pons. Activity of GABA-T (E.C. 2.6.1.19) and GAD (E.C. 4.1.1.15), and GABA concentration were measured in homogenates of these brain areas. GABA-T activity was estimated using a method based on a condensation reaction of succinic semi-aldehyde with 3-methyl-2-benzothiazolone-2-hydrazone (Vasiliev and Yeryomin, 1968). GAD activity was measured by a rise in concen-tration of the reaction product of GABA and ninhydrin (Lowe et al., 1958). GABA concentration was measured fluorometrically (Sutton and Simmonds 1974).

Portions of the dorsal hippocampus and frontal cortex were separated out for histochemical investigation and frozen in liquid nitrogen-cooled isooctane. Slices 10 microns thick were prepared in a cryostat. GABA-T activity was determined using Van Gelder's (1965) method, with changing composition of the medium ensuring maximum speed of reaction (Galustyan and Prianishnikov, 1978). The activity of mitochondrial α-glycerophosphate dehydrogenase (GPDH; E.C. 1.1.99.5), lactate dehydrogenase (LDH; E.C. 1.1.1.27), succinate dehydrogenase (SDH; E.C. 1.3.99) β-hydroxybutyrate dehydrogenase (HBDH; E.C. 1.1.1.30) and NAD-linked glutamate dehydrogenase (GDH; E.C. 1.4.1.2) was revealed according to general principals of dehy-drogenase histochemistry (Lojda et al., 1979). Enzyme activity was measured microdensitometrically by the formation of formazan. The "plug method" was used on a single beam cytophotometer at a wave-length of 546 nm while statistical assessment of the results using an "Electronica TZ-IBM" computer was by Student's t-test and factor analysis. Gastric mucosal erosions were observed visually at the end of the experimental procedure.

The following compounds were investigated: the irreversible inhibitor of GABA-T, γ-vinyl GABA (GVG) and γ-acetylenic GABA (GA-GABA) (Centre de Recherche Merrell International, Strasbourg, France), the GABA receptor agonist, muscimol (Hoffmann-La Roche, Basle, Switzerland) and the benzodiazepine tranquilizers diazepam (Seduxen[R], Gedeon Richter, Hungary) and phenazepam (Institute of Pharmacology, Moscow, USSR) together with the GABA antagonist, picrotoxin (Sigma, St. Louis, USA) and thiosemicarbazide (TSC), the blocker of GABA synthesis. All compounds were administered intraper-itoneally; GVG (750 mg/kg) and GA-GABA (50 and 100 mg/kg) 2 h before the experiment, diazepam (2.5 mg/kg) and phenazepam (1 mg/kg) 30 min before and TSC (3 mg/kg) 1 h before. Muscimol and picrotoxin were administered twice; 2 mg/kg muscimol and 1 mg/kg picrotoxin were initially administered 30 and 20 min respectively before exposure to stress, and then again 1.5 h after the onset of stress, at a dose of 1 mg/kg. An equivalent quantity of sterile water was given to the controls.

2. RESULTS AND DISCUSSION

2.1. EMOTIONAL/BEHAVIORAL AND SOMATIC SIGNS OF STRESS

The intensity of the stressor determined animals' behavioral and somatic response. Thus, IS caused scarcely any erosive damage during the post-stress period. Presumably reduced motor activity in an open field situation is a typical sign of behavioral impairment following moderate stress (Ayrapetyants et al., 1980). Behavior is increasingly suppressed by greater stress (Matlina et al., 1975; Valdman et al., 1979; Ayrapetyants and Vein, 1982). As we had shown earlier (Andreev et al., 1982; Ignatov et al., 1982), PES produced great changes in rats' emotional/behavioral status (Figure 1), while emotional response and aggression both decreased (see Figure 2). This reaction to stress may have pathological aspects, not withstanding its overtly adaptive nature. The onset of gastric mucosal erosion in rats would bear this out (see Figure 2), being somatic manifestations typical of adaptational impairment.

2.2. GABA SYSTEM'S RESPONSE TO STRESS

The changes noted in an animal's behavioral and somatic status were accompanied by changes in GABA-ergic brain processes. The changes in forebrain and brainstem, however, differed in their pattern and intensity. Furthermore, altered GABA system activity varied according to the intervals of exposure to stress (Figure 3). During

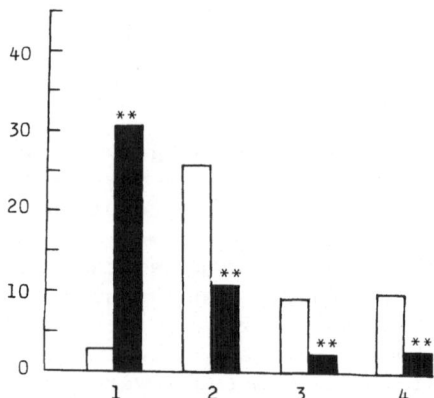

Fig. 1. The effect of painful/emotional stress on animal behavior in an open field. Ordinate: averages of 1 – latent period of escaping from the center of the arena; 2 – number of squares crossed; 3 – number of rearings; 4 – number of holes explored. Unshaded columns – control; shaded columns – test animals. **$p < 0.01$ (Van der Waerden's X test). 15 animals per group.

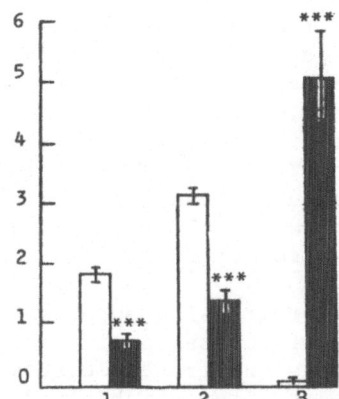

Fig. 2.   The effect of painful emotional stress on emotional
          response, aggression and gastric mucosal erosion.
          Ordinate, average ± SEM of 1 - emotional response (points
          scored); 2 - aggression (points scored); 3 - number of
          erosions. ***p<0.001 (Student's t test). Other information
          on Figure 1.

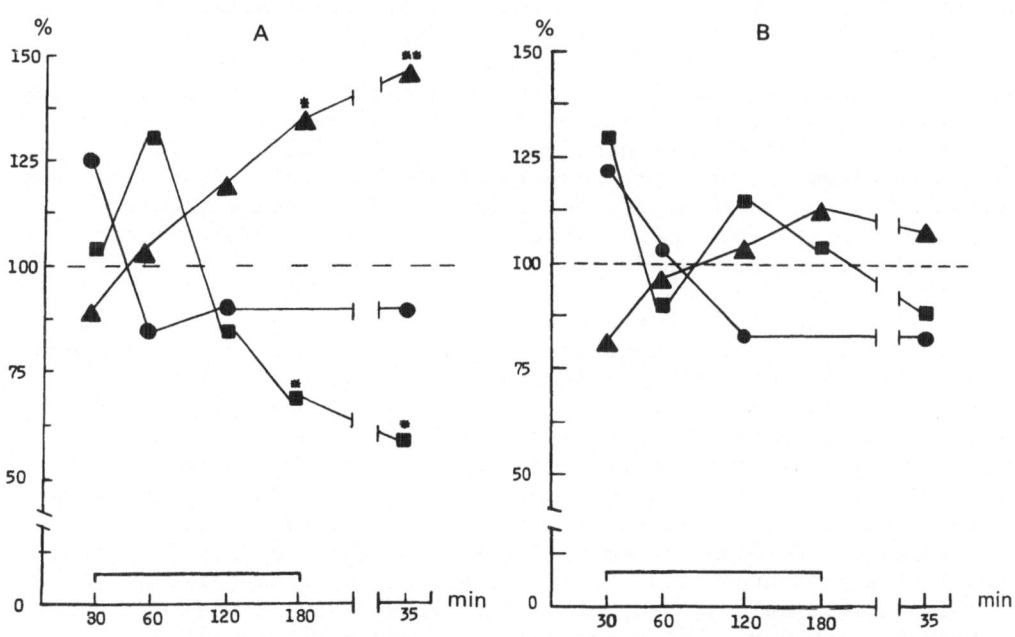

Fig. 3.   The effect of painful/emotional stress on GABA system activ-
          ity in (A) the forebrain and (B) the brainstem at different
          intervals. Ordinate: average changes in GABA concentration
          (▲); in GABA-T activity (■) and GAD (●); as a % of control
          level (taken as 100%). Abscissa: time in min. Line above
          abscissa: duration of stress. *p<0.05; **p<0.01 (Student's
          t test). 5 animals per group.

the initial stage, stress reduced GABA levels in the areas of the brain under study, while GAD and GABA-T were slightly activated; this could indicate some increase in GABA turnover. However, as stress extended from 60 to 120 min, GABA concentration in the forebrain increased together with inhibition of GABA-T and a relatively steady level of GAD. GABA concentration and the activity of its metabolizing enzymes did not change substantially in the brainstem; this could have been brought about by the greater resistance of the GABA-ergic systems of this area to stress (Volicer and Klosowitz, 1979).

We also demonstrated progressive suppression of GABA-T activity, corresponding with degree of stress, during our histochemical investigations of the hippocampus and frontal cortex (see Table 1). As this table shows, GABA-T activity is inhibited considerably more when stress reaches PES levels - this phenomenon is less marked during IS. Factor analysis also confirmed the significance of increased GABA-T inhibition on exposure to PES in the hippocampal and frontal cortex neurones.

Differences observed in the GABA system's reaction to different lengths of exposure to stress coincide with a number of other author's findings (Nilova, 1963; Hahn et al., 1975; Earley and Leonard, 1979; Volicer and Klosowitz, 1979). It is a characteristic feature that a drop in GABA level, superimposed on a rise both in GABA synthetic and catabolic ability, takes place in the early stages of stress, receding impairment of the animals' behavior and somatic condition. It can thus be assumed that the initial activation of GABA metabolism is a primary adaptive reaction. Raised and prolonged stress are accompanied by further alterations in GABA metabolism; GABA inactivation processes are retarded; it is synthesized at a relatively steady rate while its concentration rises.

In our opinion, such changes in GABA metabolism represent the subsequent stage of the adaptive response countering the development of somatic signs of stress, owing to reinforcement of inhibitory processes in the central nervous system (CNS). However, as stress continues to intensify, the extent of this adaptive reaction would appear insufficient; the fact that gastric mucosal erosion set in is proof of this. At first sight, certain recent findings would appear to run counter to our hypothesis on the adaptational, protective role of changes in the GABA system during stress. Meerson (1981) showed that intense and prolonged PES, lasting 6 h, is accompanied by GAD and GABA-T activation, together with decreased GABA concentration. We believe, however, that these findings should be viewed as experimental confirmation of the qualitatively different readjustment of GABA metabolism under more extended PES.

Long-lasting and actually increasing post-stress activation of cerebral neurochemical systems is known to be one of the signs of pathological change in the metabolism of a number of neurotrans-

Table 1. The Effect of Varying Degrees of Stress on GABA-T Activity
          in Hippocampal and Frontal Cortex Neurones

Structure	Control	GABA-transaminase activity (in units of optical density)			
		Painful/ emotional stress	% control	Immobil- izing stress	% control
Hippocampus	0.073±0.003	0.035±0.003**	47.9	0.050±0.004**	68.5
Frontal cortex	0.084±0.006	0.034±0.004**	40.5	0.065±0.004*	77.4

Results expressed as mean ± SEM.   5 rats per group.
* $p < 0.05$.
**$p < 0.001$ vs. control.

mitters (Anokhina, 1975; Matlina et al., 1975; Lapin, 1979). This
process is accompanied by a further reduction of their level in the
CNS (Anokhina, 1975; Lapin, 1970). Bearing this in mind, the activation of GAD and GABA-T, as well as the drop in GABA concentration
observed by Meeson (1981), show not the adaptational but the actual
pathological nature of changes in GABA metabolism in the CNS.

3. THE GABA SYSTEM AS A TARGET IN THE PHARMACOLOGICAL TREATMENT OF STRESS

A generalized rise in GABA level is known to reflect an increase
not only in the synaptosomol pool, but also in glial elements which
do not possess GABA receptors (Sarhan and Seiler, 1979; De Feudis,
1980). It is possible that suppressing GABA inactivation together
with a rise in GABA concentration under stress chiefly causes an
accumulation of GABA in the synaptosomal pool. It is equally possible that this constitutes a defensive process, which, however, may
be insufficient to provide full protection. If this is the case,
pharmaceuticals which produce similar but less marked changes in
GABA metabolism or direct activation of postsynaptic receptors must
presumably afford protection against stress.

4. EFFECTS OF GABA-ERGIC COMPOUNDS ON BEHAVIORAL AND SOMATIC SIGNS OF STRESS

Diazepam and muscimol substantially reduced the number of
gastric erosions in rats (see Table 2). At the same time they both
acted differently on emotional/behavioral status, as altered by
stress (Figure 3 and Table 2). Diazepam and GA-GABA reduced

post-stress behavioral inhibition. Latency of escape from the center
of the arena was decreased by diazepam, which also prevented post-
stress reduction in emotional response and aggression. Muscimol and
GA-GABA failed to produce a reduction in behavioral depression during
the post-stress period.

GABA-negative compounds, and TSC in particular, lowered capacity
to withstand stress (see Figure 4 and Table 2). The GABA-positive
action of diazepam and muscimol is known to be focused on post-
synaptic receptor zones (Costa, 1979; Enna and Maggi, 1979; Andreev

Table 2. Effect of Compounds Under Study on Emotional Response,
Aggression and Gastric Mucosal Erosion During Stress

Group	No. Erosions	Emotional response (points scored)	Aggression (points scored)
Control	0.0	1.60±0.16	2.73±0.18
γ-vinyl-GABA	0.0	0.33±0.13*	1.20±0.30*
Stress	6.69±2.61*	0.29±0.11*	1.30±0.24*
γ-vinyl-GABA + stress	1.69±0.66*+	0.31±0.13*	1.85±0.32*
Control	0.0	1.67±0.16	2.06±0.23
γ-acetylenic-GABA	0.0	0.33±0.12*	0.80±0.21*
Stress	8.93±1.81*	0.67±0.16*	1.06±0.27*
γ-acetylenic-GABA + stress	3.47±0.69*+	0.07±0.06*+	0.80±0.22*
Control	0.20±0.14	1.60±0.19	3.20±0.14
Muscimol	0.0	0.46±0.18*	1.85±0.38*
Stress	5.00±1.14*	0.47±0.16*	1.33±0.24*
Muscimol + stress	1.36±0.44*+	0.50±0.18*	1.78±0.39*
Control	0.0	1.20±0.11	1.93±0.34
Diazepam	0.0	0.20±0.11*	1.40±0.16
Stress	8.87±1.59*	0.13±0.09*	0.60±0.13*
Diazepam + stress	2.33±0.43*+	0.67±0.19*+	2.20±0.23+
Control	0.0	1.06±0.18	1.93±0.23
Picrotoxin	0.13±0.12	1.00±0.22	2.66±0.34
Stress	5.53±1.19*	0.44±0.13*	1.00±0.17*
Picrotoxin + stress	8.27±1.73*	0.60±0.16	1.53±0.47
Control	0.0	1.60±0.16	2.60±0.22
Thiosemicarbazide	0.50±0.24	1.00±0.11*	2.20±0.21

Results are expressed in the form of M ± SEM. 9-15 rats per group.
* difference statistically significant compared to the control.
+ difference statistically significant compared to the "stress"
group (p<0.05).

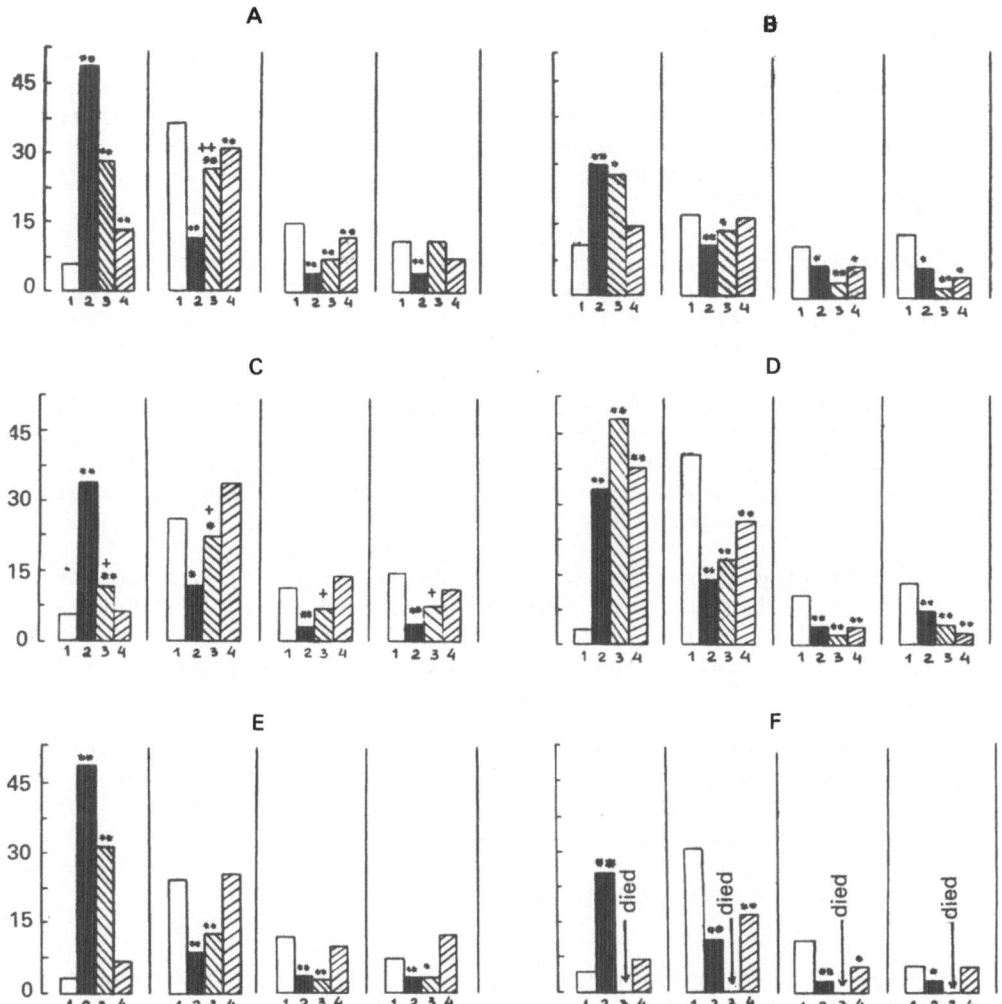

Fig. 4. The effect of various compounds on animal behavior in an
open field under stress. Ordinate: mean of parameters
tested. First group of columns: latency of escaping
from center of the arena; second group: number of
squares crossed; third group: number of holes explored.
1 - control; 2 - stress; 3 - stress+compound; 4 - compound. A - γ-vinyl-GABA (750 mg/kg); B - γ-acetylenic-GABA
(100 mg/kg); C - diazepam (2.5 mg/kg); D - muscimol (2 + 1
mg/kg); E - picrotoxin (1 + 1 mg/kg); F - thiosemicarbazide
(3 mg/kg). *p<0.05; **p<0.01 vs. control group. +p<0.05;
++p<0.01 vs. the "stress" group. (Van der Waerden's X
test). 9-15 animals per group.

et al., 1979; Andreev and Ignatov, 1981; Rayevsky, 1981). Thus, it
is possible that stimulation by these two compounds of GABA-ergic
transmission under stress could produce a level of GABA system
activity suitable for blocking the sequence of neurotransmitter
changes which occur on exposure to stress. Prevention of somatic
impairment is obviously the result. This, in its turn, might serve
as evidence of the importance of synaptic GABA-positive action of
benzodiazepines and GABA-T inhibitors in countering stress (Ignatov
et al., 1982).

The differences between the effects of these compounds on
stress-induced behavioral depression can be explained by the peculiarities of their action on GABA-ergic transmission. GVG and GA-GABA
are known to influence the compartmentation of GABA (Sarhan and
Seiler, 1979; Pericic, 1980). Unlike GA-GABA, GVG barely inhibits
GAD (Schechter et al., 1977; Andreev et al., 1982; Ignatov et al.,
1982) and produces an accumulation of GABA, mainly in the synaptosomal pool, whereas GAD does so in the extrasynaptic pool (Sarhan
and Seiler, 1979). These differences may point to a broader stress-
protective range of action for GVG.

The differing effects of diazepam and muscimol on behavior are
probably linked with the fact that muscimol activates all GABA receptors when administered systemically (Costa, 1979; Rayevsky, 1981);
this also explains its non-selective depressant action (Costa, 1979;
Enna and Maggi, 1979; Andreev et al., 1979).

Diazepam activates GABA-ergic transmission processes resulting
from decreased binding between endogenous GABA and GABA receptors
(Costa, 1979). This mechanism also characterizes the qualities of
the psychotropic action of diazepam and the special properties of its
anti-stress action, which is connected to its selection of certain
target populations of GABA receptors.

5. THE EFFECT OF GABA-ERGIC COMPOUNDS ON THE GABA SYSTEM'S REACTION
DURING STRESS

Our findings on the effect of GABA-ergic compounds on GABA
concentration and GABA-T activity in forebrain structures serve as
a convincing confirmation of the vital role played by GABA receptor
activation in preventing somatic and behavioral signs of stress
(see Table 3). This table shows that GVG and GA-GABA produced
considerably more inhibition of GAGA-T and increase in GABA concentration than exposure to stress both in rats already under stress
and in unstressed animals.

Neither muscimol nor diazepam substantially changed GABA concentration in the control animals; they also countered the stress-
induced rise in GABA (see Table 3). This fact clearly supports an

activation of postsynaptic GABA receptors which is instrumental in preventing the changes induced in GABA concentration by stress.

Picrotoxin failed to change any of the parameters of GABA metabolism, while moderately enhancing the effect of stressors on GABA system activity. TSC considerably decreased the animals' resistance to stress compared with picrotoxin. TSC's action is probably unconnected with any general changes in GABA concentration, but was thought to be caused by a drop in concentration of the transmitter

Table 3. Effect of the Compounds Under Study on GABA System Activity in Forebrain Structures During Stress

Group	GABA concentration (μmol/g)	GAD activity (μmol per 1 mg protein/h)	GABA-T activity (μmol/g)
Control	2.62±0.10	6.60±0.71	133.38±14.40
γ-vinyl-GABA	8.44±0.28*	5.29±0.72	49.91± 4.68*$_+$
Stress	3.84±0.34*	5.79±0.86	79.75± 9.91*
γ-vinyl-GABA + stress	8.86±0.21*$_+$	6.11±0.86	50.88±13.38*$_+$
Control	2.98±0.19	8.64±0.34	287.21±25.29
γ-acetylenic-GABA	10.56±0.35*	5.74±0.32*	104.04±22.00*
Stress	4.62±0.32*	8.82±0.27	190.54±25.03*
γ-acetylenic-GABA + stress	11.16±0.24*$_+$	7.07±0.39*	160.08±33.04
Control	3.48±0.13	12.89±2.92	122.78±25.20
Muscimol	3.64±0.23	11.52±1.64	105.03±20.30
Stress	5.56±0.23*	10.13±2.77	83.68±20.95
Muscimol + stress	4.10±0.18*$_+$	9.48±2.45	99.79±19.75
Control	3.28±0.42	9.75±1.30	449.58±28.87
Diazepam	3.56±0.23	9.22±0.60	286.54±55.87*
Stress	4.30±0.37	8.97±0.84	264.65±37.75*
Diazepam + stress	3.36±0.13+	8.41±0.64	298.65±61.05
Control	2.82±0.06	9.27±0.89	351.91±51.13
Picrotoxin	2.72±0.22	8.04±0.46	276.85±39.19
Stress	3.86±0.28*	8.92±0.71	230.63±57.21
Picrotoxin + stress	4.28±0.20*	9.93±0.94	192.52±42.64*
Control	3.28±0.42	12.45±1.04	385.46±80.36
Thiosemicarbazide	2.60±0.29	9.64±1.46	423.50±76.39
Stress	−	−	213.33±30.51*

Results are expressed in the form of M ± SEM. 9-15 rats per group.
* difference statistically significant compared to the control.
+ difference statistically significant compared to the "stress" group (p<0.05).

at nerve endings (Abe and Matsuda, 1977; Sarhan and Seiler, 1979) and hence, in the receptor region.

As the above findings show, the rise in GABA concentration and retardation of its enzymatic breakdown in forebrain structures form part of a defense mechanism. Similar but more pronounced changes in GABA metabolism or a direct reduction of GABA-ergic transmission brought about by pharmacologically active compounds are central to the stress-protective effect. The selective anti-stress action may be viewed as resulting less from total activation of GABA receptors than from selective action on the process of GABA compartmentation, potentiating transmitter accumulation in the synaptosomal pool or selective stimulation of particular types of GABA receptor.

6. CHANGES IN BRAIN ENERGY METABOLISM DURING STRESS

Certain extreme conditions, including PES, increase CNS energy requirements, as shown by a drop in ATP level and rise in concentration of ADP and inorganic phosphate (Vladimirova, 1956). At the same time, stressors may facilitate the oxidation of carbohydrates, which determines the state of ATP regeneration (Le Page, 1946). The rate of glycolysis, as well as its coupling to the tricarboxylic acid cycle (TAC), are of the greatest importance for a fuller appreciation of stress-induced changes in glycolysis. There is evidence of stress intensification raising LDH activity in midbrain and medulla (Marushevskaya, 1971). PES, however, as shown by Galustyan in our laboratories (Ignatov et al., 1982), produces no substantial changes in LDH activity in hippocampal and frontal cortical neurones (see Table 4), suggesting that this type of stress does not induce abrupt changes in metabolism in the forebrain. The fact that PES changes GPDH neither in the hippocampus nor the frontal cortex indicates normal functioning of the glycolytic pathway (see Table 4).

Stress disturbs the oxidizing processes of the TAC. As will be seen from Table 4, PES produced a substantial rise in SDH activity in hippocampal and frontal cortical neurones of 17-20%. HBDH activity rose by 32.7-45.5%; these changes did not occur in animals which had not been deprived of food. While the concentration of ketone bodies in the blood rose during food deprivation, they were also oxidized more rapidly during stress.

The degree of stress is also important in producing changes in metabolism. Activation of SDH in hippocampal and frontal cortical neurones is one of the features typical of the stronger, or PES type of stress (see Table 4). Factor analysis also confirmed the decisive role of stress level in raising SDH activity in various sections of the hippocampus (field CA 1 - CA 3) and dentate gyrus.

Table 4. The Effect of Varying Degrees of Stress on Enzymes of Energy Metabolism in Hippocampal and Frontal Cortex Neurones

Structure	Group	Enzyme activity (optical density units)				
		LDH	GPDH	SDH	HBDH	GDH
Hippocampus	Control	0.240±0.018	0.322±0.011	0.306±0.005	0.117±0.008	0.123±0.003
	Immobilization stress	0.242±0.008 (100.8%)	0.343±0.006 (106.5%)	0.291±0.008 (95.1%)	0.122±0.004 (104.3%)	0.147±0.007* (120.0%)
	Painful/emotional stress	0.249±0.006 (103.8%)	0.315±0.019 (97.8%)	0.361±0.016* (118.0%)	0.114±0.007 (97.4%)	0.174±0.009** (141.5%)
Frontal cortex	Control	0.193±0.008	0.333±0.004	0.290±0.002	0.095±0.003	0.154±0.012
	Immobilization stress	0.168±0.011 (87.0%)	0.332±0.015 (99.7%)	0.290±0.002 (100.0%)	0.099±0.004 (104.2%)	0.152±0.004 (98.7%)
	Painful/emotional stress	0.199±0.005 (103.1%)	0.326±0.013 (97.9%)	0.348±0.015** (120.0%)	0.106±0.013 (116.7%)	0.197±0.018 (127.9%)

Results are expressed in the form of M±SEM. 5 rats per group. Animals had free access to food prior to experiment. * $p < 0.05$; ** $p < 0.01$ compared to the control.

The opposite reaction was noted in a number of investigations, i.e. inhibition of TAC oxidation under stress (Meerson, 1981). The discrepancy between these findings and our own results might well derive from more intense and prolonged stress. Inhibition of oxidation in the TAC undoubtedly signifies greater disruption of homeostatic mechanisms.

Changes in energy metabolism occurring during stress may be reflected in the metabolism of a number of aminoacids. For example, glutamic acid concentration is known to be regulated by the activity of certain enzymes which are closely connected with TAC intermediates, primarily aspartate aminotransferase (AAT) and GDH. As will be seen from Table 4, showing the results of our investigations, PES induced GDH activation in the hippocampus and the frontal cortex, while IS was only able to do so to any noticeable extent in the dentate gyrus. IS was found to activate GDH considerably less than PES in the pyramidal neurones of the hippocampus (field CA 1 - CA 3).

Stress was thus shown to produce readjustments of the metabolic pathways closest to the GABA shunt. The inhibition of GABA-T, together with activation of SDH and GDH, may be an indication of intensified oxidation processes within the Krebs cycle. The use of glutamate as an energy substrate could produce results of the greatest importance. Under normal conditions GDH activity is lower than that of AAT, but the balance of the reaction is tilted towards the formation of α-ketoglutarate (Lai et al., 1977). The importance of this pathway is underscored by its involvement in the transport of ammonia from the brain (Weil-Malherbe, 1962). Considering our data together with the drop in glutamate concentration during stress observed by other authors it would seem that the balance of the action, catalyzed by GDH, may be tipped in the opposite direction. A similar readjustment in glutamate metabolism, related to possible impairment of detoxication mechanisms, might potentially be of pathological importance.

Blocking the GABA shunt could well have a direct effect on the regulation of metabolic compartmentation. Inhibiting GABA catabolism reduces the "drain" of GABA and its precursors from the synaptosomal pool to the cell bodies of postsynaptic neurones and neuroglia. This reduced loss of substrate promotes a rise in the efficacy of energy processes at the synaptic endings.

7. THE EFFECTS OF GABA-ERGIC COMPOUNDS ON CHANGES IN BRAIN ENERGY METABOLISM DURING STRESS

Tranquilizers of the benzodiazepine group and GA-GABA - a GABA-T inhibitor - were selected for investigation into the pharmacological involvement of GABA-T inhibition in restoring CNS metabolism impaired by exposure to stress. The choice of benzodiazepines was made ac-

cording to their ability to inhibit GABA-T, to exert a selective
action on the affective element of PES (Andreev et al., 1979;
Vasiliev et al., 1979; Galustyan, 1979; Ostrovskaya et al., 1975) and
to regulate brain metabolism during certain types of pathological
states (Carlsson et al., 1976; Berntman and Siesjo, 1978).

Table 5 shows the effect of diazepam, phenazepam and GA-GABA on
changes in enzyme activity, using hippocampal neurones as an example.
These compounds produced a similar effect on neurones of the dentate
gyrus and the frontal cortex. Diazepam, phenazepam and GA-GABA
reduced GABA-T activity. The compounds also maintained their inhibitory effect even during exposure to stress. Pretreatment with these
substances completely suppresses the changes in SDH and GDH activity
produced by stress. Diazepam and phenazepam exerted no observable
effect on SDH or GDH in stress-free animals; nor did GA-GABA
influence SDH, although it inhibited GDH activity in the hippocampal
neurones substantially.

Galustyan had shown earlier (1979) that small doses of 0.5-1
mg/kg diazepam produced considerable inhibition of GABA-T up to 50%,
primarily in the limbic structures of the brain. This being so,
findings relating to a tranquilizing element in valproate take on
additional significance (Lal et al., 1980; Kharlamov and Rayevsky,
1980); this compound inhibits GABA-T, chiefly in the structures of
the limbic system.

The effect of benzodiazepines, as distinct from that of GA-GABA,
is probably unconnected with GABA accumulation, since diazepam not
only produces a substantial rise in GABA accumulation, but also
prevents its level increasing under stress. Clearly the metabolic
rather than the transmitter function of GABA is instrumental in the
benzodiazepines achieving their normalizing effect on energy metabolism. GABA is known to play a substantial part in the compartmentation processes of aminoacid metabolism (Balazs et al., 1973).
The transport of Krebs cycle intermediates from the presynaptic
endings to the cell bodies of postsynaptic neurones and glial cells
takes place through the GABA shunt (Hertz, 1979). Thus, GABA-T
inhibition may reduce substrate loss from the presynaptic pool and
suppress oxidation of glutamic acid (properties which are uncharac-
teristic of nerve tissue under conditions of rest), apart from merely
raising GABA concentration.

The material presented enables us to put forward our own views
on the involvement of the GABA system in the regulation of stress and
its pharmacological correction. Inhibition of enzymatic breakdown of
GABA, especially in forebrain structures, resulting in an increase in
its synaptosomal fraction, is one of the mechanisms of neurochemical
adaptation to stress. Similar, but less pronounced changes in GABA
metabolism or direct reduction of GABA-ergic transmission due to the
effect of pharmacologically active compounds is central to the

Table 5. The Effect of 2.4 mg/kg Diazepam, 1 mg/kg Phenazepam and
50 mg/kg γ-acetylenic-GABA on Changes in Activity of
Hippocampal Enzymes Under Emotional Stress

| | Enzyme activity (optical density units) | | |
	GABA-T	GDH	SDH
Group			
Control	0.095±0.005	0.120±0.010	0.287±0.014
Diazepam	0.022±0.008** (23.2%)	0.137±0.004 (114.2%)	0.283±0.004 (98.6%)
Stress	0.018±0.005*** (18.9%)	0.168±0.009* (140.0%)	0.320±0.011 (111.5%)
Stress + diazepam	0.053±0.006$^{++}_{**}$ (55.8%)	0.111±0.008^{++} (92.5%)	0.271±0.005^{++} (94.4%)
Control	0.158±0.010	0.102±0.010	0.389±0.009
Phenazepam	0.107±0.005*** (67.7%)	0.106±0.006 (103.9%)	0.383±0.002 (98.5%)
Stress	0.117±0.005** (74.1%)	0.151±0.005*** (148.0%)	0.442±0.010** (113.6%)
Stress + phenazepam	0.117±0.001** (74.1%)	0.095±0.005^{+++} (93.1%)	0.370±0.010^{++} (101.3%)
Control	0.108±0.012	0.149±0.005	0.396±0.011
GA-GABA	0.046±0.006** (42.6%)	0.128±0.004* (85.9%)	0.401±0.013 (101.3%)
Stress	0.065±0.007* (60.2%)	0.188±0.003*** (126.2%)	0.476±0.021** (120.2%)
Stress + GA-GABA	0.037±0.004$^{++}_{***}$ (34.3%)	0.137±0.006^{+++} (91.9%)	0.367±0.013^{++} (92.7%)

Results are expressed in the form of M ± SEM. 5 rats per group.
* $p<0.05$; ** $p<0.01$; ***$p<0.001$ compared to the control.
++$p<0.01$; +++$p<0.001$ compared to the "stress" group.

stress-protective effect. Furthermore, it may be assumed that the
anti-stress action, suppressing behavioral as well as somatic impairment, is due not so much to total activation of GABA receptors as to
selective action on the processes of GABA compartmentation (which
produces an accumulation of the transmitter in the synaptically
active pool) or to selective stimulation of certain types of GABA
receptors.

GABA-T inhibition during stress, precluding a rise in the
activity of dehydrogenase connected with the GABA shunt, and thereby
normalizing the metabolic balance in CNS compartments, may be viewed

as one means of energetic adaptation to stress. Normalization of the brain's energy metabolism, produced by inhibiting enzymatic breakdown of GABA, under the influence of benzodiazepines, is thus one of the fundamental mechanisms in achieving their stress-protective effect. Their action on the GABA system is therefore twofold: firstly, they can reinforce its action as a neurotransmitter; secondly, they may represent a new approach to further progress in the field of pharmacological correction of stress.

REFERENCES

Abe, M., and Matsuda, M., 1977, γ-Aminobutyric acid metabolism in subcellular particles of mouse brain and its relationship to convulsions, J.Biochem., 82:195-200.

Andreev, B. V., 1978, Role of Cholinergic and serotoninergic processes for modulation of the activity of negative reinforcement system, Bull.Exp.Biol.Med., 86:pp.1462-1465 (Engl.Tr.).

Andreev, B. V., Galustyan, G. E., and Marusov, I. V., 1979, On the problem of significance of GABAergic processes for realization of diazepam anxiolytic effect, in: "Proceedings of the Leningrad V. M. Bekhterev Psychoneurological Research Institute," Vol. 91, E. A. Babayan and M. M. Kabanov, eds., Leningrad, pp.20-28.

Andreev, B. V., and Ignatov, Yu. D., 1981, Effect of GABA-negative drugs on anxiolytic and sedative effects of diazepam, Bull.Exp.Biol.Med., 91:756-758 (Engl.Tr.)

Andreev, B. V., Vasiliev, Yu. N., Ignatov, Yu. D., Kachan, A. T., and Bogdanov, N. N. 1981, Effect of electroacupuncture on manifestations of emotional stress due to pain, Bull.Exp. Biol.Med., 91:21-23. (Engl.Tr.).

Andreev, B. V., Ignatov, Yu, D., Nikitina, Z. S., and Sytinsky, I. A., 1982, Anti-stress role of the GABAergic system of the brain, J.Higher Nerv.Activ. I. P. Pavlov, 32:511-519.

Anichkov, S. V., Zavodskaya, I. S., Moreva, E. V., and Vedeneeva, Z. S., 1969, Neurogenic dystrophies and their pharmacological correction, "Meditsina", Leningrad.

Anichkov, S. V., 1974, The selective action of transmitter drugs, "Meditsina", Leningrad.

Anorhina, I. P., 1975, Neurochemical mechanisms of mental diseases, "Meditsina", Moscow.

Ayrapetyants, M. G., Khonicheva, N. M., Mekhedova, A. Ya., and Iliana Wiliar, 1980, Reactions to moderate functional loads in rats with different individual behaviour, J.Higher Nerv. Activ., 30:994-1002.

Ayrapetyants, M. G., and Vein, A. M., 1982, "Neuroses in Experiment and in Clinic," P. V. Simonov, ed., "Nauka", Moscow.

Balazs, R., Machinyama, Y., and Patel, A. J., 1973, Compartmentation and the metabolism of γ-aminobutyrate, in: "Metabolic Compartmentation in the Brain," MacMillan, London - Basingstoke, pp.57-70.

Baxter, C. F., 1970, The nature of γ-aminobutyric acid, in: "Handbook of Neurochemistry," A. Lajtha, ed., Vol.3, Plenum Press, New York, pp.289-353.

Berntman, L., and Siesjo, B. K., 1978, Brain energy metabolism and circulation in hypoxia, in: "Proc. Eur. Soc. Neurochem.," Vol.1, pp.253-265.

Biswas, B., and Carlsson, A., 1977, The effect of intracerebro-ventricularly administered GABA on brain monoamine metabolism, Naunyn-Schmiedebergs Arch.Pharmacol., 299:41-46.

Broverman, D. M., Klaiber, E. L., Vogel, W., and Kosayasini Y., 1974, Short-term versus long-term effects of adrenal hormone on behaviour, Psychol.Bull., 81:672-694.

Bunyatyan, N. Kh, 1976, Recent advances in the field of biochemistry and biochemical pharmacology of γ-aminobutyric acid, J.of D.I. Mendeleev All-Union Chem.Soc., 21:130-136.

Carlsson, C., Hagendal, M., Kaaskik, A. E., and Siesjo, B. K., 1976, The effects of diazepam on cerebral blood flow and oxygen in rats and its synergistic interaction with nitrous oxide, Anesthesiology, 45:319-325.

Costa, E., The role of gamma-aminobutyric acid in the action of 1,4-benzodiazepines, Trends Pharmacol.Sci., 1:41-44.

Curtis, D. R., and Johnston, G. A., Amino acid transmitters in central nervous system, 1974, Rev.Physiol., 69:98-199, Springer-Verlag, Berlin.

Defeudis, F. V., 1980, Binding studies with muscimol: relation to synaptic γ-aminobutyrate receptors, Neuroscience, 5:675-688.

Earley, C. J., and Leonard, B. E. Consequences of reward or nonreward conditions; run away behavior, neurotransmitters and physiological indicators of stress, Pharmacol.Biochem.Behav., 11:215-219.

Enna, G. J., and Maggi, A., 1979, Biochemical parmacology of GABA-ergic agonists, Life Sci., 24:1727-1738.

Galustyan, G. E., 1979, Quantitative histochemical study of the influence of benzodiazepine tranquilizers on the activity of GABA transaminase in rat brain, Bull.Exp.Biol.Med., 87:164-166.

Galustyan, G. E., and Prianishnikov, V. A., 1978, Cytospectrophoto-metric study of the kinetics of histochemical reaction of GABA transaminase in cryostat sections of rat cerebellar cortex, Histochemistry, 57:68-77.

Gottesfeld, Z., Kvetnansky, A., Kopin, I. J., and Jacobowitz, M., 1978, Effects of repeated immobilization stress on glutamate decarboxylase and choline acetyltransferase in discrete brain regions, Brain Res., 152:374-378.

Hann, Z., Telegdy, G., and Lissak, K., 1975, The role of the pituitary-adrenal system in changes induced by noxious stimuli in hypothalamic gamma-aminobutyric acid content in the rat, Acta Physiol.Acad.Sci.Hung., 46:325-330.

Hertz, L., 1979, Functional interactions between neurones and astrocytes, I. Turnover and metabolism of putative amino acid transmitters, Prog.Neurobiol., 13:277-323.

Ignatov Yu. D., Galustyan, G. E., and Andreev, B. V., 1982, The role
      of GABAergic mechanisms in stress-protective effect or benzo-
      diazepine tranquilizers, in: "Neurochemical Bases of Psycho-
      tropic Effect," A. V. Valdman, ed., Institute of Pharmaco-
      logy, Moscow, pp.118-126.
Kharlamov, A. N., and Rayevsky, K. S., 1980, Tranquilizing effect of
      n-dipropylacetate and other GABAergic substances in con-
      ditions of conflict situation, Bull.Exp.Biol.Med., 89:35-37.
Lai, J. C. K., Walsh, J. M., Dennis, S. C., and Clark, J. B., 1977,
      Synaptic and non-synaptic mitochondria from rat brain:
      isolation and characterization, J.Neurochem., 28:625-631.
Lal, H., Sherman, G. T., Fielding, S., Dunn, R., Kruse, H., and
      Theurer, K., 1980, Evidence that GABA mechanisms mediate the
      anxiolytic action of benzodiazepines: a study with valproic
      acid, Neuropharmacology, 19:785-789.
Lapin, I. P., Goals and possibilities of the substitution of tran-
      quilizers in the treatment of protracted neurotic abnormal-
      ities and in the prophylaxis of after-effects of chronic
      emotional stress, in: "Proceedings of the Leningrad V. M.
      Bekhterev Psychoneurological Research Institute," Vol.91,
      E. A. Babayan and M. M. Kabanov, eds., Leningrad, pp.28-37.
Le Page, G. A., 1946, Biological energy transformations during shock
      as shown by tissue analyses, Am.J.Physiol., 146:267-281.
Lojda, Z., Gossrau, R., and Schieler, T. H., 1979, Enzyme Histo-
      chemistry, A Laboratory Manual, Springer-Verlag, Berlin.
Lowe, I. P., Robins, E., and Eyerman, G. S., 1958, The fluorimetric
      measurement of glutamic decarboxylase and its distribution in
      brain, J.Neurochem., 3:8-18.
Marushevskaya N. Ya., 1971, The activity of dehydrogenases in brains
      of young and old animals in various functional states and
      during the training and retention of conditioned defensive
      reflex. Summary of Candidate Thesis, Stavropol.
Matlina, E. Sh., Baru, A. M., and Vasiliev, V. N., 1975, Emotions:
      the role of some transmitters and hormones in mechanisms of
      induction and retention of emotional states, in: "Advances of
      Science and Technique. Human and Animal Physiology," Vol.
      15, Moscow, VINITI, pp.30-93.
Maynert E. W., and Levi, R., 1964, Stress-induced release of brain
      norepinephrine and its inhibition by drugs, J.Pharmacol.
      Exp.Ther., 143:90-95.
Meerson, F. Z., 1981, Adaptation, stress and prophylaxis,
      "Meditsina", Moscow.
Nilova, N. S., 1963, The contents of free amino acids in cerebral
      hemispheres during the excitation of central nervous system,
      Doklady AN SSSR, 150:1161-1163.
Ostrovskaya, R. U., Molodavkin, G. M., Porfirieva, R. P., and
      Zubovskaya, A. M., 1975, On the mechanism of anticonvulsant
      action of diazepam, Bull.Exp.Biol.Med., 79:50-53
Palkovits, M., Brownstein, M., Kizer, J. S., Saavedra, J. M., and
      Kopin, I. J., 1976, Effect of stress on serotonin concen-

tration and tryptophan hydroxylase activity of brain nuclei, Neuroendocrinology, 22:298-304.

Pericic, D., 1980, Effect of γ-vinyl GABA on enzymes of GABA system in specific brain regions, Period.Biol., 82:19-23.

Pycock, C., and Horton, R., 1978, Regional changes in the concentrations of cerebral monoamines and their metabolites after ethanolamine-O-sulphate induced elevation of brain γ-aminobutyric concentrations, Biochem.Pharmacol., 27:1827-1830.

Rayevsky, K. S., 1981, Neurochemical aspects of the pharmacology of GABAergic drugs, Pharmacol.Toxicol., 44:517-529.

Roberts, E., 1976, Disinhibition as an organizing principle in the nervous system. The role of the GABA system. Application to neurological and psychiatric disorders, in: "GABA in Nervous System Function," E. Roberts, T. N. Chase, and D. B. Towers, eds., Raven Press, New York, pp.515-539.

Sarhan, S., and Seiler, N., 1979, Metabolic inhibitors and subcellular distribution of GABA, J.Neurosci.Res., 4:399-421.

Schechter, P. J., Tranier, Y., Jung, M. J., and Bohlen, P., 1977, Audiogenic seizure protection by elevated brain GABA concentration in mice: effect of γ-acetylenic GABA and γ-vinyl GABA, two irreversible GABA-T inhibitors, Eur.J.Pharmacol., 45:319-328.

Singh, H. G., Singh, R. H., and Udupa, K. N., 1979, Effects of electroshock on GABA glutamate content in rat brain, Indian J.Exp.Biol., 17:418-419.

Sutton, I., and Simmonds, M. A., 1974, Effects of acute and chronic pentobarbitone on the γ-aminobutyric acid system in rat brain, Biochem.Pharmacol., 23:1801-1808.

Synstisky, I. A., Soldatenkov, A. T., and Lajtha, A., 1978, Neurochemical basis of the therapeutic effect of γ-aminobutyric acid and its derivatives, Prog.Neurobiol., 10:89-135.

Tapia, R., GABAergic mechanisms and their relationship to some hormones in the central nervous system, in: "Comparative Aspects of Neuroendocrine Control of Behavior," Front.Horm.Res., (Karger, Basel), 6:86-103.

Valdman, A. V., 1975, Psychopharmacological aspects of emotional stress, Vestnik AMNSSR, No.8, pp.26-33.

Valdman, A. V., Kozlovskaya, M. M., and Medvedev, O. S., 1979, Pharmacological regulation of emotional stress, "Meditsina", Moscow.

Van Gelder, N. M., 1965, The histochemical demonstration of γ-amino butyric acid metabolism by reduction of a tetrazolium salt, J.Neurochem., 12:231-237.

Vasiliev, V. Yu, and Yeryomin, V. P., 1968, Express method for the estimation of activity of γ-aminobutyrate-α-oxoglutarate transaminase. Bull.Exp.Biol.Med., 66:125-126.

Vasiliev, Yu. N., Degtyaryova, E. P., Dmitriev, A. V., Dulinets, L. A., and Ignatov, Yu, D., 1979, The influence of benzodiazepine tranquilizers on the autonomic indices of the state of emotional tension in the therapeutic stomatological clinic, Stomatology, No.4, pp.29-33.

Vladimirova, E. A., 1956, The influence of conditioned reflex induced
        excitation of the central nervous system on the content of
        adenosine triphosphoric and adenosine diphosphoric acids in
        brain, Vopros.Med.Khimii, 2:47-52.
Volicer, L., and Klosowitz, B. A., 1979, Effect of ethanol and stress
        on gamma-aminobutyric acid and guanosine 3',5'-monophosphate
        levels in the rat brain, Biochem.Pharmacol., 28:2677-2679.
Weil-Malherbe, H., 1962, Ammonia metabolism in the brain, in:
        "Neurochemistry," Springfield, Charles C. Thomas Publishers,
        pp.321-330.

Neurochemical Pharmacology of GABA-ergic Drugs

K.S. Rayevsky, G.I. Kovalev and A.N. Kharlamov

Laboratory of Neurochemical Pharmacology, Institute of Pharmacology,
Academy of Medical Sciences of the USSR, Moscow

1. INTRODUCTION

Numerous findings have shown that γ-aminobutyric acid, or GABA, is widely distributed in mammalian brain (Fahn, 1976) and acts as a major transmitter of the central nervous system (Curtis and Johnston, 1974; Sytinsky, 1977). Much interest in the GABA system of the brain has recently been shown by research workers from a variety of disciplines: neuroanatomists, neurophysiologists, neurochemists, pharmacologists, neurologists and psychiatrists. GABA constitutes one of the most interesting fields of study on the contemporary neurobiological scene; its many aspects have been tackled using an interdisciplinary approach. This has all contributed to our knowledge of how the GABA system of the brain operates. Evidence of GABA involvement in neurotransmission in many brain structures has been found; the regional distribution of this aminoacid has been investigated and the properties and location of the enzymes of GABA biosynthesis and breakdown have been discovered, together with the specific transport systems responsible for removing GABA from the synaptic cleft following its release from nerve endings and binding sites on the postsynaptic membrane.

The significance of GABA for central nervous system function may be seen from its involvement in regulating motor activity, maintaining the convulsive threshold, shaping emotional behavior, controlling the release of some pituitary hormones and coordinating the highest forms of integrative activity of the brain, including conditioned reflexes, learning processes and memory. The GABA-ergic system interacts closely with both dopaminergic pathways and other brain transmitter systems.

261

The number of publications on GABA system involvement in the
pathogenesis of many disorders of the central nervous system continues to grow. We may therefore consider GABA involvement in the
pathogenesis of disease such as epilepsy, Huntington's chorea,
Parkinsonism and other extrapyramidal disorders as an established
fact (Enna, 1980).

The set of problems relating to the presence of GABA in the
brain has likewise continued to grow over the years. It has become
necessary to study GABA from many different angles, ranging from
transmitter/receptor interaction at the molecular level to attempts
to use specially synthesized GABA structural analogs in clinical
practice, which are able to penetrate the blood-brain barrier and
reproduce GABA-like effects under in vivo conditions (Zakusov, 1968).
Modern methods have enabled us to study some neurochemical sequences
of synaptic transmission and to determine the possibilities of pharmacological treatment capable of modifying or modulating the function
of the inhibitory GABA-ergic synapse with varying degrees of selectivity. Compounds with the property of changing GABA-ergic transmission in a given direction have become known as GABA-ergic and now
constitute a separate group of neurotropic agents.

No generally accepted concept of "GABA-ergic" exists as yet. It
is recognized, however, that such agents may be collectively defined
as substances imitating GABA's effect, changing GABA concentration in
brain tissue, augmenting or inhibiting transmitter release from the
nerve endings, inhibiting GABA reuptake by terminals and surrounding
glial tissue, preventing interaction between the aminoacid and pre-
and post-synaptic receptors and, finally, raising the latter's sensitivity to GABA (Rayevsky, 1981).

In this article we set out to examine some fundamental aspects
of the GABA problem - chiefly pharmacochemical, neurochemical and
pharmacological. Amongst the wealth of available information, we
took as our guide certain findings from our own laboratories on how
different chemical compounds act selectively at a given point in the
GABA sequence, as a result of which the experimentally observed
changes in GABA-ergic transmission occur. It seems likely that this
approach will lead to the discovery of regulators of brain GABA
function which could be applied to medical practice.

2. METHODS

2.1. BIOCHEMICAL METHODS

2.1.1. Tissue Preparation

All experiments were performed on the brain of male Wistar rats
weighing 140-180 g. The animals were sacrificed (decapitated), the

brain removed and certain structures separated, all within the course
of about 45–50 sec. The operations were all performed in cold. The
method of Hajos (1975) was used in experiments on synaptosomes.

2.1.2. Measuring GABA, Glutamate and Aspartate Levels

Rat heads were frozen in liquid nitrogen and stored at -20°C.
The structures under study were homogenized in cold (4°C) 75%
ethanol, heated in a boiling water bath and centrifuged twice at
15,000 x g for 20 min on each occasion. The supernatant was evaporated and the residue made up with an appropriate volume of distilled water and subjected to electrophoresis (6 volts/cm) on FN–11
paper (GDR) using the method of Grassman et al. (1955) for 7 hours
in 0.05% Na acetate buffer (pH = 4.0). The electrophoretograms were
dried in the air, stained with a 0.05% solution of ninhydrin in acetone, and kept at a temperature of 70-80°C for 20 min. The chromogen
was eluted with 0.005% copper sulphate in 75% ethanol and measured
colorimetrically at 530 nm. Aminoacid concentration was expressed in
μmol/g tissue.

2.1.3. Measuring GAD Activity (EC 4.1.1.15)

We assayed rat brain homogenates using the fluorescent method of
Lowe et al. (1958). Results are expressed as μmol GABA generated per
g per h (fresh tissue).

2.1.4. Measuring GABA-transaminase Activity (EC 2.6.1.19)

The method of Vasilyev et al. (1970), based on the ability of
3-methyl-2-benzothiazole-2-hydrazone (MBTH) to form a chromogen on
condensation with succinic semialdehyde (SSA), was used. In brief,
0.5 ml of a 15% homogenate of brain tissue was added to 1 ml of
substrate mixtures (50 μmol α-ketoglutarate, 50 μmol GABA, 200 μmol
tris buffer, pH 7.2). The mixture was incubated for 30 min at 38°C
and after 20% trichloracetic acid had been added, was then centrifuged for 5 min at 4,000 x g. A volume of 0.05 ml supernatant was
then poured into 0.5 ml 1% MBTH and placed for 3 min in a boiling
water bath, followed by 5 min in a cold bath at 10°. The next stage
consisted of adding 1 ml 0.25% $GeCl_3$ to each assay; the precipitate
formed was then dissolved in 4 ml acetone. The extinction of the
solutions was measured at 660 nm 20 min later. Activity was expressed as μmol SSA per g per h of fresh tissue.

2.1.5. Study of ^{3}H-GABA Uptake

For standard experiments investigating synaptosomal ^{3}H-GABA
uptake we used methods described fully elsewhere.

2.1.6. Study of ^{3}H-GABA Release

The superfusion method of Raiteri et al. (1974) was used, with minor modifications (Kovalov et al., 1982). In brief, 50 ml of a suspension of crude synaptosomes containing 7-8 mg protein per ml was incubated in 1 ml of medium (i.e. NaCl 124 mM, KCl 5 mM, MgCl$_2$ 1.5 mM, CaCl$_2$ 1.3 mM, glucose 10 mM, NaHPO$_4$ 20 mM, KH$_2$PO$_4$ 1.2 mM, aminohydroxyacetic acid 0.2 mM, pH 7.35). Two min later, ^{3}H-GABA was added (Izotop, USSR - 22.4 Ci/mmol) to a final concentration of 10^{-4}mM. The reaction mixture was incubated for 5 min at 37°C and the protein was then separated on a GFC Whatman filter in a superfusion chamber and washed with 15 ml buffer. Superfusion was performed at a rate of 0.6 ml/min. The superfusate was collected once every minute in a scintillation vial containing 8 ml Bray's fluid. Radioactivity was measured using an Intertechnique SL-4000 counter.

2.1.7. Protein Measurement

The standard Lowry et al. (1951) method was used.

3. BEHAVIORAL METHODS

3.1. ANTICONVULSANT ACTIVITY

Anticonvulsant activity was tested on white cross-bred male mice weighing 18-20 g. A convulsion model produced by GABA negative compounds such as thiosemicarbazide (TSC) or bicuculline (BCC) was used. A dose of 18 mg/kg aqueous solution of TSC was administered subcutaneously. BCC was dissolved in 0.1 HCl, made up to the required volume with distilled water (pH approx 6.0). Convulsive attacks followed by death of animals in the control group were observed in 90-95% of all cases 50-60 min following TSC administration, and up to 10 min after administering BCC. A record of the time when convulsions supervened, of the number of animals affected by attacks, and of the number of animal deaths over a 120 min period was made.

3.2. CONDITIONED ACTIVE AVOIDANCE RESPONSE

Wistar albino male rats were used, weighing 160-180 g when the experiment began. The animals were allowed food and water ad libitum, the food consisting of standard pellets. The conditioned avoidance response was tested out in an automatic shuttle box, both sections of which were lit alternately by a 60 watt electric bulb. Seven seconds after the fixed signal of switching the light, a 10-30 mA electrical current was passed through the floor as reinforcement. The animals were conditioned daily, always at the same time. Each conditioning session included 20 instances of exposure to light linked with electrical stimulation. A total of 15 or more correct responses out of 20 during one session was adopted as a criterion of successful training.

3.3. CONDITIONED PASSIVE AVOIDANCE RESPONSE

The experimental chamber was divided into two sections, consisting of a larger, illuminated compartment and a smaller, darkened one, connected by a 10x10 cm passageway. The smaller compartment contained an underfoot grid, through which a 10-30 mA electric current was passed. This served as an unconditioned stimulus.

The experimental session consisted of 2 stages. On day 1, one of the animals (male rats weighing 180-200 g) was placed in the center of the illuminated compartment and the time spent in the light and dark compartments was noted over a 3-min period. Electrical stimulation was applied to the electrode floor at the end of the first trial, when the animal was usually in the darkened compartment. Another 24 hours later the rats were once again placed in the experimental chamber, where the length of time the animals had spent in each of the compartments was again noted.

Animals belonging to the control group spent a high proportion (70%) of the total exposure time in the darkened area of the chamber by reason of their biological characteristics. But their behavior was reversed the next day; the animals then preferred the lit compartment, and only 20% of total exposure time was spent in the darkened section.

3.4. Conflict Situation Method

Albino male rats weighing 200-220 g were deprived of fluid for 2 days and each animal was then placed in an experimental chamber for 10 min, where observations on manner of drinking water from a trough were carried out. On the 6th day of the experiment, while the rat was drinking, an 8-10 mA electrical current was applied, thereby producing a conflict situation and causing fluid-deprived animals to refrain from drinking from the trough because this would have been accompanied by aversive stimulation. During an experimental session lasting 20 min the following parameters were observed: number of punished responses - i.e. the incidence of water-taking negatively reinforced by electric shock, the number of approaches made to the trough and general motor activity.

3.5. Statistical Methods

Mean, standard error and reproducibility were calculated from our results, using Wilcoxin's and Mann-Whitney's non-parametric tests as well as Student's t-test.

4. NEUROCHEMICAL MECHANISMS OF THE ACTION OF GABA-ERGIC COMPOUNDS

It is logical to seek selectively-acting GABA-ergic compounds amongst familiar naturally occurring and synthetic substances resembling GABA in structure and function. These might be investigated as possible regulators of neurochemical transmission. An approach of this kind would involve studying how these substances interact with enzymatic links in the GABA biosynthetic metabolic chain transport processes (as carriers), mechanisms for binding with the receptor complex, and so forth. The present work takes as its theme the systematic study of certain properties of a wide range of GABA structural derivatives and GABA structural analogs. The end-product of our research would be the creation of compounds with the required properties of high structural specificity, ability to penetrate the blood-brain barrier and absence of side-effects, amongst other qualities. Changes in GABA-ergic transmission may occur due either to a specific direct effect or an indirect, metabolically-mediated action. In our view the neurochemical/pharmacological study of the GABA system must settle two interrelated problems. The first consists of investigating whether the effect of certain selected or specially synthesized compounds depend either structurally or functionally on any one of the "links" in the GABA system; this would enable the contribution made by any one component to the overall pharmacological effect to be assessed separately. The second task – almost the reverse of the first – consists of discovering which link or alteration in the GABA system actually shapes the neuro-chemical range of action of a particular compound or characterizes certain pathological conditions of the brain. This approach would use available data on the pharmacological profile of the compounds concerned and knowledge of the pathological processes occurring within the nervous system.

5. ACTION ON ENZYMES

We focused our efforts on a number of GABA-ergic compounds which had already been accepted into clinical practice; firstly, some GABA derivatives and structural analogs at the experimental stage and entering the phase of clinical study; secondly, a group of compounds widely used in the study of GABA-ergic processes. The substances investigated included sodium valproate, β-phenyl-GABA (phenibut), hydroxybutyrate, calcium pantothenate (pantogam) and piracetam, amongst others.

The effect of these compounds was studied on: glutamate decar-boxylase (GAD, EC 4.1.1.15), the enzyme responsible for GABA synthesis within the brain; 4-aminobutyrate: 2-oxoglutarate-aminotrans-ferase (GABA-T, EC 2.6.1.19) – the enzyme promoting GABA degradation; the processes of active GABA transport across the nerve ending – transmitter release and reuptake. We also studied general trans-mitter aminoacid concentration in brain tissue.

As shown in our laboratories by Rayevsky and Kharlamov (1980), valproate possesses a clear-cut anxiolytic action, as well as an anticonvulsant effect. On studying the effect of valproate on transmitter aminoacid concentration, we noted that administering a dose of 200 mg/kg produced statistically significant changes in the aspartate system alone. Aspartate concentration was reduced in all 3 brain structures studied (see Table 1): by 31% in the cortex, 29% in the hippocampus and by 56% of initial level in the hypothalamus (Kharlamov et al., 1982). Our findings agree with those of Schechter et al. (1978), who noted a reduction in aspartate concentration in mouse brain. Valproate also has a known capacity for inhibiting the activity of succinic semialdehyde dehydrogenase (EC 1.2.1.16) and aldehyde reductase (EC 1.1.1.2), which catalyzes the transformation of succinic semialdehyde (SSA) into hydroxybutyrate. It remains unclear, however, how these biochemical findings relate to the underlying pharmacological effects of valproate.

By administering congeners by cyclization and also by conjugation with other physiologically active compounds, a number of lipophilic structural analogs of GABA in addition to GABA derivatives were produced. This led to the treatment of nervous and psychiatric illnesses with agents which varied widely in their structure as well as their pharmacological and neurochemical range. Phenibut, which has been accepted into clinical practice as a tranquilizer, belongs to this group of compounds (Khaunina and Lapin, 1976). Phenibut, or β-phenyl-GABA, is one of the lipophilic derivatives of GABA able to penetrate the BBB. We investigated its action on GABA system enzymes and aminoacid concentration in the brain.

A dose of 1 mmol/kg phenibut showed no effect on GABA-T or GAD activity in rats 30 min after i.p. administration, whereas it did increase GABA concentration in a cerebral cortex homogenate (see Figure 1). Baclofen, a substance with a similar structure, also failed to produce changes in GAD activity (Olsen et al., 1978). Unlike GABA, phenibut has no interaction with the reuptake systems (Morozov et al., 1977) or the bicuculline-sensitive receptors (Rägo et al., 1982). In addition, during superfusion of rat cerebral cortex synaptosomes, phenibut reinforced Ca^{2+}-dependent spontaneous release of ^3H-GABA (Kovalev et al., 1982): in a concentration of 100 μM – up to 130.7±5.9% (see Figure 2) and up to 115.9±4.3% at a concentration 50 μM. The effect of adding GABA at a concentration of 100 μM was shown by a substantial rise in its ability to release tritiated marker from the synaptosomes (268.2±20.0%) – approximately 5 times the effect of an equimolar concentration of phenibut. The effect could be produced, at least in part, by the ability of GABA to stimulate the egress of accumulated ^3H-GABA due to the mechanism of "homo-exchange" (Raiteri et al., 1975). We studied the effect of the GABA receptor antagonists, bicuculline and picrotoxin, on the release process, to help us understand the possible mechanisms underlying this interaction. The former concentration produced a mildly

Table 1. Effect of Sodium Valproate on Concentrations of GABA, Glutamate and Aspartate in Rat
Brain (Kharlamov et al., 1982)

Brain region	GABA (µmol/g tissue)		Glutamate (µmol/g tissue)		Aspartate (µmol/g tissue)	
	Control	Valproate	Control	Valproate	Control	Valproate
Cortex	2.54±0.30	2.90±0.30	10.08±0.52	10.78±1.30	3.10±0.22	2.10±0.20*
Hippocampus	1.50±0.20	1.70±0.32	8.40±0.43	9.30±1.10	2.40±0.20	1.70±0.13*
Hypothalamus	3.40±0.51	3.40±0.40	6.60±0.45	5.80±0.71	3.60±0.34	1.60±0.41*

Note: Mean ± SEM.
       *p<0.05.

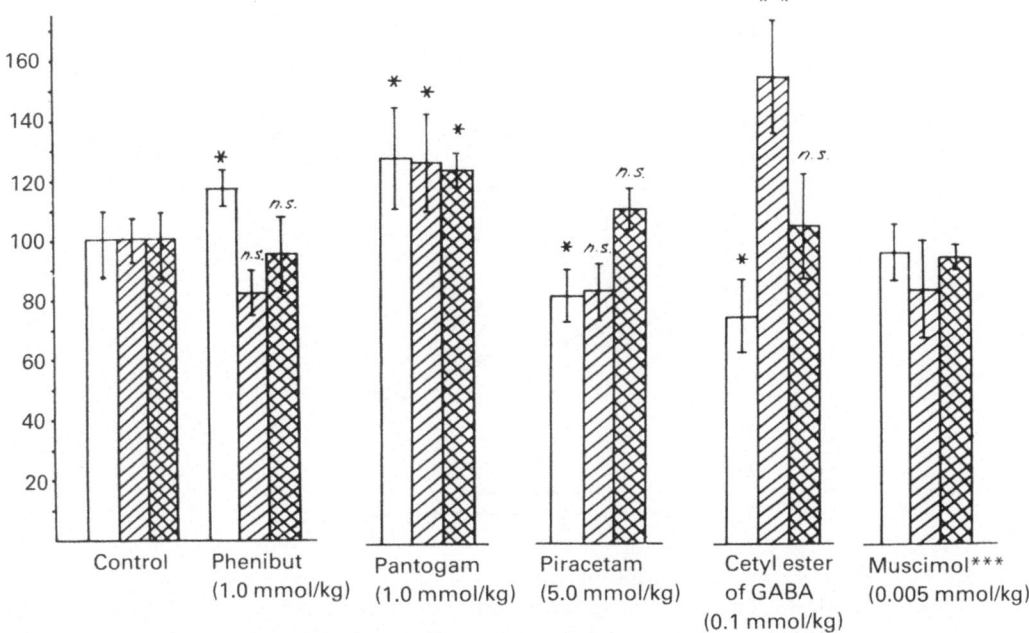

Fig. 1.   Effect of GABA structural analogs on GABA concentration and
          activity of its degrading enzymes in rat brain.   Unshaded
          columns: GABA concentration in rat cerebral cortex;   diag-
          onally striped columns: GAD activity; cross-hatched columns:
          GABA-T activity.   For baseline (100%) we took: 2.22 μmol
          GABA/g fresh tissue; 34±3 μmol GABA/h/100 mg protein (GAD);
          59±85 μmol succinic semialdehyde/h/g fresh tissue (GABA-T).
          The compounds were administered intraperitoneally 30 min
          before the start of the biochemical procedure (6-8 animals
          for each measurement taken).   *p<0.05; **p<0.01; ***data of
          Frey et al. (1979).

stimulating effect (123.6±4.9%) on spontaneous release at a concentration of 100 μM, compared with the effect of phenibut (see Figure
3a).   Picrotoxin did not alter the baseline level of GABA release
(102.3±2.3%).   The effect of BCC itself on spontaneous release may be
explained by this antagonist's ability to cause depolarization of the
cell membranes by blocking cationic permeability (Heyer et al.,
1981).   Nevertheless, when phenibut and bicuculline were introduced
simultaneously into the superfusion medium, phenibut's effect was
somewhat attenuated (by up to 113.0±6.3) whereas if phenibut is given
in association with picrotoxin the releasing effect is noticeably
potentiated, by up to 165.3±12.4%.

     Phenibut, at a concentration of 50 μM, did not alter the release
of labelled compound produced by potassium (28 mM) depolarization,
while a concentration of 100 μM substantially reinforced this effect

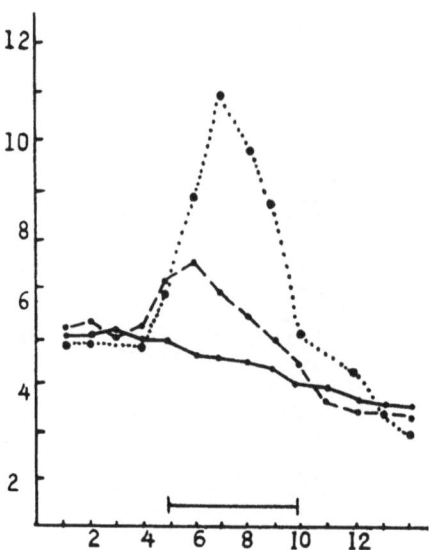

Fig. 2.   Effect of GABA and phenibut on spontaneous release of
          ^{3}H-GABA from P2 fractions of rat brain cortex.   Abscissa –
          fraction number; ordinate – percentage of radioactivity in
          each fraction from summed activity of all fractions and
          filtrate.   Unbroken line – baseline release; broken line –
          effect of 10^{-4}M phenibut; dotted line – effect of 10^{-4}M
          GABA.   Interval of from 5 to 10 min – superfusion with
          buffer with or without compound under study.   *p<0.05;
          **p<0.02.

(see Figure 3b).   GABA receptor antagonists varied in their ability
to influence the phenibut effect in relation to K^+ depolarization:
picrotoxin did not quantitatively change the effect, while bicuculline reversed it.   On the basis of data on bicuculline's antagonizing
action on the rise in GABA concentration produced in the rat brain
limbic system by 100 mg/kg phenibut (Allikmets et al., 1982), our
findings may be viewed as evidence of the existence of bicuculline-
sensitive receptors; their activation by phenibut and GABA leads to
facilitated GABA release.   A similar effect was noted earlier by
Roberts et al. (1978), who described how baclofen enhanced ^{3}H-GABA
release from synaptosomes.

     Our findings on phenibut do not appear to be comparable with
results obtained with baclofen, its structural analog.   These data
indicate the presence of bicuculline-insensitive receptors in slices
of the cerebellum, striatum and frontal cortex which are activated by
baclofen and GABA (Bowery et al., 1980).   GABA and (-)-baclofen both
bind to the same GABA$_B$ type receptors (Bowery et al., 1981) and produce a drop in Ca^{2+}-dependent NA, DA and 5-HT release from slices of

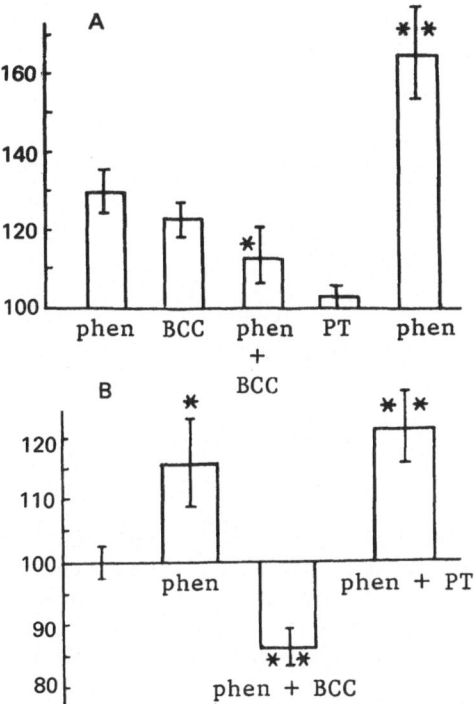

Fig. 3.   Effect of phenibut and GABA receptor agonists on spontaneous
          (A) and potassium-stimulated (B) release of ^{3}H-GABA from P2
          fractions of rat cortex.   BCC - bicuculline; PT - picro-
          toxin; phen - phenibut.   All compounds were administed at a
          concentration of 10^{-4}M.   The effect was calculated as the
          percentage increase in radioactivity over baseline level (A)
          or the action of 28 mM KCl (B), taken as 100% respectively.
          *p<0.05; **p<0.01.

the corresponding brain structures.   On the other hand (-)-baclofen
is known to inhibit K^+-dependent egress of aspartate and glutamate
from cortex slices, without influencing GABA release (Collins et al.,
1982).   Interesting results have been obtained from studying the
influence of phenibut and GABA on ^{3}H-dopamine release from synaptosomes of the rat nucleus accumbens (Kovalev, and Khetay, 1982).   It
was found that neither GABA nor phenibut alter spontaneous release of
labelled dopamine from rat nucleus accumbens at a concentration of
50 µM, indicating that neither compound exerts any influence on the
process of dopamine accumulation.   Furthermore both GABA and phenibut
reduced potassium (30 mM)-induced release of ^{3}H-DA, at a concentration of 50 µM (see Table 2), to 75.3±14.5% and 75.7 ± 21.3% respectively.   When a concentration of 50 µM picrotoxin was introduced
into the superfusion medium, it lost its inhibitory action.   Picrotoxin itself had no influence on the process of ^{3}H-DA release.   These

Table 2. Effect of GABA-ergic Compounds on K^+-Stimulated ^{3}H-Dopamine
          Release During Synaptosomal Superfusion

Compound and concentration (μM)	±	(n)	P
Control	100± 8.8%	(9)	
GABA, 50	75.3±14.5%	(4)	0.02
GABA, 50 + picrotoxin, 50	107.57± 4.56%	(3)	
Picrotoxin, 50	105.4 ± 9.1	(3)	
Phenibut, 50	75.7 ±21.3	(5)	0.05
Phenibut, 50 + picrotoxin, 50	104.1 ± 2.4	(3)	

findings point to the existence of presynatpic GABA-ergic control
over DA release in the nucleus accumbens, mediated by picrotoxin-
sensitive receptors activated by GABA and phenibut.

The synthesis of N-pantoyl-GABA (homopantothenic acid, or
"pantogam") was one of the first attempts to effect passage of this
type of compound through the BBB (Kopelevich et al., 1977). Rats
were given a single dose of 1 mmol of a calcium salt. A moderate
rise in GABA concentration, GAD and GABA-T activity in cerebral
cortex homogenates was noted 30 min later (see Figure 1). A dose of
150 μmol/kg pantogam, given continuously over 3 days, activated GAD,
repressed GABA-T and raised glutamate concentration; this appears to
indicate activation of the GABA shunt (Rosanov, 1980).

No effect was noted on synaptosomal uptake of GABA, however
(Kovalev et al., 1979). Although pantogam penetrates the BBB efficiently, it is only converted into GABA to a minor extent, since
over 90% is excreted by the organism unchanged (Rosanov, 1979).
Presumably the compound's effect is non-specific and brought about
by its influence on energy metabolism.

The cyclic form of GABA, 1-acetamidopyrrolidone-2 (piracetam),
also appears to produce a non-specific GABA-ergic effect, while its
biochemical properties show some similarities to the GABA-mimetic
cetyl ester of GABA (Ostrovskaya et al., 1972; Ostrovskaya et al.,
1982). These two compounds have the shared ability to reduce the
GABA content of the cerebral cortex. In this connection, it is
interesting to note that muscimol, another GABA-mimetic drug. also
tends to reduce GABA concentration in the brain (Frey et al., 1979).
Bearing in mind the tendency of these compounds to produce a
reduction in GAD activity, GABA-mimetics could be presumed capable
of activating feedback mechanism for retarding GABA synthesis.

Piracetam, like the series of methyl and phenyl derivatives of pyrrolidone-2, was shown to be ineffective "in vitro" with respect to the ^{3}H-GABA system (Morozov et al., 1977). Phenibut, the inverted form of 4-phenyl-pyrrolidone-2, was also inactive, as already mentioned (Morozov et al., 1977). Substituting a bulky phenyl group into the beta position seems to eliminate the molecules' ability to utilize the GABA transport system; this was successfully confirmed using GABA-L-2,4-diaminobutyric acid, an inhibitor of neuronal transport. The steric, less unwieldy methyl GABA derivatives α-, β- and γ-methyl GABA maintained their affinity for the synaptosomal carrier system of rat cerebral cortex, competitively inhibiting transport of ^{3}H-GABA (Morozov et al., 1977). These findings have been confirmed by those of other authors.

Guvacine and nipecotic acid are the best known inhibitors of neuronal uptake of ^{3}H-GABA, together with 2,4-diaminobutyric acid (Johnston et al., 1976; Kovalev et al., 1979). Nipecotic acid and GABA were both found to use the same carrier; furthermore, the former had a greater affinity for it than GABA. It is generally maintained that inhibition of GABA uptake takes place non-competitively in the case of nipecotic acid (Johnston, 1976). When we used the standard synaptosomal preparation in our experiments to investigate the dynamics of such inhibition, our findings agreed with those generally accepted (Kovalev and Rayevsky, 1981): that non-competitive inhibition takes place without any change in Michaelis constant (5.8 ± 1.6 μM) (in the control tests, without the addition of inhibitor, K_m = 5.5±1.2 M). V_{max} was greatly reduced, from 0.97±0.10 nmol/ min/mg protein in the control to 0.40±0.08 nmol/ min/mg protein. The inhibition constant was calculated as 13±3 μM; this too, corresponds with published findings, giving this value as 11±3 μM (Johnston, 1976; Beart et al., 1972).

The K_m of nipecotic acid as a GABA uptake inhibitor is also known to be the equivalent of the K_m for uptake of nipecotic acid itself (Krogsgaard-Larsen and Johnston, 1975). Furthermore, nipecotic acid uptake is inhibited and its release promoted by GABA. These two effects are reciprocal (Johnston et al., 1976). Taken together, these facts provide evidence that GABA and nipecotic acid can both use the same active transport system but are out of line with the non-competitive uptake inhibition observed by us and other research workers.

In view of the complexity of these findings, we took as our starting point the substantial involvement of ^{3}H-GABA "homo-exchange" in the process of its own synaptosomal accumulation (Raiteri et al., 1975), whereby the kinetic parameters of net transport can only be correct if this factor is taken into account.

We studied the possible role of homo-exchange, adopting the methods based on potassium depolarization (56 mM) of synaptosomes prior to incubating them with labelled substance (Ryan and Roskoski, 1977). The standard synaptosomal preparation was subjected to pre-liminary disintegration for this purpose. Our findings on the inter-dependence of "apparent" and net uptake on GABA concentration show that, at a ^{3}H-GABA concentration in the medium of up to 2 μM indirect "apparent" and net uptake coincide. Beginning at the 4-5 μM level, net uptake reaches a plateau, whereas "apparent" uptake does not reach saturation at the concentration levels under study. After 3 min incubation at a 5 μM concentration of ^{3}H-GABA, net uptake amounts to 50% of "apparent" uptake (Kovalev and Rayevsky, 1981).

Findings on the dynamics of GABA net uptake inhibition using standard and disintegrated synaptosomal preparations are given in Figure 4, showing that K_m and V_{max} of the control experiments differ in this instance from the corresponding values obtained with standard synaptosomes (K_m = 2.0±0.9 and 5.5±1.2 μM; V_{max} 0.38±0.09 and 0.97 ± 0.10 nmol/min/mg protein respectively).

When the inhibitor is present, the value of K_m is substantially increased compared with the corresponding control value, rising to

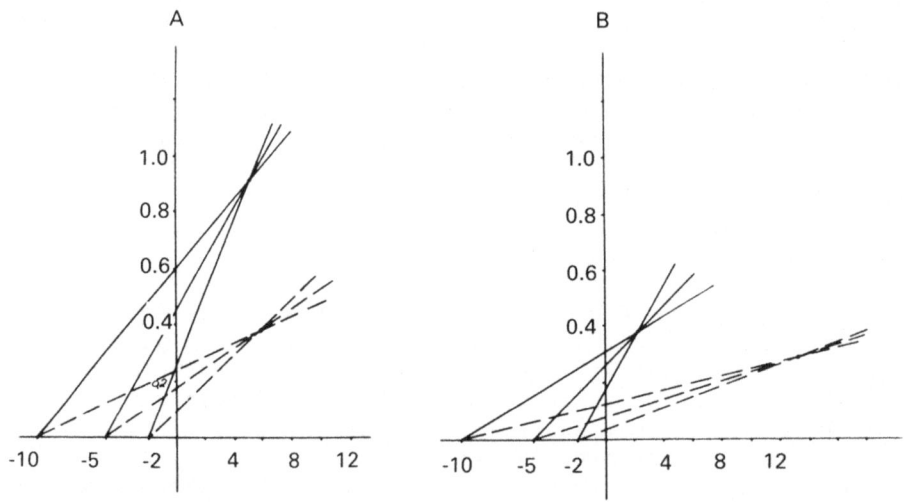

Fig. 4.   Kinetics of nipecotic acid inhibition of ^{3}H-GABA uptake by
          (A) standard preparation, and (B) previously depolarized
          synaptosomes of rat cerebral cortex, employing direct
          Cornish-Bowden coordinates (Kovalev and Rayevsky, 1981).
          Abscissa: concentration of ^{3}H-GABA in the incubation
          medium (μM); ordinate: rate of ^{3}H-GABA accumulation by
          synaptosomes (nmol/min/mg protein). Unbroken line –
          control, no inhibitor added; broken line – inhibitor
          present. Preliminary depolarization performed using a
          medium with 56 mM KCl.

12.8±3.1.µM, while the value of V_{max} remains virtually unchanged, at 0.29±0.06 nmol/min/mg protein. These findings provide clear evidence of the competitive nature of the inhibition. The inhibition constant was calculated as 3.9±0.2 µM on the basis of the above experiments.

We thus demonstrated the complexity of the accumulation process of exogenous labelled GABA, which can only be achieved in vitro with the participation of at least two mechanisms; homo-exchange and net uptake mediator involvement. The competitive nature of inhibition of nipecotic acid net uptake and the non-competitive character of GABA accumulation indicate that in determining the kinetic parameters of presynaptic GABA transport inhibition, the substantial involvement of aminoacid "homo-exchange" in this process should be borne in mind. At the same time our findings may be taken as confirming the existence of GABA net uptake and the physiological function it fulfills of complementing GABA's mediator activity within the brain.

6. THE PHARMACOLOGY OF GABA-ERGIC COMPOUNDS

6.1. ANTICONVULSANT ACTION

The connection between the development of convulsive states and disruption of brain GABA-ergic system may now be taken as proven (Meldrum, 1978). It is now generally recognized that administering GABA-negative compounds such as thiosemicarbazide, picrotoxin and bicuculline brings on convulsant attacks.

We performed a study on the anticonvulsant effect of GABA-positive compounds operating in different ways on thiosemicarbazide or bicuculline-induced model convulsions. As illustrated in Table 3, all the substances investigated, except for pantogam, were effective to some degree in suppressing thiosemicarbazide-induced convulsions. Only sodium valproate, however, was successful in forestalling the onset of bicuculline-induced convulsant attacks, and then only at the highest dose tested (see Table 3).

It was interesting to note that AHAA, the GABA-T inhibitor, which proved highly effective on the thiosemicarbazide convulsion model, was hardly active by comparison in experiments with the bicuculline model, only preventing death in 33.3% of the animals. When considering the findings, it is especially important to note how efficacy varies from one GABA-ergic compound to another in the models under study. Our data corroborate those of authors who maintain that most GABA-positive substances are effective against model convulsions connected with the use of GAD inhibitors (Khaunin, 1978). Pantogam's failure to counteract convulsions (failure as an anti-

Table 3. The Effect of GABA-ergic Substances on Thiosemicarbazide- and Bicuculline-induced Convulsions in Mice

Substances	Dose (mg/kg)	Thiosemicarbazide (18 mg/kg)			Bicuculline (3 mg/kg)		
		Latent period of onset of convulsant attacks (min)	No. of animals with convulsions	No. of animals dying	Latent period of onset of convulsant attacks (min)	No. of animals with convulsions	No. of animals dying
0.85% NaCl		61.2± 3.1	91.7	87.5	12.2±0.9	90.0	90.0
Phenibut	100	131.0±14.7*	50.0*	33.3*	13.6±2.1	100.0	100.0
Phenibut	200	146.3± 2.6*	50.0	0	16.3±3.8	80.0	80.0
Pantogam	400	67.8± 7.7	100.0	100.0	9.0±1.5	90.0	90.0
Piracetam	1400	86.2± 7.2	83.3	50.0*	14.2±4.2	87.5	87.5
AHAA	20	62.0± 4.1	16.7*	0	14.4±2.5	83.3	66.6*
Sodium valproate	200	93.4± 7.6*	83.3	16.7*	14.8±3.8	100.0	100.0
Sodium valproate	300	137.0±11.2	33.3*	33.3*	14.4±3.8	87.5	100.0
Sodium valproate	400	124.0±18.7*	16.7*	16.7*	13.2±3.0	16.6*	16.6*

Note: bicuculline administered subcutaneously, all other compounds administered intraperitoneally. *$p < 0.05$.

convulsant) might be due to administering too low a dose. This
compound was found to act as an anticonvulsant against pentylenete-
trazole-induced convulsions by Kovler et al. (1980). Piracetam's
failure to produce this effect might be explained by its influence
being concentrated on the cerebral cortex (Ostrovskaya et al., 1982),
whereas mid-brain structures appear to play a major part in controlling the spread of convulsant activity. The differing effectiveness
of the two GABA-T inhibitors, valproate and AHAA, warrants particular
attention. Findings similar to our own on the difference in anticonvulsant action of these two compounds were obtained by Iadarola
and Gale, using pentalenetetrazole and electric shock-induced
convulsions. In view of the fact that AHAA is a much more powerful
GABA-T inhibitor than valproate, and its properties more closely
resemble those of irreversible inhibitors such as γ-acetylenic-GABA
and γ-vinyl-GABA, it is interesting to compare our observations with
Kendall's et al. (1981) recent findings, that γ-vinyl-GABA exerts an
anticonvulsant effect when picrotoxin is administered, but is ineffective against bicuculline-induced convulsions.

Valproate's success, found to apply in various convulsant
models, may be interpreted with reference to this compound's ability to produce a selective build-up of GABA at the nerve terminals
(Iadarola and Gale, 1981).

One other possible explanation of sodium valproate's anticonvulsant effect might be connected with its ability to produce a
metabolic shift towards the accumulation of succinic semialdehyde
(Van der Laan et al., 1979) and hydroxybutyrate (Sneed et al.,
1980) - late products of the GABA shunt - which could in themselves
play a part in the anticonvulsant defense mechanism. Nor should
the possibility be excluded of the involvement of decreased aspartate concentration in the brain (Schechter et al., 1978; Kharlamov
et al., 1982), not to mention valproate's ability to activate GAD
(Nau and Loscher, 1982). It was also shown recently by Loscher
(1981) that valproate's metabolites may exert an anticonvulsant
effect. The part played by GABA system activation in producing an
anticonvulsant effect is given further support by findings obtained
with other GABA-positive compounds.

Thus, cetyl-GABA, progabid and SL 75 102 - compounds with good
brain penetration - show a broad range of anticonvulsant action, in
the case of convulsions produced by bicuculline, picrotoxin, pentylenetetrazole, strychnine or electrical stimulation (Ostrovskaya et
al., 1972; Worms et al., 1982). The position as regards the anticonvulsant properties of baclofen and phenibut, its closest analog,
remains unclear. Some doubts have been cast on the anticonvulsant
action of the former by Meldrum (1978). In our experiments, phenibut was inactive against bicuculline, but provided a good measure
of protection against GABA deficiency convulsions; this agrees with
the above regarding phenibut's ability to increase GABA concentration
in the brain and reinforce its release from the synaptosomes.

7. GABA-ERGIC COMPOUNDS AND CONDITIONED REFLEXES

We used several GABA-ergic compounds, which function differently, to investigate the involvement of GABA-ergic processes in
shaping complex forms of animal behavior in general and the acqui-
sition of conditioned avoidance response (CAR) in particular. A
deterioration in the performance of active avoidance response was
noted following administration of AHAA, phenibut and bicuculline (see
Figure 5). A lengthening of the latency of CAR agreed with observations made earlier (Khaunina, 1978) and with results obtained using
baclofen, a close analog of phenibut, by Delini-Stula (1977). An
increase in latency of the reflex brought about by AHAA had also been
reported by Benton and Rick (1974). The fact that bicuculline's
effect is similar to that of AHAA and phenibut may at first seem
surprising. This action appears to be non-specific, however, and is
a result of the animal's "frozen" pre-convulsant state, as confirmed
by Cuomo and Cortese's (1980) findings, indicating bicuculline's
inhibitory effect on motor activity. The fact that pantogam, which
has sedative properties, fails to produce any effect (Kovler et al.,
1980) might be due to giving too low a dose. CAR performance remained unaltered with piracetam, but this result is not contradictory
to findings on the compound's favorable influence on an animal's
conditioned reflex performance (Mashkovsky et al., 1977; Giurgea et
al., 1971), in that these studies were investigating the effect of
repeated dosing with piracetam. In view of the numerous data available on the depressant effect of most GABA-positive substances and of
GABA itself (Ostrovskaya et al., 1972; Allikmets et al., 1979), our
findings on the influence of the GABA-T inhibitors and sodium valproate on the conditioned response could be interpreted as follows:

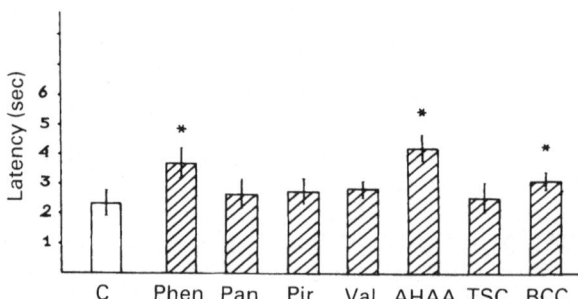

Fig. 5.   Influence of GABA-ergic compounds on performance in con-
          ditioned active avoidance response (Rayevsky and Kharlamov,
          1980). Abscissa - C - control; Phen - phenibut (100 mg/kg);
          Pan - pantogam; Pir - piracetam (1400 mg/kg); Val - sodium
          valproate (200 mg/kg); AHAA - aminohydroxyacetic acid
          (20 mg/kg); TSC - thiosemicarbazide (10 mg/kg); BCC -
          bicuculline (2 mg/kg). $*p < 0.05$.

The increase in latency of CAR following AHAA administration is obviously related to an overall increase in GABA concentration in the brain of over 200% (Wallach, 1961). Valproate, unlike AHAA, was inactive in this situation - possibly a result of its more selective action on GABA accumulation in the brain (Iadarola and Gale 1981). Experiments were performed administering bicuculline following pretreatment with AHAA in order to test the theory of a deterioration in CAR being linked with an overall rise in GABA in the brain. As shown in Figure 6, bicuculline cancelled out the increase in latency produced by AHAA. Latency of the reflex began to lengthen again (i.e. AHAA's effect was still maintained) 1 h after administering bicuculline, just when this compound had ceased to act - a confirmation of our idea of the link between AHAA's effect and the accompanying overall rise in brain GABA (Rayevsky and Kharlamov, 1980). These results corroborate the findings of Tunnicliff et al. (1976), which reveal a negative correlation between GABA concentration, the activity of its degradative enzymes and the ability of various animal strains to develop conditioned responses.

8. GABA, LEARNING PROCESS AND MEMORY

The involvement of the GABA-ergic link in the learning process and memory has hardly been studied, and such data as we do possess is highly contradictory. The most widely recognized approach to studying the processes of memory are founded on the conditioned passive avoidance response method and its various modifications. We investigated AHAA and sodium valproate - GABA-T inhibitors - using

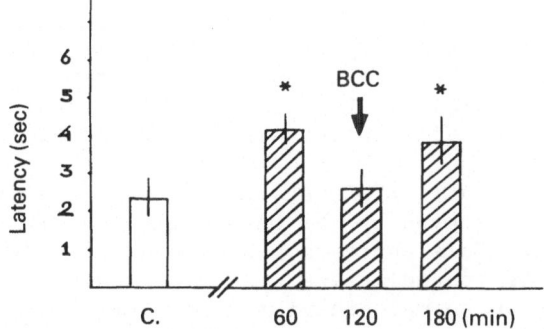

Fig. 6. Reversal of the inhibitory effect of aminohydroxyacetic acid (AHAA) on conditioned active avoidance response by bicuculline (Rayevsky and Kharlamov, 1980). Abscissa – unshaded column (C) – control; shaded column – 60, 120 and 180 min after administering AHAA. *p<0.05. M ± SEM.

this model. We showed that valproate alone, when administered systemically, and only then at the early stage of memory trace formation, had an influence on this process, which it impaired (see Figure 7). Relevant data in the literature do not agree completely with our findings. Grecksch et al. (1978), for example, showed that administering valproate to the hippocampus improved memory consolidation. Grecksch and Matthies (1981) also found an improved working of memory after administering picrotoxin into the hypothalamus. It is worth noting that systemic administration of picrotoxin produced the opposite effect, i.e. a deterioration in reproduction of the memory trace.

It will be seen from surveying the above findings that GABA-ergic compounds may produce completely opposite effects on memory processes, depending on the method of administration. Those effects produced by systemic administration of GABA-ergic compounds should apparently be taken as non-specific. The effect we noted for sodium valproate, for instance, might be considered non-specific, connected with the tranquilizing properties of the compound, as a result of which the force of the negative reinforcing stimulus may be reduced.

9. GABA INVOLVEMENT IN EMOTIONAL BEHAVIOR IN ANIMALS

Abundant data have recently accumulated on GABA system involvement in the shaping of emotional responses such as aggression, behavior in a conflict situation and defensive behavior. Thus, it was established that GABA system function broke down in aggressive isolated animals: GAD activity was reduced (Blindermann et al., 1979) and GABA concentration fell in the striatum, the hippocampus and the olfactory bulb (Delini-Stula and Vassout, 1978). Administering AHAA, sodium valproate and piracetam considerably reduced the incidence of attack in aggressive isolated animals (Krsiak et al., 1981). GABA-negative compounds, such as thiosemicarbazide, bicuculline, picrotoxin and allylglycine are apparently able to promote displays of aggression (Poshivalov, 1981). We can assume on these grounds that the GABA system of the brain is fundamentally involved in the development of aggressive behavior in animals (Krsiak et al.1981).

Creating a model conflict situation is the most suitable means of evaluating the part played by the GABA-ergic link in a compound's tranquilizing action. This approach has already revealed the benzodiazepines' success as tranquilizers. The influence of GABA-ergic compounds themselves on animal behavior in a conflict situation was thus considered worth studying.

We set out to investigate the influence of AHAA, pantogam, piracetam and valproate on rat behavior in a conflict situation. Sodium valproate was found to exert a definite tranquilizing effect (see Figure 8), comparable with that of the benzodiazepines. The

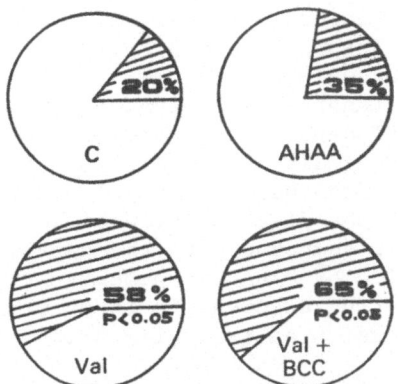

Fig. 7.   Effect of sodium valproate (Val) and aminohydroxyacetic
          acid on the acquisition of conditioned passive avoidance
          response.   Shaded segment - % time spent by animals (rats)
          in darkened compartment on 2nd day of training after
          administration of 0.85 NaCl (C); of 20 mg/kg AHAA
          (aminohydroxyacetic acid); 200 mg/kg sodium valproate (Val)
          and 2 mg/kg bicuculline, following prior administration of
          sodium valproate (Val + BCC).

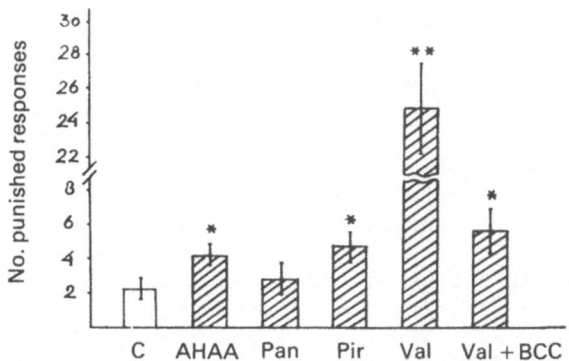

Fig. 8.   Effect of GABA-ergic substances on the behavior of animals
          in a conflict situation (Kharlamov and Rayevsky, 1980).
          Abscissa: C - control; AHAA - aminohydroxyacetic acid
          (20 mg/kg); Pan - pantogam (500 mg/kg); Pir - piracetam
          (1400 mg/kg); Val - sodium valproate (200 mg/kg); Val + BCC
          - bicuculline administered after pretreatment with 200 mg/kg
          sodium valproate. *p<0.05; **p<0.01. M ± SEM.

number of punished incidences of water-taking showed a lesser but
statistically significant increase under the influence of AHAA and
piracetam. The specificity of valproate's tranquilizing action was
also borne out, both by Kozlovakaya's et al. (1981) cat experiments

and Lal's et al. (1980) investigations. The fact that AHAA's anticonflict action is less pronounced correlates well with the poor performance of γ-acetylenic-GABA (another GABA-T inhibitor) in a similar situation (Andreev and Ignatov, 1981) and might be explained by the general depressant effect prevailing in both instances. Piracetam's mildly anxiolytic effect was demonstrated on another model, distinct from our own, by File and Hyde (1978). We therefore consider that the distinct tranquilizing effect of sodium valproate - a GABA-ergic compound with the capacity of selectively inducing the accumulation of GABA in the nerve terminal (Iadorola and Gale, 1981) - is a powerful argument in favor of the GABA system's involvement in the evolution of tranquilization.

CONCLUSION

     Reviewing data on a) the neurochemical range of the GABA struc-tural analogs and derivatives under consideration and b) their pharmacological action, it is interesting to see how these two aspects correlate.

     It may be clearly seen that overall activation of the GABA system of the brain, produced either by muscimol type GABA receptor agonists or non-selective aminoacid metabolism inhibitors, such as AHAA, γ-acetylenic GABA and gabaculine, is associated with the onset of sedation, a drop in locomotor activity and general subduing of the animals. A similar effect would appear to develop following high doses of phenibut (Khaunina 1978), baclofen (Delini-Stula, 1977), pantogam (Kovler, 1980), benzodiazepine tranquilizers, cetyl ester of GABA (Ostrovskaya et al., 1972) and progabide (Worms et al., 1982). An increase in total GABA concentration is likewise observed, due either to inhibition of its metabolic degradation, or GABA formation from its precursor "prodrugs", such as progabide and cetyl ester of GABA. The increase in latency of the conditioned active avoidance response observed in rat behavior experiments, following admini-stration of AHAA and phenibut, which is reversed by bicuculline, corroborates this view. The generalized depressant effect is pre-sumably masking the display of more subtle emotionally-related effects produced on behavior, which means that any anxiolytic effect is barely noticeable upon administering AHAA, notwithstanding a considerable increase in GABA concentration (Wallach, 1961). A similar picture is obtained with muscimol, which depresses animals generally, to the point of loss of righting reflex, although muscimol operates differently. Muscimol cannot generally be shown to exercize a tranquilizing effect (De Feudis, 1980).

     Valproate, on the other hand, does manage to raise GABA concen-tration selectively in the transmitter compartment of the nerve terminals (Iadarola and Gale, 1981); this would appear to be respons-ible for this compound's anticonvulsant effect, which is almost

missing in the case of AHAA type substances, at least with respect to
bicuculline-induced convulsions. The delayed anti-convulsant effect
of γ-vinyl-GABA correlates with this compound's ability to produce
GABA accumulation in the transmitter compartment at that particular
time (Iadarola and Gale, 1981). Findings indicating that nipecotic
acid esters exert an anticonvulsant effect which can penetrate the
brain and increase GABA concentration in the synaptosomes (Wood et
al., 1980) agree with this hypothesis.

Our discovery of a clear-cut anxiolytic effect of valproate
used on rats in a conflict situation could also be compared with
benzodiazepines' action, and likewise appears to tie up with the
accumulation of transmitter aminoacid at the nerve endings. As with
the benzodiazepines, the GABA-ergic character of valproate's effect
is supported by its bicuculline reversibility (Rayevsky and
Kharlamov, 1980). The general GABA level in the brain usually
remains unchanged at the dose level of 200 mg/kg associated with
this effect, but this is accompanied by a drop in the concentration
of aspartate, an excitatory transmitter (Schechter et al., 1978;
Kharlamov et al 1982), which may prove to be a fundamentally new
mechanism for both anticonvulsant and tranquilizer effects. Other
transmitter systems may be involved in producing the aniticonvulsant
effect, however - dopaminergic nigrostriatal and tegmento-limbic
pathways in particular (Gale and Casu, 1981; Rayevsky et al., 1982).

Piracetam was the only one of the various GABA structural
analogs we studied without a depressant action; in a conflict situation, however, this compound did show a slight anxiolytic effect
(see Figure 8); this is borne out by the findings of File and Hyde
(1978). GABA concentration in the brain and GAD activity are somewhat reduced by systemic administration of piracetam - possibly a
result of the regulatory influence exercized by a feedback mechanism.
It was recently found that piracetam, like other GABA-mimetics, can
reinforce GABA-ergic inhibition in the cortex (Ostrovskaya et al.,
1982). This is borne out by information on the selective accumulation of labelled piracetam in a particular region of the brain
(Ostrowski et al., 1975). Available neurochemical data raise
expectations of discovering selective regulators of GABA system
function amongst compounds having both a close structural resemblance
to the aminoacid itself and an affinity for a particular link in the
GABA-ergic transmission chain. Besides substances inhibiting enzymes
of GABA metabolism with a selective and reversible action they would be
the compounds mediating GABA release, competitive uptake inhibitors
which are effective when administered systemically and various types
of GABA receptor agonist. A search for substances able to regulate
the state of the transmitter compartment at GABA-ergic nerve endings
also appears a promising line to pursue.

REFERENCES

Allikmets, L., Kh., Polevoy, L. G., Tsaryeva, T. A., and Zharkovsky,
    A. M., 1979, Dopaminergic component in the mode of action of
    GABA derivatives and structural analogs, Farmakol.Toksikol.,
    42:(6) 603-606.
Allikmets, L. Kh., Rägo, L. K., and Nurk, A. M., 1982, Influence of
    the GABA-receptor blocker bicuculline on the effects of
    phenibut and diazepam, Byull.Eksp.Biol.Med., 93:(5) 64-65.
Andreev, B. V., and Ignatov, Yu. D., 1981, Effect of GABA-negative
    drugs on the anxiolytic and sedative effects of diazepam,
    Byull.Eksp.Biol.Med., 91:(6), 685-687.
Beart, P. M., Johnston, G. A. R., and Uhr, M. L., 1972, Competitive
    inhibition of GABA uptake in rat brain slices by some GABA
    analogues of restricted conformation, J.Neurochem.,
    19:1855-1861..
Benton, D., and Rick, J. T., 1974, The effect of aminohydroxyacetic
    acid on brain gamma-aminobutyric acid levels, acetylcholin-
    esterase activity and shuttle-box learning in rats. - IRCS
    (Research on: Neurobiology and Neurophysiology; Pharmacology;
    Psychology), 2:1221-1231.
Blindermann, J. M., Maitre, M., and Mandel, P., 1979, Apoenzyme
    concentration and turnover-number of L-glutamate decarboxy-
    lase in some regions of rat brain, J.Neurochem., 32:245-246.
Bowery, N. G., Hill, D. R., Hudson, A. L., Doble, A., Middlemiss, D.
    N., Shaw, J., and Turnbull, M., 1980, - Baclofen decreases
    neurotransmitter release in the mammalian CNS by an action at
    a novel GABA receptor, Nature, 283:92-94.
Bowery, N. G., Hill, D. R., and Hudson, A. L., 1981, ^{3}H-GABA and ^{3}H-
    baclofen are ligands for the same bicuculline-insensitive
    site on mammalian CNS synaptic membranes, Brit.J.Pharmacol.,
    74:222-223.
Collins, G. G., Anson, J., and Kelly, E. P., 1982, Baclofen: effects
    on evoked field potentials and amino acid neurotransmitter
    release in the rat olfactory cortex slice, Brain Res.,
    238:371-383.
Cuomo, V., and Cortese, I., 1980, Effects of bicuculline and
    chlordiazepoxide on locomotor activity and avoidance per-
    formance in rats, Experientia, 36:1208-1210.
Curtis, D. R., and Johnston, G. A. R., 1974, Amino acid transmitters
    in the mammalian CNS, Ergebn.Physiol.Biol.Chem.Exp.
    Pharmacol., 69:97-188.
De Feudis, F. V., 1980, Physiological and behavioural studies with
    muscimol, Neurochem.Res., 1047-1068.
Delini-Stula, A., 1977, Baclofen-induced modification of conditioned
    discriminative avoidance behaviour and contraversive turning
    in the rat, Eur.J.Pharmacol., 46:265-274.
Delini-Stula, A., and Vassout, 1978, Modulatory effects of baclofen,
    muscimol and GABA on interspecific aggressive behaviour in
    the rat, Neuropharmacology, 17:1063-1065.

Enna, S. J., 1980, GABA and neuropsychiatric disorders, Can.J. Neurol.Sci., 7:257-259.

Fahn, S., 1976, Regional distribution studies of GABA and other putative neurotransmitters and their enzymes, in: "GABA in Nervous System Function," E. Roberts, T. N. Chase, and D. B. Tower, eds., Raven Press, New York, pp.169-186.

File, S. E., and Hyde, J. R. G., 1978, Piracetam, a non-sedative anxiolytic drug, Br.J.Pharmacol., 62:425-426.

Frey, H.-H., Popp, C., and Löscher, W., 1979, Influence of inhibitors of the high affinity GABA uptake on seizure thresholds in mice, Neuropharmacology, 18:581-590.

Gale, K., and Casu, M., 1981, Dynamic utilization of GABA in substantia nigra: regulation by dopamine and GABA in the striatum, and its clinical and behavioural implications, Molec.Cell.Biochem., 39:369-405.

Giurgea, C., Lefevre, D., Lescrenier, C., and David-Remacle, M., 1971, Pharmacological protection against hypoxia-induced amnesia in rats, Psychopharmacologia, (Berl.), 20:160-168.

Grabmann, W., Hannig, K., and Plöckl, M., 1955, Eine Methode zur quantitativen Bestimmung der Aminosäurezusammensetzung von Eiweisshydrolysaten durch Kombination von Elektrophorese und Chromatographie, Hoppe-Seyler's Z.Physiol.Chem., 299:258-276.

Grecksch, G., and Matthies, H., 1981, Differential effects of intra-hippocampally or systemically applied picrotoxin on memory consolidation in rats, Pharmac.Biochem.Behav., 14:613-616.

Grecksch, G., Wetzel, W., and Metthies, H., 1978, Effect of n-dipropylacetate on the consolidation of a brightness discrimination, Pharmacol.Biochem.Behav., 9:(2) 269-271.

Hajos, F., 1975, An improved method for the preparation of synaptosomol fractions in high purity, Brain Res., 93:485-489.

Heyer, E. J., Nowak, L. M., and Macdonald, R. L., 1981, Bicuculline: a convulsant with synaptic and nonsynaptic actions, Neurology, 31:1381-1390.

Iadarola, M. J., and Gale, K., 1981, Cellular compartments of GABA in brain and their relationship to anticonvulsant activity, Molec.Cell.Biochem., 39:305-330.

Johnston, G. A. R., 1976, Physiologic pharmacology of GABA and its antagonists in the vertebrate nervous system, in: "GABA in Nervous System Function," E. Roberts, T. N. Chase, and D. B. Tower, eds., Raven Press, New York, pp.395-412.

Johnston, G. A. R., and Stephenson, A. L., 1976, Uptake and release of nipecotic acid by rat brain slices, J.Neurochem., 26:83-87.

Kendall, D. A., Fox, D. A., and Enna, S. J., 1981, Effect of γ-vinyl GABA on bicuculline-induced seizures, Neuropharmacology, 20:351-355.

Kharlamov, A. N., and Rayevsky, K. S., 1980, Tranquilizing effect of n-dipropyl acetate and other GABA-ergic substances in a conflict situation, Byull.Eksp.Biol.Med., 89(7):35-37.

Kharlamov, A. N., Kovalev, G. T., and Shumkova, O. V., 1982, Effect

of sodium valproate on the concentration of transmitter amino
acids in rat brain, Pharmacol.Toxicol., 45(5):31-34.

Khaunina, R. A., 1978, The neurotropic activity of phenylpyrroli-
dones-2 (P-2), Byull.Eksp.Biol.Med., 85(3):301-304.

Khaunina, R. A., and Lapin, I. P., 1976, Phenibut - a new tranquilizer, Khim.Pharm.Zh., 10(12):125-127.

Kopelevich, V. M., Evdokimova, G. S., Mariyeva, T. D., and
Shmuilovich, L. M., 1971, The synthesis of D-homopantothenic
acid, Khim.Pharm.Zh., 5(9):21-22.

Kovalev, G. I., Kopelevich, V. M., Bulanova, L. N., Bursky, R. N.,
Gunar, V. I., and Rayevsky, K. S., 1979, Synthesis of some
γ-aminobutyric acid derivatives and study of their effects on
accumulation of labeled (^{3}H)-γ-aminobutyric acid by brain
synaptosomes, Khim.Pharm.Zh., 13(10):18-24.

Kovalev, G. T., and Rayevsky, K. S., 1981, Nipecotic acid, competitive inhibitor of ^{3}H-GABA net-uptake by rat brain synaptosomes, Byull.Eksp.Biol.Med., 91(6):692-694.

Kovalev, G. T., Prikhozhan, A. V., and Rayevsky, K. S., 1982, Presynaptic component in the mode of action of phenybut,
Byull.Eksp.Biol.Med., 94(11):59-61.

Kovalev, G. T., and Hetey, L., 1983, GABA and phenybut inhibit
potassium-stimulated release of ^{3}H-dopamine from rat brain
n. accumbens synaptosomes, Neurokhimia, 2(3) 315-318.

Kovler, M. H., Avakumov, V. M., and Kruglikova-Lvova, R. P., 1980,
Pantogam - new psychopharmacological drug, Khim.Pharm.Zh.,
14(9):118-122.

Kozlovskaya, M. M., Kharlamov, A. N., Rayevsky, K. S., and Valdman,
A. V., 1981, GABAergic link in development of tranquilizing
effects in cats, Byull.Eksp.Biol.Med., 91(1):45-48.

Krogsgaard-Larsen, P., and Johnston, G. A. R., 1975, Inhibition of
GABA uptake in rat brain slices by nipecotic acid, various
isoxazoles and related compounds, J.Neurochem., 25:797-802.

Krsiak, M., Sulcova, A., Tomasikova, Z., Dlohozkova, N., Kosar, E.,
and Masek, K., 1981, Drug effects on attack, defense and
escape in mice, Pharmacol.Biochem.Behav., 14 Suppl.1:47-52.

Lal. H., Shearman, G. T., Fielding, S., Dunn, R., Kruse, H., and
Theurer, K., 1980, Evidence that GABA mechanisms mediate the
anxiolytic action of benzodiazepines: a study with valproic
acid, Neuropharmacology, 19:785-789.

Löscher W., 1982, Anticonvulsant activity of metabolites of valproic
acid, Arch.Int.Pharmacodyn.Ther., 249:158-163.

Lowe, J. P., Robins, E., and Eyerman, G. S., 1958, The fluorometric
measurement of glutamic decarboxylase and its distribution in
brain, J.Neurochem., 3:8-18.

Lowry, O. H., Rosebrough, N. J., Farr, A. L., and Randall, R. J.,
1951, Protein measurement with the Folin phenol reagent,
J.Biol.Chem., 193:265-275.

Mashkovsky, M. D., Roschina, L. F., and Polezhayeva, A. I., 1977,
Some features of the pharmacological action of piracetam,
Farmakol.Toksikol., 40(6):676-683.

Meldrum, B. S., 1978, Gamma-aminobutyric acid and the search for new
        anticonvulsant drugs, Lancet, 2:304-306.
Morozov, I. S., Kovalev, G. I., Maisov, N. I., Kovalev, G. V., and
        Rayevsky, K. S., 1977, The effects of γ-aminobutyric acid and
        analogs on uptake of (^{3}H)-γ-aminobutyric acid by rat brain
        synaptosomes, Khim.Pharm.Zh., 11(1):13-15.
Nau, H., and Löscher, W., 1982, Valproic acid: brain and plasma
        levels of the drug and its metabolites, anticonvulsant
        effects and γ-aminobutyric acid (GABA) metabolism in the
        mouse, J.Pharmacol.Exp.Ther., 220:654-659.
Olsen, R. W., Ticku, M. K., Van Ness, P. C., and Greenlee, D., 1978,
        Effects of drugs on γ-aminobutyric acid receptors, uptake
        release and synthesis in vitro, Brain Res., 139:277-294.
Ostrovskaya, R. U., Parin, V. V. and Tsybinqu, N. M., 1972, The
        comparative neurotropic potency of gamma-aminobutyric acid
        and its cetyl ester, Byull.Eksp.Biol.Med., 73(1):51-55.
Ostrovskaya, R. U., Molodavkin, G. M., and Kovalev, G. T., 1982,
        GABAergic cortical component in the action of piracetam and
        cetyl ester of GABA, Byull.Eksp.Biol.Med., 93(4):62-64.
Ostrowski, J., Keil, M., and Schraven, E., 1975, Autoradiographische,
        Untersuchungen zur vertellung von Piracetam ^{14}C bei Ratte und
        Hund, Arzneimittel Forsch., 25:589-596.
Poshivalov V. P., 1981, Pharmaco-ethological analysis of social
        behaviour of isolated mice, Pharmac.Biochem.Behav. 14, Suppl.
        1, pp.53-59.
Poshivalov, V. P., 1981, GABAergic correlates of aggression and
        intraspecies sociability in isolated mice, Byull.Eksp.Biol.
        Med., 91(5):584-587.
Raiteri, M., Angelini, F., and Levi, J., 1974, A simple apparatus for
        studying the release of neurotransmitters from synaptosomes,
        Eur.J.Pharmacol., 25:411-414.
Raiteri, M., Federico, R., Coletti, A., and Levi, J., 1975, Release
        and exchange studies relating to the synaptosomal uptake of
        GABA, J.Neurochem., 24:1243-1250.
Rayevsky, K. S., 1981, Neurochemical aspects of the pharmacology of
        GABAergic substances, Farmakol.Toksikol., 44(5):517-529.
Rayevsky, K. S., and Kharlamov, A. N., 1980, Effects of GABAergic
        substances on conditioned avoidance reflex in rats, Farmakol.
        Toksikol., 43(3):284-288.
Riago, L. K., Nurk, A. M., Korneyev, A. Ya., and Allimets, L. Kh.,
        1982, Phenibut binds to bicuculline-insensitive GABA
        receptors in the rat brain, Byull.Eksp.Biol.Med.,
        94(11):58-59.
Roberts, P. J., Gupta, H. K., and Shargill, N. S., 1978, The inter-
        action of baclofen (β-(4-chlorophenyl) GABA) with GABA
        systems in rat brain: evidence for a releasing action, Brain
        Res., 155:209-212.
Rozanov, V. H., 1979, Studies in (^{14}C)-GABA and (^{14}C)GABA-pantoyl
        metabolism in the mouse, Ukr.Biokhim.Zh., 51(6):629-633.

Rozanov, V. H., 1980, The effects of pyridoxal phosphate and panto-
        thenate derivatives on the γ-aminobutyrate shunt in mouse
        brain, Vopr.Med.Khim., 26(1)42-46.
Ryan, L. D., and Roskoski, R., 1977, Net uptake of γ-aminobutyric
        acid by a high affinity synaptosomal transport system,
        J.Pharmacol.Exp.Ther., 200:285-291.
Schechter, P. J., Tranier, Y., and Grove, J., 1978, Effect of
        n-dipropylacetate on amino acid concentrations in mouse
        brain: correlations with anti-convulsant activity,
        J.Neurochem, 31:1325-1337.
Sneed, O. C., III, Bearden, L. J., and Pegram, V., 1980, Effect of
        acute and chronic anticonvulsant administration on endogenous
        γ-hydroxybutyrate in rat brain, Neuropharmacology, 19:47-52.
Sytinsky, I. A., 1977, Gamma-aminobutyric Acid as an Inhibitory
        Neurotransmitter, Leningrad, Nauka.
Tunnicliff, Y., Wimer, C., and Wimer, R. E., 1976, Brain γ-amino-
        butyric acid metabolism and behaviour in the mouse and rat,
        Gen.Pharmacol., 7:67-69.
Van der Laan, J. V., Boer, de Th., and Bruinvels, J., 1979, Di-n-
        propylacetate and GABA degradation. Preferential inhibition
        of succinic semialdehyde dehydrogenase and indirect inhi-
        bition of GABA-transaminase, J.Neurochem., 32:1769-1779.
Vasil'ev, V. Yu., Sytinsky, I. A., and Nicolaeva, Z. K., 1970,
        Further investigation of the properties of γ-aminobutyrate-
        glutamate-aminotransferase, Biokhimiya, 35:556-561.
Wallach, D. P., 1961, Studies on the GABA pathway, I. The inhibition
        of γ-aminobutyric acid - α-ketoglutaric acid transaminase in
        vitro and in vivo by U-7524 (aminohydroxyacetic acid),
        Biochem.Pharmacol., 5:323-331.
Wood, J. D., Schousboe, A., and Krogsgaard-Larsen, P., 1980, In vivo
        changes in the GABA content of nerve endings (synaptosomes)
        induced by inhibitors of GABA uptake, Neuropharmacology,
        19:1149-1152.
Worms, P., Deportere, H., Durand, A., Morselli, P. L., Lloyd, K. G.,
        and Bartholini, G., 1982, γ-aminobutyric acid (GABA) receptor
        stimulation. I. Neuropharmacological profiles of progabide
        (SL-76002) and SL-75102, with emphasis on their anticonvul-
        sant spectrum, J.Pharmac.Exp.Ther., 220:660-671.
Zakusov, V. V., ed., 1968, "Sodium Hydroxybutyrate," Meditsina,
        Moscow.

Analysis of Dopamine Receptor Supersensitivity After Chronic Neuroleptic Treatment in Rats

A.M. Zharkovsky and L.H. Allikmets

Department of Pharmacology, Tartu University, Estonian SSR

1. INTRODUCTION

Continuous treatment of laboratory animals with neuroleptics produces tolerance to some of their pharmacological effects and hypersensitivity of dopamine (DA) receptors. Most of the evidence for these phenomena has been based on behavioral or biochemical studies following systemic administration of neuroleptics. Schelkunov first demonstrated in 1967 that withdrawal of rats from long-term haloperidol treatment produced increased amphetamine stereotypy. Following withdrawal from chronic neuroleptic therapy, threshold doses of the direct DA-receptor agonist, apomorphine, required to induce stereotyped behavior in the guinea pig (Klawans and Rubovits, 1972) and cage-climbing in the mouse (Von Voigtlander et al., 1975), were decreased whilst the intensity of apomorphine-induced stereotyped behavior in the rat was increased (Tarsy and Baldessarini, 1974). Moreover, the binding of tritiated agonists or antagonists to striatal preparations of animals withdrawn from chronically or subchronically administered neuroleptics was enhanced (Burt et al., 1976; Muller and Seeman, 1977; Goldstein et al., 1980). Kinetic studies have shown that the maximal number of DA antagonist binding sites in the striatum rose after long-term treatment with neuroleptics (for review, see Seeman, 1980). The significance of these adaptive changes in the dopaminergic system for therapy with or side effects of neuroleptics is not clear. Some authors have speculated that mechanisms underlying chronic movement disorders (tardive dyskinesia) seen in some patients on long-term neuroleptics may be related to an induced supersensitivity of postsynaptic receptors in the basal ganglia (Tarsy and Baldessarini, 1977). Some neuroleptics, however, classed as atypical on the basis of their anomalous properties, are now receiving widespread clinical acceptance.

289

In the present study, we compared the typical neuroleptic, haloperidol, with drugs of the substituted benzamide group, sulpiride and metoclopramide. Metoclopramide, at antiemetic doses, has been shown to induce extrapyramidal signs without exhibiting antipsychotic activity in man (Peringer et al., 1976). Sulpiride has been reported to differ from other neuroleptics by not producing major extrapyramidal side effects. In animal experiments, sulpiride seems to be a weak blocker of apomorphine stereotypy and does not produce catalepsy even at rather high doses (Jenner et al., 1978). Because it is generally accepted that the supersensitivity of DA receptors after long-term neuroleptic treatment is central to the pathophysiology of tardive dyskinesia or withdrawal dyskinetic syndrome (Klawans et al., 1977; Baldessarini, 1980), we attempted to check the effect of various neurotropic drugs on neuroleptic-induced DA receptor supersensitivity.

2. MATERIALS AND METHODS

ANIMALS

Male Wistar rats, initially weighing 150-175 g, were used in all experiments. Animals were housed 4-5 per cage and given free access to food and drink.

DRUGS

In the first experiment, animals were given 1 mg/kg haloperidol (Gedeon Richter, Hungary), 50 mg/kg sulpiride (Delagrange, France), 10 mg/kg metoclopramide (A.H. Robins Co., Richmond, U.S.A) and saline once a day for 14 consecutive days. All drugs were administered intraperitoneally.

In the second experiment, animals were divided into groups of 16-18 animals and given haloperidol in combination with various neurotropic drugs (Table 1). Physostigmine, artane, piracetam and diazepam were obtained from commercial sources, phenibut was a gift from Prof. V. Perekalin.

Two hours after acute or chronic treatment, some of the animals in each group (5-6 rats) were sacrificed to determine the DA metabolite level. The remaining animals were withdrawn from chronic drug or saline treatment. On the fifth day of withdrawal, 4-5 animals from each group were given apomorphine hydrochloride (commercially available solution) 1 mg/kg s.c. and stereotypy was scored 15 min and 30 min later by a "blind" observer, according to the method of Allikmets et al. (1979). The remaining animals were used for binding studies in vitro.

Table 1.   Chronic Drug Treatment Schedule (all drugs were
           administered over a 14 day period)

Group	Neuroleptic dose (mg/kg)	Drug, dose (mg/kg)	Frequency of drug treatment per day
I	Saline	Saline	
II	Haloperidol 1.0	Saline	
III	Haloperidol 1.0	Physostigmine 0.5	twice
IV	Haloperidol 1.0	Artane 3	once
V	Haloperidol 1.0	β-Phenyl-GABA 50 (phenibut)	once
VI	Haloperidol 1.0	Piracetam 1000	once
VII	Haloperidol 1.0	Diazepam 2.5	once

Note: In control experiments, neuroactive drugs were administered
        separately (without neuroleptic) according to the schedule
        shown.

Biochemical Studies

In order to determine DA metabolite, brains of sacrificed rats
were rapidly removed and striatal and limbic tissues (tuberculum
olfactorium plus nucl. accumbens) dissected out on ice, according to
Glowinski and Iverson (1966) and pooled for each rat. The DA metabolites, homovanillic acid (HVA) and 3,4-dihydroxyphenylacetic acid
(DOPAC) were separated on Sephadex G-10 (Pharmacia Fine Chemicals,
Uppsala, Sweden) columns and determined spectrofluorimetrically using
the method of Earley and Leonard (1978).

Adenylate Cyclase Activity in Vitro

Adenylate cyclase activity in homogenates of caudate nuclei was
measured by the method of Kebabian et al. (1972). Animals were
sacrificed by decapitation. Striatal and limbic tissues of each rat
were homogenized in 50 vol (striatum) or 25 vol (limbic) of 4 mM
Tris-maleate buffer, pH 7.4, and 4 mM EGTA in a Teflon glass homogenizer (Braun-Melsungen Co., West Germany). Aliquots (50 μl) of this
homogenate were added to the standard assay mixture (final volume
0.5. ml) for measuring adenylate cyclase activity. This assay
mixture contained (final concentrations in mM): Tris-maleate, 80;
$MgCl_2$, 4; EGTA, 0.6; theophylline, 10; ATP, 0.5), and the indicated
concentrations of dopamine. The assay mixture was incubated in the
absence on ATP for 20 min at 0°C, then for 1 min at 30°C. The
adenylate cyclase reaction was then initiated by the addition of ATP.
The reaction was allowed to proceed for 3 min at 30°C and was termin-
ated by boiling for 2 min.

Samples from each rat were incubated in duplicate, and duplicate 50 µl aliquots of each sample were assayed for cyclic AMP in both the presence and absence of 5 mM (striatum) or 10 mM (limbic tissue) of dopamine HCl (Ferak, Berlin, W. Germany) by the method of Brown et al. (1972) using a cyclic AMP assay kit (Radiochemical Centre, Amersham, England). A linear standard curve was obtained between 0.5 - 8.0 pmol.

Binding Studies

Membrane preparations and binding studies were performed according to a slight modification of the method described by Creese et al. (1975). Tissue was homogenized in 50 vol cold Tris HCl buffer (50 mM, pH 7.7) using a Teflon-glass homogenizer (20 passes, 500 rpm). The homogenate was centrifuged twice at 50,000 x g for 10 min with rehomogenization of the intermediate pellet in the fresh buffer. The final pellet was resuspended in freshly prepared 50 mM Tris HCl buffer containing 0.1% ascorbic acid, 10 µM pargyline, 120 mM NaCl, 5 mM KCl, 2 mM $CaCl_2$ and 1 mM $MgCl_2$ to give a final pH of 7.1 at 37°C. The tissue preparation was preincubated for 5 min at 37°C and then placed on ice. ^{3}H-Spiperone (Radiochemical Centre, Amersham; specific activity 21 Ci/mmol) was diluted on the day of use with 0.1% ascorbic acid solution. Incubation tubes in triplicate received 100 µl of ^{3}H-spiperone (0.5 nM), 100 µl of displacer (cold spiroperidol, 1 µM) or 0.1% ascorbic acid and 800 µl tissue suspension. The mixtures were incubated at 37°C for 15 min and rapidly filtered under vacuum through Whatman GF/B filters with three 5 ml rinses of cold 50 mM Tris HCl buffer. Radioactivity was counted by liquid scintillation spectrometry in 8 ml of Redy-Solv (Beckman) at a counting efficiency of 37%. Specific ^{3}H-spiperone binding was defined as that displaced by 1 µM of cold spiroperidol and represented 65-70% of total. Protein was determined by the method of Lowry et al. (1951).

3. RESULTS

EXPERIMENT I.   DA-RECEPTOR SUPERSENSITIVITY IN RATS AFTER CHRONIC
              HALOPERIDOL, SULPIRIDE AND METOCLOPRAMIDE TREATMENT

The animals in haloperidol and metoclopramide groups showed sedation and catalepsy after drug administration throughout the experiment. The rats receiving repeated sulpiride exhibited only mild sedation. After withdrawal of chronic injections, rats pretreated with neuroleptics exhibited a significantly greater degree of stereotyped behavior after apomorphine than saline treated rats (Table 2). The intensity of stereotypy was maximally increased 15-30 min after apomorphine injection. The duration of stereotypy was also increased in all neuroleptic pretreated groups of rats.

Repeated neuroleptic treatment also resulted in an increase in ^{3}H-spiroperidol binding (Table 3). However, there were considerable differences between haloperidol, sulpiride and metoclopramide pretreated groups in this respect. Thus, in the haloperidol pretreated group, an increase in ^{3}H-spiroperidol binding was observed in both striatum (+69%) and limbic structures (+52%). After chronic sulpiride administration, an increase in ^{3}H-spiroperidol occurred in limbic structures only without significant changes in striatum. The opposite effect was seen in the case of metoclopramide, increased binding in striatum but no changes in limbic tissues (Table 3). Cyclic AMP production determined in rats withdrawn from neuroleptics did not differ significantly from the control group.

Table 2. Apomorphine-induced Stereotypy in Rats on the 5th day after Withdrawal from Repeated Saline or Neuroleptic Treatment. Apomorphine hydrochloride (0.5 mg/kg i.p.) was injected and stereotypy was scored 15 and 30 min later. The duration of stereotypy was also measured.

Drug, dosage (mg/kg/day)	Stereotypy score		Duration of stereotypy (min)
	15 min	30 min	
Control: saline	2.1±0.2	1.3±0.2	38.8±5.8
Haloperidol, 1	2.8±0.3*	2.2±0.2*	52.4±7.1*
Metoclopramide, 10	2.6±0.4	1.9±0.2*	68.8±9.2*
Sulpiride, 50	2.6±0.2	2.0±0.3*	59.0±2.0*

The data represent means ± SEM from 5 animals.
*p<0.05. (Mann-Whitney U-test).

Table 3. Specific Binding of ^{3}H-Spiroperidol (0.5 nM) after Repeated (14 days) Administration of Neuroleptics. Five days after termination of drug or saline treatment, animals were sacrificed and striata or limbic systems from two rats were pooled and assayed in triplicate.

Drug, dosage (mg/kg/day)	Striatum fmoles/mg protein	%	Limbic system fmoles/mg protein	%
Control - saline	24.6±22	100.	14.9±2.1	100
Haloperidol, 1	41.6±4.3**	169.1	22.7±3.6**	152.3
Sulpiride, 50	26.1±1.7	106.1	39.1±4.9**	262.4
Metoclopramide, 10	61.4±9.8**	249.6	19.0±2.3*	127.5

The data represent means from three experiments ± SEM. Statistical comparison was made by Student's t-test.
* p<0.05; **p<0.01.

In all cases, however, a rise in cyclic AMP in the presence of DA was evident (Table 4).

EXPERIMENT II.   THE EFFECT OF VARIOUS NEUROTROPIC DRUGS ON
                 HALOPERIDOL-INDUCED TOLERANCE AND DA-RECEPTOR
                 SUPERSENSITIVITY IN RATS

As shown in Table 5, acute injection of haloperidol caused a substantial rise in striatal concentration of the DA metabolites, HVA and DOPAC. After chronic treatment with haloperidol, a significant reduction in HVA and DOPAC concentration was observed. Thus, DA metabolism displays the development of tolerance to the effect of neuroleptics.

Of the cholinergic drugs studied, neither physostigmine nor artane influenced this effect of haloperidol. Nor did the neurotropic agent, piracetam, or the tranquilizer, diazepam, change the effects of acute or chronic haloperidol on striatal HVA and DOPAC levels, at least in the doses used (Table 5). Only phenibut, the phenyl derivative of GABA, was capable of preventing the development of tolerance to the effect of the neuroleptic on DA metabolism in some measure. Moreover, after acute injection of haloperidol in combination with phenibut, HVA concentration was significantly higher than after haloperidol alone (Table 6).

Table 4. Dopamine-Sensitive Adenylate-Cyclase Activity in Striatum
          and Limbic System of Rats after Withdrawal from Chronic
          Neuroleptics or Saline. Animals were sacrificed four days
          after termination of chronic neuroleptic treatment and the
          brain regions analyzed for cyclic AMP formed in the
          presence and absence of dopamine.

| Drug, dosage (mg/kg/kay) | Cyclic AMP formed (pmol/mg protein) | | | |
| | Striatum | | Limbic system | |
	- DA	+ DA (5 μM)	- DA	+ DA (10 μM)
Control	21.8±0.9	30.8±2.5	18.0±1.1	37.0±3.4
Haloperidol, 1	24.0±1.5	35.8±3.0	20.5±1.3	29.3±4.8
Sulpiride, 50	20.0±1.5	37.0±3.3	19.8±1.3	29.0±3.0
Metoclopramide, 10	19.2±2.1	27.0±3.8	17.3±1.5	31.8±2.8

Each value is the mean of 5-6 determinations in triplicate ± SEM.

Table 5. Effect of Haloperidol or Haloperidol in Combination with
          Physostigmine and Artane on Striatal HVA and DOPAC Level
          after Acute or Repeated (14 day) Injection in Rats. Rats
          were sacrified 2 h after acute or repeated injection of
          haloperidol. All acute treatments showed a significant
          increase in DA metabolite formation above control level.

Drug, dose (mg/kg)	Duration of treatment (days)	HVA µg/g wet tissue	%	DOPAC µg/g wet tissue	%
Control: saline		0.46±0.04	100	0.83±0.04	100
Haloperidol, 1	1	3.48±0.26	756.5	3.26±0.27	392.7
	14	2.14±0.14*	465.2	2.11±0.14*	254.2
Haloperidol + physostigmine, 0.5	1	3.25±0.11	706.5	3.70±0.18	445.8
	14	2.20±0.14*	478.3	2.15±0.24*	259.0
Haloperidol + artane, 3	1	2.96±0.09	649.5	3.29±0.42	396.4
	14	1.86±0.14*	404.3	2.44±0.17*	293.9

Each data point represents mean ± SEM from 6-8 determinations.
*$p < 0.05$ chronic vs. acute (Student's t-test).

This effect of phenibut may be due to the ability of the drug
to increase HVA and DOPAC concentration (Table 7) indirectly via
agonistic action on Ca^{2+} sensitive GABA receptors (Rago et al., in
press). It has been suggested that these receptors are localized on
DA-ergic terminals and regulate DA release (Bowery et al., 1980).

Chronic administration of haloperidol in combination with physostigmine enhanced behavioral supersensitivity: the stereotypic
response to apomorphine on the fifth day of withdrawal was increased
as compared with haloperidol (Table 8). When diazepam was given
chronically with haloperidol, the intensity of apomorphine stereotypy
was reduced.

Phenibut also showed a tendency to reduce stereotypy but the
values were not significant (Table 8). Neither artane nor piracetam
altered the signs of behavioral supersensitivity. A good correlation
between behavioral data and binding studies was obtained (Table 9).
Thus, simultaneous treatment with haloperidol and physostigmine
resulted in significant elevation of striatal ^{3}H-spiroperidol binding
as compared with chronic haloperidol alone. Diazepam on the other
hand reduced ^{3}H-spiroperidol binding which had been increased by
chronic haloperidol (Table 9).

Table 6. Effect of Haloperidol and Haloperidol in Combination with β-Phenyl-GABA (phenibut), Piracetam and Diazepam after Acute or Repeated Administration in Rats on Striatal HVA and DOPAC Concentration. Rats were sacrificed 2 h after acute or repeated (14 days) injection of drugs. DA metabolites were separated and measured spectrofluorimetrically.

Drug, dose (mg/kg)	Duration of treatment (days)	HVA µg/g wet tissue	%	DOPAC µg/g west tissue	%
Control: saline	14	0.68±0.04	100	0.90±0.04	100
Haloperidol, 1	1	3.21±0.20	472	3.48±0.32	387
	14	1.78±0.12^a	262	2.29±0.16^a	254
Haloperidol + phenibut 50	1	4.18±0.26	615	3.30±0.10	367
	14	3.63±0.24^b	534	3.82±0.23^b	424
Haloperidol + piracetam, 1000	1	3.71±0.23	546	3.81±0.21	423
	14	2.28±0.27^a	335	2.41±0.12^a	268
Haloperidol + diazepam, 2.5	1	3.65±0.18	537	3.66±0.32	407
	14	1.75±0.21^a	257	2.39±0.18^a	266

Each data point represents m ± SEM from 5-6 determinations.
a - $p < 0.05$ - chronic vs. acute; b - $p < 0.05$ - chronic haloperidol + drug vs. chronic haloperidol alone (Student's t-test). All acute treatments showed highly significant ($p < 0.01$) increases in metabolite level above control values.

Table 7. Effect of Phenibut, Piracetam and Diazepam after Repeated Administration on Striatal HVA and DOPAC Concentration in Rats. Animals were sacrificed 2 h after repeated (14 days) injection of drugs.

Drug, dose (mg/kg)	HVA µg/g wet tissue	DOPAC µg/g wet tissue
Control: saline	0.50±0.02	0.90±0.04
Phenibut, 50	0.89±0.14*	1.27±0.5*
Piracetam, 1000	0.63±0.07	0.97±0.07
Diazepam 2.5	0.75±0.08	1.38±0.23

Each data point represents m ± SEM from 5-6 determinations.
*$p < 0.05$ (Student's t-test).

Table 8. Apomorphine-Induced Stereotypy in Rats on the Fifth Day
          after Withdrawal from Chronic Administration of Haloperidol
          or Haloperidol Combined with Neurotropic Drugs. Apomor-
          phine hydrochloride was injected i.p. at a dose of 0.5
          mg/kg and stereotypy was scored at 15 and 30 min.

Drug, Dosage (mg/kg/day)	Stereotypy score	
	15 min	30 min
Control: saline	1.29 ± 0.2	1.90 ± 0.3
Haloperidol, 1	1.9 ± 0.2*	2.5 ± 0.2*[+]
Haloperidol + physostigmine, 1	2.4 ± 0.2*	3.4 ± 0.3*[+]
Haloperidol + artane, 3	2.0 ± 0.4*	2.7 ± 0.2*
Haloperidol + β-phenyl-GABA, 50	1.4 ± 0.3	2.0 ± 0.4
Haloperidol + piracetam, 1000	2.2 ± 0.3*	2.7 ± 0.3*
Haloperidol + diazepam, 2.5	1.3 ± 0.2[+]	1.9 ± 0.2[+]

Chronic treatment with these drugs for 14 days failed to influence
apomorphine-induced stereotypy. The data points show means ± SEM from
5-6 animals.
* $p < 0.05$ compared with control.
[+] $p < 0.05$ compared with chronic haloperidol (Mann-Whitney U-test).

Table 9. Specific Binding of ^{3}H-Spiroperidol (0.5 nM) after Repeated
          Administration (14 days) of Neuroleptics in Combination
          with Physostigmine, Artane and Diazepam. Five days after
          termination of drug or saline treatment, rats were sacri-
          ficed, striata from two rats pooled and assayed immediately
          (see "Methods"). Each assay was performed in triplicate.

Drug, dose (mg/kg/day)	0.5 pM ^{3}H-spiroperidol binding (fmoles/mg protein)	%
Control: saline	31.4 ± 28	100
Haloperidol, 1	46.9 ± 1.3*[+]	149.4
Haloperidol + physostigmine, 1	55.3 ± 2.8*[+]	176.8
Haloperidol + artane, 3	50.1 ± 2.1*[+]	159.6
Haloperidol + diazepam, 2.5	39.0 ± 1.4[+]	124.2
Diazepam, 2.5	27.0 ± 2.4	86.0

Chronic (14 days) artane and physostigmine treatment given alone did
not differ from saline control. The data points represent means from
4 experiments ± SEM. Statistical evaluation by Student's t-test.
* $p < 0.05$ - compared with saline control.
[+] $p < 0.05$ - compared with chronic haloperidol alone.

4. DISCUSSION

Behavioral experiments clearly indicate that all neuroleptics
studied induce a state of behavioral supersensitivity in response to
apomorphine when administered chronically. Direct binding studies,
using ^{3}H-spiroperidol as ligand, however, showed considerable regional differences between the neuroleptics studied. Thus, in
the haloperidol pretreated group, ^{3}H-spiroperidol binding was
increased equally in both the structures studied (striatum and
limbic structures). This indicates that dopamine receptor hypersensitivity in these brain regions occurs during withdrawal. These
data confirm earlier findings of Muller and Seeman (1977) that
chronic haloperidol treatment enhances ^{3}H-agonist and ^{3}H-antagonist
binding in striatum and limbic system of the rat brain. In contrast,
sulpiride-treated rats showed an increase in binding only in limbic
structures and not in the striatum. The opposite was true of the
metoclopramide group: elevation of binding in striatum and binding
in the limbic system were observed. These data indicate that enhancement of stereotypy observed after chronic neuroleptic treatment
is mediated not only via the nigro-striatal system but also via the
mesolimbic dopaminergic system. Moreover, an increased apomorphine
stereotypic response may be obtained even in the case of mesolimbic
DA-receptor supersensitivity without changing the sensitivity of DA
receptors of striatal origin as seen in sulpiride-pretreated rats.
It should be mentioned that behavioral supersensitivity may not
necessarily be due exclusively to changes in these two dopaminergic
systems. Acetylcholine, serotonin, noradrenaline and other neurotransmitters and receptors or other brain regions might be involved
(Muller and Seeman, 1977). Our data also indicate that supersensitivity after neuroleptic treatment is a characteristic of the
D_2 type of receptors that are not linked to adenylate cyclase, since
we did not observe any changes in adenylate cyclase activity after
withdrawal from chronic neuroleptics. Nor could other authors find
any changes in adenylate cyclase activity after chronic haloperidol
or chlorpromazine treatment (Von Voigtlander et al., 1975). These
data have not been confirmed by others (Kaneno et al., 1978). Although we have not presented data in this work on kinetic changes in
^{3}H-spiroperidol binding, our unpublished results, as well as data
obtained by others (for review see Seeman, 1980; Bacopoulus, 1981),
show an increase in D_2 receptor density (obtained by Scatchard plot)
in both striatum and limbic system after chronic haloperidol or
sulpiride respectively.

If our experimental data are compared with the clinical profile of the neuroleptics studied, it seems that drugs with high
extrapyramidal potential, such as haloperidol and metoclopramide,
induce DA-receptor supersensitivity in the nigro-striatal system.
In contrast, sulpiride, which produces relatively weaker extrapyramidal disturbances, induces supersensitivity in the limbic
system. These data may indicate that various dyskinetic reactions

seen after chronic neuroleptic treatment are mediated via induction of DA-receptor supersensitivity in the nigro-striatal system.
When sulpiride is given in effective antipsychotic doses, it too
may produce extrapyramidal disturbances (Cassano et al., 1975).

Indeed, if animals are chronically treated with a higher dose of
sulpiride (100 mg/kg) than in the present study, an increase in
^{3}H-spiroperidol binding in rat striatum may be observed (Hall et al.,
1981).

The significance of DA receptor supersensitivity seen in the
limbic system after chronic neuroleptic treatment is not fully understood. Based on our data and data obtained by others (see review
of Seeman 1980; Allikmets et al., 1979), we may conclude that the
ability of neuroleptics to induce DA receptor supersensitivity is
a common feature of all neuroleptics and this phenomenon develops
very rapidly even after acute treatment (Schwartz et al., 1978).
These data lead us to suggest that the supersensitivity of DA
receptors seen after chronic neuroleptic treatment is necessary for
their antipsychotic action. This proposal might be especially true
for the limbic system because, apart from classical neuroleptics,
atypical antipsychotics such as clozapine, sulpiride and thioridazine
preferentially affect limbic DA-ergic mechanisms (this study;
Bartholini, 1977). If this speculation is so, the commonly-involved
DA hypothesis of schizophrenia, postulating an abnormal increase in
activity of central DA neurones and/or over-activity of DA receptors
as an etiologic factor in schizophrenia, would have to be revised.
It should be admitted, however, that our hypothesis requires further
investigation before it can be proved or disproved.

Tardive dyskinesia and other hyperkinetic disturbances which
follow chronic neuroleptic treatment may well be due to striatal
DA receptor supersensitivity (Baldessarini and Tarsy, 1978;
Baldessarini, 1980; Klawans et al., 1977). It was therefore of
interest to study the effects of various neurotropic drugs when
given chronically to rats in combination with haloperidol on tolerance phenomena and DA receptor supersensitivity developing after
chronic haloperidol treatment.

As shown by the above data, only phenibut combined with haloperidol was able to prevent the development of tolerance in the
dopaminergic system, apparently through its action on the GABA-ergic
system (Rägo et al., in press). As seen in behavioral and binding
studies, the inhibitor, physostigmine, produced an additional increase in sensitivity in striatal DA-receptors. Previous investigations had shown that physostigmine was able to suppress DA-ergic
system activity, apparently through activation of the cholinergic
system (Millington and Wurtman, 1982). This additional suppression
of the DA-ergic system by chronic treatment may result in more rapid
development of DA receptor supersensitivity following withdrawal.

It has recently been reported that lithium, administered chronically in combination with haloperidol, effectively countered DA
receptor supersensitivity in behavioral and binding experiments
(Klawans et al., 1977; Pert et al., 1978; Allikmets et al., 1979).
On the basis of these studies, the authors suggested that lithium
might be effective in prophylaxis of dyskinetic disorders.

Of the drugs studied, diazepam was highly effective in countering DA-receptor supersensitivity. This effect of diazepam is not
fully understood. Most probably, it is mediated through the activation of the GABA-benzodiazepine complex which, in turn, modulates
DA-ergic transmission (Costa et al., 1975). The successful prevention of haloperidol-induced supersensitivity in the striatum by
the simultaneous administration of diazepam suggests that this drug
might be used clinically to prevent the development of dyskinetic
disorders seen after chronic haloperidol treatment in man.

REFERENCES

Allikmets, L. H., Stanley, M., and Gershon, S., 1979, The effect of
    lithium on chronic haloperidol enhanced apomorphine aggression
    in rats, Life Sci., 25:165-170.
Allikmets, L. H., Zarkovsky, A. M., and Nurk, A. M., 1981, Changes in
    catalepsy and receptor sensitivity following chronic neuro-
    leptic treatment, Eur.J.Pharmacol., 75:145-147.
Bacopoulos, N. G., 1981, Antipsychotic drug effects on dopamine and
    serotonin receptor: in vitro binding and in vivo turnover
    studies, J.Pharmacol.Exp.Ther., 219:708-714.
Baldessarini, R., 1980, Dopamine and the pathophysiology of dy-
    skinesias induced by antipsychotic drugs, Ann.Rev.Neurosci.,
    3:23-41.
Baldessarini, R. I., and Tarsy, D., 1978, Tardive dyskinesia, in:
    "Psychopharmacology: a Generation of Progress," M. A. Lipton,
    A. Di Mascio, and K. F. Killam, eds., Raven Press,
    pp.993-1004.
Bowery, N. G., Hill, D. R., Hudson, A. L., Doble, A., Middlemiss,
    Shan I., and Turnbull, M., 1980, Baclofen decreases neuro-
    transmitter release in mammalian CNS by an action at a novel
    GABA receptor, Nature, 283:92.
Brown, B. L., Ekins, R. P., and Albano, I. D. M., 1972, Saturation
    assay for cyclic AMP using endogenous binding protein,
    Adv.Cyclic Nucleotide Res., 2:25-40.
Burt, D. R., Creese, I., and Snyder, S. H., 1977, Antischizophrenic
    drugs: chronic treatment elevates dopamine receptor binding in
    brain, Science, 196:326-328.
Cassano, G. B., Castrogiavanni, P., Conti, L., and Bonollo, I., 1975,
    Sulpiride versus haloperidol in schizophrenia, a double blind
    comparative trial, Curr.Ther.Res., 17:189-201.

Costa, E., Guidotti, A., Mao, C. C., and Suria, A., 1975, New concepts on the mechanism of action of benzodiazepines, Life Sci., 17:167-186.

Creese, I., Burt, D. R., and Snyder, S. H., 1975, Dopamine receptor binding. Differentiation of agonist and antagonist states with ^{3}H-dopamine and ^{3}H-haloperidol, Life Sci., 17:992-1002.

Earley, C. I., and Leonard, B. E., 1978, Isolation and assay of noradrenaline, dopamine, 5-hydroxytryptamine and several metabolites from brain tissue using disposable Bio-Rad columns packed with sephadex G-10, J.Pharmacol.Meth., 1:67-79.

Goldstein, M., Lew, I. J., Asano, T., and Ueta, K., 1980, Alterations in dopamine receptor: effects of lesion and haloperidol treatment, Comm.Psychopharmac., 4:21-25.

Hall, M. D., Jenner, P., Marsden, C. D., Murugaiah, K., Rupniak, N. M. I., and Theodorou, A. E., 1982, Repeated sulpiride administration to rats, like repeated haloperidol, induces cerebral dopamine receptor suspersensitivity, Brit.J.Pharmacol., 74:229-230.

Jenner, P., Clow, A., Reavill, C., Theodorou, A., and Marsden, C. D., 1975, A behavioural and biochemical comparison of dopamine receptor blockade produced by haloperidol with that produced by substituted benzamide drugs, Life Sci., 23:545-550.

Klawans, H. L., and Rubovits, R., 1972, An experimental model of tardive dyskinesia, J.Neural.Transmiss., 33:235-246.

Kebabian, I. W., Petzold, G. L., and Greengard, P., 1972, Dopamine sensitive cyclase in caudate nucleus of rat brain and its similarity to the "dopamine receptors", Proc.natl.Acad.Sci., (Wash.), 69:2145-2149.

Klawans, H. L., Weiner, W. J., and Nausieda, P. A., 1977, The effect of lithium on an animal model of tardive dyskinesia, Prog.Neuropsychopharmacol., 53-60.

Lowry, O. H., Rosebrough, N. I., Farr, A. L., and Randall, R. I., 1951, Protein measurement with the Folin phenol reagent, J.biol.Chem., 193:265-272.

Millington, W. R., and Wurtman, R. J., 1982, Choline and physostigmine enhance haloperidol induced HVA and DOPAC accumulation, Eur.J.Pharmacol., 80:431-434.

Muller, P., and Seeman, P., 1977, Brain neurotransmitter receptors after long term haloperidol: dopamine, acetylcholine serotonin, noradrenergic and naloxone receptors, Life Sci., 21:1751-1758.

Peringov, E., Jenner, P., Donaldson, I. M., and Marsden, C. D., 1976, Metoclopramide and dopamine receptor blockade, Neuropharmacology, 15:463-469.

Pert, A., Rosenblatt, I. E., Sivit, C., Pert, C. B., and Bunney, W. E., 1978, Long-term treatment with lithium prevents the development of dopamine receptor supersensitivity, Science, 201:171-173.

Rägo, L. K., Nurk, A. M., Kornejev, A. I., and Allikmets, L. H., 1982, Phenibut binds to bicuculline-insensitive GABA receptors in rat brain, Bull.exp.Biol.Med., in press.

Schelkunov, E. L., 1967, Adrenergic effect of chronic administration
       of neuroleptics, Nature, 214:1210-1212.
Schwartz, I. C., Costentin, I., Martres, M. P., Protais, P., and
       Baudry, M., 1978, Modulation of receptor mechanisms in the
       CNS: hyper- and hyposensitivity to catecholamines,
       Neuropharmacology, 17:665-685.
Seeman, P., 1980, Brain dopamine receptors, Pharmacol.Rev.,
       32:229-313.
Tarsy, D., and Baldessarini, R. J., 1974, Behavioural super-
       sensitivity to apomorphine following chronic treatment with
       drugs which interfere with the synaptic function of catechol-
       amines, Neuropharmacology, 13:927-940.
Tarsy, D., and Baldessarini, R. J., 1977, The pathophysiologic basis
       of tardive dyskinesia, Biol.Psychiat., 12:431-450.
Von Voigtlander, P. E., Losely, E. G., and Triezenberg, H. I., 1975,
       Increased sensitivity to dopaminergic agents after chronic
       neuroleptic treatment, J.Pharmacol.Exp.Ther., 193:88-94.

Neuroleptics – Allosteric Tyrosine Hydroxylase Regulators

M.F. Mineyeva, V.S. Kudrin and A.Yu. Shemanov

Laboratory of Neurochemical Pharmacology, Institute of Pharmacology,
Academy of Medical Sciences of the USSR, Moscow

1. INTRODUCTION

The action of psychotropic compounds on catecholaminergic
processes is a key factor in their action on the function of the
central and peripheral nervous system (Zakusov, 1973; Valdman et
al., 1979; Breese et al., 1978; Bunney and Aghajanian, 1975),
since impairment of these processes may be an important stage in
the onset and development of schizophrenia and other psychoses
(Owen et al., 1978; Snyder, 1974; Lideman et al., 1980; Le Fur et
al., 1979).

Neuroleptics act largely through their effect on dopaminergic
brain processes. They block pre- and post-synaptic receptors when
administered acutely (Snyder et al., 1974; Di Chiara et al., 1978;
Le Fur et al., 1979), while long-term administration renders these
receptors hypersensitive. In addition, tyrosine hydroxylase helps
to shape how neuroleptics act on dopaminergic processes; α-methyl-
para-tyrosine (AMT), a competitive tyrosine hydroxylase (TH; E.C.
1.14.16.2) inhibitor, potentiates the effect of neuroleptics when
administered systemically (Antelman et al., 1976). Acute administration of antipsychotic compounds produced an increase in turnover
of dopamine (DA) and an increase in TH reaction rate – at least
in the nigro-striatal and mesolimbic DA systems of the brain
(Zivkovic et al., 1975). These experiments also brought to light
a reduction in Michaelis' constant (K_m) for the pterin co-factor.
TH, by virtue of the way in which it is physiologically regulated
(Levitt et al., 1965; Murrin et al., 1977; Bunney and Aghajanian,
1975; Di Chiara et al., 1978), makes a major contribution in its
turn towards the regulation of catecholaminergic processes and is
directly involved in maintaining the physiological function of the

nervous system. TH may be linked with a number of behavioral
responses (Valdman et al., 1981; Kudrin et al., 1981; Ahlenius,
1976). An analysis of how its action is regulated could pave the
way for investigations clarifying the workings of psychopharmacological compounds in general, and neuroleptics in particular.

Our research focused on regulation of brain TH by neuroleptics
as compared with other types of psychopharmacological compounds.
The experiments performed enabled us a) to establish that the drug's
effect on the enzyme's affinity for tyrosine makes an essential
contribution to the regulation of its activity, b) to form a hypothesis as to how neuroleptics exert an inhibitory effect on TH
regulation by presynaptic receptors. In our view, this mechanism
may involve not just attenuation of the signal from presynaptic
receptors to TH, but also a transformation of this enzyme such as
to make it refractory to this signal.

2. METHODS

Both rat and bovine brains were used in these experiments.

Cross-bred or Wistar male rats weighing 200±40 g were decapitated. Their brains were rapidly removed over ice. The striatum
and hypothalamus were separated by Glowinski and Iversen's (1966)
method. The nucleus accumbens septi (adjoining septal nuclei) were
separated by Horn's (1974) procedure. Bovine brains were obtained
from the slaughter-house and processed in the same manner as rat
brain. Tissues were homogenized in a Potter-Elvehjem homogenizer
in a 1:10 homogenate, in a medium depending on the nature of the
experiment. Soluble and membrane-bound TH was separated by Kuchensky
and Mandell's (1972) method, closely observing the author's recommendations in the interests of obtaining standard preparations of
the enzyme, the kinetic properties of which had been carefully
identified. Purified TH was obtained from the microsomal fraction
of the appropriate rat or cattle brain structure by affinity chromatography on diiodothyronine-sepharose 4B by the method of Mineyeva
et al. (1979), described above. The P_2 fraction was obtained by Gray
and Whittaker's (1962) procedure, and homogenized in Krebs phosphate
buffer solution. Synaptosomes were produced using Hajos' (1978) or
Gray and Whittaker's (1962) method.

The following experiments on TH regulation by presynaptic receptors were then performed. Synaptosomes were suspended in Krebs
pH 7.35 phosphate buffer - composition as follows: NaCl - 124 mM,
KCl - 5 mM, Na_2HPO_4 - 20 mM, $MgCl_2$ - 1.3 mM, KH_2PO_4 - 1.2 mM,
$CaCl_2$ - 0.75 mM and D-glucose - 10 mM. The buffer was saturated
with carbogen before suspending the synaptosomes. The suspension
having been divided up into several portions, assays with different

additives were incubated simultaneously:

   a) control, without additive
   b) assay with dopamine
   c) assay with dopamine immobilized with fluorphenazine (Fluphenazin)
   d) assay with a combination of dopamine and fluorphenazine
   e) assay with fluorphenazine.

If the synaptosomal protein concentration in the assay was greater than 3.5 mg/ml, that of the additives amounted to $10^{-5}M$, in accordance with Venter et al's (1972) findings on how the concentration of a receptor agonist or antagonist in experiments involving the suspension of isolated cells should be selected according to total surface area - i.e. concentration in the incubation medium. Incubation took place for 30 min at a temperature of 37°C. The synaptosomes were centrifuged at 20,000 x g for 30 min and the soluble and membrane-bound enzyme was separated by Kuchensky and Mandell's (1972) method. The P_2 fraction was used in experiments with des-tyr-gamma-endorphin (DTGE) and Tyr-D-Ala-Gly-Phe-amide (TAE).

Tyrosine hydroxylase activity in the crude synaptosomal fraction was ascertained by Nagatsu et al's (1964) method, adapted to synaptosomes. Two modifications of this method were employed. The first involved measuring TH activity by the rate of ^{3}H-DOPA accumulation in the presence of 3-hydroxybenzylhydrazine (NSD-1015), a dopa decarboxylase inhibitor, at a concentration of 1 mM, using 2,3,5,6-^{3}H-tyrosine (76 Ci/mol) from Amersham (Kuchensky and Segal, 1974). The incubation medium was made of synaptosomal suspension: 0.2-0.3 mg protein; compound under study; NSD-1015 - 1 mM; ^{3}H-tyrosine, 2-10 μM. Final volume of incubation mixture: 110 μl. The incubation lasted 30 min at a temperature of 30° and 3-iodo-tyrosine (5 mM) was added to blanks. Nagatsu et al's (1964) procedure was used to isolate and measure ^{3}H-DOPA. Radioactivity was measured on an Intertechnique SL-4000 liquid scintillation counter.

The second variation consisted of measuring TH activity according to rate of accumulation of tritiated water, in the presence of NSD-1015 to reduce the formation of labelled products of DA metabolism. The 3,5-^{3}H-tyrosine (2.8 Ci/mmol) was obtained from the Institute of Isotopes, Hungary. The incubation mixture consisted of: synaptosomal suspension, 0.1-0.2 mg of protein, compound under study, NSD-1015 (1 mM) and ^{3}H-tyrosine (0.2-6 μM). Final volume of incubation mixture: 200 μl. Incubation lasted 15 min at a temperature of 37°C. Nagatsu et al's (1964) method was used for measuring tritium-labelled water.

The accumulation of reaction product was linear with time when the incubation period was less than 20 min. It also depended on protein concentration in the assay of up to 2 mg/ml. Before use 3,5-^{3}H-tyrosine was purified chromatographically on aluminum oxide, Dowex 50B x 4 (Coyle, 1972).

The initial speed of the TH reaction was monitored by a direct spectrophotometric method (Mineyeva-Vyalikh, 1976) which measured increase in absorption at 335 nm; this was caused by the formation of the oxidized form of the pterin cofactor. The compound 6,7-dimethyl-5,6,7,8-tetrahydropterin (DMPH$_4$) was the substance used as cofactor. Measurements were made on a DW-2 UV VIS differential spectrophotometer (Aminco).

Protein level was ascertained by the method of Lowry et al. (1951), using the following reagents: tris base, tyrosine from Sigma, recrystallized maleic acid from Yerevan Chemicals, DMPH$_4$ from Calbiochem - also synthesized by ourselves using the method of Mager and Addink (1967). Flupentixol isomers were kindly provided by Dr. Moller-Nielssen, Lundbeck and Co. A/S, Copenhagen, Denmark. Des-Tyr-gamma-endorphin and Tyr-D-Ala-Gly-Phe-NH$_2$ were synthesized in the Laboratory of Peptide Synthesis (Director, M. I. Titov), National Cardiological Research Center, Acad. Med. Sci., USSR.

3. RESULTS

3.1. DIRECT ACTION OF NEUROLEPTICS ON THE KINETICS OF THE TYROSINE HYDROXYLASE REACTION

The effect of azabutyrone, a neuroleptic of the butyrophenone group, on the kinetics of the TH reaction was investigated using the soluble and membrane-bound enzyme. The azabutyrone had been synthesized and its pharmacology studied at the Institute of Pharmacology of the Acad. Med. Sci. of the USSR by Rayevsky et al. (1976). As shown in Figure 1, a 10 mg/kg dose of azabutyrone did not alter the kinetics of the reaction with tyrosine and DMPH$_4$. When added to soluble or membrane-bound TH from the hypothalamus, azabutyrone produced a rise in K_m for tyrosine, without altering the kinetics of the reaction with DMPH$_4$.

Our findings suggest that azabutyrone exerts an indirect influence on the K_m for the pterin cofactor, while its effect on the K_m for tyrosine when administered systemically could be brought about by this neuroleptic acting directly on TH. One fact drew our attention; when azabutyrone is present, the reaction rate with inhibitory concentrations of substrate is maintained at a high level, corresponding to a near-saturating concentration of tyrosine; thus azabutyrone has been able to suppress substrate-induced inhibition of TH; this is particularly noticeable in the TH of the hypothalamus. Out of a number of azabutyrone analogs, only those members of the group which had shown neuroleptic action in behavioral experiments (Nazarova et al., 1976) produced any effect on substrate-induced inhibition of the enzyme (see Table 1). These data would appear to indicate that the ability of a neuroleptic to influence the K_m for tyrosine and eliminate substrate-induced inhibition of TH may be

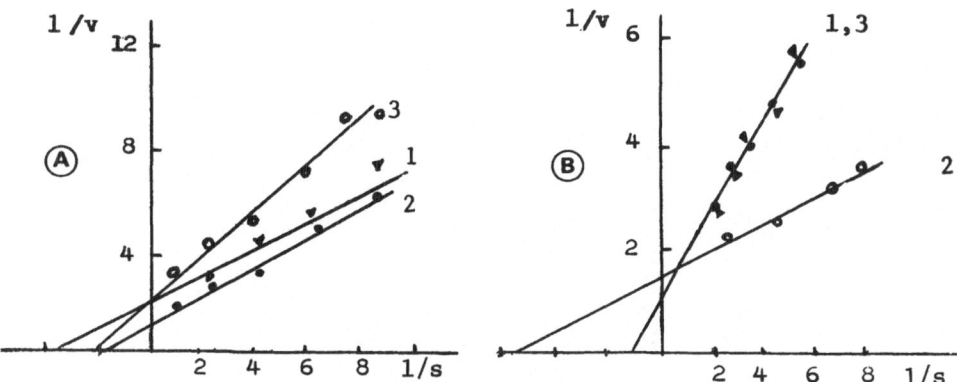

Fig. 1.   Effect of azabutyrone on the kinetics of the tyrosine
hydroxylase reaction. (A) - tyrosine, (B) - DMPH$_4$; 1 -
control, 2 - azabutyrone, 10 mg/kg i.p., 3 - azabutyrone,
10^{-5}M, in vitro. Experimental conditions, buffer - 0.1 M
tris-maleate, pH 6.25, 30°C, membrane-bound tyrosine
hydroxylase from rat hypothalamus 30 mg/ml. (A) - tyrosine
concentration 0.187 mM, (B) - DMPH$_4$ concentration 100 µM
(0.100 mM).

central to neuroleptic action. Such a view finds support from the
fact that the suppression of substrate-induced inhibition only occurs
with pharmacological compounds classed as neuroleptics (Mineyeva and
Rayevsky, 1976; Rayevsky and Mineyeva, 1978). As will be seen from
Table 2, neither tranquilizers, stimulants, antidepressants nor
morphine suppress this phenomenon, which distinguishes them from both
typical and atypical neuroleptics. It was our team which first
showed how neuroleptics can act directly on TH, a finding manifested
as an increase in K_m for tyrosine and suppression of substrate-
induced inhibition of TH from the hypothalamus. This effect operates
selectively, in that compounds belonging to other pharmacological
categories do not act in this way - although certain substances may
have an indirect action on the rate of the TH reaction up to a
saturating concentration of tyrosine. Cocaine activates the TH
reaction in vitro, for example - this effect is related to accelerated disassociation of the reaction product. When administered
systemically, cocaine inhibits the TH reaction without altering the
K_m for tyrosine (Mineyeva et al. (1978)).

3.2.   STEREOSPECIFIC NATURE OF NEUROLEPTIC INFLUENCE ON SUBSTRATE
        INHIBITION OF HYPOTHALAMIC TH

        Our assumption that neuroleptics selectively suppress substrate-
induced inhibition of TH was subsequently confirmed by experiments

Table 1. Effect of Azabutyrone Analogs on TH Reaction Rate

Compound	Reaction rate, as % of control tyrosine concentration	
	0.1 mM	0.3 mM
Control	100	16.7
No. 1, azabutyrone $m=3$; $Z=CH_2-C-$; $X=F$	35.7	97.6
No. 2, $m=4$; $Z=CH_2-C-$; $X=F$	102.4	95.2
No. 3, $m=5$; $Z=CH_2-C-$; $X=F$	61.9	69.0
No. 4, $m=3$; $Z=CH_2-C-$; $X=H$	42.9	21.4
No. 5, $m=3$; $Z=CH_2-C-$; $X=Cl$	71.4	23.8

Assay consisting of: 0.1 M tris-maleate, pH 6.25; membrane bound TH from rat hypothalamus, 30 mg protein/ml; $DMPH_4$ - 18.3 μM. Initial reaction rate, taking control as 100% at a tyrosine concentration of 0.1 mM, consisting of 200 μmol/mg protein/min.

Table 2. The Effect of Various Types of Psychopharmacological Agents on TH from Rat Hypothalamus

Experimental variable	Reaction rate as % of control	Experimental variable	Reaction rate as % of control
Control 1	100	Amitriptyline	52
Control 2	40	Pyrazidol	53
Haloperidol	100	Imipramine	40
Fluorphenazine	100	Fluacizin	54
Aminazine	100	Cocaine	47
Carbidine	80	Dopamine	20
Cis-flupentixol	90	Apomorphine	27
Clozapine	79	Morphine	40

Assay consisting of: buffer, 0.05 M tris-maleate, pH 6.0; isolated TH from rat hypothalamus, fraction III, processed by norit A: 1 μg/ml; $DMPH_4$ 146 μM. Tyrosine concentration: 100 μM in the "control 1" assay; 350 μM in the assays with added pharmaceutical agents and "control 2". Compounds added at a concentration of 10^{-5}M. Initial reaction rate in control 1 taken as 100%, consisting of 292 nmol/mg protein/min.

with flupentixol isomers illustrating the stereospecific nature of
the relationship between neuroleptics and TH, and showing how they
suppress the above phenomenon. The experiments were performed on TH
isolated from bovine hypothalamus using diiodothyronine-Sepharose 4 B.
These experiments enabled us to show that the cis-isomer exerts a
concentration-dependent action on substrate-induced inhibition, which
is completely suppressed at a concentration of 10^{-4}M. This amount of
flupentixol is no higher than the tyrosine concentration required for
substrate inhibition to occur. The trans-isomer exerts an inhibitory
effect over a concentration range of 10^{-6}-10^{-4} and does not eliminate
the substrate-induced inhibition of TH (Mineyeva et al., 1982).
Flupentixol only displays a stereospecific effect when concentrations
of tyrosine are inhibitory (see Figure 2). It will be seen from
Figure 2 that the curve showing the interrelationship between reaction rate and tyrosine concentration of beef hypothalamic enzyme
(curve 1) has a similar maximum velocity as that obtained for the
enzyme from rat brain hypothalamus. The 100-140 μM concentration
range of tyrosine (saturation point) overlaps into the substrate
inhibition concentration of greater than 140 μM. A concentration of
5×10^{-5}M trans-flupentixol does not noticeably affect the nature of
the curve for reaction rate plotted against tyrosine concentration.
The same concentration of cis-flupentixol, represented by curve 3,

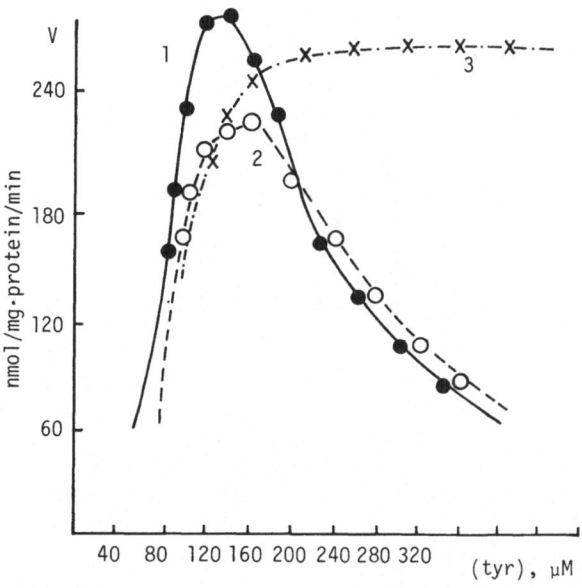

Fig. 2. Stereospecificity of the suppression of substrate inhibition
         of tyrosine hydroxylase by flupentixol. Experimental con-
         ditions: 0.1 M tris-maleate, pH 6.0. $DMPH_4$ - 140 μM;
         tyrosine - 40-360 μM; flupentixol: 5×10^{-5}M. 1 - control;
         2 - trans-flupentixol; 3 - cis-flupentixol.

strongly influences this relationship, increasing the saturating
concentration of tyrosine and extending the saturation range to such
an extent that, at least up to a concentration of 360 μM tyrosine,
it forms a plateau. It may thus be seen that in the presence of cis-
flupentixol the TH reaction with tyrosine conforms with Michaelis-
Menten kinetics without displaying substrate inhibition, at least for
a tyrosine concentration range of up to 360 μM. A comparison of
curves 2 and 3 reveals clear-cut differences between the influences
of flupentixol's isomers to the right of the curve, corresponding
with the substrate inhibition range, as against an almost identical
influence to the left. This confirms our assumption concerning the
presence of two areas of tyrosine binding on the TH molecule, one
situated at the catalytic and the other at the non-catalytic center
of the enzyme. This assumption is borne out by the results of
experiments with AMT, which inhibits the enzyme, competing with
tyrosine at the catalytic center. It will be seen from Figure 3
that cis-flupentixol, like fluorphenazine, fails to reverse AMT's

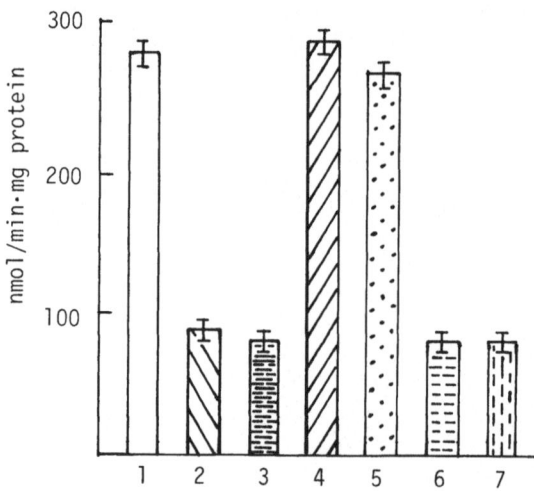

Fig. 3.   The effect of fluorphenazine and cis-flupentixol on
          substrate inhibition of tyrosine hydroxylase in the
          presence of α-methyl-paratyrosine (AMT).
          1 - control 1, 0.1 mM tyrosine;
          2 - control 2, 0.36 mM tyrosine (substrate inhibition);
          3 - 10^{-4} AMT;
          4 - 10^{-5}M fluorphenazine   All in 0.36 nM
          5 - 10^{-5}M cis-flupentixol   tyrosine
          6 - 10^{-4}M AMT plus 10^{-5}M fluorphenazine   All in 0.1 nM
          7 - 10^{-4}M AMT plus 10^{-5}M cis-flupentixol tyrosine
          Experimental conditions: buffer - 0.05 M tris-maleate,
          pH 6.0, 30°C. DMPH$_4$ concentration - 0.140 mM; purified
          tyrosine hydroxylase from rat hypothalamus - 1 μg/ml.

inhibitory effect. Our findings lead us to believe that neuroleptics act as allosteric regulators of TH, reacting directly on the enzyme, and thereby preventing the interaction between tyrosine and TH at the site of regulation (Mineyeva et al., 1982).

The part played by substrate inhibition of TH was unclear, however. Furthermore, many authors have expressed the view that substrate-induced inhibition of TH could not occur in vivo, since tyrosine concentration in brain structures is below the saturating concentration of tyrosine (Mandell, 1978). Hence the need to find some means of clarifying the physiological role played by substrate-induced inhibition of TH, to further our understanding of this phenomenon and its contribution to the mechanism of neuroleptic action.

4. HOW PRESYNAPTIC RECEPTORS REGULATE HYPOTHALAMIC TH ACTIVITY

The suppression of substrate inhibition of TH by neuroleptics might in some way be connected with their ability to act as dopamine receptor antagonists, in view of their known ability to antagonize presynaptic dopamine receptors to varying degrees (Karobath, 1975). Experiments investigating the mechanism of the regulation of TH activity by presynaptic receptors were performed on synaptosomes subjected to a) prior electrical stimulation, imitating the passage of a "spike", and b) the influence of the dopamine receptor agonists, dopamine (DA) and IDA, the immobilized form of dopamine, and c) fluorphenazine, a DA receptor antagonist.

Electrical stimulation of the synaptosomes was shown experimentally to produce depolarization of the plasma membrane (Mineyeva et al., 1977). Depolarizing the synaptosomal membrane produces an increase in DA release, which could interact with presynaptic receptors - although the conditions prevailing during this experiment would have made it difficult for a sufficiently high concentration of DA to be released onto the synaptosomal surface by virtue of its dispersal throughout the incubation medium. The dynamic characteristics of synaptosomal TH which had been subjected to the action of the depolarizing current were observed to change profoundly, revealing a reduction in K_m for tyrosine, a reduction in semi-saturating concentration for $DMPH_4^m$ and a raised maximum velocity (Mineyeva et al., 1977; Mineyeva et al., 1978). Our findings agreed with those of Murrin et al. (1976): the striatum and nucleus accumbens were subjected to cauterizing unilateral electrical stimulation and the properties of the soluble TH from these structures were then measured. Murrin et al. viewed their findings as a reduction in K_m for tyrosine and pterin cofactor in terms of activation of the enzyme. The changes we noted in the kinetic constants of TH might also be interpreted as enzyme activation. Both in Murrin et al.'s (1976) experiments and our own, only "kinetic activation of the enzyme," or the possibility of this occurrence, could be claimed, however, and not a real increase in reaction rate, for when the K_m for tyrosine decreases the enzyme may be subject to substrate inhibition, while the optimum tyrosine concentration for activating it remains unchanged at nerve endings. Kinetic activation, linked with a reduced K_m for tyrosine, could thus bring about a reduction in true reaction rate.

Stimulating presynaptic dopamine receptors in vivo inhibits the accumulation of DOPA in brain structures, as Di Chiara et al. (1978) have shown in the nigro-striatal system. This fact is generally viewed as a consequence of the direct inhibitory action of "taken up" DA on TH. These findings led us to perform experiments on synaptosomes which had been incubated with naturally-occurring low molecular weight dopamine and dopamine immobilized on a high molecular weight polymer of 70,000 daltons, which was unable to penetrate the synaptosomal membrane at a measurable rate (Mineyeva et al., 1980). IDA maintained its ability to inhibit TH on coming into direct contact with the enzyme; this indicates that IDA is biologically active. As illustrated in Table 3, the membrane-bound enzyme from the control assay exhibits marked cooperativity as regards tyrosine binding. TH from synaptosomes, once exposed to the action of DA or IDA, is less subject to this effect. The enzyme from the assay with DA, and from that with IDA in particular, reveals a major reduction in semi-saturating concentration of tyrosine, of a magnitude characterizing the affinity of the enzyme for the substrate, similiar to the K_m, but calculated using Hill's coefficient (Kurganov, 1979). The apparent K_m for $DMPH_4$ is also reduced. Similar but less pronounced changes may be noted in the soluble enzyme (see Table 4). These results are consistent with the kinetic activation of TH and are reminiscent of kinetic changes in the TH reaction observed following synaptosomal depolarization. The results given in the table confirm that IDA displayed the same action as low-molecular weight DA which can penetrate the nerve ending via the reuptake mechanism. Dopamine also performs efficiently in its transmembrane action, apparently changing the kinetics of the reaction in comparison with tyrosine and reducing the K_m for pterin cofactor, which is not typical of DA's direct effect on TH. Extra-synaptosomal dopamine could therefore be assumed to act as a trigger for the regulation of TH by presynaptic receptors, initiating a chain of events culminating in modification of the enzyme and alteration of its kinetics properties. It should also be noted that a reduction in semi-saturating concentration of tyrosine brings the enzyme to a state of substrate inhibition at a tyrosine concentration optimal for monitoring. In these circumstances substrate inhibition of TH in the tissues of the brain can develop at a concentration of tyrosine usual for brain structures, due to a change in the enzyme's properties. A steady tyrosine concentration in the tissues can in itself be an essential pre-condition for achieving substrate inhibition of TH under these conditions. Substrate inhibition of TH could thus be a focal point of TH regulation and control as far as the presynaptic receptors are concerned.

Table 3. Kinetic Parameters of Membrane-bound TH from Hypothalamic
Synaptosomes, Incubated with Dopamine (DA) and Immobilized
Dopamine

Assay	Hill's coefficient	V_{max}tyr.	$S_{0.5}$ tyr.	K_m DMPH$_4$ app.
Control	4.3±0.2	25.6±1.5	86.5±4.3	1.56±0.17
DA	1.5±0.02	28.2±2.2	30.0±2.1	0.94±0.11
IDA	1.3±0.5	38.3±2.3	18.1±0.9	0.96±0.11

Experimental conditions: buffer – tris-maleate, pH 6.0, 30°C. On
measuring the semi-saturating concentration of tyrosine ($S_{0.5}$) the
concentration of DMPH$_4$ was 0.2 mM. On measuring the apparent K_m for
DMPH$_4$ the tyrosine concentration was optimal for each experiment.

Table 4. Kinetic Parameters of Soluble TH from Rat Hypothalamic
Synptosomes, Previously Incubated with Dopamine (DA) and
Immobilizing Dopamine (IDA)

Assay	K_m DMPH$_4$ mM	V_{max}DMPH$_4$ nmol/mg protein/min	K_m tyr. mM	V_{max}tyr. nmol/mg proteim/min
Control	0.23±0.05	46.5±3.7	0.08±0.001	24.0±2.0
DA	0.08±0.01	42.0±3.4	0.065±0.001	43.6±4.5
IDA	0.08±0.001	43.0±3.4	0.070±0.001	30.8±3.2

Experimental conditions: buffer – tris-maleate, pH 6.0, 30°C. On
measuring the K_m of DMPH$_4$ tyrosine concentration was optimal for each
experimental variant. On measuring the apparent K_m for tyrosine
DMPH$_4$ concentration was 0.2 mM.

We made this finding when working on synaptosomes from the
hypothalamus, which play an important part in the regulatory processes shaping behavioral response and the control of endocrine
function (Tuomisto, 1978). The hypothalamic dopamine systems are
known to be distinctive. Hence, both Annunziato et al. (1980) and
Demarest et al. (1981) showed that TH-containing neurones from the
tuberoinfundibular region are not regulated by presynaptic receptors.
Our experiments, however, clearly revealed the presence of nerve
endings in the hypothalamus, where TH is indeed regulated by presynaptic receptors which react to dopamine. The possibility should
not be excluded that these terminals may not belong to the group of
hypothalamic neurones but are a projection of the nigro-striatal
(George et al., 1982) or some other dopamine system. Hypothalamic

TH is more sensitive to the action of atypical neuroleptics than
striatal TH (Rayevsky et al., 1978), indicating the major function
fulfilled by the hypothalamic dopamine system in the mechanisms of
neuroleptic action.

5. EFFECT OF FLUORPHENAZINE ON CONTROL OF HYPOTHALAMIC SYNAPTOSOMAL TH BY PRESYNAPTIC RECEPTORS

Figure 4 shows TH reaction rate plotted against tyrosine concentration for membrane-bound synaptosomal enzyme, incubated with a)
dopamine, and b) a combination of dopamine and fluorphenazine. As
the figure shows, the addition of DA together with fluorphenazine to
the incubation mixture leaves the TH reaction kinetics observed
during the assay with DA unchanged. However, fluorphenazine is found
to produce a slight but noticeable action when added separately to
the synaptosomes (without dopamine). This graphs show a rise in the
concentration of tyrosine optimal for TH, which suggests that neuro-

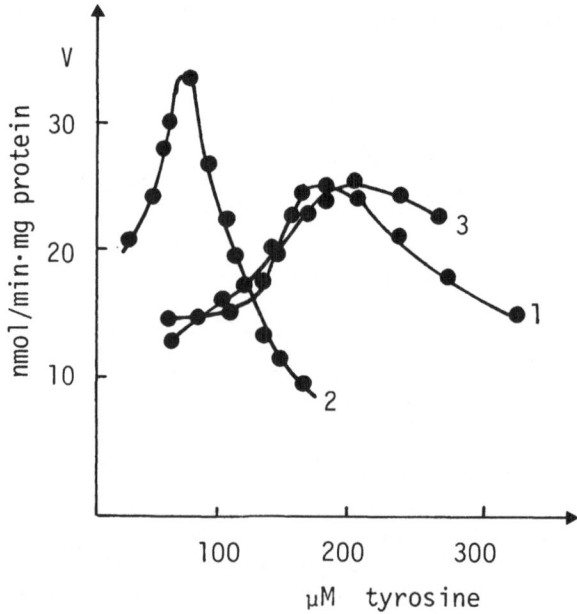

Fig. 4. The antagonistic effect of fluorphenazine to changes in the
         kinetics of the tyrosine hydroxylase reaction induced by
         dopamine action on the outer synaptosomal membrane. Experi-
         mental conditions: 0.05 M tris-maleate, pH 6.1, 30°C. DMPH₄
         - 0.140 µM; tyrosine - 30-350 mM; membrane-bound tyrosine
         hydroxylase from rat hypothalamus - 30 µg/ml. 1 - control;
         2 - test incubated with DA; 3 - test incubated with DA and
         fluorphenazine.

leptics could act as presynaptic receptor antagonists by influencing TH directly, as well as acting on the receptor as such, thereby pre-empting its reception of the signal from the presynaptic receptors.

6. EFFECT OF DES-TYR-GAMMA-ENDORPHIN ON THE KINETICS OF TYROSINE HYDROXYLATION IN SYNAPTOSOMES OF THE STRIATUM AND NUCLEUS ACCUMBENS SEPTI

Our investigations into the workings of des-tyr-gamma-endorphin (DTGE), an endogenous peptide, also helped to establish the fact that compounds exerting a neuroleptic action directly affect the kinetics of the TH reaction with tyrosine. A number of DTGE's properties relate to the neuroleptics (De Wied et al., 1978). The dopaminergic systems of the brain are involved in the mechanics of DTGE's action, as in the case of neuroleptics (De Wied, 1979). The peptide's influence on TH was investigated following the first version of the method of Rayevsky et al. (1982), so as to shed light on how DTGE might be involved in regulating DA biosynthesis in striatal synaptosomes. Hydroxylation was measured at tyrosine concentrations of 2 µM and 10 µM (see Table 5). This table shows how DTGE failed to affect the speed of tyrosine hydroxylation in the synaptosomes when tyrosine was present in the incubation mixture at a concentration 2 µM, and significantly increased the reaction rate ($p<0.05$) at a tyrosine concentration of 10 µM. These factors led us to believe that DTGE could be reducing the affinity of synaptosomal TH for its substrate (Rayevsky et al., 1982). These findings are in line with the results of experiments with fluorphenazine on hypothalamic synaptosomes, in which the enzyme was separated in the form of a membrane-bound fraction, after the synaptosomes had been incubated with antagonist. Reaction rate was measured by a different method, using synthetic $DMPH_4$ as cofactor. The extent of DTGE's effect on hydroxylation in synaptosomes of the striatum and of the nucleus accumbens differed, however. The experiments were again performed using the second version of the method for ascertaining TH activity, incubating the assays for 15 min and using a protein concentration of 0.5 mg/ml The linearity of accumulation of the reaction product with time and protein concentration in the assay is maintained under these circumstances. Incubation time was greater during the first set of experiments with DTGE and striatal synaptosomes, since we were interested in the possibility of DTGE acting on TH or the potential effects on the enzyme of this peptide's active fragments, which are formed when it is incubated with synaptosomes (Burbach et al., 1980). As shown in Figure 5, a 10^{-6}M concentration of DTGE increased the K_m for tyrosine in striatal synaptosomes by more than 50%, but was able to exert a comparable effect on the synaptosomes of the neighboring septal nucleus at a concentration of only 10^{-4}M (see Figure 6). These findings show that the action of DTGE, to which the attributes of an endogenous neuroleptic are often ascribed, is marked by its own

Table 5. Effect of Des-tyr-γ-endorphin on Hydroxylation of Tyrosine
          in Striatal Synaptosomes

Tyrosine concentration µM	Tyrosine hydroxylation rate, pmol ^{3}H-DOPA/mg protein/min		
	Control	DTGE 10^{-6}M	DTGE 10^{-4}M
2.0	0.087±0.022	0.097±0.029	–
10.0	0.146±0.062	0.266±0.088*	0.223±0.018*

Difference from control for corresponding concentration of tyrosine
*p<0.05 (Student's test).

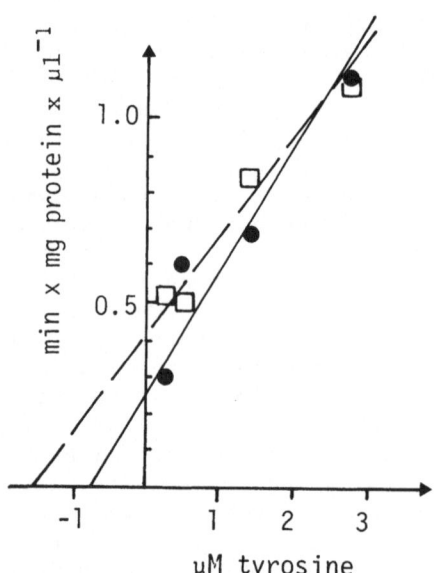

Fig. 5. The effect of des-tyr-γ-endorphin (DTGE) on hydroxylation
         of tyrosine in rat striatal synaptosomes. Abscissa –
         tyrosine concentration (µM). Ordinate – relationship
         between tyrosine concentration (µM) and initial reaction
         rate (pmol/mg protein/min). ●——● control; □ – –□ DTGE
         (10^{-6}M). Experimental conditions: Krebs phosphate buffer,
         pH 7.35, 37°C, 15 min. Each concentration of tyrosine
         represents the mean of three separate measurements.
         Reaction rate is determined by the amount of ^{3}H-labelled
         water formed.

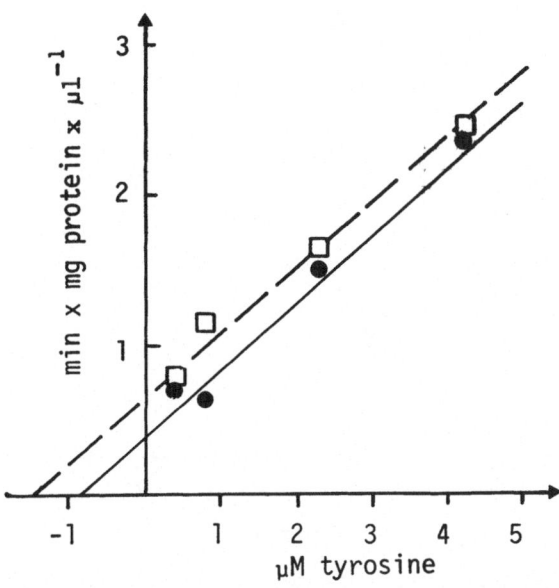

Fig. 6. The effect of des-tyr-γ-endorphin on hydroxylation of
         tyrosine in the synaptosomes of the nucleus accumbens
         septi in rat brain. Abscissa - tyrosine concentration
         (μM). Ordinate - relationship between tyrosine concen-
         tration (μM) and initial reaction rate (pmol/mg
         protein/min). ●——● control; ▯ - - -▯ DTGE (10^{-4}M).
         Reaction conditions as in Figure 5.

peculiarities - viz. influencing the hydroxylation of tyrosine in the
striatum and in the adjoining septal nucleus to a varying extent.
Taking Tyr-D-Ala-Gly-Phe-amide (TAE) for purposes of comparison,
which has pronounced opioid traits (McGregor et al., 1978; Zakusov et
al., 1981), this synthetic tetrapeptide analog of the enkephalins
exerts a less potent action than DTGE on the hydroxylation of tyrosine in the striatum at a concentration of not less than 10^{-4}M, and
has no effect on this process in synaptosomes of the neighboring
septal nucleus (see Figure 7). This provides a means of differentiating between the DTGE effect and the action of opioid peptides.

     In the course of research into the properties of DTGE and TAE as
presynaptic dopamine receptor antagonists, applied to synaptosomes
under identical conditions, DTGE revealed traits distinguishing it
from classical neuroleptics (see Tables 6, 7 and 8). Neither DTGE
nor TAE impeded DA's transmembrane effect; i.e. they did not antagonize presynaptic dopamine receptors. At the same time the actions of
DTGE and DA did not appear to be additive.

     The interaction between these substances could perhaps be ex-
plained as follows: changes mediated by presynaptic receptors alter

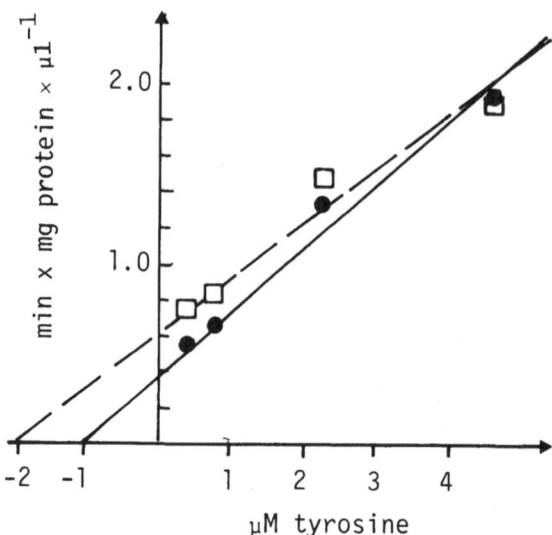

Fig. 7. Effect of the tetrapeptide analog of the enkephalins (TAE) on the hydroxylation of tyrosine in rat striatal synapto-somes. Abscissa – tyrosine concentration (μM). Ordinate – relationship between tyrosine concentration (μM) and initial reaction rate (pmol/mg protein/min). ●——● control; □ – – □ TAE (10^{-4}M). Reaction conditions as in Figure 5.

the kinetics of TH isolated from the synaptosomes of rat hypothalamus exposed to DA; these changes manifest as a reduction in semi-saturating concentration of tyrosine, equivalent to that of the K_m for tyrosine. Assuming that an equivalent trans-membrane effect of DA on the apparent Michaelis constant for tyrosine is also seen in the striatum, this factor could shift the range of the DTGE-induced inhibitory reaction towards a lower tyrosine concentration and there-by interfere with the manifestation of its effect at the concen-tration of tyrosine used.

These findings point to the complexity of the mechanisms of TH regulation by presynaptic receptors; more research is required to understand how they operate. Our results also illustrate the dif-ference between DTGE and typical neuroleptics, in common with the findings of Weinberger et al. (1979). These authors discovered that, unlike haloperidol, DTGE administered to animals does not accelerate DA biosynthesis in striatal slices when measured under in vitro conditions. The way in which DTGE acts at dopamine receptor level still remains somewhat unclear. This peptide does not displace ^{3}H-spiroperidol from its binding site on striatal membranes in vitro, although one of its metabolites does appear to possess this ability (Reisine et al., 1979). Data both supporting and denying its capacity to do so in vivo have been put forward respectively by

Table 6. Effect of Des-tyr-γ-endorphin on Dopamine-induced
          Inhibition of Tyrosine Hydroxylation in Rat Striatal
          Synaptosomes

| Experimental variable | Rate of tyrosine hydroxylation as % of control (M ± SEM) | |
	^{3}H-tyrosine, 0.8 μM	^{3}H-tyrosine, 4.4 μM
Control	100.0±26.9[+] (5)	100.0±12.7[++] (8)
DTGE 1 μM	67.7±31.1* (7)	89.9±15.9** (8)
DA, 2 μM	45.2±25.3* (5)	44.3± 9.2** (8)
DTGE 1 μM + DA, 2 μM	45.8±25.5* (8)	40.0±20.8** (8)

[+] speed of tyrosine hydroxylation - 1.058 pmol/mg protein/min.
[++] speed of tyrosine hydroxylation - 3.002 pmol/mg protein/min.
Incubation period: 15 min. Protein concentration in assays -
0.5-1 mg/ml. Difference from control (Student's test) -
* $p<0.01$; ** $p<0.002$.

Table 7. Tyrosine Hydroxylation in the Striatal Synaptosomes:
          Effect of Peptides on Inhibitory Effect of Dopamine when
          Potassium Ions (30 mM) and Cocaine (40 mM) are Present

| Experimental variable | Rate of tyrosine hydroxylation (tyrosine concentration 1.7 M) as % of control; no added dopamine | |
	Without addition	Dopamine, 3 μM
Control	100.0±18.0 (8)	72.2±14.4** (4)
DTGE, 1 μM	96.0±20.5 (6)	75.8±12.2* (7)
DTGE, 100 μM	92.1±20.1 (8)	71.7± 8.4** (6)
TAE, 1 μM	90.3±33.7 (8)	75.1±11.5** (6)
TAE, 100 μM	90.1±10.8 (7)	71.2±16.4 (6)

Conditions of incubation - as in Table 6.
Reaction rate in control (no addition) taken as 100%;
this corresponds to 1.232 pmol/mg protein/min.
Difference from the control, without dopamine (Student's t-test).
* $p<0.05$; ** $p<0.02$; *** $p<0.01$. No. of experiments given in brackets.

Pedigo (1979) and Lachti et al. (1982). Our data indicate another
possible route for inhibiting dopaminergic transmission as influenced
by DTGE - viz. a drop in synthesis rate of the transmitter when the
affinity of TH for tyrosine decreases (Rayevsky et al., 1982) due to
the DTGE-induced changes in TH kinetics. The fact that DTGE, like
the neuroleptics, produces an action contrasting with DA's transmembrane effect on the kinetics of tyrosine hydroxylation is also
worth noting.

Table 8. Hydroxylation of Tyrosine in the Synaptosomes of the
          Adjoining Nucleus; Effect of Peptides on the Inhibitory
          Effect of Dopamine in the Presence of Potassium (30 mM)
          and Cocaine (40 µM)

Experimental variable	Rate of tyrosine hydroxylation (tyrosine concentration expressed as % of control, without added dopamine) $M \pm m$ (where $m = t_{0.975} \times$ SEM)	
	No added dopamine	Dopamine, 3 µM
Control	100.0±19.4 (8)	68.0±14.0** (6)
DTGE µM	85.9±18.1 (8)	68.5±32.7* (4)
DTGE, 100 µM	86.9±18.4 (8)	68.0±16.6** (8)
TAE, 1 µM	94.9±23.4 (8)	70.4± 5.3***(8)
TAE, 100 µM	90.1±10.8 (7)	71.2±16.4 (6)

Conditions of incubation, as in Table 6. Reaction rate in control
(no tyrosine added); concentration: 1.181 pmol/mg protein/min.
Difference from control, with no dopamine added: *$p<0.04$ (one-tailed
Wilkinson's test); **$p<0.02$ (Student's test); ***$p<0.01$ (Student's
test). No. of experiments given in brackets.

6. DISCUSSION

As shown above, influencing the kinetics of the TH reaction with
respect to tyrosine in the hypothalamus, striatum and nucleus
accumbens septi is characteristic of compounds possessing neuroleptic
properties. We began by showing that substrate inhibition of TH
might be involved in the mechanics of TH regulation by presynaptic
receptors, while neuroleptics (dopamine receptor antagonists) could
be suppressing the regulatory signal from the presynaptic receptors
which causes inhibition of synthesis of the neurotransmitter, thereby
eliminating substrate inhibition of TH.

We proceeded to show that the mechanism by which dopamine regulates TH cannot merely be attributed to direct inhibition of TH by
dopamine, which is subject to reuptake; influence at presynaptic
receptor level clearly plays an important part here. This mechanism
may involve modifications in the enzyme, alterations in its kinetics
and a reduction in true rate of dopamine biosynthesis without the
enzyme losing any of its potential activity. Our findings show that
substrate inhibition of TH might operate by a fundamentally different
mechanism, whereby the phenomenon could depend on enzymes involved in
TH modification rather than factors determining intracellular concentration of tyrosine. It is difficult to see how the enzyme is modified under these circumstances, but it is plain that such modification must be rapidly reversible. TH modification could very well

occur as a result of its phosphorylation (Lazar, 1982). This supposition could resolve the paradox that activating the receptors
brings about the activation of protein kinase and a rise in protein
phosphorylation rate, whereas phosphorylation of isolated TH leads to
kinetic activation of the enzyme; in fact, TH inhibition is actually
observed under in vivo conditions. TH reaction rate in vivo is
determined by the rate of DOPA accumulation in the presence of endogenous concentrations of tyrosine and tetrahydrobiopterin; reaction
rate is thus assessed at only one level of tyrosine. As the semi-
saturating concentration of tyrosine for TH has been reduced, the
endogenous tyrosine concentration will now be inhibitory. The mechanism suggested for presynaptic receptor regulation of TH in no way
conflicts with the concept of dopamine's physiological role as a
direct regulator of the TH reaction. According to McMillan et al.
(1980) it is the dopamine present in the storage granules which plays
a key role in dopamine control of TH.

6.1. REGULATION OF TYROSINE HYDROXYLASE OF NEUROLEPTICS - POSSIBLE
      MECHANISM OF ACTION

     The conclusion drawn from our experimental findings that two
tyrosine binding sites are present on the TH molecule, together with
evidence that compounds like fluorphenazine and DTGE, with a neuroleptic action, produce an identical effect on the enzyme's kinetic
properties by direct and transmembrane modes of action on TH suggest
that neuroleptics act as follows: The TH molecule, with several
sites of regulation, can be seen on Figure 8, which shows how the
conformation of the active center is maintained by endogenous ligand
at the allosteric site. Tyrosine, at a concentration exceeding
saturation level, occupies a site 3', which has less affinity for TH
than the catalytic center. Obstacles are thereby created for interaction between the endogenous ligand and TH at the site regulated by
the endogenous ligand. The ligand disassociates from the enzyme; the
conformation of the catalytic center thereby changes and the enzyme
loses activity. Neuroleptics interact with TH when the regulatory
site is occupied by tyrosine, reducing TH's affinity for tyrosine,
and forcing tyrosine out of the site. The site with a higher affinity for tyrosine is accordingly occupied by tyrosine, but the
regulatory site cannot be properly occupied by it under these circumstances. At the same time interaction between enzyme and neuroleptic interferes with modification of TH, due to the action of
factors which display their activity during activation of presynaptic
receptors. This diagram is based on the assumption that neuroleptics
penetrate through the plasma membrane and interact with TH directly.
Our findings also point to the presence in the TH molecule of
recognition sites for dopamine receptor antagonists, as shown by
neuroleptics' selectivity of suppression of TH substrate inhibition.
If this is so, there may be some structural resemblance between such
recognition sites on the TH molecule and the otherwise distinct

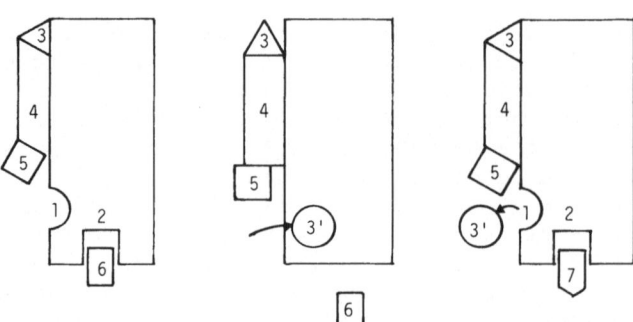

Fig. 8. Scheme of tyrosine hydroxylase regulation with tyrosine and
        neuroleptics. 1 and 2: regulatory sites for tyrosine and
        endogenous ligand, respectively; 3, 4 and 5: sites for
        tyrosine, pterin cofactor, oxygen; 6 and 7: endogenous
        ligand and neuroleptic, respectively; 3': tyrosine in
        regulatory site.

complex receptor molecule. There is reason to believe that a
naturally-occurring antagonist of presynaptic dopamine receptors
could be serving as an endogenous ligand for TH, stabilizing the
active conformation of the enzyme. If these concepts correspond to
the facts, then isolated TH could be used as a model of the receptor
contact site. This would facilitate investigations into molecular
mechanisms of interaction with the antagonists. TH, and that from
the hypothalamus in particular, could also be used as a test system
for identifying neuroleptics which, as already shown, can act as
allosteric regulators of TH.

REFERENCES

Ahlenius, S., 1974, Reversal by DOPA of the suppression of locomotor
        activity induced by inhibition of tyrosine hydroxylase and
        dopamine β-hydroxylase in mice, Brain Res, 79:57-65.
Annunziato, L., Cerrito, F., and Raiteri, M., 1981, Characteristics
        of dopamine release from isolated nerve endings of the tubero-
        infundibular neurones, Neuropharmacology., 20:727-731.
Antelman, S. M., Szechtman, H., Chin, P., and Fisher, A. E., 1976,
        Inhibition of tyrosine hydroxylase but not dopaminehydroxylase
        facilitates the action of behaviourally ineffective doses of
        neuroleptics, J.Pharm.Pharmacol., 28:66-68.
Bradford, H. V., 1970, Metabolic response of synaptosomes to elec-
        trical stimulation release of amino acids, Brain Res.,
        19:239-247.
Breese, G. R., Mueller, R. A., Hollister, A., and Mailman, R., 1978,
        Importance of dopaminergic pathways and other neural systems
        to behaviour and action of psychotropic drugs, Fed.Proc.,
        37:10:2429-2433.

Coyle, J. T., 1972, Tyrosine hydroxylase in rat brain - cofactor
    requirements, regional and subcellular distribution, Biochem.
    Pharmacol., 21:1935-1944.
Di Chiara, G., Corsini, G. U., Mereu, G. P., Tissari, A., and Gessa,
    G. L., 1978, Self-inhibitory dopamine receptors: their role in
    the biochemical and behavioural effects of low doses of apo-
    morphine, in: "Dopamine 2. Advances in Biochemical Psycho-
    pharmacology," P. J. Roberts, G. N. Woodruff, and L. L.
    Iversen, eds., vol. 19, Raven Press, New York, pp. 273-295.
Demarest, K. T., and Moore, K. E., 1979, Comparison of dopamine
    synthesis regulation in the terminals of nigrostriatal,
    mesolimbic, tuberoinfundibular and tuberohypophyseal neurons,
    J.Neural Transmiss., 46:263-277.
George, S. R., and Van Loon, G. R., 1982, Characterization of high
    affinity dopamine uptake into the dopamine neurones of the
    hypothalamus, Brain Res., 234:339-355.
Glowinski, J., and Iversen, L. L., 1966, Regional studies of
    catecholamines in the rat brain, I. The disposition of
    ^{3}H-norepinephrine, ^{3}H-dopamine and ^{3}H-DOPA in various regions
    of the brain, J.Neurochem., 13:655-669.
Gray, E. G., and Whittaker, V. P., 1962, The isolation of nerve
    endings from brain: an electron-microscopic study of cell
    fragments derived by homogenization and centrifugation,
    J.Anat., 96:79-87.
Hajos, T., 1975, An improved method for the preparation of synap-
    tosomal fractions in high purity, Brain Res., 93:485-489.
Horn, A. S., Cuello, A. S., and Miller, R. J., 1974, Dopamine in the
    mesolimbic system of the rat brain: endogenous levels and the
    effects of drugs on the uptake mechanism and stimulation of
    adenylate cyclase activity, J.Neurochem, 22:265-270.
Karobath, W. E., 1975, Dopamine-Receptor Blockade, ein moglicher
    Wirkungsmechanismus antipsychotisch wirksamer Pharmaka, Pharm-
    acopsychiatrie, 8:152-161.
Kuchenski, R., and Mandell, A. J., 1972, Regulatory properties of
    soluble and particulate rat brain tyrosine hydroxylase,
    J.Biol.Chem., 247:3114-3122.
Kuchenski, R., and Segal, D. S., 1974, Intrasynaptosomal conversion
    of tyrosine to dopamine as an index of brain catecholamine
    biosynthetic capacity, J.Neurochem. 22:1039-1044.
Lachti, N. A., and Gay, D. D., 1982, Lack of effect of des-tyr-
    endorphin on in vivo ^{3}H-spiperone binding, Eur.J.Pharmacol.,
    80:127-131.
Lasar, M. A., Mefford, J. N., Barchas, J. D., 1982, Comparison of in
    vitro phosphorylation and in vivo administration of haloperi-
    dol, Biochem.Pharmacol., 31:2599-2609.
Le Fur, G., Guillouz, F., and Uzan, A., 1980, In vivo blockade of
    dopaminergic receptors from different rat brain regions by
    classical and atypical neuroleptics, Biochem.Pharmacol.,
    29:267-270.
Levitt, M., Spector, S., Sjoerdsma, A., and Udenfriend, S., 1965,

Elucidation of the rate limiting step in noradrenaline biosynthesis in the perfused guinea pig heart, J.Pharmacol.Exp.Ther., 148:1-8.

Lowry, O. H., Rosebrough, N. J., Farr, A. L., and Randall, R. J., 1951, Protein measurement with Folin phenol reagent, J.Biol.Chem., 193:263-275.

Mager, H. I., Addink, R., and Berends, W., 1967, Coupled and decoupled processes in the autooxidation of partially reduced pteridines and flavins, Rec.Trav.chim.Pays-Bas, 86:833-851.

Mandell, A., 1978, Redundant mechanisms regulating brain tyrosine and tryptophan hydroxylases, Ann.Rev.Pharmacol.Toxicol., 18:461-493.

McGregor, W. H., Stein, I., and Beluzzi, J. D., 1978, Potent analgesic activity of the enkephaline-like tetrapeptide H-tyr-D-Ala-Gly-PheNH$_2$ Life Sci., 23:1371-1378.

Mineyeva, M. F., Tummler, D., Kuznetsova, E. A., Vassiliev, A. E., and Rayevski, K. S., 1982, Change in kinetic properties of tyrosine hydroxylase due to dopamine effect on the external synaptosomal membrane, Ann.Ist.Super.Sanita 18:45-48.

Murrin, L. C., Morgenroth III V. H., and Roth, R. H., 1976, Dopaminergic neurons: effects of electrical stimulation on tyrosine hydroxylase, Molec.Pharmacol., 12:1070-1081.

Nagatsu, T., Levitt, M., and Udenfriend, S., 1964, A rapid and simple radioassay for tyrosine hydroxylase activity, Analyt.Biochem., 9:122-126.

Owen, F., Crow, T. J., Poulter, M., Cross, A. J., Longden, A., and Riley, G. J., 1978, Increased dopamine receptor sensitivity in schizophrenia, Lancet ii:223-226.

Pegigo, N. W., Schallert, T., Overstreet, D. H., Ling, N. C., Ragan, P., Reisine T., and Yamamura, H. I., 1979, Inhibition of in vitro ^{3}H-spiperone binding by the proposed antipsychotic des-tyr-endorphin, Eur.J.Pharmacol., 60:359-364.

Poshel, B. P., and Nitelman, F. W., 1966, Hypothalamic self-stimulation: its suppression by blocade of norepinephrine biosynthesis and reinstatement by methamphetamine, Life Sci., 5:11-16.

Reisine, T. D., Pedigo, N. W., Ragan, P., Ling, N., and Yamamura, H. I., 1979, Abnormal brain opiate mechanisms in schizophrenia, in: "Endogenous and Exogenous Opiate Agonists and Antagonists," Proc.Int.Narcotic Res., North Falmouth, Mass, 1980, pp.117-120.

Snyder, S., 1976, The dopamine hypothesis of schizophrenia: focus on the dopamine receptor, Am.J.Psychiatry, 133:197-202.

Tuomisto, J., 1978, Neurophysiological intervention on the pituitary-hypothalamic relationship, Ann.Clin.Res., 10:120-132.

Venter, J., Diton, J. S., Maroko, P. R., and Kaplan, N. O., 1975, Biologically active catecholamines covalently bound to glass beads, Proc.Nat.Acad.Sci.USA., 69:1141-1145.

Wied, De, D., Kovacs, G. L., Bohus, B., van Ree, J. M., and Greven, H. M., 1978, Neuroleptic activity of the neuropeptide γ-LPH

62-67 (des-Tyr-γ-endorphin: DT E), Eur.J.Pharmacol., 49:427-426.

Wied, De, D., 1979, Schizophrenia as an inborn error in the degradation of γ-endorphin - a hypothesis. Trends Neurosci. 49:79-82.

Zivkovic, B., Guidotti, A., Revetta, A., and Costa, E., 1975, Effect of thioridazine, clozapine and other antipsychotics on the kinetic state of tyrosine hydroxylase and on the turnover rate of dopamine in striatum and nucleus accumbens, J.Pharmacol.Exp.Ther., 194:37-46.

Correlation Between the Affinity of Antidepressants for Membranes and Their Influence on Monoamine Transport

N.A. Avdulov and A.V. Valdman

Institute of Pharmacology, Academy of Medical Sciences
of the USSR, Moscow

1. INTRODUCTION

Antidepressants belonging to different chemical groups have been found to inhibit monoamine reuptake by rat brain slices (Dolzhenko and Komissarov, 1981) as well as rat brain synaptosomes (Jirkovsky and Lippman, 1978). Many of the drugs in this group are also known to increase monoamine release (Raiteri et al., 1977). Antidepressants may thus be divided into three groups: the first functions mainly by intensifying monoamine release from axon terminals, the second acts similarly to the first while also inhibiting neurotransmitter reuptake, and the third group chiefly affects monoamine reuptake (Dolzhenko and Komissarov, 1981).

From a study of the literature we are unable to identify impaired function of any one neurotransmitter or enzyme as the cause of depression, in that clinically effective antidepressants influence the transport of a variety of neurotransmitters and the function of a number of enzyme systems. Hence it has been suggested that these antidepressants' neurochemical effects are centered on components of neuronal membranes affecting a variety of membrane enzymes and transport systems and processes rather than specific proteins (transmitters and enzymes).

Publications clarifying the part played by lipids in membrane processes have recently been claiming the attention of research workers (Loh and Law, 1980; Antonov, 1982; Ivkov and Berestovsky, 1982). The function of membrane lipids may vary, from reinforcing the membrane (Day and Levy, 1969) to acting on membrane protein activity - especially mitochondrial enzymatic activity (Coleman, 1973) - as well as Na,K-ATPase activity (Grisham and Barnett, 1973),

influencing the binding of various ligands to membrane proteins, as
when thiamine phosphate binds with pyruvate oxidase (Cunningham and
Hager, 1971), or the ionic permeability of membranes (Antonov, 1982).
Since most neurotransmitters and pharmaceutical substances are
present in solution as cations when the pH is at physiological level,
membrane lipids may be serving as binding sites for these compounds
(Loh and Law, 1980).

Inhibition of neurotransmitter reuptake by presynaptic neuronal
membranes is one important action of tricyclic and atypical antidepressants. It has also been observed that most antidepressants
have an effect on membrane enzyme activity (Jirkovsky and Lippman,
1978). The absence of any precise selectivity in most antidepressants' overall neurochemical effect would suggest that they exert a
nonspecific membrane action. Hence it is interesting to compare the
affinity for membranes of a selection of well-known tricyclic and
atypical antidepressants with their action on neurotransmitter
accumulation by rat brain synaptosomes. Assessing the specificity
of the neurochemical and biophysical properties of the compounds
under study also appeared worthwhile.

2. MATERIAL AND METHODS

2.1. SYNAPTOSOMAL SEPARATION

Experiments were performed on white cross-bred male rats
weighing 180–200 g. The brain was removed shortly after decapitation
and placed in a Petri dish previously chilled over ice. Brain tissue
was washed free of blood with a 0.32 M sucrose solution (0–4°C).
The washed brain tissue was then homogenized in a glass homogenizer
with a Teflon pestle in a medium containing 0.32 M sucrose, 0.2. mM
EDTA and 0.05 M tris-HCl, pH = 7.4. The ratio of brain weight in g
to volume of homogenate in ml equalled 0.1. The homogenate was
centrifuged at 1000 g for 10 min. The supernatant was collected
and centrifuged at 11,000 x g for 20 min before being poured off.
The pellet, containing synaptosomes, mitochondria and myelin, was
resuspended in 0.7 ml 0.32 M sucrose/g original brain weight.

Table 1. Michaelis Constant (K_m) Values for
          Neurotransmitter Accumulation by Crude
          Synaptosomal Fraction of Rat Brain

Transmitter	K_m, 1 x 10^{-6} M
Noradrenaline	0.6
Dopamine	0.06
Serotonin	0.2
GABA	30.0
Glutamate	30.0

2.2. PRODUCTION OF LIPOSOMES

Model membrane particles were obtained by rapid injection of
an ethanolic solution of total phospholipid fraction of hen's egg
(containing an average of 50 mg lipid per ml) into a tris-HCl
buffer solution (pH = 7.4), using a magnetic stirrer. The liposomes
produced had a single bilayer membrane (Bratzri and Korn, 1973).

2.3. NEUROTRANSMITTER ACCUMULATION BY CRUDE SYNAPTOSOMAL FRACTION

A volume of 50 µl of synaptosomal suspension containing an
average of 0.5 mg protein per ml was incubated, during the course of
standard experiments, with different concentrations of neurotransmitters in 1 ml of incubation medium containing 100 mM NaCl, 6 mM KCl,
2 mM $CaCl_2$, 1.14 mM $MgCl_2$, 5 mM Na_2HPO_4, 10 mM glucose, 100 mM
sucrose, 0.125 mM pargyline, 0.54 mM EDTA, 1.14 mM ascorbic acid and
30 mM tris-HCl buffer (pH-7.4). This incubation took place at 37°C
for 20 min with continuous stirring. Snyder and Coyle's (1969)
method was used to separate the synaptosomes from the incubation
medium and to determine neurotransmitter concentration. Michaelis
constant values obtained for transmitter build-up agree with findings
from the literature (Snyder and Coyle, 1969; Rayevsky and Maisov,
1975).

Experiments were subsequently performed at standard neurotransmitter concentrations: 0.2 µM for noradrenaline, 0.02 µM for
dopamine, 0.08 µM for serotonin, 10 µM for GABA and 10 µM for
glutamate. Synaptosomal accumulation occurred at 0.17 of maximum
rate at these concentrations; this enabled us to record both competitive and noncompetitive inhibition of the accumulation of the above
neurotransmitters by the compounds tested. The synaptosomes incubated at 37°C for 20 min were found to take up 0.3±0.03 nmol noradrenaline, 2±0.3 pmol dopamine, 9±0.2 nmol serotonin, 3.5±0.4 µM
GABA and 3±0.3 µM glutamate. These amounts were taken as representing 100% accumulation of the respective neurotransmitter by the
synaptosomal accumulation process. A statistical evaluation was made
of the results, using an HP-33E calculator, establishing means and
standard deviations - p = 0.05.

Radioactivity was measured on an "Intertechnique" SL-4000 liquid
scintillation counter (France).

2.4. KINETICS OF SYNAPTOSOMAL SEROTONIN REUPTAKE

Experiments were performed on a crude synaptosomal fraction of
rat brain. Different concentrations of ^{3}H-serotonin were incubated
for 3 min at 37°C, mixing continuously, in an incubation medium
composed as described above; 50 µl synaptosomes were present, con-

taining an average of 0.5 mg protein per ml. Incubation period was
selected once the linearity of serotonin accumulation had been ascertained (see Figure 1). The final stages of the procedure consisted
of filtering the synaptosomes on the Whatman G-45 filters, using a
vacuum pump and washing them with 8 ml of incubation medium, at a
temperature of 22°C. The filters were dried at room temperature;
in 10 ml of scintillator containing 3 ml methylcellosolve and 7 ml
toluene, in which 0.5% PPO and 0.01% POPOP were dissolved. Radioactivity was measured on a "Mark 1" Nuclear Chicago (USA) liquid
scintillation counter. Findings were displayed on a Lineweaver-Burk
graph, using a double-reciprocal plot.

2.5. INTERACTION BETWEEN ANTIDEPRESSANTS AND LIPOSOMES

A compound's affinity for membranes was determined by the
fluorescent probe method. We used the universal probe 1-anilino-
naphthalene-8-sulfonate (ANS), which attaches itself to the polar
head group zone of lipid membranes. Fluorescence of the probe was
measured at an excitation wave length of 360 nm and recorded at
480 nm using an "Opton" spectrofluorometer (West Germany). We
proceeded to a 2-stage titration procedure in order to ascertain how
ANS interacts with model membranes, and 0.2 mg/ml membranes were
titrated with the probe. An analysis of our findings by double
reciprocal plot produced a K_c value of 0.02 μM^{-1} and an N_u of 69.1
for ANS.*

On the basis of these findings, we investigated the interaction
between antidepressants and membranes at standard membrane and probe
concentration - i.e. 0.2 mg/ml liposomes and 10 μM ANS - and a probe/
lipid molecular ratio of 1:25. Approximately 50% of the probe had

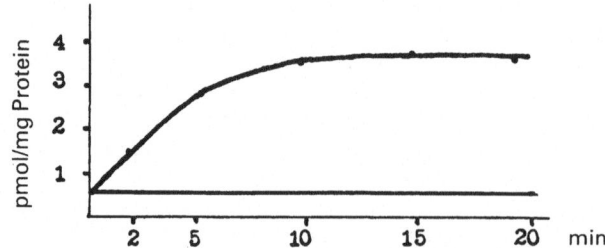

Fig. 1. Accumulation of serotonin by crude synaptosomal fraction
         of rat brain plotted against incubation time. Abscissa:
         incubation time in min; ordinate: accumulation of
         ^{3}H-serotonin creatinine sulfate in pmol/mg protein.

---

*(Translator's Note: for definition and discussion of these
  constants, see Section 3.2.)

failed to bind to membrane under these experimental conditions, thus
enabling us to identify either reduced or increased bindings, as
indicated by enhanced or reduced fluorescence (in our procedure,
any fluorescence seen was produced by bound probe). Fluorescence
was recorded following each addition of varying amounts of a 1 mM
solution of the antidepressant under study after a 5 min interval.

2.6. RATING A COMPOUND'S RELATIVE HYDROPHOBICITY

We studied the partition of these compounds in a two-phase
water-organic system, n-octanol/water, in order to assess their
relative hydrophobicity.

Aqueous solutions of a compound were introduced into the system,
the volume of the aqueous phase brought to 1 ml and 1 ml octanol was
added. The resulting phases were first mixed for 2 min using a
laboratory mixer, then separated out on a table model centrifuge at
4000 rpm for 20 min. Concentrations of drug distributed in each
phase were determined by spectrophotometry. An HP-33E (USA) electronic calculator was used for processing the results.

Standard programs were used for regression analysis and calculating correlation coefficients. The Lowry method was applied for
protein assay.

2.7. MATERIALS

The following compounds were used in our experiments: ^{3}H-nor-
adrenaline (specific activity, 12.4 Ci/mmol), ^{3}H-dopamine, with a
specific activity (SA) of 12.5 Ci/mmol, ^{3}H-serotonin-creatinine-
sulfate (SA 12), ^{3}H-GABA (SA 10) and ^{14}C-glutamate (SA 50 Ci/mmol) –
all from Amersham (UK). All reagents were of analytical grade. The
following pharmaceuticals were used: 1,8-ANS (Sigma), imipramine
(CIBA-Geigy), desmethylimipramine (CIBA-Geigy), chlorimipramine
(CIBA-Geigy) (tricyclic antidepressants were used in form of hydro-
chlorides), pirazidol, inkazan and azaphen (USSR) and befuraline
(synthesized at the Institute of Pharmacological Research of the
Acad. Med. Sci. of the USSR).

3. RESULTS

3.1. DRUG EFFECTS ON SYNAPTOSOMAL NEUROTRANSMITTER ACCUMULATION

The results of our experiments, given in Table 2, show that
tricyclic antidepressants of the imipramine group substantially
inhibit synaptosomal neurotransmitter and GABA accumulation at the
concentration employed, whereas atypical agents only inhibit the
accumulation of up to two of the monamines studied (one in the case

Table 2. The Effect of 50 μM Concentration of Antidepressants on Synaptosomal Neurotransmitter
          Accumulation in Rat Brain

Drug	Neurotransmitter accumulation (as % of control)				
	Noradrenaline	Dopamine	Serotonin	GABA	Glutamate
Control	100±10	100±10	100±9	100±10	100±9
Imipramine	57±6*	69±7*	31±2*	78±9*	86±4
Desmethylimipramine	67±7*	61±7*	21±3*	77±9*	96±10*
Chlorimipramine	46±5*	64±7*	26±4*	55±7*	78±10*
Befuraline	75±8*	102±11	86±10	121±15	95±10
Trazodone	92±10	104±10	33±4*	96±10	101±11
Pirazidol	59±6*	87±9	71±8*	99±15	95±10
Inkazan	104±11	99±10	35±6*	88±10	81±10
Azaphen	102±11	90±10	77±8*	70±10*	78±8*

* Differs significantly from control (p = 0.05).

of befuraline, trazodone and inkazan; two in that of pirazidol and
azaphen). The relatively weak action of all these drugs on synaptosomal glutamate accumulation is particularly striking, although the
tricyclic antidepressant, chlorimipramine, did have a significant
action on this system. Our results basically agree with those found
in the literature (Mashkovsky et al., 1979).

3.2. INTERACTION BETWEEN ANTIDEPRESSANTS AND MODEL LIPID MEMBRANES

Some data on the interaction between drugs and model membranes
are given in Table 3, including drug–membrane binding constants (K_c),
number of drug binding sites per unit of membrane material (N_u),
total affinity of drug for the membrane ($K_c N_u$) and finally the
drug's effect on "density of electric charge on the membrane surface
(f_s)". The latter was calculated on the basis of Gouy-Chapman's
theory (Vladimirov and Dobretsov, 1980). These data show that the
bicyclic antidepressant befuralin binds most effectively to membranes, while pirazidol, a tetracyclic drug, proved least effective
in this respect. Tricyclic drugs affected the density of the electric charge on the membranes' surface most. Chlorimipramine was the
most successful of these; it also had the greatest overall affinity
for the membranes.

3.3. COMPARISON BETWEEN ANTIDEPRESSANTS' AFFINITY FOR MEMBRANES AND THEIR EFFECT ON SYNAPTOSOMAL NEUROTRANSMITTER ACCUMULATION

By comparing and contrasting our findings, we were able to
illustrate that a close correlation exists between antidepressants'
effect on synaptosomal neurotransmitter accumulation and these drugs'
influence on the density of the surface charge on the lipid phase of
the membranes (see Table 4). No correlation is to be seen, however,
between the drugs' neurochemical properties and a) their binding
constant with model membranes, b) the number of antidepressant binding sites per unit of membrane material and c) the overall affinity
of the drug for these membranes.

Table 3.   Some Characteristics of the Interaction Between
           Antidepressants and Model Membranes

Drug	K_c, μM^{-1}	N_u	$K_c N_u$, μM^{-1}	f_s, %
Imipramine	0.137	65	9.1	0.5
Desmethylimipramine	0.116	56	6.7	0.5
Chlorimipramine	0.234	82	18.8	0.625
Befuralin	0.279	22	6.2	0.25
Pirazidol	0.041	115	4.6	0.375

Table 4. Values of Correlation Coefficients (r ± SEM) Between Neurochemical and Biophysical Properties of the Antidepressants Under Study

Neurochemical properties	Biophysical properties			
	K_c	N_u	$K_c N_u$	f_s
Noradrenaline	0.17±0.48	-0.7 ±0.25	-0.8 ±0.18*	-0.8±0.18*
Dopamine	0.7 ±0.25	-0.3 ±0.45	-0.5 ±0.46	-0.9±0.08**
Serotonin	0.15±0.49	-0.15±0.49	-0.5 ±0.36	-0.9±0.08**
GABA	0.1 ±0.49	-0.4 ±0.43	-0.8 ±0.22*	-1.0±0.0002**
Glutamate	-0.32±0.44	-0.19±0.48	-0.96±0.04**	-0.8±0.16*

Neurochemical properties relate to the effect of concentration of imipramine, desmethylimipramine, chlorimipramine, befuraline and pirazodol on synaptosomal neurotransmitter accumulation. Biophysical properties relate to the affinity of these antidepressants for membranes.
* p = 0.05; ** p = 0.01; n = 5.

3.4. THE EFFECTS OF IMIPRAMINE AND TRAZODONE ON SEROTONIN REUPTAKE

Our investigations showed that imipramine and trazodone are noncompetitive inhibitors of synaptosomal serotonin reuptake (Figure 2), in that they change V_{max} without effecting K_m. Trazodone is twice as active as imipramine in this respect (K_i = 1.3±0.2 µM, as against k_i = 2.8±0.4 µM).

3.5. RATING OF RELATIVE HYDROPHOBICITY OF THE ANTIDEPRESSANTS
      UNDER STUDY

Values for partition coefficients of the compounds under study in a biphasic aqueous-organic system of n-octanol-water are given in Table 5, together with their natural logarithms. Befuraline and chlorimipramine were shown to be the most hydrophobic of the drugs tested (K_p>1), while the other antidepressants exhibit a greater affinity for water than for octanol (K_p<1).

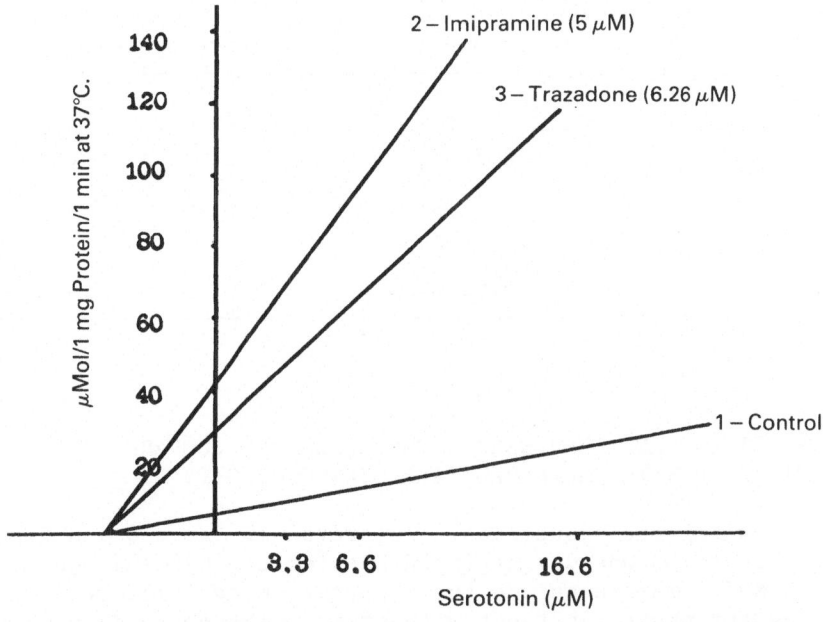

Fig. 2. Non-competitive inhibitors of serotonin reuptake: influence
         on the kinetics of the process shown by Lineweaver-Burk
         plot. Abscissa: serotonin concentration in µM; ordinate:
         accumulation of ³H-serotonin creatinine sulfate, in µmols/
         1 mg protein/1 min at 37°C.

Table 5. Some Properties of the Partition of Antidepressants in an
         n-octanol-water system

Drug	K_p	$\ln K_p$
Imipramine	0.163±0.023	−1.814±0.127
Desmethylimipramine	0.168±0.024	−1.784±0.125
Chlorimipramine	1.45 ±0.11	0.372±0.03
Befuralin	3.79 ±0.3	1.332±0.09
Trazodone	0.38 ±0.03	−0.967±0.09
Inkazan	0.016±0.001	−4.135±0.37
Pirazidol	0.69 ±0.06	−0.371±0.09

Table 6. Correlation Coefficients (and their standard errors)
         Between Relative Hydrophobicity of Known Antidepressents
         and Some of Their Biophysical Properties

Biophysical properties	Properties of relative hydrophobicity	
	K_p	$\ln K_p$
K_c	0.8 ±0.22*	0.66±0.32
N_u	−0.6 ±0.37	−0.26±0.54
$K_c N_u$	−0.01±0.58	0.18±0.56
f_s	−0.63±0.35	−0.45±0.46

*$p = 0.05$; $n = 5$.

3.6. AFFINITY OF ANTIDEPRESSANTS FOR MEMBRANES COMPARED AND CONTRASTED WITH THEIR RELATIVE HYDROPHOBICITY

By comparing the properties illustrated in Table 6 and Figure 3,
a close correlation was demonstrated between the binding constants of
drugs with model membranes and their partition coefficient in the
n-octanol-water system, but not with other membranotropic properties
of these antidepressants. This is corroborated by the other reports
in the literature to the effect that interaction between drugs and
lipid bilayer membranes should not be viewed merely as the passage
from aqueous to organic solvent (Simon et al., 1979; Vladimirov and
Dobretsov, 1980).

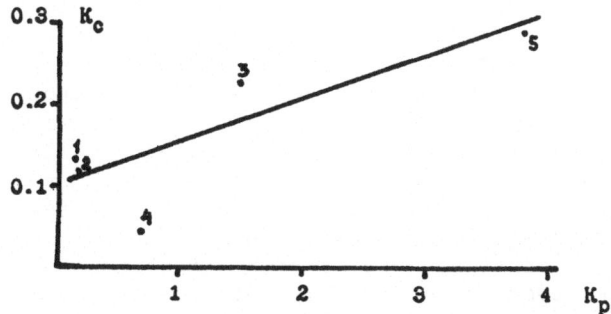

Fig. 3. Correlation between relative hydrophobicity of antidepressants and their binding constants with model
phospholipid membranes. Abscissa: values for the binding
constant of antidepressants with model phospholipid
membranes in μM; ordinate: values of partition coefficients
of the antidepressants tested in a two-phase aqueous-organic
system of n-octanol-water, in fixed units. (1) imipramine;
(2) desmethylimipramine; (3) chlorimipramine; (4) pirazidol;
(5) befuraline.

4. RESULTS AND DISCUSSION

Our investigations show that the influence exerted on the lipid
phase of presynaptic membranes by the drugs tested makes a major
contribution to their neurochemical effects. Our assessment of how
they influence synaptosomal accumulation of neurotransmitters and
neuroactive aminoacids confirmed the findings of others that drugs of
the imipramine group impair catecholamine, serotonin and GABA transport, while atypical antidepressants only affect the transport of one
or two of the monoamines under study. This fact would appear to be
connected with the different ways in which these compounds affect the
density of the surface charge of the lipid phase of the presynaptic
membrane, and is borne out in particular by the observed correlation
between the compounds' neurochemical effects and influence on the
membrane surface charge.

The noncompetitive nature of serotonin reuptake inhibition by
trazodone and imipramine would indicate that these compounds do not
interact directly with the active center of the protein carrier of
the transmitter, but become established in the membrane alongside the
active center, possibly in the bound lipid or the area where this
carrier is situated. This produces changes in the carrier's surroundings, modifying the structure of its active center and thereby
reducing its affinity for substrate.

The fact that a correlation exists between the compounds' binding constant with model lipid membranes and the partition coefficients of antidepressants in an n-octanol-water system confirms the reports of others as to the relatedness of these properties (Vladimirov and Dobretsov, 1980). Some aspects of the interaction between these compounds and membranes may thus be modelled by partitioning the antidepressants between aqueous-organic systems, especially that of their behavior in the hydrophobic area of lipid molecule tails (Ivkov and Derestovsky, 1982). The lack of any correlation between the other aspects of these compounds' affinity for membranes, however, together with their relative hydrophobicity, showed that, overall, the interaction between the compounds and membranes constitutes a much more complex process than simple passage from aqueous phase to organic solvent. Here, the processes occurring at the membrane surface and in the lipid polar head zone must be taken into account; hence, the way in which the compounds act on the surface of membranes, the number of binding sites per unit of membrane material and other factors must also be considered. Since the drugs tested are present in solution in ionized form at physiological pH values, they are hardly likely to penetrate into the hydrophobic area of the bilayer. The drugs tested are therefore to be found in the membrane in the lipid polar head zone, as our experiments show.

It should be added that the antidepressants tested inhibited neurotransmitter accumulation at one tenth of the strength required to inhibit Na,K-ATPase activity (Maisov et al., 1976), without reacting directly on the active center of the hypothetical carrier. This finding suggests that it is these very hypothetical carriers which constitute the weakest link in the neurotransmitter transport system through presynaptic membranes. The accumulation of antidepressant cations in the lipid bilayer of presynaptic membranes would appear to produce such shifts in surface charge of the membranes as to modify the active center of the carrier of neurotransmitters through the membrane, but insufficient to interfere with Na,K-ATPase. Impairment of neuronal neurotransmitter transport under the action of the compounds tested appears to be connected with their indirect action on the hypothetical neurotransmitter carriers.

Our investigations point to the need for broadening the range of biological investigation of test compounds at the stage of preclinical study.

REFERENCES

Antonov, V. F., 1982, Lipids and Ion Permeability of Membranes, Nauka, Moscow.
Batzri, S., and Korn, E. D., 1973, Single bilayer liposomes prepared without sonification, Biochim.Biophys.Acta., 298:1015-1019.

Cunningham, C. C., and Hayer, L. P., 1971, Crystalline pyruvate oxidase from Escherichia coli, II. Phospholipid as an allosteric effector for the enzyme, J.Biol.Chem., 246:1583-1589.

Day, C. W., and Levy, R. S., 1969, A generalised functional role for phospholipids, J.Theor.Biol., 22:541-542

Dolzjenko, A. T., and Komissarof, I. V., Peculiarities of presynaptic action of different antidepressants, Personal communication.

Grisham, C. M., and Barnett, R. E., 1973, The role of lipid-phase transitions in the regulation of the (sodium + potassium) adenosine triphosphatase, Biochemistry, 12:2635-2637.

Ivkov, V. G., and Berestovsky, G N., 1982, Lipid Bilayer of Biological Membranes, Nauka, Moscow.

Jirkovsky, I., and Lippman, W., 1978, Antidepressants, Annual Reports Med.Chem., 13:1-10.

Loh, H. H., and Low, P. Y., 1980, The role of membrane lipids in receptor mechanisms, Ann.Rev.Pharmacol.Toxicol., 20:201-234.

Maisov, N. I., Sandalov, Yu. G., Glebov, R. N., and Raevsky, K. S., 1976, Effects of psychotropic drugs on synaptosomal uptake of ^{3}H-GABA and Na,K-ATPase activity, Bull.Exper.Biol., No. 1, 45-47.

Raiteri, M., Del Carmine, R., Bertollini, A., and Levi, G., 1977, Effect of desmethylimipramine on the release of ^{3}H-norepine-phrine induced by various agents in hypothalamic synaptosomes, Mol.Pharmacol., 13:746-758.

Simon, S. A., Stone, W. L., and Bennett, P. B., 1979, Can regular solution theory be applied to lipid bilayer membranes? Biochim.Biophys.Acta., 550:38-47.

Snyder, S. H., and Coyle, J. T., 1969, Regional differences in ^{3}H-norepinephrine and ^{3}H-dopamine uptake into rat brain homogenates, J.Pharmacol.Exper.Ther., 165:78-86.

Drug Dependence
Edited by Yu. V. Burov

Psychophysiological and Genetic Factors Responsible for Alcohol and Morphine Dependence in Animals

Yu. V. Burov, S.A. Borisenko, A.B. Kampov-Polevoi
and A.I. Maisky

Laboratory for Research into Drug Addiction Prevention and Treatment,
Institute of Pharmacology, Academy of Medical Sciences
of the USSR, Moscow

1. INTRODUCTION

Clinical observations and experimental findings leave no doubt as to the variety and profusion of alcohol's and other narcotics' effects on the organism. This explains why so many hypotheses have been produced explaining the development of alcoholism and drug addiction in terms of impairment of neurotransmitter regulatory systems (Anokhina, 1979; Tabakoff and Ritzman, 1979), of the characteristics of metabolism of xenobiotics (Belknap et al., 1972), of CNS sensitivity (Kakihana et al., 1966) or other contributory factors.

Clinicians studying the reasons for ethanol dependence have long been concentrating attention on a number of psychological peculiarities and disorders common in alcoholic patients during the pre-morbid phase, which they class as depressive (Kannabikh, 1914; Desyatnikov and Sorokin, 1981, and others). There is evidence of susceptibility to alcohol amongst individuals with certain types of personality. In others, however, alcohol has not been observed to produce any positive reaction; it actually reduces general buoyancy and well-being and even produces physical and mental discomfort. Such subjects usually seek to avoid alcoholic beverages at all costs (Strelchuk, 1973) - hence, alcohol's apparent ability to produce psychopharmacological effects which differ according to mental state, promoting a feeling of comfort in some and discomfort in others.

There is thus reason to believe that the development of drug dependence within the organism is determined by a complex set of cause/effect relationships, comprising both interaction between

341

heterogenous genetically-determined biological elements and environmental (e.g. social) factors. The latter determine the multiple
variations in the course of the disease, the prospects for recovery
and the usefulness of pharmacological treatment. These differences
in how individual organisms react to alcohol and narcotics have a
biological basis - namely the biochemical and biophysical polymorphism of the population - shaping general resistance to noxious
influences while allowing a survival pattern to emerge. Those
less adapted to existing conditions are known to respond less
favorably to test situations, other things being equal. It follows that alcoholism and drug addiction, in common with the vast
majority of illnesses, are disorders affecting those already predisposed.

     It also follows from these considerations that any investi-
gation into the cause and development of drug dependency and alcoholism should take the biological propensity of individual
laboratory animals for consuming drugs and ethanol into account.
This approach would a) help to identify any idiosyncratic pathological response to these substances and b) strengthen the basis
for extrapolating relevant findings to man.

     Using such models, the present work sets out to explore i)
the possible interrelationships between alcohol and morphine's
tranquilizing, antidepressant and stimulant effects (i.e. activating
the positive reinforcement brain structures) and ii) individual
susceptibility to these substances in particular laboratory animals.
We would add that although research workers have generally been
quick to recognize the syndrome of potential addictiveness in the
pathogenesis of alcoholism and drug addiction, the special effects
of individual narcotics in shaping motivation have been somewhat
neglected by experimental pharmacologists.

2. MATERIALS AND METHODS

2.1. THE POSITIVE EMOTIONAL STATE PRODUCED BY ETHANOL

     Self-stimulation is the reaction used as experimental model for
studying the effects of psychotropic substances on those brain
structures involved in shaping a positive emotional state in animals.

     Our experiments were performed on 22 male rats weighing
250-300 g. Electrodes were implanted, according to the coordinates
of De Groot's (1959) brain atlas, in the medial forebrain bundle,
level with the lateral hypothalamus (A - 4.4 mm, L - 1.5. mm,
H - 8.5. mm) in 12 animals, or the diagonal septal bundle
(A - 8.5 mm, L - 0.7 mm, H - 8.0 mm) in the remaining 10 animals.

     The operation was performed using 45 mg/kg nembutal i.p. as
anesthetic. The rats were afterwards housed in individual cages

measuring 40 x 12 x 15 cm, on a standard diet, with free access to
food and water. Seven days after the operation the rats were trained
in the self-stimulation procedure until a stable response threshold
and a steady incidence of self-stimulation set in following 5 min
after the application of current of various strengths. The experiments took place between 9 am and 12 noon. Lever-pressing was accompanied by stimulation of the brain by sinusoidal impulses lasting
0.02 sec with a frequency of 50 Hz. A 15% ethanol solution administered intraperitoneally at doses of 0.5, 0.75, 1.2 and 4 g/kg, and
acetaldehyde at doses of 25, 50, 75, 100 and 200 mg/kg were also
given i.p. 30 min prior to the experiment. The effect of ethanol and
acetaldehyde on self-stimulation response having been established for
each rat, the animals were kept for 14 days in conditions allowing
them to choose between imbibing water or 15% ethanol. It had already
been found that this is the concentration favored by rats when
allowed to choose between either water or solutions of 5, 10, 15 or
20% ethanol. Average incidence of self-stimulation and average daily
ethanol consumption were calculated and compared with controls, using
Fisher's criterion (Plokhinsky, 1970).

2.2. HOW TOLERANCE IS DEVELOPED TO THE ACTIVATING EFFECTS OF MORPHINE, AMPHETAMINE AND ALCOHOL ON RATS' POSITIVE REINFORCEMENT SYSTEM

This set of experiments was performed on 11 male rats weighing
250-300 g with electrodes implanted in the medial forebrain bundle,
level with the lateral hypothalamus. For coordinates, operating
techniques and housing of the animals, see previous section. After
training the animals in steady self-stimulation an initial level was
established - lowered threshold and increased incidence of self-
stimulation, under the influence of equivalent doses of amphetamine
(0.5 mg/kg), morphine (3 mg/kg) and ethanol (0.5 g/kg, 15% solution),
all of which were administered intraperitoneally 30 min prior to the
experiment. Animals were used in the experiments which had shown
enhanced self-stimulation on first receiving these substances.

The above dose levels are the minimum at which the substances
tested show an almost equally strong activating effect on self-
stimulation in one particular animal. (These had been established
in previous experiments). Following a single test dose of each of
the substances, one only was given daily at the above dose until
tolerance was achieved and the substance failed to promote self-
stimulation.

Once tolerance to the main substance had developed, we proceeded
to test with starting dosages of two other substances at the same
time as the original substance, while doubling the dose of the
latter. Fisher's criterion was again used for statistical evaluation
of our findings (Plokhinsky, 1970).

2.3. MEASURING ETHANOL'S TRANQUILIZING EFFECTS

A total of 116 male white rats, weighing 230–250 g, were split into one control and two experimental groups. They were trained in a chamber with an electrode floor to escape from painful 60–70 V stimulation of the paws by jumping into one of two spaces, one larger than the other. The size of the smaller, measuring 6 x 8 cm, was the smallest area in which two rats, having escaped together, could be contained. The experiments of Burov and Salimov (1975) had already shown that shared escape was only possible in the absence of any competitive interaction between the two animals. This could only be achieved after administering tranquilizers to the animals. The other (larger) space was 1.5 times the size of the smaller and could accommodate animals escaping together on any occasion without the need for tranquilizers. Once all the animals had been trained to escape, they were divided into pairs at random. The effect of a 0.25–1 g/kg dose of ethanol administered i.p. 30 min prior to the experiment on shared escape into the larger and smaller spaces was investigated. A shared escape into the smaller space was interpreted as alcohol exerting a tranquilizing effect on the animals, while reduced escapes into the larger area showed increased conflict between two animals (aggressive interaction) under the influence of ethanol.

2.4. THE ANTIDEPRESSANT EFFECTS OF ETHANOL

These experiments were performed on 220 cross-bred male white rats, each weighing 200–250 g.

The method described by Porsolt et al. 1978) was used to assess rats' propensity for developing depressive-like states (DLS). Rats were made to swim for 600 sec in a plastic container 50 cm deep, containing water up to a depth of 25 cm, at a temperature 20°C. They were then kept in a cage heated to 32°C and finally returned to the vivarium. This process was repeated 24 h later. Total time when the animals remained frozen in one position was recorded, during which they swam around passively, with the head held vertically and inclined slightly forward, hardly emerging above the surface. According to Porsolt et al. (1978) the degree of DLS is directly proportional to total time spent immobile.

Ethanol's effect on the intensity of DLS was investigated as follows. Doses of 0.5 and 1.0 g/kg ethanol were given intraperitoneally, with the control receiving physiological saline (1.0 g/kg), 30 min before renewed testing.

Competitiveness was assessed between rats in a conflict situation whereby pairs of rats in a pool compete for a dry space only big enough for one animal, as described previously (Kampov-Polevoi, 1978). In this procedure, rats were placed in a container measuring 32 cm across and 50 cm deep, containing water 19 cm deep at a temper-

ature of 20°C. A round-shaped wooden shaft measuring 4 cm across and
10 cm high was fastened to the bottom of the basin in a central
position, and secured to a base measuring 13 x 13 x 10 cm, so as to
protrude 1 cm above the surface. Trials lasted 30 min. After
20 min the victorious animal was removed from the pool for 1 min and
then re-introduced into the continuing experiment (in a control
encounter). The number of a) attempts made by each animal to oust
its partner from the dry spot and b) number of successful encounters
was recorded. The ratio of b) to a) measured success in ousting a
partner (SOP).

Animals victorious in this competition for a dry spot, which had
ousted a rival during a control encounter and shown above 0.6 SOP,
were classed as dominant. Those which had given up the struggle
early were unable to hold their ground on the dry space and scored
less then 0.3 SOP were considered submissive.

The rats were then separated into twos, pairing dominant and
submissive animals. Each pair was housed in a separate cell until
the procedure was repeated 24 h later. Any pair in which no clear
dominant/subordinate relationship had been established during the
first trial - i.e. where the animal's behavior failed to correspond
to the above criteria - was excluded from subsequent experiments.

The effect of alcohol on dominant/subordinate relationships was
investigated by administering doses of 0.5 and 1.0 g/kg i.p. to
either the dominant, the submissive or both rats 30 min prior to
trials.

Statistical evaluation was then performed on the data obtained
using Meddis' (1975) non-parametric criterion. Significance of differences was also assessed using Fisher's method (Plokhinsky, 1975).

2.5. INVOLVEMENT OF THE PRINCIPAL HISTOCOMPATIBILITY SYSTEM IN THE
REACTION OF THE ORGANISM TO MORPHINE AND ETHANOL

This research used 2,398 5-6 months old mice of congenitally
resistant strains, weighing 20-35 g, which were fed a standard dry
laboratory diet. CR strains, bred on the stock A/Sn, C3H/Sn and
G57B1 10/Sn with the H-2 haplotypes shown on the next page, served
as models for this research.

The strains of mice differ in their H-2 system, either in
various areas or in its entirety; all other genes in these breeds
remain identical.

The CR strains, bred on the stock A/Sn (A/Sn; A.SW) and C3H/Sn
(C3H.OH; C3H.SW), were used to throw light on any possible association
between the H-2 system and the reaction of the organism to chronic
administration of morphine. Tolerance and physical dependence
induced by a 4-day program of morphine administration were modelled

Strain	Haplotype H-2
A/SnY	$H-2^a$
A.SW	$H-2^s$
A.Ca	$H-2^f$
C3H/SnY	$H-2^k$
C3H.OH	$H-2^o$
C3H.JK	$H-2^{ja}$
C3H.SW	$H-2^b$
C3H.NB	$H-2^p$
C57BL/ Sny	$H-2^b$
B10(R 101)	$H-2^g$
B10(R 103)	$H-2^g$
B10.CNB	$H-2^p$
B10.M	$H-2^f$
B10.D2	$H-2^d$

following the method of Saelens et al. (1971). The degree of morphine tolerance developed was assessed by observing how repeated administration reduced morphine's analgesic effect. The hot plate method (surface temperature 56°C) was used to measure the strength and degree of analgesis produced by a single 20 mg/kg dose of morphine. Sudden withdrawal having been induced by administering a 100 mg/kg dose of nalorphine (i.p.), physical dependence was rated by a count of stereospecific jerks. Open field trials were used to study changes in motor activity and orientating reaction following chronic morphine administration.

The tendency of the CR line of mice to become "alcoholic" was monitored by changes in locomotor activity in chambers with a photo-cell by the method advocated by Schuster et al. (1975). During these experiments ethanol was administered at a dose of 1.0 g/kg body weight, while control animals received physiological saline. In cases of chronic alcohol dependence, the amount of locomotion was measured at the same time for a total period of 4-5 days. Chronic alcohol dependence was induced both voluntarily, by allowing the animals to choose between a 10% solution of ethanol and water, and forcibly, by providing 10% alcohol solution as the only available form of fluid. The duration of ethanol narcosis was judged according to the sideways position observed after administering a 4.0 g/kg dose of ethanol. Volume of solution administered: 0.01 ml/g body weight. Fisher-Student's criterion was used for statistical evaluation of results.

3. RESULTS AND DISCUSSION

3.1. EFFECT OF ETHANOL ON SELF-STIMULATION RESPONSE (SS)

According to our experiments a 0.25 g/kg dose of ethanol does not influence SS of the hypothalamus or the septum. Doses of 0.5, 0.75 and 1 g/kg of ethanol lowered the SS threshold and increased its incidence as compared with original level in five rats with electrodes implanted in the hypothalamus. A dose of 2.5 g/kg failed to influence SS threshold, but was 38% successful in this respect on average when the dose was raised to 4.0 g/kg. As shown in Table 1, the incidence of SS under these circumstances showed a tendency to rise (2 g/kg) or fall (4 g/kg) 2-4 times compared with controls. Absolute changes in the SS values are given in Table 2.

In three rats none of the ethanol doses tested affected SS threshold, while a tendency towards an increased incidence of SS was observed. In the four remaining animals with electrodes implanted in the hypothalamus small doses of ethanol failed to facilitate SS, while raised doses inhibited this response.

In three out of the 10 animals with electrodes implanted in the septum, ethanol promoted SS; maximum effect was reached at a dose level of 1 g/kg. It failed to do so in the seven remaining rats at

Table 1. The Effect of Ethanol and Acetaldehyde on Hypothalamic and Septal Self-stimulation Response

Substance	Hypothalamus		Septum	
	SST	ISS	SST	ISS
Ethanol, g/kg:				
0.25	0	0	0	0
0.5	↓ 14	↑ 12	0	↑ 7*
0.75	↓ 18	↑ 25	↓ 9	↑ 14
1	↓ 10	↑ 17	↓ 12	↑ 20
2	0	↑ 6*	↓ 7	↑ 18
4	↑ 38	↓ 2-4 counts	↑ 18	↓ 2 counts
Acetaldehyde, mg/kg:				
25	0	0	0	0
50	0	↓ 8*	0	0
75	0	↓ 7*	0	0
100	↑ 40	↓ 52	↑ 24	↓ 30
200	↑ 2 counts	↓ 64	↑ 82	↓ 75

SST – self-stimulation response threshold; ISS – incidence of self-stimulation response. Upward and downward pointing arrows – reduction or increase (as % of initial values, $p < 0.05$). *Result not significant.

Table 2. A. The Effect of Ethanol on Hypothalamic and Septal Self-stimulation Response

Brain structure			Hypothalamus										Septum					
No. animals			2		5		6		10		12		14		19		21	
Ethanol dose (g/kg) 15% solution (%96)	Values for SS reaction		SS threshold	No. SS responses (counts)	SS threshold	No. SS responses (counts)	SS threshold	No. SS responses (counts)	SS threshold	No. SS responses (counts)	SS threshold	No. SS responses (counts)	SS threshold	No. SS responses (counts)	SS threshold	No. SS responses (counts)	SS threshold	No. SS responses (counts)
0.25 (0.04)	baseline		25	720	60	350	14	890	30	670	120	220	40	540	8	1120	75	420
	ethanol		25	740*	60	340*	14	900*	30	650*	120	230*	40	560*	8	1150*	75	400*
0.5 (0.078)	baseline		25	690	60	370	14	910	30	640	120	270	40	580	8	1180	75	380
	ethanol		22	750	48	420	12	1090	24	710	105	300	40	620*	8	1140*	75	410*
0.75 (0.117)	baseline		25	700	60	320	14	860	30	640	120	260	40	570	8	1160	75	360
	ethanol		20	850	45	420	12	1040	24	820	100	320	30	650	8	1310	60	430
1 (0.156)	baseline		25	740	60	340	16	900	30	660	120	230	40	520	10	1210	75	410
	ethanol		22	860	55	410	14	1020	28	790	110	280	25	610	8	1430	65	520
2 (0.31)	baseline		30	660	60	310	16	930	30	690	140	200	50	560	10	1100	80	390
	ethanol		30	690*	60	320*	16	980*	30	710*	140	250	45	650	10	1290	75	450
4 (0.625)	baseline		30	680	60	290	16	870	30	650	140	210	50	520	10	1190	80	370
	ethanol		45	410	80	80	22	320	50	180	200	120	60	280	12	610	95	170

Table 2. B. The Effect of Acetaldehyde on Hypothalamic and Septal Self-stimulation Response

Brain structure		Hypothalamus										Septum					
No. animals		2		5		6		10		12		14		19		21	
acetaldehyde dose (mg/kg)	Values for SS reaction	SS threshold	No. SS responses (counts)	SS threshold	No. SS responses (counts)	SS threshold	No. SS responses (counts)	SS threshold	No. SS responses (counts)	SS threshold	No. SS responses (counts)	SS threshold	No. SS responses (counts)	SS threshold	No. SS responses (counts)	SS threshold	No. SS responses (counts)
25	baseline	30	630	60	260	16	820	30	620	140	210	50	520	15	1050	80	350
	acetaldehyde	30	610*	60	270*	16	800*	30	610*	140	200*	50	530*	15	1030*	80	350*
50	baseline	30	890	70	270	18	840	35	600	140	220	55	490	15	1040	80	330
	acetaldehyde	30	550*	70	250*	18	770*	35	560*	140	210*	55	480*	15	1060*	80	340*
75	baseline	40	570	70	240	20	790	40	580	140	210	55	500	20	960	90	300
	acetaldehyde	40	530*	70	220*	20	730*	40	530*	140	190*	55	510*	20	970*	90	280*
100	baseline	40	540	75	260	20	770	40	570	140	200	55	480	25	920	90	320
	acetaldehyde	55	260	105	140	30	400	50	300	200	110	70	340	30	640	115	250
200	baseline	40	550	75	230	20	760	40	580	140	220	55	480	25	940	90	300
	acetaldehyde	80	350	160	150	40	490	75	380	300	140	100	130	45	280	160	70

*p>0.05 - results not significant.

No. of SS responses - mean of 3 values obtained for self-stimulation at optimal current strength.

any of the doses tested, however. Nor did acetaldehyde at doses of
25, 50 or 75 mg/kg affect SS of the hypothalamus or septum, either in
those animals showing a release of SS resulting from ethanol or those
in which SS remained uninfluenced by small doses of ethanol at both
these brain structures (see Table 2B). Acetaldehyde raised SS threshold and reduced the incidence of SS with increasing doses. Quantitative changes in SS values after administering ethanol and acetaldehyde are shown in Table 2.

Rats implanted with hypothalamic and septal electrodes, in which
ethanol had promoted SS, daily consumed 7.8 and 6.3 ml respectively
when allowed to choose freely between ethanol and water over a 14-day
period. This level differed significantly (p<0.001) from the 0.7 and
1 ml alcohol consumed by rats in which SS had not been promoted under
the influence of alcohol.

These investigations point to a connection between ethanol's
ability to activate positive reinforcement and the development of a
predilection for this substance. By activating these sites ethanol
may be producing an emotional state peculiar to rats resembling
euphoria in man, responsible for shaping addiction. Unlike alcohol,
acetaldehyde exerts a basically inhibitory effect on positive reinforcement sites, and would appear to produce a condition resembling
dysphoria in man. The data of Alkabane (1960) showing how alcohol-
induced psychosomatic symptoms differ from those produced by acetaldehyde support our findings. Blood alcohol level becomes high
relatively soon after alcohol consumption, and euphoria, together
with other positive emotions, is among the chief signs observed.
Subsequently acetaldehyde's effects begin to prevail as alcohol
becomes converted to acetaldehyde. Thus Mayfield and Allen (1967)
rightly concluded that the urge to drink through a hangover
represents a need to superimpose alcohol's effects over those of
acetaldehyde.

We therefore have reason to believe that alcohol's stimulation
of positive brain reinforcement sites and connected feelings of
enjoyment are decisive factors in shaping partiality for alcohol.

Clinically observed cases of multiple drug addiction, together
with the fact that one stimulant can be replaced by another if need
be, further demonstrate that these agents' effects are mediated by
the same brain mechanisms, neurochemical processes included.
Published findings on the uni-directional shifts in brain biogenic
amine levels produced by addictive drugs with particular reference to
alcohol (Anokhina, 1978), morphine (Maynert and Klingman, 1962) and
amphetamines (Leonard, 1972) support our view. Tolerance developed
to the drug consumed, including its stimulant effect, goes hand in
hand with euphoria itself to shape drug dependence. We would therefore suggest that cross tolerance is likely to exist between different substances producing euphoria via a common pathway.

3.2. TOLERANCE TO THE ACTIVATING EFFECTS OF MORPHINE, AMPHETAMINE AND ALCOHOL ON THE POSITIVE REINFORCEMENT SYSTEM IN RATS

Initial values of SS threshold varied from one animal to another, ranging between 20 and 140 μA. SS rate (220–560 pedal pressings per 10 min) was inversely proportional to the initial level of SS threshold. The threshold was reduced by 5–40 μA (p<0.05) by doses of 0.5 mg/kg amphetamine, 3 mg/kg morphine or 0.5 g/kg ethanol, while the number of SS increased at threshold current by 20–130 bar pressing per 10 min (<0.05) compared with initial values. A more pronounced reduction in SS threshold, established according to the absolute magnitude of the intensity of the stimulating current, was observed in animals with comparatively higher original SS threshold levels. The period required for the different rats to develop tolerance varied from one individual to another and averaged 3 days for amphetamine, 3.2 for ethanol and 7.3 days for morphine (see Figure 1). Treatment with two compounds at initial doses after tolerance had developed to the main substance failed to promote SS; in most cases the facilitation in SS already initiated was maintained when the test substances were given. On doubling the dose of the main substance, a stepping-up of SS was usually observed, occurring briefly after administering 1 mg/kg amphetamine and 1 g/kg ethanol or for a longer period (of 2–4 days) in various animals treated with 6 mg/kg morphine. SS continued to decline thereafter until reaching the level corresponding to tolerance. Thus reduced SS, produced by doubling the dose of the substance used, lasted virtually the same time as that required for tolerance to develop to that substance. Doubling the dose of amphetamine used in one instance and of ethanol in a further two cases accelerated the decline in SS.

Apart from alcohol, other compounds with a euphoric effect, especially stimulants, hallucinogens, narcotic analgesics and tranquilizers, are also known to produce feelings of enjoyment.

The above data lead us to conclude that:

a) repeated administration of amphetamine, ethanol and morphine, at the lowest doses required to promote SS, initially activate the positive reinforcement system
b) this system's response to these agents then declines and
c) finally tolerance to their effects is developed.

The fact that an equivalent dosage of test drugs, given after tolerance has been developed to the main substance, produces no activating effect on the positive reinforcement system (i.e. cross-tolerance is developed), proves that these substances are achieving their effects via common neurophysiological and neurochemical mechanisms.

Adrenergic structures are known to play a leading part in activating the hypothalamic positive reinforcement system (Stein, 1968).

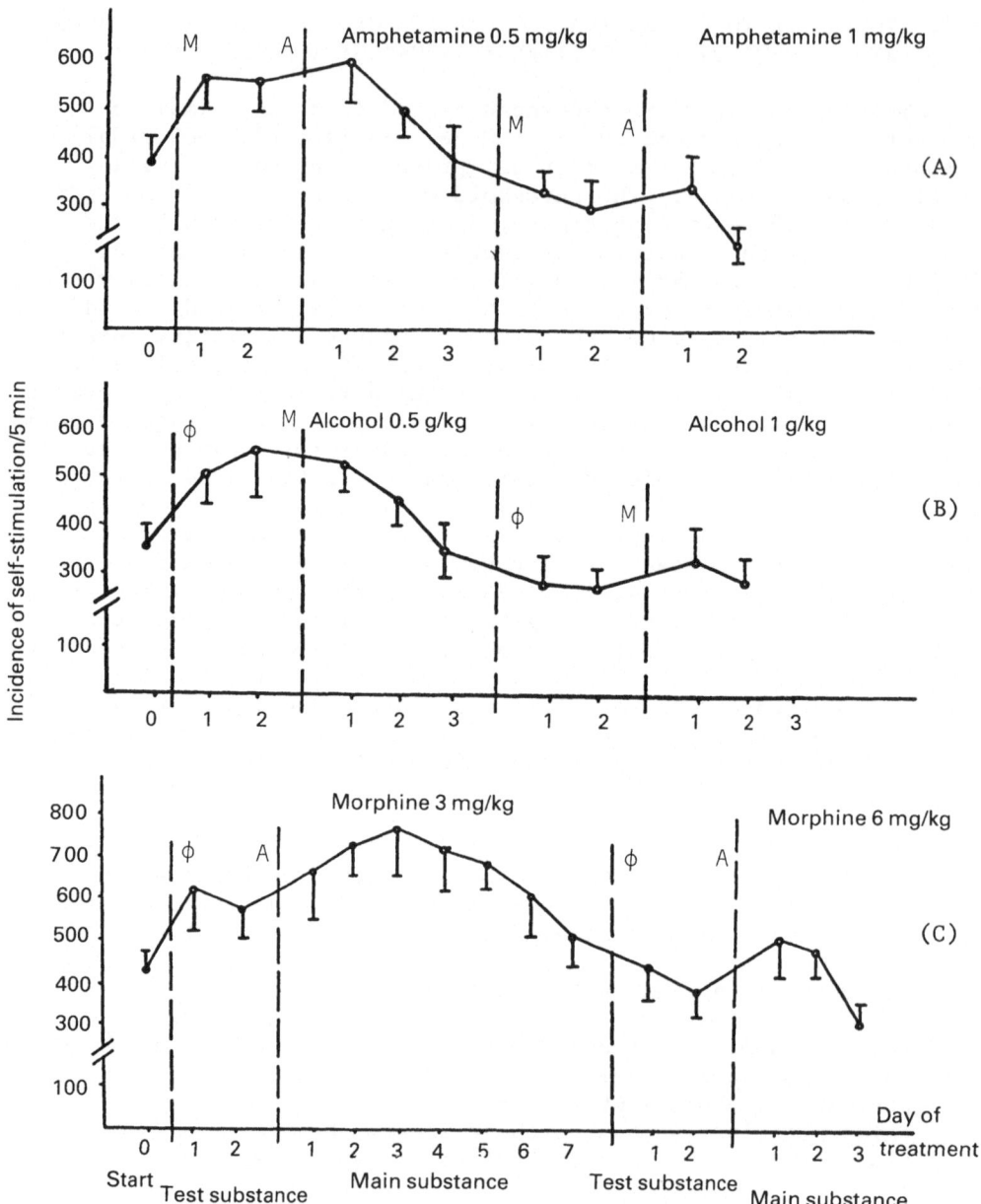

Fig. 1. How tolerance develops to the activating effect of (A)
         amphetamine, (B) alcohol, and (C) morphine on positive
         reinforcement structures. Abscissa – day on which main or
         test substance is administered. Ordinate – incidence of
         self-stimulation per 5 min. φ – amphetamine; A – alcohol;
         M – morphine. Shows average episodes of self-stimulation
         observed after treatment of all test animals used (see
         section on methods).

It has also been shown that alcohol, amphetamine and morphine have a unidirectional action on adrenergic structures, causing increased discharge of adrenergic transmitters and changing reuptake processes by reducing and ultimately exhausting supplies of transmitter. (Maynert and Klingman, 1962; Leonard, 1972; Anokhin, 1980). It may therefore be maintained that the effects of the test drugs on the hypothalamic positive reinforcement system are produced vfs adrenergic mechanisms. This would provide an explanation of the cross-tolerance to alcohol, amphetamine and morphine observed during our experiments. The unequal periods required for tolerance to a particular substance to develop could reflect its action at different points in the process of adrenergic transmission.

3.3. THE TRANQUILIZING EFFECTS OF ALCOHOL

Doses of 0.25-1 g/kg alcohol did not affect an animal's ability to escape singly into the larger or smaller space. But the onset of aggression was observed in the case of shared escape into the larger space, in the form of fighting between the incumbent and the animal attempting to jump into the space. In some cases, the incumbent even leapt off the space in order to engage in combat with its fellow-animal. No such signs of aggression were observed in the control animals. The ED_{50} for provoking aggression (i.e. escape into the larger space) was 0.5 g/kg. If the alcohol dose was either doubled or halved, this effect decreased to 20 and 30% respectively. In the case of shared escape into the smaller space, treatment with 0.5 g/kg alcohol resulted in 50% successful shared escapes.

The total of 31 rats in which alcohol had produced a tranquilizing effect consumed an average of 22 ml alcohol daily when given the choice between water and a 15% ethanol solution. The 85 rats made aggressive by alcohol consumed less than 2 ml alcohol solution per day.

Our findings show that alcohol has a tranquilizing effect on a certain portion of the rats, proportional to their consumption of (substantial) amounts of alcohol when given the choice of alcohol or water. Burov's research confirms the tranquilizing nature of this effect; he showed that the tranquilizers benactyzine and chordiazepoxide produce a drop in alcohol consumption during the stress of painful electric shock. This could indicate that the agents used are acting along the lines of a therapy substitute in this instance; i.e. tranquilizers and alcohol have converging effects. Burov and Speranskaya (1977) showed that the effective dose of alcohol, of 32 mg/kg, required to suppress emotional tension in rats behaving defensively towards their partners, according to Preobrazhenskaya and Simonov's (1970) criterion, was only 1/70th of the 2.25 g/kg dose required to inhibit the conditioned defensive reflex. The effect described is typical of tranquilizers.

3.4. ALCOHOL'S ANTIDEPRESSANT EFFECT

Our experiments showed that rats became highly active when first
placed in a situation where forced to swim. They swam vigorously and
tried to leap out of the container or dive under its sides. The
animals soon began to abandon vigorous activity, however, which then
alternated with increasing periods of immobility, when they merely
swam around passively, leaning forward in an upright position with
the head barely protruding above the surface. The rats could be
divided into two groups on the basis of their performance, consisting
of highly active (HA) animals, which spent 186±39.7 sec immobile, and
relatively inactive (RI) rats, which remained immobile for an average
of 353±19.9 sec. When the RI group were placed repeatedly in a
situation obliging them to swim, the time spent immobilized rose
substantially to 447±29.7 sec. This could indicate a rise in their
DLS. No such effect occurred in the case of HA rats, however – in
fact the overall time spent immobilized fell to 45±7.6 sec, showing
that they had adapted to the conditions of the experiments.

A dose of 0.5 g/kg alcohol substantially reduced the intensity
of DLS in RI rats; the overall time spent immobile fell to 175±32.2
sec in these animals, a difference which was statistically significant (p< 0.01). A dose of 1.0 g/kg produced no noticeable effect on
the time spent immobile by either animal group.

When the two groups of rats were again introduced to a choice
between water and 15% ethanol solution, HA rats showed a pronounced
predilection for alcohol, with a total daily 15% ethanol consumption
of 51±5.1 ml/kg, while RI animals virtually abstained, with a consumption of 3±3.1 mg/kg per day.

The trial, involving competition between pairs over a dry space
in a pool of water, gave the following results: the pairs of animals
hotly contested the dry space when first introduced into the test
situation, with the defeated rat then giving up its claims to the
space, which had been taken by the dominant animal. A total of 90
animal pairs, each consisting of a clearly dominant and a submissive
rat, were selected for the experiment. The animals were further
subdivided into nine groups of 10 pairs each.

Three of these animal groups were given 0.1 ml physiological
saline per 100 g body weight and served as controls. In one of the
control groups all animals were given saline, while dominant rats
only received it in a further group and submissive partners only in
the remaining group. All three sets of animals behaved in the same
way: the rat defeated on the first day only made a few weak attempts
to oust the victorious rat, although its vigor must have been restored during the previous 24 hours. The victorious rats' efficacy
of pushing attempts (EPA) easily surpassed both that of the defeated
rat and its own initial EPA (p<0.01).

Animals belonging to groups 4-6 and 7-9 received doses of 0.5
and 1.0 g/kg ethanol respectively, along the same lines: all rats
were treated in groups 6 and 9 but dominant partners only in groups
4 and 7 and submissive animals only in groups 5 and 8.

No difference was observed between the behavior of dominant
rats treated with 0.5 g/kg ethanol and that of the control. The
same dose of ethanol given to submissive rats caused a fall in the
EPA of the dominant rat, without affecting that of the submissive
rat, however. If the dose given to the dominant rat was doubled
no typical rise in EPA was observed, which usually accompanies the
establishing of dominant/submissive relationships. Treating the
submissive rat with the same dose produced a similar picture:
treating both rats in a pair with 1.0 g/kg leads to a drop in EPA
level in the dominant animal. Dominant/submissive relationships
and dominance itself decline in 50% of cases – i.e. the rat which
had got to the dry space first could hardly stand up to its partner's
attacks and makes no attempt to reach the dry space in a trial
encounter. Our findings illustrate how small doses of 0.5 g/kg
ethanol revive rivalry in submissive rats without substantially
affecting the behavior of their dominant partners. The latter,
however, become less competitive when treated with a higher dose of
1.0 g/kg ethanol, which was also less successful in restoring the
submissive rat's behavior.

Subsequent experiments were devised to ascertain what the submissive state, which arises from conflict within an animal group and
forms the basis of animal behavior, might have in common with the
inactivity resulting from the development of DLS. They involved
placing both types of animal in a situation forcing them to swim,
within 24 h of taking part in the above test of competing for a dry
spot. Results showed that submissive rats remained immobile for
448±13.1 sec – almost exactly the same time as for RI rats on the
second day of trials. The same was true of HA animals and dominant
rats, which remained immobile for 182±0.17 secs. Further trials,
whereby animals were to choose between water and 15% ethanol
solution, gave results very similar to the previous set of experiments – dominant (HA rats) consumed 6±0.23 and submissive animals –
46±9.2 ml/kg per 24 h of 15% ethanol solution.

3.5 INVOLVEMENT OF THE MAIN HISTOCOMPATIBILITY SYSTEM IN THE
ORGANISM'S REACTION TO MORPHINE AND ETHANOL

In our opinion the best model to use when seeking biological
correlates determining the organism's predisposition to alcohol
addiction is the human (HLA) or mouse (H-2) main histocompatibility
system. This system is so constructed as to set the level of the
organism's sensitivity threshold to noxious agents. The presence of
certain H-2 and HLA antigens in mouse and human phenotype is evidence

of hereditary susceptibility to disorders of the vascular supply of
bone and the endocrine, neuro-psychiatric and other systems, as well
as heightened susceptibility to a range of viruses (Snell et al.,
1976).

     The discovery that the H-2 system regulates the level of cyclic
nucleotides in the liver, spleen and sexual organs is central to an
understanding of the biological role of this system. The same may be
said of the sex organs' reaction to exogenous administration of
hormones and the encoding of a) high or low testosterone levels and
b) specific androgen binding protein in the plasma. Goldman et al.
(1977) showed that the final path of the H-2 system is also the
corticosteroid receptor. The main histocompatibility system makes
a contribution to neurotransmitter as well as endocrine system regulation (Smeraldi and Scorza-Smeraldi, 1976). A structural similarity
has been noted between HLA-AI, a lymphocytic antigen in man (and, to
a lesser extent, other factors), and dopaminergic and β-adrenergic
receptors. Preincubating chlorpromazine and some of its metabolites
with HLA-AI was found to suppress this antigen's sensitivity to the
cytotoxic action of anti-HLA-AI antibody and complement. The 7-OH-
analog of chlorpromazine proved to have the strongest protective
action of the various derivatives of the compound. It exerted an
antipsychotic neuroleptic-type action in clinical trials. This
finding illustrates the major part played by certain HLA-antigens in
the individual response to neuroleptics (Smeraldi et al., 1980).
Dissociation in the sensitivity of strains differing in their H-2
system with regard to hormones and neurotransmitters might be caused
by the previously noted fluctuation in basal level of cyclic AMP,
since the H-2 system was shown to modulate the cyclic nucleotides'
response to hormonal stimulus. Noticable fluctuations in the action
of glucagon, insulin and prostaglandips on adenylate cyclase activity
were observed in mice of $H-2^a$ and $H-2^k$ halotype; this would confirm
that specific interaction between hormones and other membrane-active
agents with a cellular surface may be regulated by main system histocompatibility (Meruelo and Edidin, 1975). Svejgaard and Ryder
(1976) explain the connection between the main histocompatibility
system and "non-immune" diseases by this same similarity between
certain transplantation antigens and hormone neurotransmitter receptors. According to the latter hypothesis, membrane-building and
soluble transplantational alloantigens bear a structural resemblance
to receptors of some endogenous biologically active compounds. These
ligands' affinity for true receptors is immeasurably higher. The
antigen receptors, however, are numerically superior. These are
situated in virtually all organs and tissue, above the true receptors. The latter are generally in the minority in target organs and
may cause pathological binding of hormones and neurotransmitters,
thereby creating conditions conducive to a number of non-immune
diseases. In view of the importance of immune, endocrine and neurotransmitter factors in shaping sensitivity to narcotics of varying
structure in man and animals, we suggested that the histocompatibility system may be involved in genetic control of the organism's

response to the narcotic (Maisky and Vedernikova, 1978). Investigations were carried out on congenitally resistant (CR) strains of mice in order to test this view experimentally. Experiments on CR strains of mice bred from G57B/10 failed to show any noticeable differences in response to morphine when tested for sensitivity to pain. In the case of CR strains bred from A/Ns and C3H/Sn, however, which respond to morphine by a vastly increased analgesic reaction, strains which differed in their sensitivity to the narcotic could be discerned. Thus the duration of strain ACA's response showing pain to heat stimulation increased to 474% compared with the control, and to 386 and 294% for A/Sn and ASW mice - i.e. high, medium and low response level respectively. The latter rating was used exclusively to describe morphine response level found in CR stains of one group, since only the strain bred from C57BL/10 proved to be completely resistant to the narcotic amongst those investigated. A high morphine response level in the pain sensitivity test was found in C3H.OH and C3H.NB mice (425 and 408% respectively), while a medium or low response level was shown by strains C3H.YK (365%), C3H/Sn (301%) and C3H.Sw (256%). Since mice belonging to each of these groups only differ in their H-2 gene complex and the chromosome portion, one of the loci which determines morphine sensitivity must either be situated in the H-2 complex or be closely linked to it.

We obtained F_1 hybrids by crossing high response and low response strains bred from A/Sn and C3H/Sn in order to determine how far morphine sensitivity in mice is inherited. It was found that all the hybrids' response resembled the highly responsive parental strain and hence that heightened morphine sensitivity, as determined by analgesic test, is genetically dominant (Maisky et al., 1978). We then proceeded to study the possibility of a link between the H-2 system and the organism's response to chronic administration of morphine (see Table 3).

Our research showed that the main tissue compatibility system exercises control over tolerance, and to a lesser extent over physical dependence (see Table 4). Since no correlation between the sensitivity of the strain with respect to these two stages of experimentally-produced drug addiction has been discovered, we presume that polygenic control is involved in this pathology.

The use of the open field test to reveal changes in locomotor and exploratory activity in mice during morphine treatment showed that narcotic drugs alter these types of behavior in all strains investigated, and that these effects are dose-related (see Figure 2). The reaction of CR rats on morphine, however, was clearly shown to be disproportionately strong in the locomotion trial. A marked tendency for the jerk response to relate to previous level of motor activity was identified on comparing these findings with the results obtained from abstinence syndrome models. The H-2 system would appear to be involved in the control of a complex item of behavior such as loco-

Table 3. The Effect of Morphine on Pain Sensitivity in Congenitally
Resistant Strains of Mice and their F_1 Hybrids

Strain	No. test animals	Pain sensitivity		
		Control (sec)	Experiment	
			sec	% of control
A/Sn	8	10.9±1.06	42.1±5.32	386
A.CA	10	10.0±0.41	47.4±4.08	474
A.SW	9	9.0±0.45	26.5±2.8	294
C3H/Sn	12	11.6±0.34	35.0±3.45	301
C3H.OH	7	14.1±1.12	57.1±1.43	425
C3H.NB	6	14.7±1.0	54.3±3.12	408
C3H.Sw	8	12.8±0.79	32.8±0.24	256
C3H.JK	7	12.6±0.39	46.0±4.91	365
C57BL10/Sn	11	10.5±0.59	13.7±0.86	130
BIO.D2	7	8.0±0.54	11.4±0.61	126
BIO.M	6	11.6±0.62	13.6±0.55	117
BIO.CNB	6	9.8±0.63	12.0±0.54	122
RIO3	6	8.6±0.5	9.2±0.70	107
RIO1	7	10.0±0.36	10.5±0.87	101
(A.CAxASw) F_1-hybrid (C3H.OHxC3H/Sn)	7	11.0±1.32	45.1±0.51	410
F_1-hybrid (C3H.NBxC3H/Sw)	6	13.6±0.81	54.6±2.9	415
F_1-hybrid	8	14.6±0.77	51.4±0.84	351

motion which, in turn, is produced by the same mechanism as the jerk
reaction following precipitate nalorphine withdrawal (Vedernikova et
al., 1978). The connection between the principal histocompatibility
system of mice and susceptibility to alcohol was investigated using a
similar approach (Shoshina and Maisky, 1982). The standard pharmacological test for motor activity, using photocells to record the
number of movements made, was applied to CR strain mice after a
single 2.5 g/kg dose of ethanol. Over a given period of time it
revealed statistically significant differences in the locomotor
response of animals with different H-2 systems (see Figure 3).
Findings on the degree of alcohol consumption in CR strains of mice
presented the greatest interest, correlations having subsequently
been made with available clinical material. Transitory changes in
the animals' fluid consumption over a protracted experimental period
were corrected as follows. The ratio between ethanol consumed per
kg body weight in the test group, using 3-5 mice per data point, and
the amount of water consumed by the control group, housed under
identical conditions, was adopted as an index of consumption. As
with morphine, the CR strains reacted along the lines of the standard
inbred strain from which they had been bred. Hence R 107 and R 111
mice (original strain: C57BL10/Sn) were prone to a higher alcohol

Table 4.   Differences Between Mice of Various Strains During the
           Development of Experimentally-produced Drug Addiction

Mouse strain	Stage of experimentally-produced drug addiction					
	Tolerance (length of pain reaction)					Physical dependence No. jerks per 10 min
	Control	Day				
		I	II	III	IV	
A.Sw	13±1.6	13±1.1	60*	60*	25± 6.8	38±12.3
A/Sn	14±3.0	14±3.1	60	60*	33± 5.8	9± 3.2
C3H.OH	11±0.5	60*	60*	60*	23± 2.9	12± 2.0
C3H.Sw	12±0.8	60*	46±6.5	60*	40±10.0	24± 3.5

*Statistically significant difference from control
 (maximum observation period: 60 sec).

consumption than the CR strains A/Sn and A.Sw (standard A/Sn strain).
This corresponds to the original (G57BL10/Sn) strain's higher alcohol
consumption compared with the A/Sn strain. Although the dynamics of
ethanol consumption are similar in all four strains investigated
during the period of a) free access to ethanol, b) ethanol withdrawal, and c) post-withdrawal syndrome, a highly significant difference is to be seen in consumption level between the CR strains in
question at practically all stages of alcohol addiction (see Figure
4). We would therefore maintain that the H-2 system is involved in
polygenic control of congenital ethanol susceptibility in mice.

CR strains of mice have been found particularly valuable biological models for experimentation for two reasons. Firstly, the
principal histocompatibility system of mouse tissue bears structural
and other resemblances to that of man, and secondly, it is possible
to discern HLA halotype by typing human transplantational antigens,
using specific anti-HLA transplantational sera. Transplantational
antigens may therefore serve as an objective criterion in assessing
an individual's predisposition to a given disease, in circumstances
where principal system histocompatibility is associated with a given
pathology. By identifying specific marker antigens, peculiar to a
given diagnostic category, the risk of a particular individual developing the disorder can be assessed, and the diagnosis of the ailment
more easily established.

The reasoning behind pharmacological experiments on CR strains
of mice led on to the hypothesis that the above system could shed
light on drug addiction in man under similar circumstances (Maisky et
al., 1978a). The possibility, mentioned earlier, of some connection
operating between the principal histocompatibility system and a
number of psychiatric diseases, particularly schizophrenia and manic-
depressive conditions, would support our view.

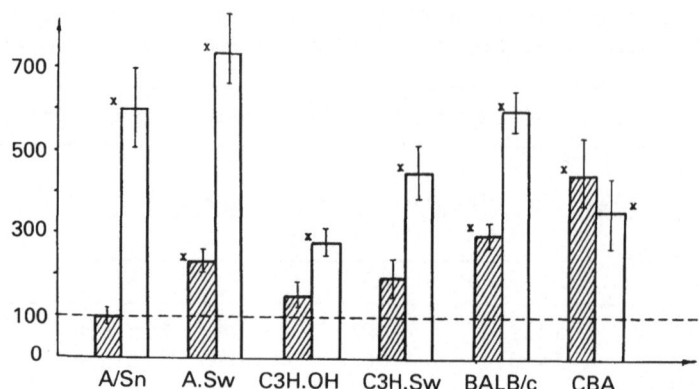

Fig. 2.   Inter-strain differences in mouse response to chronic treat-
          ment with morphine in the open field test.   Shaded columns -
          total dose of 100 mg morphine.   Unshaded columns - total
          dose of 500 mg morphine.

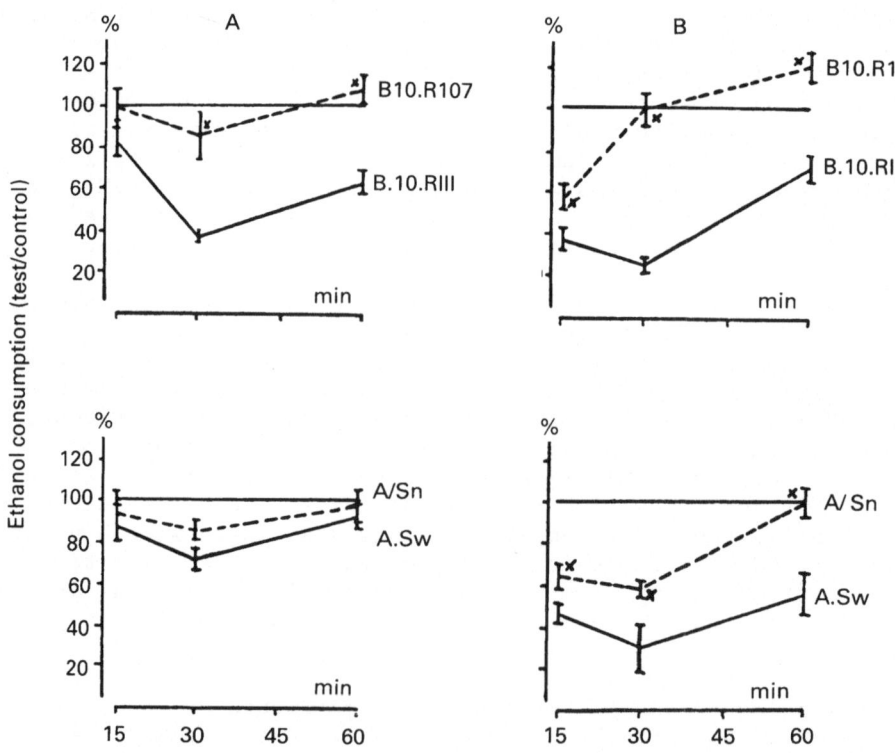

Fig. 3.   Locomotor changes produced in congenitally resistant strains
          of mice by acute administration of ethanol.   A - 1 g/kg
          dose; B - 2.5 g/kg dose; x - statistically significant
          intraspecies differences.

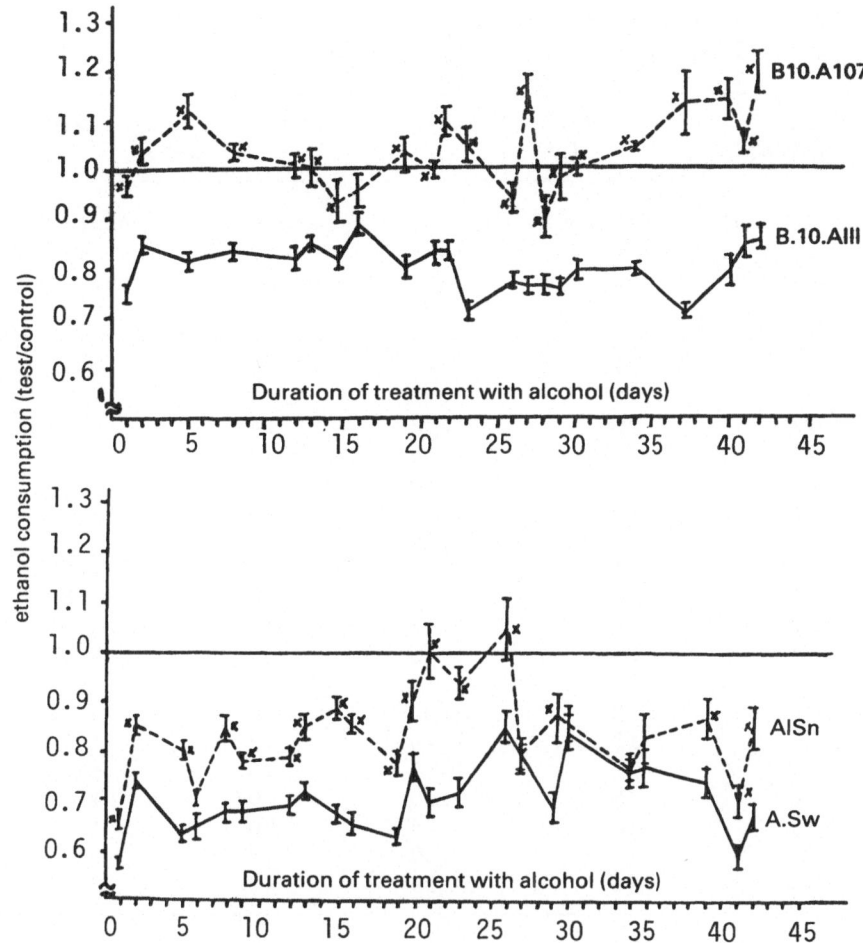

Fig. 4.   Ethanol consumption of mice of congenitally resistant
          strains. x – statistically significant intraspecies
          differences.

        Tissue typing of patients of Russian nationality, suffering from
multiple drug addiction or alcoholism, helped to bring out a number
of special features in the distribution of transplantational antigens
in the two groups of patients (see Table 5) (Maisky et al., 1980).
The information included in the table demonstrates that the frequency
with which locus A antigens occur is close to the control value in
both drug addicts and alcoholics. Of the locus B antigens, HLA-B5,
HLA-B12 and HLA-B14 occur the most frequently in alcoholics (coefficient of relative risk = 2.44, 1.65 and 2.02 respectively). In the
multiple drug addiction group the following locus B and C antigens
appear most frequently: HLA-B5, HLA-B14, HLA-BW17 and HLA-CW4, with
coefficient of relative risk of 4.00, 3.26, 3.14 and 2.97 respec-

Table 5. Characteristics of Distribution of HLA Antigens in
          Alcoholics and Multiple Drug Addicts

Antigens	Control (n=126) Frequency of ocurrence %	Alcoholism (n=117) Frequency of occurrence %	Coefficient of relative risk	Multiple drug addiction (n=59) Frequency of occurrence %	Coefficient of relative risk
HLA-A1	30	27	0.84	25	0.79
HLA-A2	48	49	1.01	49	1.03
HLA-A3	17	20	1.22	14	0.74
HLA-A9	21	24	1.15	22	1.04
HLA-A10	22	24	1.10	29	1.42
HLA-A11	18	20	1.10	20	1.14
HLA-A28	8	6	0.74	5	0.62
HLA-B5	12	25	2.44	37	4.00
HLA-B7	24	23	0.96	20	0.82
HLA-B8	12	9	0.77	10	0.84
HLA-B12	17	25	1.65	12	0.67
HLA-B13	14	16	1.16	15	1.08
HLA-B14	4	8	2.02	12	3.26
HLA-B15	11	15	1.46	12	1.00
HLA-BW17	5	7	1.47	14	3.14
HLA-B27	6	5	0.80	9	1.37
HLA-BW35	19	18	0.93	20	1.09
HLA-B40	15	14	0.89	12	0.76
HLA-CW4	20	26	1.39	42	2.97

tively. Only HLA-B5 (in both of patients) and HLA-CW4 (group of
patients on multiple drugs) differ significantly from the normal
distribution (χ^2 test).

The raised incidence of HLA-B14 would suggest the presence of
cirrhotic changes in both groups of patients, in view of the association observed between this antigen and alcohol-induced cirrhosis of
the liver (Betancor et al., 1977). It should also be mentioned that
attempts are now being made to use transplantational antigen analysis
as a means of establishing the homogeneity of the diagnostic forms of
diseases. In some cases, notably in schizophrenia, the arbitrary
manner in which diseases are sub-defined according to ill-divided
criteria is only fully brought out by HLA-antigen typing. Hence, the
connection between schizophrenia and the principal histocompatibility
system discovered by tissue typing the entire patient population does
not become apparent from investigating patients sub-divided according
to the clinical manifestations of the disease which they exhibit.
Indeed, the fact that the same marker antigens are associated both

with alcoholism and multiple drug addiction would favor a single
etiology for both disorders and common mechanisms involving genetic
predisposition for addiction to substance with different chemical
structures. An attempt should thus be made to identify the actual
physiopathological mechanisms which link these conditions.

An increased incidence of the HLA-B5 antigen had previously
been noted in diseases of different etiology. The association
between the HLA-B5 antigen and Burger's disease (a disorder of the
peripheral vasculature) is frequently met with in heavy smokers; it
is viewed as an illustration of the genetic link between this antigen
and a pathological indulgence in tobacco smoking (McLoughlin et al.,
1976). The substantially raised incidence of HLA-B5 in schizo-
phrenics and children suffering from psychotic disturbances (Golse
et al., 1977) offers a great deal of interest as far as these investigations are concerned. In this country, the latter disorder is
classified as childhood schizophrenia. In view of the well-known
pattern of drug-dependent patients' individual character and personality traits, it may well be that genetic markers can be identified for increased susceptibility to drug addiction and personality
deterioration.

Some findings recently obtained from joint investigations
(with Treskov) of alcoholic patients at a drug addiction treatment
center marked a new departure. Taking alcoholism as a multifactorial
disorder, it could be established clinically that in a randomly
selected group of 104 diagnosed male alcoholics, most showed only a
relatively slight impairment of mental activity, and this is one of
the decisive factors in the clinical picture of the disease. The
patients could be divided into 5 groups on the basis of our study.
No pathological mental disturbance was observed in the 1st group,
consisting of 22 male patients, at the time of investigation. The
2nd group contained 11 subjects with the mildest type of impairment,
such as mild depression, and the 3rd - 20 patients with reactive or
near endogenous manic-depressive disorders. Group 4 consisted of
12 patients suffering from psychopathic disorders - predominantly
emotional outbursts accompanied by mental tension and anxiety. The
5th and last group was made up of patients with atypical affective
disorders, manifesting in a variety of ways, perhaps with apathy,
tiredness, lethargy, etc. HLA-antigen tissue-typing of these
patients enabled us to establish that the chance of carriers of
HLA-CW4 antigen in group 3 was 11.3 times greater and that of
HLA-B5 carriers - 5.38 times higher. Those at most risk in the
5th group investigated were carriers of HLA-B5 antigen (7.8 times
greater), followed by HLA-A (3.0 times) and HLA-CW4 (2.5 times).
These findings show that the HLA-antigens which we have proposed as
genetic markers of susceptibility to alcoholism may help to decide
whether a disputed psychiatric disorder is actually present, and
whether a combination of traditional therapy and psychotropic agents
should be used.

We have thus shown how certain pharmacological effects of ethanol and morphine can play a major part in the mechanisms shaping weaknesses for drug and alcohol. These include their ability to activate positive reinforcement sites in animals, induce feeling of euphoria in man and exert tranquilizing and anti-depressant effects, which largely determine the functional state of the CNS. Genetically-determined factors would appear to be decisive in coding for differences in predisposition to ethanol and alcoholism. Such differences emerged between individuals in our population of cross-bred white rats. Response to alcohol differed according to "emotional" status. The fact that the organism's sensitivity to ethanol and predisposition for alcoholism are polygenically determined certainly does not render the search for possible genetic markers of alcoholism and drug addiction superfluous. This search could bear fruit if biochemical mechanisms and the physiological means by which they direct emotions are fully studied. Our data on involvement of the principal histocompatibility system in mice (H-2) in shaping genetic markers might be sought in loci of this system. Individuals with borderline psychiatric disorders are at particular risk of becoming alcoholic or addicted to drugs. Hence the proposed use of genetic markers for prophylaxis of these disorders, for objective assessment of their severity in individual patients and prescribing rational therapy with psychotropic agents, based on a sound grasp of the underlying pathological process.

REFERENCES

Akabane, J., 1960, Pharmacological aspects of manifestation of the acute after-effects of alcoholic beverages: role of acetaldehyde in alcoholism, Med.J.Shinshu Univ., 5:113.

Anokhina, I. P., 1974, Neurochemical aspects in pathogenesis of alcohol and drug dependence, Drug Alcohol Dependence, 4:215-273.

Anokhina, I. P., 1978, Neurochemical aspects of pathogenesis of alcoholism and narcomania, 1978, Proc. Conf. on clinical picture, prophylaxis and treatment of alcoholism and narcomania, Moscow, pp.69-72.

Belenky, M. L., 1963, Elements of quantitative valuation of pharmacological effect, Gosmedizdat, Leningrad.

Belknap, J. K., MacInnes, J. D., and McClearn, G. E., 1972, Ethanol sleep times and hepatic alcohol and aldehyde dehydrogenase activities in mice, Physiol.Behav., 9:453-457.

Betancor, L. P., Lomax, G., Schuller, P. A., and Rey, C. J., 1977, Antigenos, HLA en la porfiria hepatocutanca tardia alcoholica, Rev.Clin.Espan., 146:607-608.

Burov, Yu, V., and Salimov, R. M., 1975, The effect of drugs on intraspecies aggression, Bull.Exp.Biol.Med., 79:64-67.

Burov, Yu. V., and Speranskaya, Ya. N. P., 1977, The effect of psychotropic drugs on the development of defense conditioned

reflex during experiemental neurosis, Bull.Exp.Biol.Med.,
    No.6, 696-698.
De Groot, J., 1959, The Rat Forebrain in Stereotaxic Coordinates,
    N. V. Noord-Hollandsche, Uitgevers, Maatschappij, Amsterdam.
Desyatnikov, V. F., and Sorokina, T. T., 1981, Occult depression in
    practice, Minsk, High Sch.Publ.House.
Goldman, A. S., Katsumata, M., Yaffe, S. J., and Gasser, D. L., 1977,
    Cleft palate: a molecular mechanism linked to the H-r locus,
    Pediat.Res., 11:4:456.
Gobse, B., Debray-Pirev, P., Dansset, J., Lipinski, M., and Hors, J.,
    1977, Etude des groups HLA dans les psychous infantiles de
    développement et d'une enzymopathie, C.R.Acad.Sci.,
    D 284:1733-1735.
Kampov-Polevoy, A. B., 1978, Effect of pharmacological agents on
    domination and subordination relations in a pair of rats,
    Bull.Exp.Biol.Med., No. 9, 306-308.
Kannabikh, Yu. V., 1914, Cyclothymia, Its Symptomatology and Course,
    Moscow.
Kakihana, R., Brown, D. R., and McClearn, G. E., 1966, Brain
    sensitivity to alcohol in inbred mouse strains, Science,
    154:1574-1575.
Leonard, B. E., 1972, Effect of four amphetamines on brain biogenic
    amines and their metabolites, Biochem.Pharmacol.,
    21:1289-1297.
Maisky, A. I., and Vedernikova, N. N., 1974, Analysis of the factors
    that take part in the regulation of sensitivity to narcotics,
    "32nd Internat. Congr. on Alcoholism and Drug Dependence,"
    Warshawa, p.94.
Maisky, A. I., and Vedernikova, N. N., 1979, Genetic control over
    sensitivity of experimental animals to narcotics,
    Succ.Contemp.Biol., No. 2, 199-214.
Maisky, A. I., Vedernikova, N. N., and Volov, V. A., 1978, System H-2
    participation in the regulation of mouse sensitivity to mor-
    phine, Bull.Exp.Biol.Med., No.6, 688-690.
Maisky, A. I., Vedernikova, N. N., Vedernikov, A. A., and Shoshina,
    S. V., 1978a, Genetic regulation of sensitivity to narcotics:
    role of the main system of histocompatibility of the mouse and
    man, Proc.XIV Int.genet.Congress, M., p.351.
Maisky, A. I., Vedernikova, N. N., Vedernikov, A. A., Shoshina, S.
    V., Borovkova, N. K., Naidenova, N. G., and Krivov, L. I.,
    1980, Relationship between the main system of histocompat-
    ibility of the mouse and man and narcotic dependence
    liability, Bull.Exp.Biol.Med., No. 11, 596-598.
Mayfield, D., and Allen, D., 1967, Alcohol and affect: a psycho-
    pharmacological study, Am.J.Psychiat., 123:1346.
Maynert, E., and Klingam, G., 1962, Tolerance to morphine, I.
    Effects on catecholamines in the brain and adrenal glands,
    J.Pharmac.Exp.Ther., 135:285-295.
Meddis, R., 1975, A simple two-group test for matched scores with
    unequal cell frequences, Brit.J.Psychol., 66:225-227.

Meruelo, D., and Edidin, M., 1965, Modulation of hormonal effects on adenyl cyclase by the histocompatibility-2(H-2)-locus, Fed. Proc., 34:979.

McLoughlin, G. A., Helslig, C. R., Evans, C. C., and Chapman, D. M., 1976, Association of HLA-A9 and HLA-B5 with Burger's disease, Brit.Med.J., 2:1165-1166.

Plokhinsky, N. A., 1970, Biometry, Moscow Univ., Moscow.

Porsolt, R. D., Anton, G., Blavet, N., and Jalfee, M., 1978, Behavioural despair in rats: a new model sensitive to anti-depressant treatment, Eur.J.Pharmacol., 47:379-391.

Preobrazhenskaya, L. A., and Simonov, P. V., 1970, Conditioned avoidance reactions to a pain stimulation of another animal, J.Higher Nerv.Activ., 20:379-385.

Russell, J. A., and Bond, C. R., 1980, Individual differences in beliefs concerning emotions conducive to alcohol use, J.Studies Alcohol., 41:753-759.

Saelens, J. K., Granat, F. R., and Sawler, W., 1971, The mouse jumping test, a simple screening method to estimate the physical dependence capacity of analgesics, Arch.Int.Pharmacodyn.Therap., 190:213-218.

Shoshina, S. V., and Maisky, A. I.,1982, Participation of the major histocompatibility mouse system (H-2) in the response to ethanol, Bull.Exp.Biol.Med., No. 7, 55-58.

Shuster, L., Baran, A., and Eleftheriou, B. E., 1975, Opiate receptors and analgesic response in mice: basis for genetic difference, Fed.Proc. 34:713.

Smeraldi, E., and Scorza-Smeraldi, R., 1976, Interference between anti-HLA antibiotics and chlorpromazine, Nature, 260:532-533.

Smeraldi, E., Scorza-Smeraldi, R., and Fabio, F., 1980, Interference between antiHLA antibodies and CPZ-metabolites, Psychopharmacology, 67:87-91.

Snell, G. D., Dansset, J., and Nathenson, S., 1976, Histocompatibility, Acad.Press, New York.

Stein, L., 1968, Chemistry of reward and punishment, in: "Psychopharmacology. A Review of Progress," Washington, pp.105-123.

Strelchuk, I. V., 1973, Alcoholic acute and chronic intoxication, Moscow, Medicine.

Svegaard, A., and Kuder, L. P., 1976, Interaction of HLA molecules with non-immunological ligands as an explanation of HLA and disease associations, Lancet, 2:547-549.

Tabakoff, B., and Ritzman, R. F., 1979, Acute tolerance in inbred and selected lines of mice, Drug and Alcohol Dependence, 4:87-90.

The Role of Catecholamines and Serotonin in the Formation and Maintenance of Susceptibility to Alcohol

V.N. Zhukov, Yu.V. Burov and N.A. Khodorova

Institute of Pharmacology, Academy of Medical Sciences
of the USSR, Moscow

1. INTRODUCTION

No simple explanation has yet been offered of how the serotoninergic and noradrenergic systems function in alcohol dependence. Pharmacological studies on the involvement of these systems in the regulation of voluntary alcohol consumption have failed in this respect. Investigations of the effect of ethanol on the two systems, whether administered chronically or acutely, have given contradictory results, moreover. As described previously (Burov et al., 1981), we were able to sort cross-bred rats into those susceptible or not susceptible to alcohol, according to how long the effects of this narcotic were maintained. This enabled us to explore differences between the animal's inborn characteristics, concentrating on metabolic, neurochemical, hormonal and behavioral systems.

Our research aimed to throw light on a) the interaction between central serotoninergic and noradrenergic processes under the influence of alcohol and b) any contribution they may make to initiating and maintaining ethanol addictiveness in animals varying in their potential for voluntary alcohol consumption.

2. MATERIAL AND METHODS

All our experiments were performed on cross-bred male white mice.

2.1. SORTING ANIMALS INTO THOSE PRONE – AND NOT PRONE – TO VOLUNTARY ALCOHOL CONSUMPTION, ACCORDING TO THE DURATION OF ALCOHOL'S EFFECTS

Experiments were performed on animals weighing 200-300 g. The rats were given 4.5 g/kg of 25% ethanol solution at 10 a.m. The duration of alcohol's effects were noted, from how long the "sideways position" was held. Ambient temperature: 20-22°C. From the entire population of rats treated in this way, only those adopting a sideways position of short duration (62±18 min) or long duration (196 ± 23 min) under the influence of alcohol were selected. The former were considered prone and the latter not prone to alcohol.

The above divisions were based on the following findings of Kampov-Polevoi. Evidence of metabolic tolerance to alcohol was adopted as a differentiating sign. This is determined by how fast alcohol is metabolized. The duration of the sideways position induced in rats by 4.5 g/kg of 25% ethanol solution (4.0 g/kg for mice) was taken as an index of rate of metabolism. As shown previously by Burov et al. (1981) duration is inversely proportional to alcohol's rate of metabolism.

The duration of ethanol's effects on strain C57BL and CBA male mice weighing 30-32 g was measured in this manner during our first set of experiments. After testing, the mice were housed in single cells fitted with a drinking trough containing water and a 10% ethanol solution. It was discovered that ethanol's effects lasted 74±7.4 min in C57BL strain animals, with an inherited susceptibility to alcohol. In mice belonging to the CBA strain which had rejected ethanol when allowed to choose, as described above, reaction to ethanol lasted 147±11.3 min.

We were thus led to conclude that a genetically-determined susceptibility to alcohol is associated with an accelerated rate of ethanol metabolism in these animals.

We proceeded to study the duration of ethanol's effects on 84 cross-bred male rats weighing 180-220 g. A total of 24 heavily sedated (HS) and 18 lightly sedated (LS) animals were then selected. The effects of ethanol had lasted an average of 222±11.6 min in HS as compared with 67±4.5 for LS. The rats were then grouped in fours and housed for 10 days in cages measuring 32 x 47 x 16 cm containing a feeding rack and two equal-sized drinking troughs, one containing water and the other a 15% ethanol solution. Daily ethanol consumption by animals of each group was recorded A rate of alcohol consumption of 35.6 ± 5.79 and 10.8 ± 6.48 ml/kg 15% solution (p < 0.05) was observed in LS and HS rats, respectively, during the first three days of testing. Daily alcohol consumption in LS rats dropped thereafter to 14.3 ml/kg, however, and fell to an insignificant level of 5 ml/kg 15% solution in HS animals.

The animals were then housed in individual cages, each equipped
with a feeding track and equal-sized drinking troughs, one containing
water and the other 15% ethanol solution. They were observed over a
10-day period, during which the LS animal's daily alcohol consumption
increased to 39 ml/kg and that of HS to 38 ml/kg (see Table 1).

The above experiments show that LS animals are a) inherently
predisposed to alcohol, b) this tendency is displayed by the animals
even when housed in congenial conditions and c) it is accentuated by
stress, such as a move to a new cage or the effects of isolation.

The HS rats appear to show no inherent need for alcohol, but
this need does arise in response to stress. In fact, HS animals
given access to a 3rd trough containing a 13% saccharine solution,
which lacks ethanol's calorific content and pharmacological effects,
show a substantially reduced ethanol consumption. No such effect was
seen in LS animals (see Table 1). The HS group was found to consume
less alcohol than LS animals when returned to single cages and subsequently tested (2nd to 4th month - see Table 1). During the 3rd
month of trials, HS ethanol consumption began to draw level with LS.
Consumption level was hardly affected, even when animals were given
access to saccharine at the end of the 8th month of testing.

Our findings show that the stress of prolonged isolation may
create a strong need for alcohol. Initially this need fluctuates.
It may be countered by some other pleasurable sensation, such as the
taste of saccharine. These animals' need for alcohol then begins to
stabilize, apparently entering into a second phase and taking on
features of physical dependence.

We finally concluded that two distinct forms of experimental
alcoholism could be discerned in cross-bred white mice.

Form I is produced by ethanol in LS animals with an inherently
rapid rate of ethanol metabolism. It is an indication of the

Table 1. Consumption of 15% Ethanol Solution (in mg/kg) by HS and
LS Rats under Different Experimental Conditions

No. of months	1			2	3	4	5	6	7	8
	ten-day periods									
Condi-tions	group rear ing	single cages	access to sac-charine	Single cages						
LS rats	36± 5.79	39± 6.32	41± 8.17	41± 5.25	46± 4.69	37± 4.49	34± 4.10	24± 3.30	29± 1.98	19± 3.53
HS rats	11± 6.48	38± 5.30	17± 6.74	35± 5.24	35± 5.69	28± 4.40	48± 5.87	30± 6.97	30± 2.55	19± 6.74

similarity of this form of experimental alcoholism and that of mice
in which need for alcohol is genetically determined. LS rats are
also distinguished by their lasting need for alcohol, observed even
when the animals are housed under congenial conditions.

The type of experimentally-induced alcoholism seen in HS animals
is referred to as form II. An inherently low rate of ethanol metabolism is typical of form II. Animals lose their innate need for
alcohol when housed under congenial conditions. It only reappears
under stress, in an initially unstable form.

In subsequent experiments these animals were not used before 2
weeks of the alcohol treatment process had elapsed.

2.2. HOUSING OF ANIMALS; CHOICE OFFERED BETWEEN ALCOHOL AND WATER, NOTING CONSUMPTION OF BOTH FLUIDS; STATISTICAL EVALUATION OF DATA

Animals were kept continuously in single cages measuring
40 x 10 x 15 cm, lit by daylight; ambient temperature, 20-24°C. They
had free access to two equal-sized troughs (100 ml each). The left-
hand container was kept filled with water, and the other with 15%
ethanol solution. The rats had continuous access to standard dry
food pellets, composed of wheat flour, oatmeal, ground barley, sun-
flower seed oilcake, fat-free fish meal, powdered milk, molasses and
irradiated hydrolyzed yeast. The amount of the two fluids consumed
by each rat was noted each day at 10 am. All rats were weighed daily
during the experimental period. Two sets of experiments were performed, depending on how long animals had access to alcohol when
offered the alternatives described above. The first series set out
to investigate the effect of certain substances in shaping the need
for alcohol. These substances were administered when animals were
first offered the choice between alcohol and water. Average daily
consumption of these two fluids per kg/body weight was calculated
over a 10-day period of treatment with the substances and a further
week after treatment had ended. Findings were compared with those
from the control group, using Student's test (Belenky, 1963). The
second series of experiments concerned these product's effects on
shaping desire for and dependence on alcohol. The animals' initial
reactions to the choice described were set aside when long-term
treatment with the test drugs began. Average daily consumption of
water and alcohol per kg body weight was then calculated separately
for each animal a) for 1 week preceding, b) for 2 weeks during and
c) for 1 week following treatment with the test substances (i.e.
withdrawal). These were administered i.p. at a volume of 1.0-1.5 ml
twice daily (at 10 am and 4 pm) every 2 weeks. The same schedule was
followed for the control group, which received distilled water (i.p).
Statistical evaluation of our findings was performed according to the
method of Meddis (1976).

2.3. a) DESCRIPTION OF OPERATION. b) LOCAL MICROINJECTION OF
      SUBSTANCES INVOLVED IN THE NEUROCHEMISTRY OF THE BRAIN INTO
      THE CEREBRAL VENTRICLE

     Rats were sufficiently anesthetized, in an atmosphere contain-
ing fluothane, to undergo endotracheal intubation. The endotracheal
tube was then connected to the anesthetic apparatus, which was thereby providing controlled fluothane anesthesia using a semi-circuit,
while the animal breathed independently. The rats were secured in a
stereotactic frame and a cannula surgically inserted into the left
cerebral ventricle according to the following coordinates : A 5.8;
L 1.5; H 2.0 (Pellegrino et al., 1979). Single doses of the test
compounds were administered into the cerebral ventricle via a capil-
lary inserted through the cannula, stretching its entire length and
joined by a polyethylene tube to a microinjector. The substance was
administered for 30 sec and the cannula removed 5 min after the
microinjection. The wound was stitched tightly, the area treated
with an iodine solution and sprinkled with "bitsillin-5", and the
rats finally taken out of the stereotactic frame. All the test
substances were dissolved in physiological saline and administered
into the cerebral ventricle in a volume of 20 μl. A similar operation was performed on control animals, the only difference being
that physiological saline alone was used. When the cannula was used
chronically, it was secured to the plates of the skull after implan-
tation in the cerebral ventricle, using "protacryl" plastic and
stainless steel hooks. In the course of the operation, accuracy in
placing the cannula was checked by keeping in view a meniscus of
cerebrospinal fluid. The cannulas used came in the form of stainless
steel tubes with an outer and inner diameter of 0.6 and 0.3 mm respectively. Each had a removable trocar, stretching the length of the
cannula, and which would close it with a screw-on cap during the
periods between injections in experiments involving chronic treatment. The insertion of the cannula in these animals was also checked
at the end of the experiment by injecting 20 μl ink into the cerebral
ventricle under fluothane anesthesia and eventually observing stain-
ing of the cerebral ventricular cavity after removing the brain from
the animal's skull.

2.4. DETERMINING SEROTONIN CONCENTRATION IN RAT BRAIN, LIVER,
      INTESTINE AND BLOOD

     Serotonin concentration in brain, liver and intestine were
determined, closely following the method of Cox and Perhach (1973),
using an "Opton" spectrofluorometer. The method of Kulinsky et al.
(1969) was used for spectrofluorometric determination of serotonin
concentration in the blood, using the following reagents:

1. n-butanol, washed with a 1 N NaOH, then 1 N HCl solution, three
    times with distilled water and once with double-distilled water.

2. Heptane, washed with 1 N NaOH solution, 1 N HCl solution, three
    times with distilled water and once with double-distilled water.
3. A 1 N $HClO_4$ solution, containing 7 mg/ml ascorbic acid.
4. Borate buffer, 0.5 M, pH 10 (31.4 g boric acid dissolved in 1
    liter water, with the addition of 67 ml of 30% NaOH solution).
5. A 0.1% solution of thymol blue in 20% ethanol solution.
6. A 10% and a 0.4% NaOH solution.
7. Borate buffer, pH 10, saturated with NaCl and butanol.
8. Phosphate buffer, 0.05 M, pH 7.
9. A 0.1 M ninhydrin solution (1.78 g in 100 ml solution).
10. Crystalline NaCl.
11. Ascorbic acid.
12. Serotonin creatinine sulfate: 23 mg reagent, containing 10 mg
    serotonin-free base, dissolved in a 40 ml solution of 0.01 N HCl
    – stock solution, containing 250 µg/ml serotonin. The solution
    is used for preparing a working solution containing 5 µ/ml
    serotonin – i.e. by dissolving 0.5 ml of the stock solution in
    25 ml 0.01 N HCl. These solutions are then frozen. To obtain a
    product for intravenous administration the 5 µg/ml solution was
    unfrozen and 0.1 ml taken off; i.e. 0.5 µg serotonin.

2.5. PROCEDURE FOR SEROTONIN DETERMINATION

Blood is mixed with an equal volume of 1 N perchloric acid
solution, shaken for 15 min and centrifuged for 30 min at 6,000 rpm.

The clear supernatant is transferred to a flask with a ground-
glass stopper and a capacity of 50 ml; 0.5 ml of 0.5 M borate buffer
and 2 drops of thymol blue are then added. A 10% NaOH solution is
first titrated until a light blue shade shows, after which a few
drops of 0.4% NaOH solution is added until a pH of 10 is obtained.
After adding NaCl (1 g/ml) and shaking, 15 ml n-butanol is rapidly
poured in. The mixture is again shaken for 10 min and the organic
phase then taken off, using a separating funnel, and transferred to
another flask, where 2 ml of 0.1 M borate buffer is added. After
repeated shaking for 3 min, it is left to stand for a further 10 min.
The 10 ml butanol is separated off, transferred to a third vessel
containing 2.6 ml of 0.05 M phosphate buffer and 15 ml heptane and
shaken for 10 min. The entire contents are then poured into a 35 ml
centrifuge tube and centrifuged for 10 min at 1,000 rpm. A volume
of 2.4 ml aqueous phase is then separated off, to which 0.1 ml of a
0.1 M solution of ninhydrin is added. After mixing, the solution is
incubated for 30 min at 75°C. One to two hours, later, it is examined fluorometrically at an excitation wavelength of 360 nm and
emission wavelength of 470 nm.

The following formula was used for calculating results:

$$C = \frac{2 \times 0.62 \times X}{0.62 \times K \times Y} = \frac{2 \times X}{K \times Y} \; \mu g/ml$$

where 2 – dilution of blood with perchloric acid; X – reading in
fluorometric units for this assay; K – reading in units given by 1 g
serotonin, taken through all stages of the procedure; Y – volume of
perchloric acid extract taken for measurement;

$$\frac{10}{15} \times \frac{2.4}{2.6} = 0.62$$ is the correction for volume of butanol and

phosphate buffer. Before the removal of tissues for examination the
animals were anesthetized with ether and decapitated. Both venous
and arterial blood were collected in equal-sized vessels with a
chilled solution of 1 N $HClO_4$ in which 7 mg/kg ascorbic acid had been
dissolved. Various parts of the brain – hypothalamus, thalamus,
brain stem and striatum – were separated in a glass dish.

2.6. DETERMINING CATECHOLAMINE CONCENTRATION IN BRAIN TISSUE

Concentrations were determined of noradrenaline (NA) in the
hypothalamus and dopamine (DA) and its metabolite, homovanillic acid
(HVA), in the striatum, combing the brain structures of two animals
and using a modified version of Westering and Korf's (1976) method.
This involved removing the cerebral cortex of rats, separating the
above-mentioned structures, weighing and then freezing them in liquid
nitrogen, where they remained prior to analysis. Brain structures
from two animals were homogenized in 1.5 ml of 0.4 N perchloric acid.
Perchloric acid was then precipitated out, using a solution of KOH
and formic acid, bringing the pH of the homogenate to 2-3. The
solution was centrifuged at 4,000 g for 20 min at 4°C. The supernatant was transferred into a Sephadex G-10 column (Farmacia Fine
Chemicals). The columns, measuring 70 mm in length, were prepared
from Pasteur pipettes measuring 5 mm across. The columns were re-
generated before use with successive 30 ml portions of 0.01 N ammonia
and formic acid, which were discarded. Following this, 2 ml of the
solution were passed through and the eluate used to determine NA and
DA concentration and then 3 ml of 0.005 M phosphate buffer, pH 8.5;
the eluate was used to determine HVA content. NA and DA concentrations were measured fluorometrically (Chang, 1964) using a
0.2 M iodine solution as oxidizing agent. HVA concentration was
determined using the method of Anden et al. (1963). A 0.01% solution
of $K_3Fe(CN)_6$ was the oxidizing agent employed on this occasion. The
extraction of NA, DA and HVA from brain tissue was performed with
internal standards, adding 100 g HVA, 50 ng NA and 1 µg DA to the
cerebral cortex homogenate. Recovery was 70% for HVA, 92% for NA
and 76% for DA. The measurement was performed on a "Hitachi"
spectrofluorometer at wavelengths of 320.390 for DA, 420.480 for NA
and 320.450 nm for HVA.

2.7. MEASURING CONCENTRATION OF ^{14}C-SEROTONIN IN RAT SEPTUM, HYPOTHALAMUS, THALAMUS, BRAIN STEM AND CAUDATE NUCLEUS

Male mice weighing 280-300 g were used in these experiments.
They had previously been tested for predisposition (or the reverse)

to alcohol, according to how long they remained sedated by its effects.   Two weeks after these tests the animals were housed in single
cages for 10 days and given free access to a 15% ethanol solution and
water.   These rats were matched by a control group which had also
been tested during the same period as to the duration of alcohol-
induced sedation. They had been housed under the same conditions, but
without provision of the 15% ethanol solution. On the 11th day the
animals were injected in the tail vein with a dose of 20 µCi per kg
body weight ^{14}C-serotonin.   They were decapitated 10 min later and
the brain removed and freed from superficial vessels.   The brain
structures to be studied were then separated, weighed and frozen at
-4°C.   Next came the process of determining the amount of ^{14}C-sero-
tonin in these structures.   Small pieces of tissue weighing 10-30 mg,
dissected from the appropriate brain region, were dissolved in 0.5 ml
dimethylammonium hydrochloride from "Serva".   When completely dissolved, 10 ml toluene scintillator was added.   Radioactivity was
measured in the tissues using a Nuclear Chicago liquid scintillation
spectrometer.   Its efficiency was checked with an external standard.
A statistical evaluation was made by calculating mean and reliability
limits (Belenky, 1963).

3. RESULTS

3.1.   NA, DA AND HVA CONCENTRATION IN RAT BRAIN.   a) PREDISPOSED OR
       b) NOT PREDISPOSED TO ALCOHOL, AND c) UNDER THE EFFECTS OF
       ALCOHOL

     Table 2 shows our findings on hypothalamic NA and striatal DA
and HVA in rats both predisposed and not predisposed to alcohol - as
follows:

  a) having had no contact with alcohol
  b) 30 min after i.p. administration of a 2.5 g/kg dose of a 25%
     ethanol solution
  c) following a 10-day period of consumption of 15% ethanol solution
     (chosen in preference to water).

Initially our experiments revealed no difference between NA, DA and
HVA concentrations when animals with and without the inclination to
consume alcohol were compared. Ethanol produced the considerable
effect of reducing NA concentration by 30% (p<0.05) when administered
acutely to predisposed animals, a possible indication of increased
release and utilization of this neurotransmitter in such animals when
strongly under the influence of alcohol. Alcohol failed to produce
any changes in concentration of NA, DA or HVA in non-disposed rats
under equivalent conditions. Both groups of animals showed an increase in DA concentration of 25% (p<0.05) over a 10-day period of
elective alcohol consumption. The basic similarity of these changes

Table 2. Concentrations of Noradrenaline, Dopamine and Homovanillic Acid in the Brain of Rats Predisposed and Non-predisposed to Alcohol

Experiment	Hypothalamic NA (ng/g)		Striatal DA (ng/g)		Striatal HVA (ng/g)	
	Predisposed	Non-predisposed	Predisposed	Non-predisposed	Predisposed	Non-predisposed
Intact rats	1416± 93	1256±218	7015±1186	7227±972	612±51.7	616±57.9
Ethanol 2.5 g/kg, acute, i.p.	1015± 35	1145± 80	6385± 260	6285±472	555±49.0	576±62.4
10 Day consumption 15% ethanol solution	1392±102	1550± 89	9205±865	9026±865	540±88.2	643±69.5

in predisposed animals, consuming 39 ml/kg of 15% ethanol solution
per 24 h as against 9 ml/mg for non-predisposed rats, confirms their
non-specific nature. Some differences were found, on the other hand,
in NA concentration. Non-predisposed rats show an increase in NA
concentration of 19% ($p<0.05$) compared with initial level after an
elective 10-day period on alcohol, whereas non-predisposed animals
do not. In view of the substantial drop in NA concentration following acute administration of alcohol to predisposed rats, stabilization of NA concentration after a 10-day period of alcohol consumption would appear to occur as a result of its compensatory synthesis
under circumstances of increased release and utilization. In other
words, activity of the NA system is intensified during this period.

3.2.   SEROTONIN (5-HT) CONCENTRATION IN a) HYPOTHALAMUS, b) THALAMUS
        c) BRAIN STEM, d) STRIATUM, e) LIVER, f) INTESTINE AND g) BLOOD
        OF INTACT RATS. i) predisposed, ii) non-predisposed to alcohol
        consumption and iii) under the action of ethanol

        We investigated 5-HT concentration in brain and periphery in
predisposed and non-predisposed intact animals a) 30 min after a
single 2.5 g/kg i.p. dose of 25% ethanol solution and b) after a
10-day period of 15% ethanol consumption (chosen in preference to
water - see Table 3). According to our findings, obtained from a
population of white cross-bred rats, serotonin level is higher in
hypothalamus and brain stem and lower in thalamus and striatum in
rats predisposed to alcohol than in those animals refraining from
alcohol.

        We had already established that the duration of "sideways
position" under the influence of ethanol and predisposition for
alcohol consumption are inversely proportional. Our findings now
suggested that within the serotoninergic system an initially
heightened need for alcohol is bound up with how serotonin is distributed in various parts of the brain - i.e. concentrated more in
hypothalamus and brain stem and relatively less in thalamus and
striatum. A reversal in these serotonin levels in the same brain
sites would accompany a potentially smaller need for alcohol.

        Other evidence of differences in serotonin concentration between
cross-bred rats belonging to predisposed and non-predisposed categories is seen in the peripheral serotoninergic system; LS animals
have a higher serotonin concentration in intestine and a lower level
in liver and blood.

        One-way movement is observed in both groups in the course of
ethanol's influence on the peripheral serotoninergic system - namely,
increased serotonin concentration in liver and blood. The fall in
concentration in the intestine of animals predisposed to alcohol
constitutes an exception to this.

Table 3. Serotonin Concentration in Areas of Brain, Liver and Intestine (in μg/kg) and in Blood (μg/ml). How Ethanol Influences these Levels in Rats Predisposed and Non-predisposed to Alcohol Consumption

Parts of the body	Duration of alcohol's effects					
	62 + 18 min			196 + 23 min		
	Controls	Ethanol 2.5 g/kg i.p.	10-day period on 15% alcohol solution	Controls	Ethanol 2.5 g/kg i.p.	10-day period on 15% alcohol solution
		Predisposition to alcohol +				
Hypothalamus	1.58 ±0.07***	1.23 ±0.3***	1.31 ±0.05*	1.00 ±0.017	1.498±0.125**	1.57 ±0.03***
Thalamus	0.86 ±0.04**	1.08 ±0.01***	0.76 ±0.03	1.012±0.025	0.84 ±0.03**	1.3 ±0.03***
Brain stem	0.53 ±0.008***	0.65 ±0.02***	0.57 ±0.009**	0.43 ±0.01	0.632±0.036***	0.41 ±0.009
Striatum	0.47 ±0.01***	0.314±0.02***	0.57 ±0.01***	0.59 ±0.008	0.41 ±0.018***	0.69 ±0.009***
Liver	0.51 ±0.01***	0.71 ±0.09***	0.59 ±0.01***	0.62 ±0.007	0.74 ±0.02***	0.76 ±0.009***
Intestine	0.67 ±0.01***	0.61 ±0.02	0.97 ±0.001***	0.53 ±0.01	0.66 ±0.03**	0.74 ±0.01***
Blood	0.115±0.004***	0.222±0.004***	0.136±0.005*	0.159±0.007	0.237±0.003***	0.326±0.007***

The results of differences a) between two control groups and b) between control group and "action of ethanol" groups (marked with an "*").

* $p < 0.05$.
** $p < 0.01$.
*** $p < 0.001$.

Five animals per group were used in these experiments.

An alcohol-induced accumulation of serotonin in the periphery thus seems likely. Furthermore, the intestine, as the main source of peripheral serotonin, could be putting out increased quantities of this biogenic amine.

Acute ethanol administration likewise produces unidirectional changes in serotonin concentration, which increases in the brain stem and decreases in the striatum in predisposed and non-predisposed animals. The changes produced in these structures are thus non-specific. The opposite is true of those occurring in hypothalamus and thalamus; ethanol has diverse effects, reducing serotonin concentration in the hypothalamus and raising it in the thalamus in those rats with a predisposition for alcohol, these effects being reversed in non-predisposed animals. The changes in serotonin value correlations produced by ethanol under the circumstances described may be viewed as a connection between the changes occurring and the a) positive or b) negative attitude which causes animals to either welcome or reject alcohol.

During the 10-day period of elective alcohol consumption changes in 5-HT concentration in predisposed rats followed the same pattern as after acute administration. Of the structures tested the striatum, where 5-HT level increased, constituted an exception. It also exceeded the control value in the thalamus and striatum and was restored in the brain stem in non-predisposed animals following elective alcohol consumption. Hypothalamic serotonin concentration rose considerably in these animals compared with control.

3.3. ^{14}C-SEROTONIN PENETRATION OF THE BLOOD-BRAIN BARRIER (BBB) IN RATS WITH AND WITHOUT PREDISPOSITION FOR ALCOHOL a) UNDER NORMAL CIRCUMSTANCES AND b) AFTER A 10-DAY PROGRAM OF ELECTIVE ALCOHOL CONSUMPTION

Table 4 shows the results of experiments on the accumulation of ^{14}C-serotonin (^{14}C-5-HT) in various areas of the brain in i) predisposed and non-predisposed alcohol-free rats and ii) equivalent groups following a 10-day program of elective alcohol consumption.

Our findings thus indicate that the degree of accumulation, reflecting BBB penetration, of ^{14}C-5-HT in the hypothalamus is initially lower in predisposed than in non-predisposed animals. After a 10-day program of 15% ethanol solution (chosen in preference to water), predisposed animals show increased penetration of ^{14}C-5-HT through the BBB in the hypothalamus. Penetration is roughly the same as in non-predisposed rats which had not touched alcohol, having preferred water. Animals opting against alcohol, however, show a higher hypothalamic ^{14}C-5-HT BBB penetration at the outset. This is halved at the end of the 10-day elective alcohol consumption program. Amongst the brain areas studied, changes in ^{14}C-5-HT BBB

Table 4. Initial Penetration through BBB of ^{14}C-serotonin and its Pattern in Predisposed Rats after a 10-day Program of Elective Alcohol Consumption

Categories of predisposition for alcohol	^{14}C-serotonin accumulation (counts/min per g tissue) in brain structures				
	Septum	Hypothalamus	Thalamus	Brain Stem	Caudate nucleus
Non-predisposed:	37	53 ↑	21	48	51 ↑
No contact with alcohol	30÷44	47÷59	11÷31	32÷64	40÷62
After 10 days of elective alcohol consumption	50	23 ↓	25	32	18 ↓
	32÷68	13÷33	19÷31	20÷44	13÷23
Predisposed:	36	21 ↓	21	49	26 ↓
No contact with alcohol	20÷52	16÷26	16÷26	38÷60	19÷33
After 10 days of elective alcohol consumption	32	49 ↑	17	35	26
	21÷43	40÷58	14÷20	26÷44	19÷33

penetration of a similar pattern were noted in the caudate nucleus,
the only difference being that none were seen in predisposed animals
after the 10-day program with alcohol.

3.4.   THE EFFECTS ON ELECTIVE ALCOHOL CONSUMPTION OF ADMINISTERING
        NA, 3,5-DIMETHYLPYRAZOLE AND DA INTO THE CEREBRAL VENTRICLE

        The results of these investigations, presented in Table 5, show
that NA and 3,5-dimethylpyrazole, an inhibitor of dopamine β-hydroxylase, proved capable of reducing voluntary alcohol consumption
a) when the need first arose (on initial contact with 15% ethanol
solution during the 10 days, when a choice of fluids was offered) and
b) during the period of alcohol addiction, after 8 months of elective
alcohol consumption. Dopamine did not affect the latter type of
consumption under the same circumstances, however.

3.5.   EFFECTS OF 5,6-DIHYDROXYTRYPTAMINE (5,6-DHT) ADMINISTERED
        INTRACEREBROVENTRICULARLY ON ELECTIVE ALCOHOL CONSUMPTION
        DURING EVOLUTION AND DISSOLUTION OF ALCOHOL ADDICTION

        Alcohol consumption was studied in animals which had never met
with alcohol and in rats which had already become dependent. These
rats were given 5,6-DHT (a neurotoxin) intracerebroventricularly,
thereby producing a selective degeneration of serotoninergic endings
and reducing serotonin concentration in the central nervous system
(Baumgarten et al., 1971). An outline of the experiments performed
is given in Figures 1 and 2.

3.6.   THE EFFECTS OF SUBSTANCES INVOLVED IN CATECHOLAMINERGIC PROCESS
        ON VOLUNTARY ALCOHOL CONSUMPTION OF RATS AS THEIR NEED FOR
        ALCOHOL DEVELOPS

        The following substances were used in our research: the cate-
cholamine precursor L-DOPA (50 mg/kg, twice daily); α-methyldopa,
which reduces activity of the catecholaminergic system (50 mg/kg,
twice daily); cocaine - a noradrenaline reuptake inhibitor (5 mg/kg
twice daily); phentolamine and pyrroxane - α-adrenergic-blockers
(both at a dose of 30 mg/kg once daily); clonidine - a direct agonist
of presynaptic α-adrenoreceptors (0.05 mg/kg, twice daily); propranolol - a β-adrenergic blocker (30 mg/kg, once daily), haloperidol
- a post-synaptic dopaminergic receptor blocker (0.1 mg/kg, twice
daily) and apomorphine, a direct dopaminergic receptor agonist
(0.1 mg/kg, twice daily) - a dose which acts primarily on presynaptic
receptors. All these substances were administered intraperitoneally.

        As our experiments show, L-DOPA raises and α-methyldopa re-
duces rats' alcohol consumption at the formative stage of desire for

Table 5. The Effects of Noradrenaline, Dopamine and 3,5-Dimethylpyrazole on the Alcohol
          Consumption of Rats at Various Stages of Experimentally-induced Alcohol Dependence

| | Consumption of 15% alcohol in ml/kg per 24 h | | | | |
| | Stage I | | | Stage II | |
Compound	Administered 5 days	Withdrawal 5 days	10 days' background	Administered 5 days	Withdrawal 5 days
Noradrenaline 100 µg, twice per 24 h	4.6*	6.3*	41.1	24.4*	15.7**
Dopamine 100 µg, twice per 24 h	15.5	21.4	50.4	44.9	37.2
3,5-dimethylpyrazole 90 µg once per 24 h	4.2	18.8	44.5	24.5	31.8
NaCl, 0.9% solution	13.7	22.9	42.4	35.3	29.8

Difference statistically significant compared with control.
* p<0.05.
**p<0.01.

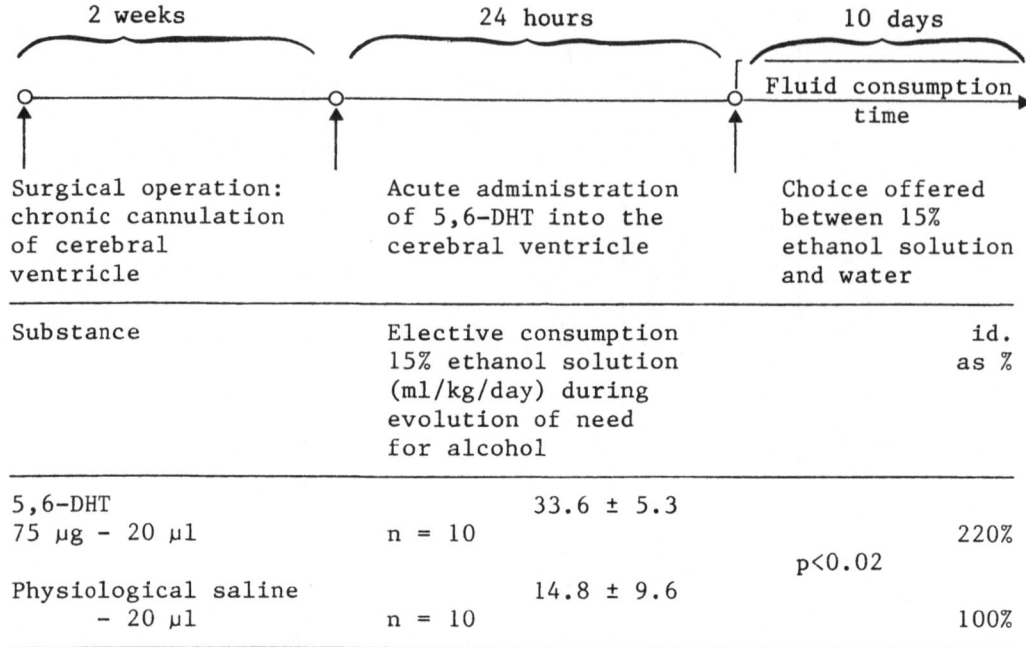

Substance	Elective consumption 15% ethanol solution (ml/kg/day) during evolution of need for alcohol	id. as %
5,6–DHT 75 μg – 20 μl	33.6 ± 5.3 n = 10	220%
		p<0.02
Physiological saline – 20 μl	14.8 ± 9.6 n = 10	100%

Fig. 1. Effects of 5,6–DHT, administered intracerebroventricularly,
on voluntary consumption of 15% ethanol solution during
development of alcohol dependence.

alcohol. In other words, increased or reduced alcohol consumption
are produced respectively by activation or inhibition of the catecholaminergic system. Further differentiation of activity within
the catecholaminergic system showed that cocaine (in that it renders
the noradrenergic system more active) produces a steep rise in alcohol consumption. The inhibitory activity of α-receptors of the
noradrenergic system produces marked inhibition of alcohol consumption in rats, as mediated by phentolamine, pyrroxane and clonidine
(see Table 6).

    Propranolol – the β-adrenergic blocker – and the two dopamin-
ergic system inhibitors, haloperidol and apomorphine hardly affect
alcohol consumption. Our findings show that the α-receptors of the
noradrenergic system play a substantial part in developing the desire
for alcohol.

4. DISCUSSION

    Our experiments show that the functional organization of the
catecholaminergic and serotoninergic systems and their reaction to
the effects of alcohol differ between rats predisposed to alcohol

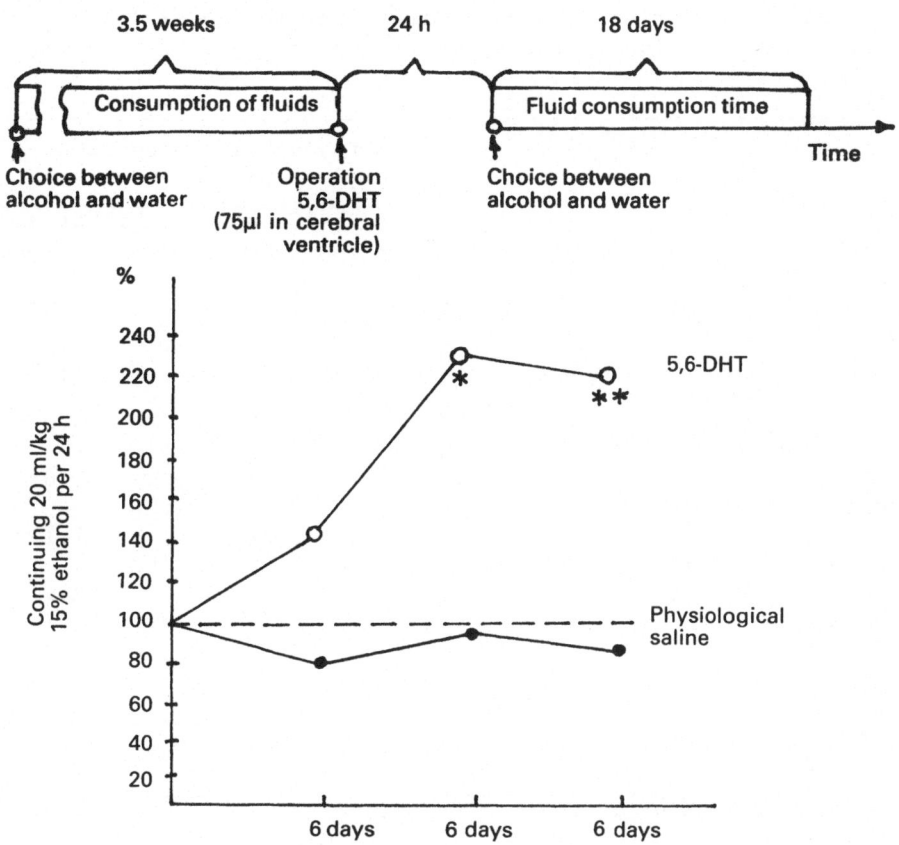

Fig. 2.   Effects of 5,6-DHT on voluntary consumption of 15% ethanol
          solution in rats already dependent on alcohol. Results
          shown as % of baseline intake (20 ml/kg solution per 24 h,
          taken as 100%). Significantly different from baseline:
          *p<0.05; **p<0.001.

consumption and those not so predisposed. It affects predisposed
rats, when administered acutely, by substantially reducing NA concentration in the hypothalamus – the area mainly responsible for this
motivation – perhaps a sign of intensified production and utilization
of this neurotransmitter under the influence of ethanol. It will be
remembered that alcohol potentiates self-stimulation response in
predisposed animals (Burov and Borisenko, 1979), together with its
related basic catecholaminergic phenomena (Hunt et al., 1976; Herberg
et al., 1976). This leads us to suppose that the noradrenergic
system is directly involved in ethanol's effects on the emotions.
The NA system remains more stable in non-predisposed animals as far
as alcohol's effects are concerned; acute administration does not
alter NA concentration in these animals. It is also significant that

Table 6. Effects of Various Drugs on Rats' Consumption of 15%
          Alcohol Solution During the Development Stage of Need for
          Alcohol

Compounds	Period of alcohol consumption 0-10 Days		
	1-5 Days	6-10 Days	Withdrawal
1. L-DOPA - 50 mg/kg twice daily	162.0*	77.0	141.0*
2. α-Methyldopa - 50 mg/kg twice daily	26.8*	28.7*	22.8*
3. Cocaine - 5 mg/kg twice daily	194.0*	183.0*	410.0**
4. Phentolamine - 30 mg/kg once daily	4.2**	14.0**	8.0**
5. Pyrroxane - 30 mg/kg once daily	24.1**	3.1**	17.1**
6. Clonidine - 0.05 mg/kg once daily	35.2*	20.7*	62.4
7. Propranolol - 30 mg/kg once daily	58.2*	53.8**	102.0
8. Apomorphine - 0.1 mg/kg twice daily	54.3*	72.8	76.2
9. Haloperidol - 0.1 mg/kg twice daily	53.2*	62.8	75.1

Results expressed as % of control. $*p < 0.05$; $**p < 0.01$.

ethanol fails to attenuate the self-stimulation response in non-
disposed animals (Burov and Borisenko, 1979). It is most likely,
therefore, that ethanol exerts no positive emotiogenic effect on
these animals. From the results of experiments investigating the
effects of substances involved in central catecholaminergic processes
it was seen that when either NA or 3,5-methylpyrazole (a dopamine-β-
hydroxylase inhibitor which prevents the transformation of DA into
NA) are given intracerebroventricularly, voluntary alcohol consumption decreases substantially. When NA is administered in this way
the observed falling off in the need for alcohol is probably connected with intensified NA system activity. This, in turn, is linked
with positive reinforcement mechanisms - i.e. the NA output produced
by alcohol is being imitated, and the need to consume ethanol in
order to achieve a "positive" state decreases. The falling off in
desire for alcohol occurring under the influence of 3,5-dimethyl-
pyrazole would appear to be connected with a waning in the activity
of the NA mechanisms of positive reinforcement, since this compound
is able to reduce the concentration of central NA. If this is so,
alcohol would probably not be capable of promoting sufficient output
of NA to produce an emotionally positive state. Our findings point
to the possible involvement of NA mechanisms in ethanol achieving its

emotiogenic effects. Pharmacological investigations have indicated
that compounds involved in catecholaminergic processes affect voluntary alcohol consumption. They further established that facilitating
the activity of the NA system using L-DOPA (a precursor of NA) or
cocaine (an NA reuptake inhibitor) leads to an increase in ethanol
consumption, and α-methyl-DOPA to a decrease. Furthermore, α-adre-
nergic receptors of the NA system are probably playing a major part
in the emotiogenic effects revealed by ethanol in the framework of
the catecholaminergic system. This is likely because when these
receptors are blocked by using phentolamine, pyrroxane or clonidine,
pronounced inhibition of voluntary alcohol consumption is observed.
It should be added that propranolol, a β-adrenoreceptor blocker,
together with haloperidol and apomorphine (compounds inhibiting DA
system activity), fail to exert any noticeable effect on alcohol
consumption.

The discovery of differences in 5-HT concentration in the hypothalamus - the major motivational region - is of prime importance
for the 5-HT-related mechanisms of the need for alcohol. As already
mentioned, a clear-cut relationship was pointed out earlier between
ethanol's capacity for activating NA positive reinforcement structures and the arousal of desire for alcohol in rats. In view of
evidence from the literature that an autonomous system of serotoninergic neurones restraining self-stimulation is present in the hypothalamus (Van der Kooy, 1977), there is reason to suppose that this
restraint is especially powerful in positive reinforcement structures
in animals which are predisposed to alcohol consumption and show a
high 5-HT concentration in the hypothalamus. Ethanol would appear to
attenuate such inhibitory effects, as shown by the drop in hypothalamic 5-HT concentration under these circumstances. Consequently,
the positive reinforcement systems are activated and a "positive"
emotional state is produced. The positive reinforcement system in
non-predisposed animals would probably be less inherently susceptible
to inhibitory influences originating from the hypothalamic serotoninergic system; serotonin concentration is lower in this region.
Inhibitory influences do appear to be increased in these animals by
ethanol, however. Consequently a "positive" emotional state would
not arise under the circumstances described.

On the basis of our findings, we would therefore suggest that
the hypothalamic serotoninergic system could be involved in modulating ethanol's "positive" and "negative" effects in white cross-
bred rats, both disposed and non-disposed to alcohol consumption.

The 5-HT systems of the brain stem likewise make an important
contribution to regulating ethanol's positive and negative effects.
These systems promote self-stimulation response (Van der Kooy, 1977).
Concentration of 5-HT in the brain stem increases in non-predisposed
animals both receiving acute ethanol and during a 10-day program of
elective alcohol consumption; this could be interpreted as further

promoting alcohol's enhancing effects on the positive reinforcement
system. When alcohol is administered acutely to non-predisposed
rats, brain stem 5-HT appears to heighten its "releasing" effect,
side by side with its raised inhibitory control over the positive
reinforcement system, mentioned earlier. Serotonin concentration
in the brain stem then rises in addition. This balanced type of
inhibitory and potentiating influence, exerted by 5-HT on the NA
positive reinforcement system as the result of acute ethanol administration, is unlikely to change the level of its activity. A positive emotional state does not then develop in response to alcohol.
Brain stem 5-HT level drops when non-predisposed animals are subjected to a 10-day voluntary alcohol consumption program - an
indication that the brain stem 5-HT system may be losing some of its
activating effect on the positive reinforcement system. This causes
ethanol to produce a negative emotional state in these animals.

The BBB may be involved in the control exerted over the NA
mechanisms of positive reinforcement by 5-HT. Its role differs in
predisposed and non-predisposed animals as far as 5-HT formed at
the periphery penetrating thence into the brain is concerned. We
established that the main differences in ^{14}C-5-HT accumulation in
these animals involved the hypothalamic region; accumulation of 5-HT
from the periphery was initially lower in disposed than in non-
disposed animals. This is one feature which is characteristic of
both types of rats studied. Further light is shed by the findings of
Shaskan and Snyder (1970), that when the 5-HT of the brain increases,
it penetrates into the NA-neurones, including their endings. It then
encroaches on their function, operating as false transmitter and
reducing the functional activity of the NA system. Bearing these
suggestions in mind, the function of the BBB, as far as 5-HT is concerned, is unlikely to potentiate the NA positive reinforcement
mechanisms initially. Peripheral 5-HT, as mediated by BBB function,
may play a vital part during the creation of the alcohol habit, in
view of the fact that ethanol increases 5-HT concentrations in the
blood of both predisposed and non-predisposed rats.

We have stated our findings on initial concentrations of NA and
5-HT in the brain, the changes occurring with acute and chronic
alcohol administration and the pharmacological involvement of NA and
5-HT processes during the period of voluntary consumption of 15%
ethanol solution. We were able to make assumptions on how the NA and
5-HT brain systems interact and the part they play in shaping and
maintaining the desire for alcohol.

The "positive emotional state" occurring under the influence of
alcohol is linked with the NA system. This system is activated by
the effects of ethanol in animals predisposed to alcohol consumption.

It is the α-receptors which contribute most within the NA system
towards achieving the positive state triggered by ethanol.

The level of activity of the NA system is governed either by
inhibitory influences originating from hypothalamic 5-HT or by activating effects deriving from the brain stem 5-HT system. This
control operates differently in predisposed and non-predisposed
animals, helping to activate the NA positive reinforcement mechanisms
in the former and inhibit these mechanisms in the latter.

REFERENCES

Anden, N. E., Roos, B. E., and Werdinius, B., 1963, On the occurrence
    of homovanillic acid in brain and its determination by a
    fluorimetric method, Life Sci., 2:448-458.
Baumgarten, H. G., Bjorklund, A., Lachermayer, L., Nobin, A.,
    and Stenevi, U., 1971, Long-lasting selective depletion of
    brain serotonin by 5,6-dihydroxtryptamine,
    Acta physiol.Scand., Suppl.373, pp.1-15.
Belenky, M. L., 1959, Elements of Quantitative Valuation of
    Pharmacological Effect, Acad. Sci. Latv. SSR, Riga.
Burov, Yu. V., 1982 Searching for pharmacological agents to treat
    alcoholism, Bull.USSR Acad.Med.Sci., 5:72-77.
Burov, Yu. V., Absava, G. I., Kampov-Polevoy, A. B., and Klyuev,
    S. M., 1981, Elimination of ethanol in white rats with
    different levels of alcohol motivation, Pharmac.Toxicol.,
    1:50-52.
Burov, Yu. V., 1979, Effect of ethanol and acetaldehyde on the struc-
    tures of positive reinforcement in the rat, Pharmac.Toxicol.,
    17:291-293.
Chang, C. C., 1964, A sensitive method for spectrophotofluorimetric
    assay of catecholamines, J.Neuropharmacol., 3:643-649.
Cox, R. H., and Perhach, J. L., 1973, A sensitive, rapid and simple
    method for the simultaneous spectrophotofluorometric deter-
    minations of norepinephrine, dopamine, 5-hydroxytryptamine and
    5-hydroxyindoleacetic acid in discrete areas of brain,
    J.Neuro- chemistry, 20:1770-1780.
Herberg, L. J., Stephens, D. N., and Franklin, K. B. J., 1976,
    Catecholamines and self-stimulation: evidence suggesting a
    reinforcing role for noradrenaline and a motivating role for
    dopamine, Pharmacol.Biochem.Behav., 4(5):575-582.
Hunt, G. E., Artens, D. M., Chester, G. N., and Becher, F. T., 1976
    α-Noradrenergic modulation of hypothalamic self-stimulation:
    studies employing clonidine, 1-phenylephrine and α-methyl-p-
    tyrosine, Europ.J.Pharmacol., 37(1):105-111.
Kulinsky, V. I., and Kostiukovskaya, L. S., 1969, Determination of
    serotonin in human blood and in whole blood of experimental
    animals, Lab.Buiss., No.7, 390-394.
Meddis, R. 1975, A simple two-group test for matched scores with
    unequal cell frequencies, Eur.J.Psychol., 66:225-227.
Pellegrino, L. J., Pellegrino, A. S., and Cushman, A. J., 1979,
    Stereotaxic Atlas of the Rat Brain, Plenum Press, New York -
    London.

Shaskan, G. E., and Snyder, S. N., 1970, Kinetics of serotonin
        accumulation into slices from rat brain: relationship to
        catecholamine uptake, J.Pharmacol.Exp.Ther., 175:404-418.
Van Der Koog, D., Fibiger, H., and Phillips, A. G., 1978, An analysis
        of dorsal and median raphe self-stimulation: effects of para-
        chlorophenylalanine, Pharmacol.Biochem.Behav., 8:441-445.
Westering B., and Korf, J., 1976, Regional rat brain levels of
        3,4-dihydroxyphenylacetic acid; concurrent fluorimetric
        measurement and influence of drugs, Eur.J.Pharmacol.,
        38:281-291.

The Importance of Some CNS Peptides in the Development of Experimental Alcoholism

Yu.V. Burov, R. Yukhananov and A.I. Maisky

Institute of Pharmacology, Academy of Medical Sciences
of the USSR, Moscow

1. INTRODUCTION

The neurochemical mechanisms of ethanol's action have not yet
been fully explored. Various authors have shown that ethanol affects
neurotransmitter accumulation, adenylcyclase activity (Kalant, 1975),
NaK-ATPase (Kalant, 1981), the enzymes of catecholamine metabolism
(Carlson et al., 1973) and transmission processes (Tewari, 1981).
There is reason to suppose that the effects of acute and possibly
chronic treatment with alcohol could be mediated by alterations in
the physico-chemical properties of biological membranes. In fact,
ethanol has been shown to increase the mobility of layers in cell
membrane (Goldstein et al., 1981) and to produce changes in the
activity of membrane-bound enzymes. It also alters the conformation
of receptor complexes situated on nerve cell membranes and the
disposition of neurotransmitters in the synapses (Ciofalo, 1980).
The broad spectrum of ethanol's effects means that considerable
difficulties are encountered when identifying cause and effect
relationships in the various disturbances produced by alcoholism.
It may be argued that of its many effects on various systems within
the organism, this biologically active compound exerts its strongest
influence on emotions. We would therefore recommend assessing
changes in the neurochemical systems most closely connected with
regulation of emotional state.

In recent years, the discovery of enkephalins was followed by
identification of the part they play in controlling complex forms
of behavior and determining how the organism responds to stress,
together with a number of other data on the interaction between
alcohol and opiates when addiction sets in (Blum, 1977). These
discoveries provide a theoretical framework for evaluating enkephalin
and endorphin economy in animals with varying propensity for alcohol.

In view of the importance of sleep disturbance in the pathogenesis of alcoholism, we made an analysis of DSIP (delta sleep peptide) concentrations – the only endogenous regulator of sleep within the organism yet described (Schoenenberger et al., 1972). Evidence coming to light that DSIP serves as an antidepressant and regulates circadian rhythms within the organism provided additional reasons for studying concentrations of this peptide (Kastin et al., 1980; Grat et al., 1981).

Experimental models of chronic alcoholism created in our laboratories enabled us to study neurochemical aspects of the formation of alcohol dependence (Burov, 1982). Using this model, we were able to study the differences between two groups of animals. The animals of one group prefer an ethanol solution, while the others reject it. In that ethanol acts differently in rats disposed to alcoholism compared with those not so disposed (Lumeng, 1982), the model used enables us to make a distinction between those alterations occurring within the organism as a result of dependence developing, and lesser changes merely due to the toxic action of ethyl alcohol. We set out to study biochemical differences observed (a) between cross-bred rats which had initially shown a different degree of predisposition for alcohol consumption, and (b) pure bred mice, some of which choose alcohol and others reject it when offered a choice. We hoped to find how certain systems react during the development of alcoholism.

2. MATERIALS AND METHODS

Cross-bred male white rats weighing 250–280 g and congenitally resistant mouse strains weighing 25–30 g were used in these experiments.

Rats were sorted into ethanol-predisposed and non-predisposed by finding out how long the drug's effects had persisted, as described by Burov (1981). Animals sedated for 85.3±19.1 min were sorted into a lightly-sedated group, and those asleep for 191.7±27.3 min were classified as heavily sedated. Fifteen days after this process, an acute dose of 25% ethanol in 0.9% physiological saline was administered i.p. Controls were given physiological saline alone. Most animals were decapitated 60 min after injection, apart from the exceptions mentioned below. Rats housed in individual cages were given a choice between water and 15% ethanol solution (fully described elsewhere by Burov, 1982), in order to simulate chronic alcoholism in rats. "Heavy drinkers" consumed 56.3±8.7 ml and "light drinkers" 6.2±1.4 ml of 15% ethanol solution per kg body weight over a 24 h period, after which ethanol was withdrawn. Mice were also used to model chronic alcoholism. They were given a choice between water and 10% alcohol. Mice having a dropping bottle containing water only were used as controls in the case of "drinker" strains of animals. The A/Sn and ASW "non-drinking" strains completely rejected the ethanol solution.

2.1. TISSUE PREPARATION

The animals were decapitated and the brains rapidly divided into
sections following the method of Glowinsky et al. (1966). The cerebral cortex, striatum and thalamus were set aside for investigation
(the mid-brain was separated by a frontal incision in front of the
tubicles of the lamina quadrigemina). Each section of the brain was
rapidly frozen in liquid nitrogen, weighed and mixed with 0.1 M
acetic acid, previously heated to 90-95°C in a boiling water bath.
The brain fragments were coarsely minced, heated in the water bath
for 5-7 min and chilled over ice. After chilling, the brain tissue
was homogenized with a Teflon pestle and centrifuged for 15 min at
10,000 g. The supernatant was saved and the precipitate rehomogen-
ized and recentrifuged. The supernatant was decanted, lyophilized,
and then dissolved in RIA buffer, at a ratio of 1 g original tissue
per 2 ml buffer. It was centrifuged at 8,000 g for 5 min, poured off
into aliquots and frozen. The sample was kept at a temperature of
−70°C until use. A volume of 100 μl of extract per assay was used to
determine DSIP concentration.

2.2. DETERMINING PEPTIDE CONCENTRATIONS BY RADIOIMMUNE ASSAY

DSIP antibodies were obtained using a conjugate of ovalbumin
and DSIP. The conjugate, obtained using glutaraldehyde, was freed of
reaction products with Sephadex G-15 and used to immunize cross-bred
grey rabbits. Blood serum was collected from the immunized rabbits
1, 4, 10 and 12 months later, during which time they were repeatedly
immunized with Freund's adjuvant. A DSIP analog having tyrosine in
position 6 instead of alanine was used for purposes of radioimmune
assay (RAI).

Tyr[6]- DSIP was iodinated using lactoperoxidase and hydrogen
peroxide. The iodine-labelled peptide was purified over DEAE-
cellulose column using gradient solution with an ammonium acetate
buffer, pH 6.8, 0.1-0.5M. A Tris-HCl buffer, pH 8, containing
0.5% albumin and 0.05% sodium azide, was used to carry out the RIA
at T=4°C (RIA buffer). The incubation mixture contained $8-10 \times 10^3$
cpm ^{125}I-tyr[6]-DSIP, 100 μl of the sample under study or DSIP standard and 100 μl antiserum at a final concentration of 1:25,000. The
volume was brought up to 0.5 ml using RIA buffer. Incubation lasted
40 h. The antigen-antibody complex was separated off from unbound
peptide using ammonium sulfate and the precipitate examined for
radioactivity. This method provided a sensitivity of 5-10 fmol per
assay. Cross-reaction between antiserum and other DSIP analogs did
not exceed 5% (0.01 in the case of all other peptides investigated).
Iodinated peptides and antisera, kindly provided by A. D. Dmitriev of
the National Institute for Psychiatric Research of the Acad. Med.
Sci., were used to identify leu- and met-enkephalins. RIA was performed as for DSIP. Antiserum was used at a final concentration of

1:2000 for met-enkephalin and 1:5000 for leu-enkephalin. This method provided a sensitivity of 15-20 fmol per test in the case of leu-enkephalin and 100-120 fmol for met-enkephalin. Cross-reaction between met- and leu-enkephalin did not exceed 1.5%.

A 10 µl aliquot of extract was added to each assay and the volume was made up by RIA buffer. A standard reagent kit from Amersham was used to determine β-endorphin concentration. Brain extract was diluted to 10 times its volume and 50 µl at a time was used to measure peptide concentration. DSIP and its analogs were synthesized at the laboratory of Peptide Chemistry of the Biochemical Institute of the USSR Acad. Sci.

3. RESULTS

3.1. THE EFFECT OF AN ACUTE DOSE OF ETHANOL ON NEUROPEPTIDE CONCENTRATION

A 1 g/kg i.p. injection of ethanol failed to exert an anxiolytic effect and produced no noticeable changes in met- and leu-enkephalin concentration. It did, however, increase DSIP level in all areas of the brain (see Figure 1). It will be seen that striatal DSIP concentration rose by almost 200% from 0.99 to 2.64 fmol/mg tissue, while its level increased by 140% in thalamus and medulla, from 1.33 to 3.35 fmol/mg tissue. As peptide concentrations were measured only shortly after ethanol treatment, it is hardly likely that the observed increases stemmed entirely from de novo synthesis. It is more likely to have been released from its larger molecular weight precursor, as described in the case of a number of other neuropeptides (De Cloet et al., 1981). If the acute dose of ethanol was increased to 2.5 g/kg, the level promoting a sedative effect, a smaller increase in DSIP concentration was seen to occur. The medulla, where DSIP concentration remained the same as for lower doses, constituted an exception. When ethanol was administered at the anesthetic dose of 4.5 g/kg, DSIP concentration did not differ significantly from control in any of the brain areas studied. These findings led us to suppose that when large doses of ethanol are administered, its effects on DSIP concentration develop earlier and have already disappeared by the time measurements are taken. We therefore checked the concentration of this neuropeptide in the striatum 20, 60 and 120 min after treatment with 1 g/kg and 4.5 g/kg dose of ethanol, in order to test this possibility. Our findings (Figure 2) show that 4.5 g/kg dose of ethanol does not alter DSIP concentration compared with control when testing takes place at these intervals. When a dose of 1 g/kg was given, however, a maximum increase in concentration of this peptide was seen after an interval of 60 min and continued to exceed control values throughout the duration of the experiment.

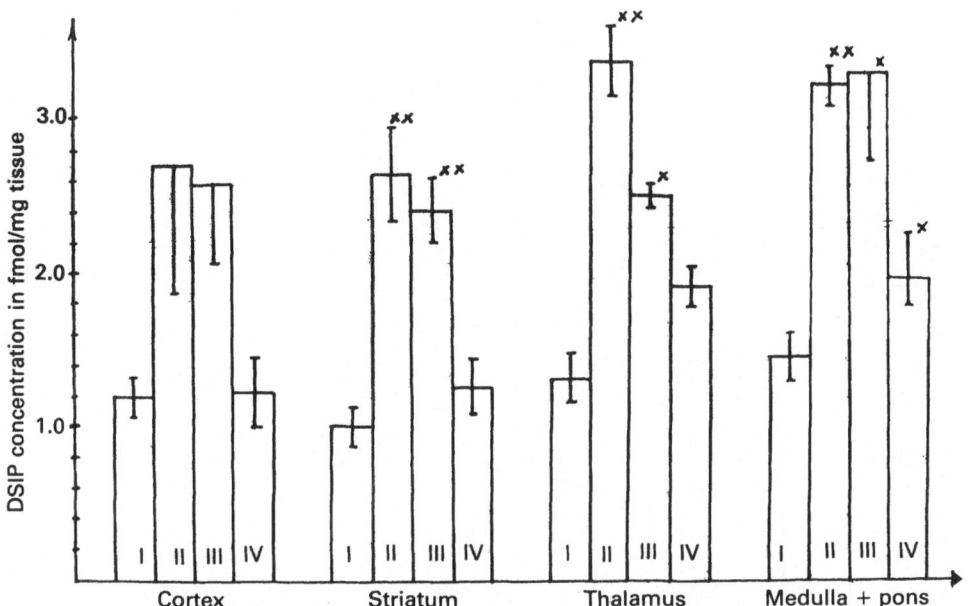

Fig. 1.   Changes in DSIP concentration in rat brain after acute
          ethanol administration.   I - physiological saline; II -
          ethanol, 1 g/kg; III - ethanol, 2.5 g/kg; IV - ethanol,
          4.5 g/kg.   *p<0.05; **p<0.01.

        When an injection of 2.5 g/kg ethanol was given, the concen-
tration of leu-enkephalin was noticeably reduced in the cortex, while
remaining unchanged in thalamus, striatum and medulla (see Figure 3).
Met-enkephalin concentration, however, rose significantly in the
medulla and showed an upward trend in striatum, cortex and thalamus
when the same dose of ethanol was given.   If the ethanol dose was
increased, the concentration of leu-enkephalin dropped considerably
in the cerebral cortex (from 57.2 to 27.3 fmol/mg tissue - see Figure
3).   No difference in peptide concentration compared with control was
seen in striatum, thalamus or medulla (see Figure 2).   The level of
met-enkephalin increased considerably in the medulla (from 540 to 960
fmol/mg tissue) when anesthetic doses of ethanol are given.   At the
same time, striatal met-enkephalin concentration dropped to 1,200
fmol/mg tissue - a considerable decrease, although an upward trend in
the level was noted at this brain site when lower doses of ethanol
were given (see Figure 4).   We then proceeded to assess peptide
concentrations in the brain of animals whose initial ethanol-induced
sedation differed in duration.

        DSIP concentration was found to be 6.87±0.8 fmol/mg protein in
whole brain homogenate of lightly-sedated animals as against the

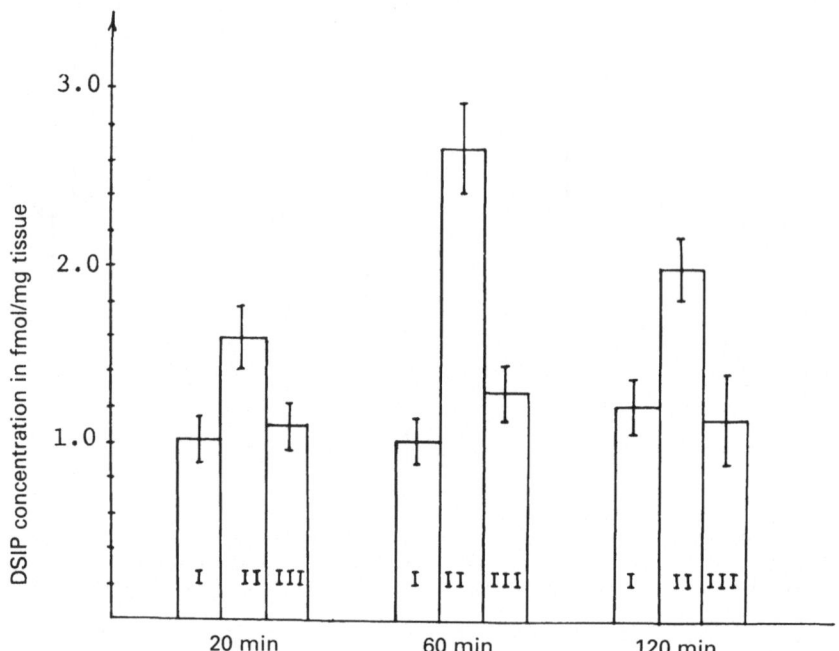

Fig.   DSIP concentration in rat striatum at various intervals
          after ethanol administration. I – physiological saline;
          II – ethanol, 1 g/kg; III – ethanol, 4 g/kg. *p<0.05
          compared with I.

considerably higher level of 11.59±1.56 in the case of heavily
sedated rats. Measurements of DSIP concentration through the various
regions of the brain demonstrated that alcohol-predisposed rats
showed reduced concentrations in cortex and striatum and slightly
raised levels in the thalamus; they resembled non-predisposed rats in
medullary DSIP level (see Table 1). A different picture emerged on
measuring the concentration of met-enkephalin. Alcohol-predisposed
animals showed a reduction in striatal and medullary DSIP levels
(Table 1). A downward trend in met-enkephalin concentration was
observed in the hypothalamus. It is not easy to deal with such
specific differences in met- and leu-enkephalin concentrations in the
brains of animals varying in desire for alcohol. The reasons for the
existence of two structurally related opioid peptides with similar
function within the brain remain unclear as yet. Their localization
in different areas of the brain are known to change in parallel.
Furthermore, their effects are almost identical when administered
intraventricularly . Met- and leu-enkephalin have been thought to
differ in their affinity for mu- and delta-opiate respectively by
Goodman (1980). No convincing experimental confirmation of this has
yet been provided, however. We ourselves support the view that the

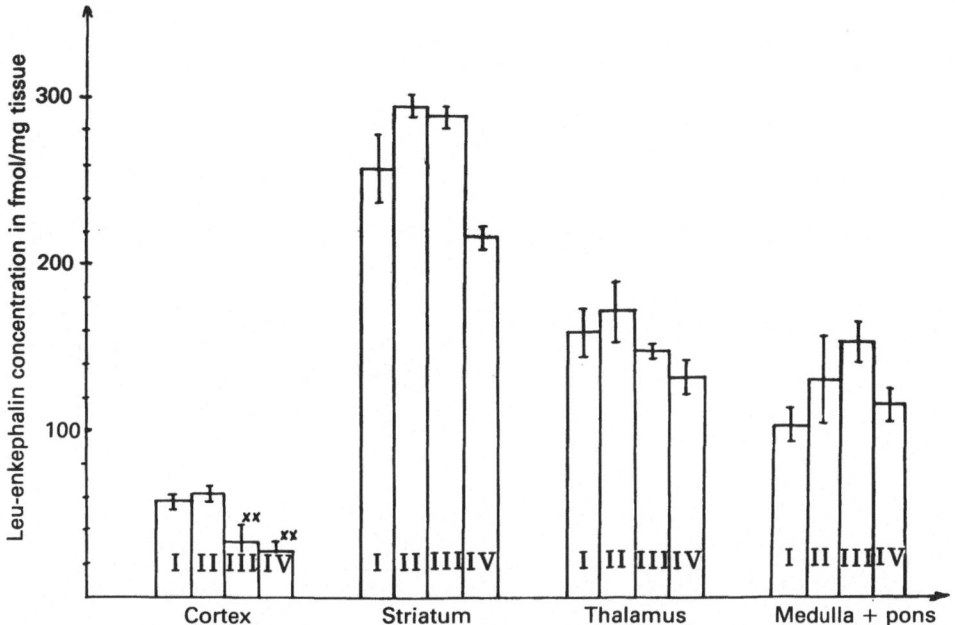

Fig. 3.   Leu-enkephalin concentration in rat brain after acute
          ethanol administration.   I - physiological saline; II -
          ethanol, 1 g/kg; III - ethanol, 2.5 g/kg; IV - ethanol,
          4.5 g/kg.   *p<0.01.

specificity of the difference between alcohol-predisposed and non-
predisposed animals in met- and leu-enkephalin levels might well
indicate that the two peptides are fulfilling different functions
within the organism.

Our analysis of β-endorphin concentration revealed a significant
rise in its cortical level in lightly-sedated rats as compared with
heavily sedated ones (see Table 1).   Furthermore, no difference
between these two groups of animals was found in striatum, thalamus
or medulla, as shown in the table.   The level of β-endorphin is known
to be considerably lower than of met- and leu-enkephalins in the
brain, although the larger peptide's functional importance as a
neuromodulator would appear to be crucial.   For instance, it produces
a wide range of effects - cataleptic, hypothermic and analgesic
(Beleslin et al., 1982; Wei et al., 1977) and is more active than the
enkephalins.   In addition, the presence of β-endorphin Σ-receptors in
the brain is thought likely (Akil et al., 1980).   The β-endorphin
present in the brain is synthesized rapidly in situ.   Its non-
pituitary origin has fostered a number of investigations employing
hypophysectomy (De Kloet, 1981).   The possibility of this peptide
being introduced from the blood can be excluded on the grounds that

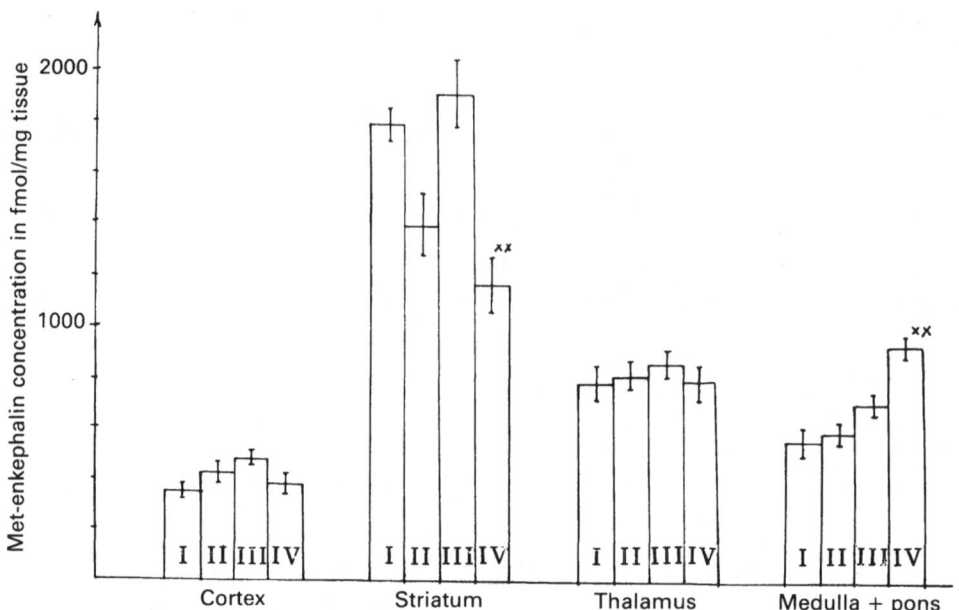

Fig. 4.   Met-enkephalin concentration after acute ethanol administration.   I - physiological saline; II - ethanol, 1 g/kg;
1II - ethanol, 2.5 g/kg; IV - ethanol, 4.5 g/kg.   *p<0.05;
**p<0.01.

its concentration does not exceed 0.5 fmol/μl plasma.   Furthermore,
if the β-endorphin measured in the brain had originated from the
plasma its concentration would be raised in all brain areas and not
just the cortex.   β-endorphin concentration in the plasma of lightly-
sedated rats was twice as high as in heavily sedated animals, at 207±
26 and 82±16 fmol/ml plasma respectively.

The duration of animals' reaction to ethanol is reflected in a
number of their neurochemical and behavioral characteristics (Brick,
1982).   This consideration led us to investigate the effects of an
acute dose of ethanol on met-enkephalin and β-endorphin concentrations in each of the above groups.   As our experiments showed, a
2.5 g/kg dose of ethanol reduced the level of met-enkephalin in the
cerebral cortex in heavily-sedated rats and left its concentration
unchanged in the same area of lightly-sedated animals (see Figure 5).
Ethanol raised met-enkephalin concentration in the thalamus and
medulla of lightly-sedated animals only.   As shown in Figure 5, the
same was true of striatum.   It could well be that acute ethanol
administration and the use of cross-bred intact animals are the
factors responsible for this complex pattern of changes in met-
enkephalin concentration.   A different picture is seen in the case of
β-endorphin.   Concentrations of this peptide increased significantly

Table 1. Neuropeptide Concentrations in Brains of Rats Divided
          According to Duration of Ethanol-induced Sedation

Peptide	Area of brain	Lightly sedated rats		Heavily sedated rats	
Leu-enkephalin	Cortex	43.5 ± 5***	(4)	19.5 ± 3.2	(5)
	Striatum	161±19	(5)	137±12.3	(5)
	Thalamus	66± 8	(5)	62.9 ±13.5	(5)
	Medulla + pons	59.7 ± 7.4	(5)	55.6 ±13	(5)
Met-enkephalin	Cortex	262±23.1	(4)	308±18	(4)
	Striatum	1520±68*	(4)	1892±103	(4)
	Thalamus	612±28.8*	(4)	772±19.4	(4)
	Medulla + pons	325±27.3**	(4)	544±16.3	(4)
β-Endorphin	Cortex	8.4 ± 0.97*	(4)	5.4 ± 0.2	(4)
	Striatum	5.4 ± 0.6	(4)	6.2 ± 0.7	(4)
	Thalamus	33.7 ± 5.0	(4)	27.4 ± 4.5	(4)
	Medulla + pons	5.7 ± 0.37	(4)	7.4 ± 1.6	(4)
DSIP	Cortex	0.94± 0.06**	(5)	1.23± 0.04	(5)
	Striatum	0.85± 0.11*	(5)	1.23± 0.06	(5)
	Thalamus	1.18± 0.22	(5)	0.85± 0.15	(5)
	Medulla + pons	1.32± 0.08	(5)	1.36± 0.13	(5)

Values given in Table – M ± SEM. No. of animals given in brackets.
Peptide concentrations in fmol/mg tissue.
*$p < 0.05$; **$p < 0.01$; ***$p < 0.005$ compared with group of animals not
predisposed to consuming alcohol.

in the cortex and remained unchanged in medulla after ethanol administration. It decreased, however, in the thalamus of heavily
sedated animals (Figure 6), while rising steeply in lightly-sedated
rats, where its level more than doubled. No change in concentration
occurred in cerebral cortex, medulla or striatum, as illustrated in
Figure 6. We noted with interest that, while differences between
lightly- and heavily-sedated animals are only seen in the cerebral
cortex before administering ethanol, this is no longer the case after
an acute dose, when β-endorphin concentration in lightly-sedated
rats rises above the value noted in heavily-sedated animal groups
(Figure 6). It is thus safe to say that differences occur between
the two groups of animals which do not appear in rats receiving
physiological saline. As far as met-enkephalin is concerned, however, the differences observed between lightly- and heavily-sedated
groups following treatment with ethanol proceeded to level out.
Likewise, leu-enkephalin concentration dropped in the cerebral cortex
after an acute dose of ethanol, as shown in Figure 3. The reverse
effect was seen in lightly-sedated animals, as may be seen by refer-

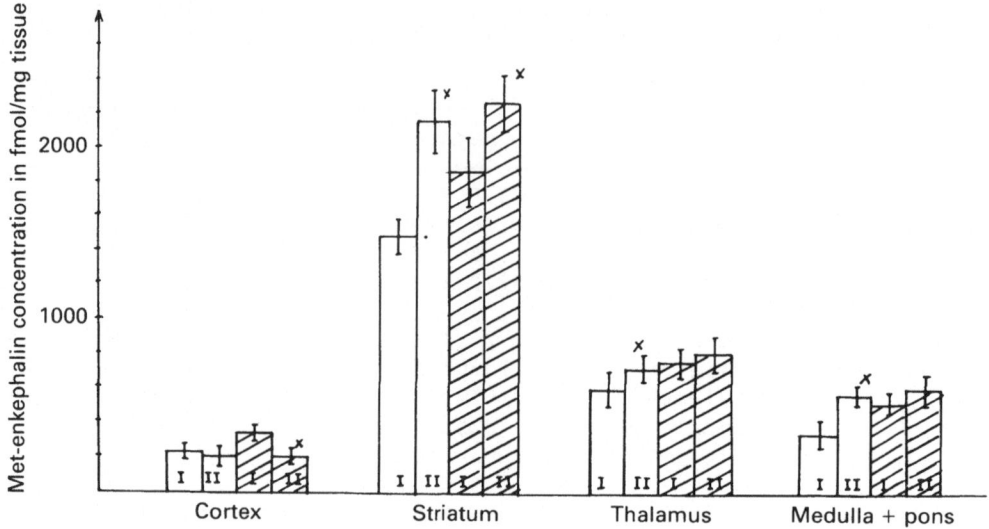

Fig. 5. Met-enkephalin concentration in rats varying in pre-
         disposition for alcohol following acute ethanol adminis-
         tration. □ - lightly sedated rats. ▨ - heavily sedated
         rats. I - physiological saline; II - ethanol, 2.5 g/kg.
         *p<0.05, compared with I.

ring to Table 1. No differences in leu-enkephalin level were found
in the other brain regions studied following acute treatment with
ethanol. The picture is the same for DSIP concentration, which rises
after a single dose of ethanol, but falls in lightly-sedated animals
(see Table 1).

3.2. MEASURING NEUROPEPTIDE CONCENTRATION IN CHRONIC ALCOHOL
         ADDICTION

     This set of experiments was concerned with the stage of chronic
experimentally-produced alcoholism (Burov, 1982). At the initial
stages of determining the need for alcohol, i.e. between the 1st and
10th day of treatment with alcohol, the rats were divided according
to whether they opted for or against a 15% ethanol solution. Those
whose ethanol solution consumption consisted of under 10% of their
total fluid requirement were considered "light" drinkers; the "heavy"
drinkers were those whose fluid consumption consisted of over 70%
alcohol solution. No differences were seen in the concentration of
DSIP, leu- and met-enkephalin or β-endorphin in these experiments for
the first 10 days after ethanol was introduced, with one minor exception (somewhat raised met-enkephalin level in the striatum of
"light" drinkers). It should be mentioned that both leu- and met-
enkephalin concentrations in the striatum, thalamus and medulla were

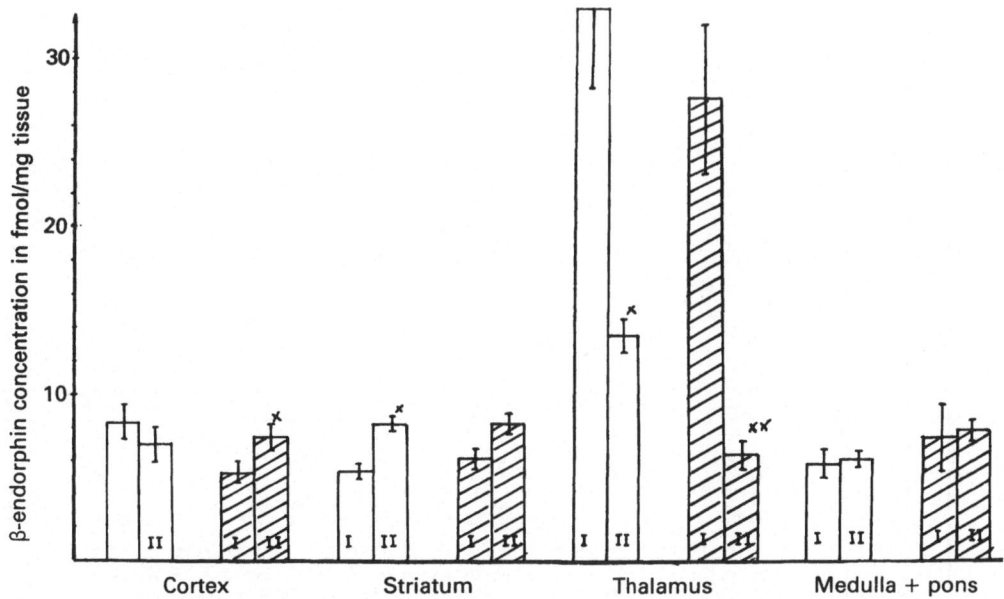

Fig. 6.   The effects of acute ethanol administration on β-endorphin
concentration in animals varying in predisposition for
ethanol. □ - lightly sedated rats; ▨ - heavily sedated
rats. I - physiological saline; II - ethanol, 2.5 g/kg.
*p<0.05; **p<0.01.

raised in each set of animals, compared with the control, while
met-enkephalin had also increased in the cerebral cortex. This could
well be due to the stress of being kept in isolation. It should be
added that hardly any difference in DSIP level was noted between the
control (without access to ethanol) and the two groups which had had
access to ethanol for 10 days.

The 2nd stage in the development of chronic alcoholism consists
of forming a preference for alcohol which is not yet physical
(alcohol) dependence (Burov, 1982). Withdrawal symptoms when deprived of alcohol are not observed in animals with a high consumption
rate from the 2nd to the 5th month of a program of elective alcohol
consumption. Our research concentrated on peptide concentrations in
animals which had been consuming alcohol for 3 months. Their consumption was recorded throughout. Rats were sorted as for stage I -
into those whose alcohol solution consumption fell below 10% or
exceeded 70% of their total fluid requirement over a given period
(i.e. "light" and "heavy" drinkers respectively). On measuring
concentrations of DSIP, β-endorphin and met-enkephalin in these two
groups of animals, no differences were found in the cortex, striatum,
thalamus or medulla (see Table 2). Leu-enkephalin level, however,
had increased significantly in the cortex of light drinkers, while no

Table 2. Peptide Concentrations in Rat Brain at Various Stages in the Evolution of Ethanol Dependence

Peptide	Region of the brain	Period of alcohol consumption	"Heavy-drinker" rats	"Light-drinker" rats
Leu-enkephalin	Cortex	10 days	37.3 ± 6.3 (4)	35.0 ± 6.5 (4)
		3 months	35.5 ± 1.2** (3)	58.5 ± 3.2 (3)
		10 months	50.5 ± 2.1** (6)	29.5 ± 3.1 (6)
		withdrawal	41.6 ± 6.2 (3)	
	Striatum	10 days	296±24 (4)	222±26.8 (4)
		3 months	105.5 ±11.8 (3)	150±19.8 (3)
		10 months	102±13.4* (6)	192.4 ±11.3 (6)
		withdrawal	173±12.1 (4)	
	Thalamus	10 days	99.2 ± 5.6 (4)	131.8 ±22.7 (4)
		3 months	55±12.1 (3)	63±10.4 (3)
		10 months	82± 5.2** (6)	123.2 ± 4.9 (6)
		withdrawal	86.5 ±10.4* (3)	
	Medulla + pons	10 days	90.5 ±15.7 (4)	134±22 (4)
		3 months	81.5 ± 3.8 (3)	93.5 ±10.7 (3)
		10 months	87.5 ± 3.7** (6)	111.5 ± 2.1 (6)
		withdrawal	94± 1.3 (3)	
Met-enkephalin	Cortex	10 days	450±23 (4)	444±13 (4)
		3 months	370±16* (4)	310±20 (4)
		10 months	450±30 (6)	390±38 (5)
		withdrawal	475±27 (3)	
	Striatum	10 days	2540±126* (4)	3200±180 (4)
		3 months	750±40 (3)	610±76 (3)
		10 months	630±14** (6)	1250±155 (7)
		withdrawal	950±80 (3)	
	Thalamus	10 days	930±23 (4)	1210±125 (4)
		3 months	470±35 (3)	495±11 (3)
		10 months	400±35** (7)	850±56 (6)
		withdrawal	530±72** (3)	

Peptide	Region			
	Medulla + pons	10 days	940±77 (4)	1130±98 (4)
		3 months	540±32 (3)	600±55 (3)
		10 months	485±28** (7)	810±48 (7)
		withdrawal	640±72 (3)	
β-Endorphin	Cortex	3 months	6.6 ±0.75 (4)	5.7 ± 0.61 (4)
		10 months	2.2 ±0.22** (6)	4.6 ± 0.6 (5)
		withdrawal	1.4 ±0.1*** (3)	
	Striatum	3 months	6.6 ±0.75 (4)	5.7 ± 0.61 (4)
		10 months	4.7 ±0.65** (5)	12.2 ± 1.6 (5)
		withdrawal	5.9 ±0.8* (3)	
	Thalamus	3 months	18.5 ±2.2 (3)	15.3 ± 3.2 (3)
		10 months	12.9 ±1.48* (5)	8.5 ± 0.74 (5)
		withdrawal	17.0 ±3.2* (3)	
	Medulla + pons	3 months	7.1 ±0.26 (3)	8.2 ± 0.87 (3)
		10 months	1.9 ±0.2* (5)	3.0 ± 0.4 (3)
		withdrawal	1.2 ±0.2* (3)	
DSIP	Cortex	10 days	1.0 ±0.12 (4)	1.03± 0.23 (4)
		3 months	0.92±0.1 (3)	1.2 ± 0.17 (3)
		10 months	0.75±0.16 (6)	1.45± 0.14 (6)
		withdrawal	0.69±0.13** (3)	
	Striatum	10 days	0.75±0.14 (4)	1.35± 0.11 (4)
		3 months	1.62±0.17 (3)	1.31± 0.14 (3)
		10 months	0.54±0.05* (7)	1.0 ± 0.03 (8)
		withdrawal	0.61±0.06* (3)	
	Thalamus	10 days	1.04±0.04 (4)	1.14± 0.03 (4)
		3 months	0.8 ±0.24 (3)	0.9 ± 0.21 (3)
		10 months	0.68±0.13* (7)	1.23± 0.05 (8)
		withdrawal	0.72±0.16* (3)	
	Medulla + pons	10 days	1.10±0.15 (4)	1.68± 0.2 (4)
		3 months	1.16±0.19 (3)	1.4 ± 0.1 (3)
		10 months	0.68±0.13 (7)	1.23± 0.05 (8)
		withdrawal	0.72±0.13* (7)	

Peptide concentrations in fmol/mg tissue. Values given in table – M ± SEM. No. of animals given in brackets. *p<0.05; **p<0.01, compared with "light drinkers"; ***p<0.05, compared with "heavy drinkers". Peptide concentration during withdrawal was measured 24 hours after ethanol had been removed from "heavy drinking" animals.

difference was observed in the striatum, the thalamus or the medulla
in either group, as Table 2 illustrated. A third stage of experimentally-induced chronic alcoholism could be distinguished in our
experimental model - that of confirmed addiction to alcohol together
with physical dependence. During this stage, which set in after a
7-month program of alcohol consumption, animals showed signs of
withdrawal when deprived of ethanol. Disturbed sleep pattern was one
of the clearest withdrawal symptoms (Viglinskaya, 1980). Our experiments in this area used animals which had had access to ethanol for
10-11 months. They were divided into light and heavy drinkers, as
before. Peptide levels were measured 24 h after withdrawal from
ethanol, when withdrawal symptoms were at their height (Viglinskaya,
1980). Reduced DSIP concentration was observed in the striatum and
thalamus of heavy drinkers as compared with light drinkers after a
10-month period of alcohol consumption (see Table 2). As seen from
this table, some increases were also noted in DSIP concentration
during withdrawal, but levels hardly varied between the two sets of
animals in any of the brain regions tested. DSIP concentration in
the striatum during withdrawal reached a level significantly below
that of light drinkers. DSIP concentration in light drinkers, in
fact, remained the same as in the earlier stages of the development
of ethanol dependence for this group. At this 3rd stage of development of experimentally-induced alcoholism, leu-enkephalin concentration was significantly reduced in medulla, striatum and thalamus
and substantially increased in cerebral cortex in heavy drinkers. A
significant rise in concentration of this peptide in the striatum was
also observed during withdrawal compared with the level in heavy
drinkers (see Table 2). In this group, no statistically significant
difference in leu-peptide concentration compared with heavy drinkers
was seen in cerebral cortex, thalamus or medulla. At this 3rd stage
of experimentally-induced alcoholism, met-enkephalin concentration
was much reduced in striatum, medulla and thalamus of heavy drinkers,
while remaining virtually unchanged in both groups in the cerebral
cortex (see Table 2). A considerable increase in met-enkephalin
occurs in the striatum and medulla during withdrawal as compared with
heavy drinkers. At the same time, the level of this peptide in the
striatum of heavy drinkers during withdrawal hardly differs from that
of light drinkers. In the medulla, the level remains low in relation
to that of light drinkers. No significant increase in concentration
occurs in the hypothalamus during withdrawal.

    The similarity between changes in met- and leu-enkephalin in the
striatum and medulla cannot be attributed to crossed immunorespon-
siveness between met- and leu-enkephalin, which does not, in fact,
exceed 1.5%. It should also be noted that leu-enkephalin concentration is raised in the cerebral cortex in heavy drinkers, while
this group shows a met-enkephalin level comparable to that of light
drinkers. Similarly, β-endorphin concentration is significantly
reduced in the cerebral cortex, striatum and medulla of heavy
drinkers. Furthermore, withdrawal from ethanol is not accompanied by

an increase in this peptide's concentration in any of the brain sites listed earlier; in fact, it drops considerably in both cortex and medulla (see Table 2). β-endorphin concentration is significantly raised in the thalamus of heavy drinkers and continues to rise during withdrawal until reaching double the level to be seen in light drinkers.

It has already been established that the degree of craving for alcohol may depend on how the main histocompatibility system in mice (H-2) and man (HLA) is composed. We now set out to find whether neuropeptide levels could be governed by this system. Congenitally resistant strains of mice served as models in our studies of neuro-peptide concentrations. Two pairs of mice of the congenitally resis-tant strains B 10. R 107 and B 10. R III were consuming ethanol when a choice was made available, while another two strains (A/Sn and A.SW rejected it. Strain B 10. R 107 consumed considerably more alcohol than B 10. R III (see Figure 7). Peptide concentrations were measured in all four strains. Considerably higher β-endorphin con-centrations were found in the drinking than in the non-drinking strains, but no difference in this respect between the two drinking strains (see Table 3). Leu-enkephalin concentration was the same in the brains of all four strains investigated, while the B 10. R 107 (heavy drinking) strain showed a reduced met-enkephalin level com-

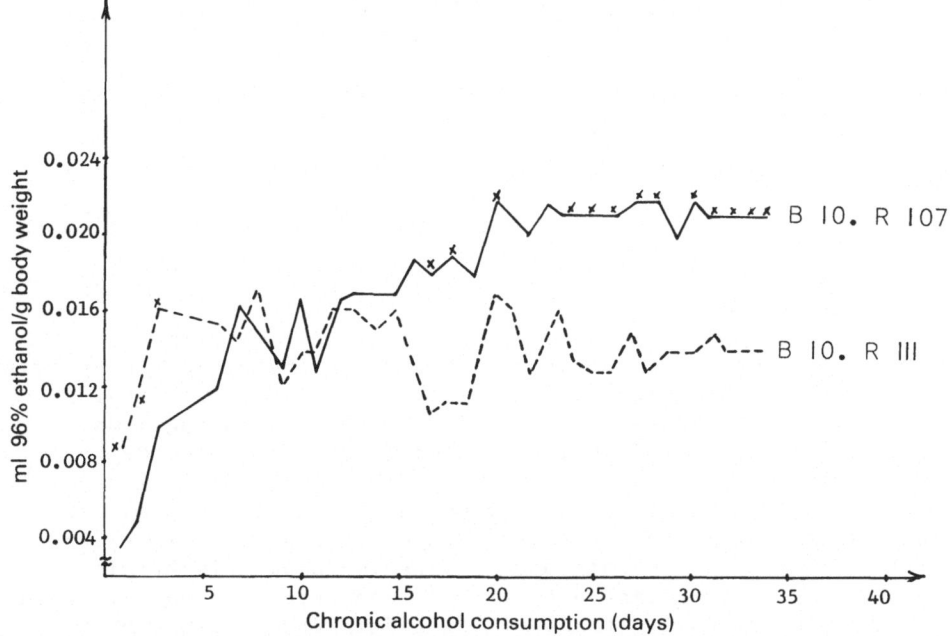

Fig. 7.   Alcohol consumption of mice belonging to congenitally-resistant strains in a state of voluntary chronic alcohol dependence.

Table 3. Neuroleptic Concentrations in Whole Brain of Pure-bred Mice
after 3 Months with Access to Alcohol

	B 10. R107	B 10. RIII	A.SW	A/Sn
Leu-enkephalin				
control	88.6 ±10.2 (5)	81.2± 6.0 (5)	76.3±14.5 (4)	81.5±12 (4)
3 months ethanol	53.5 ± 9.5 (4)	118.9±8 (4)	–	–
DSIP following 7 day treatment with ethanol	132±24* (4)	109±10.5 (4)	–	–
Met-enkephalin				
control	670.3 ±102 (5)	997±87* (5)	820±60 (4)	961±85 (4)
ethanol	880±80 (4)	1180±75 (4)	–	–
ethanol + DSIP	920±85 (4)	1110±86 (4)	–	–
β-Endorphin				
control	2.4 ± 0.15 (4)	1.9± 0.11 (4)	5.5± 0.36* (4)	4.3±0.9* (4)
ethanol	1.9 ± 0.85 (4)	2.1± 0.15 (4)	–	–
DSIP				
control	1.01± 0.04 (4)	2.1± 0.06 (4)	2.5± 0.02* (6)	2.3±0.05 (5)
ethanol	1.9 ± 0.1 (4)	1.5± 0.08 (4)	–	–

Above values – M ± SEM, in fmol/mg tissue. No. of animals given in
brackets. *p<0.05 between strains.

pared with the light drinkers B 10. R III or the non-drinkers A/Sn.
DSIP concentration was significantly higher in the non-drinking strain
A.SW compared with drinking stains, as shown in Table 3. After a
3-month period with access to ethanol, the heavy-drinking mouse
strain B 10. R 107 displayed a considerable drop in leu-enkephalin
concentration, which rose, however, in strain B 10. R III (Table 3).
Concentration of met-enkephalin did not differ significantly from
control in mice of either drinking strain. β-endorphin level was
reduced in B 10. R 107 strain and remained unchanged in strain
B 10. R III. A significant drop in DSIP concentration was likewise
noted in the B 10. R III strain of light drinkers. When a 60 nmol
dose was administered i.p. over 7 days to drinking animals which had
had access to ethanol over a 3-month period, a drop in consumption of
10% ethanol solution ensued. When peptide levels were measured after

7 days' DSIP administration, the B 10. R 107 strain of heavy drinkers
showed a considerable increase in leu-enkephalin concentration,
which remained unchanged in the brain of B 10. R III strain, however.
These factors may indicate that DSIP is changing peptide concentrations, as well as exerting an effect on catecholamine and
serotonin metabolism. It is likewise possible that the increase
in leu-enkephalin concentration is mediated through the catecholaminergic system, since activating dopaminergic neurones in the
substantia nigra is known to produce a rise in striatal enkephalin
concentration (Schwartz et al., 1961). It was also noted that
7 days' treatment with DSIP did not affect met-enkephalin concentration in either heavy or light drinkers.

4. DISCUSSION

Marked changes in peptide concentrations were found in rat brain
after both acute and chronic treatment with ethanol. This suggests
that these biologically active molecules help to shape predilection
for and dependence on ethanol. The considerable rise in DSIP concentration occuring after acute administration could be associated with
the increase in slow-wave activity produced by ethanol (Lehtinen et
al., 1981). DSIP is also known to affect noradrenaline, dopamine and
serotonin concentrations (Graf et al., 1981). It was shown to increase the activity of mitochondrial monoamine oxidase type A. This
enzyme in its turn is largely responsible for metabolizing serotonin
(Ashmarin et al., 1980). It is highly likely that a number of ethanol's effects on the concentration of monoamines and serotonin are
achieved via changes in DSIP concentration. The view that ethanol
and DSIP interact with the monoaminergic system is supported by the
finding that both are able to reverse the increase in locomotor
activity produced by d-amphetamine (Yebudaeta, 1980; Duncan. 1981).
Amphetamine supposedly achieves its effects by increasing dopamine
and noradrenaline release. DSIP cancels out this effect, while
activating the serotoninergic system.

There is now much evidence to show that DSIP can exert an antidepressant action in addition to its effects on slow-wave activity
(Barbely et al., 1980; Kastin et al., 1980). The fact that DSIP
concentration is reduced in animals predisposed to alcohol consumption means either that DSIP supplies have been exhausted due to a
lightly-sedated animal's greater susceptibility to stress, or that
the animals were deficient in DSIP at the outset and compensated for
the deficit by consuming alcohol. The fact that concentrations are
higher in strains of mice refusing alcohol than in alcohol-drinking
strains when a choice is available would support the second of these
suggestions (see Table 3). The drop in striatum and cortex of
lightly sedated rats - i.e. the very parts of the brain most involved
in the nervous system's response to stress - may be a significant
factor in the pathogenesis of alcoholism. It was actually found that

DSIP activated type A monoamine oxidase, particularly in the striatum (Ashmarin et al., 1980). DSIP concentration evens out between drinkers and non-drinkers when the animals are brought into contact with ethanol during the first two stages of experimentally-induced alcoholism - possibly a compensatory effect. Concentration then decreases considerably as substantial dependence on ethanol develops, which may be due to the inhibiting action of ethanol on protein synthesis (Poto et al., 1980) or a rise in the compensatory capacity of the organism during the 10-month period of alcohol consumption. Since DSIP produces an increase in EEG slow wave activity (Kafi et al., 1979), and sleep impairment is one of the most clear-cut symptoms of alcohol withdrawal (Viglinskaya, 1980), a reduction in concentration during withdrawal could trigger the appearance of impaired sleep, which drinking animals compensate for by consuming alcohol. A drop in DSIP concentration accompanying prolonged alcohol consumption may constitute one of the mechanisms by which tolerance develops, even though the animals may be consuming considerable quantities of alcohol. While an acute dose of ethanol sharply increases DSIP concentration, its level is reduced by chronic administration of alcohol solution. It should be mentioned that similar changes in DSIP concentration were noted to a greater or lesser extent in all parts of the brain studied. Its action would therefore appear to be equal for different areas of the CNS. No significant differences in DSIP concentrations were observed when distribution of this peptide in rat brain was investigated (Kastin et al., 1978). It may therefore be assumed that DSIP most affects serotonin metabolism in the brain, the involvement of which in sleep processes and shaping alcohol dependence has often been demonstrated (Burov,1982). The possibility should not be excluded of some link between reduced serotonin level in rats predisposed to alcohol (Zhukov et al., 1982),the rise in its concentration and the drop in 5-HIAA level produced by chronic ethanol administration (Sytinsky, 1980) and accompanying changes occurring in DSIP concentration. It can be partially confirmed that DSIP is involved in shaping alcohol dependence, most probably due to the anti-stress effects which ethanol also possesses (Burov, 1982). The involvement of enkephalins and endorphins in the mechanisms of stress has been fairly fully discussed. Evidence had been produced that naloxone eliminates the analgesia induced by stress (R. Przewlocki et al., 1980), but that stress increases β-endorphin concentration in plasma and various areas of the brain (Barta et al., 1981). While on the subject of enkephalinergic system involvement in the pathogenesis of alcoholism, it should be recalled that leu- and met-enkephalin possess a very wide range of functions, amongst other things influencing emotional state and positive reinforcement systems, learning processes, memory and other complex elements of behavior (Olson et al., 1978). The effects of opiate peptides depend on the part of the brain in which they function. If β-endorphin is administered locally into the stratum, for instance, an increase in motor activity is produced, whereas it induces catalepsy when administered into the aqueductal

gray matter. A similar phenomenon occurs in the case of enkephalins.
Met-enkephalin produces analgesia when administered into the thalamus
but characteristically brings on hyperalgesia when injected into the
periaqueductal gray matter (Nicoll et al., 1977; Frenk et al., 1978).
The contradictory nature of enkephalins' effects in different parts
of the brain are attributed to the existence of several different
types of opiate receptor (Zukin et al., 1981). An acute dose of
ethanol brings about a reasonably well differentiated response to
enkephalin concentrations seen in different parts of the brain.
Leu-enkephalin level only drops in the cerebral cortex, where delta
opiate receptors predominate. These are involved in stimulating
motor activity and producing an epileptiform EEG (Dzolyic et al.,
1982). It is interesting to note that leu-enkephalin concentration
is more than doubled in the cortex of rats disposed to alcohol, a
fact which may be connected with a higher response level in this
group of animals. It has also been shown that ethanol reacts preferentially with β-receptors (Hiller et al., 1981). Differences in
leu-enkephalin concentrations produced by chronic alcohol consumption
in heavy and light drinkers is again only noted in the cerebral
cortex, in the form of a reduced level after 3 months with access to
alcohol and a rise after a 10-month period of alcohol consumption.
In the remaining parts of the brain investigated, however, a decrease
in concentration of this peptide occurs after the 10-month program;
this could simply be tied up with inhibition of protein synthesis
during the period of chronic alcohol administration. Similarly, the
concentration of met-enkephalin is considerably reduced in the
medulla and striatum following 10 months with access to ethanol, but
somewhat raised in the cortex. After ethanol is withdrawn, the level
of both peptides rises - slightly more so in the striatum. This
could be caused by the onset of stress associated with recent withdrawal. In addition, leu-enkephalin in the medulla and thalamus and
met-enkephalin concentration in the thalamus remain significantly
reduced in animals during withdrawal, in comparison with light
drinkers. The fairly specific increase in neuropeptide concentration
in the striatum itself may be connected with its particular role in
the shaping of withdrawal. The enkephalins are known to regulate the
state of dopaminergic terminals in the striatum, thus acting as presynaptic modulators (Pollard et al., 1978), and the state of withdrawal is closely connected with activation of the dopaminergic
systems (Apokhina, 1979).

   When rats with varying susceptibility to chronic alcoholism are
subjected to acute administration of ethanol, the concentration of
met-enkephalin in the medulla, thalamus and striatum rises considerably in rats predisposed to ethanol. Met-enkephalin level also
rises in non-predisposed animals, but a good deal less. Treatment
with diazepam also produces a rise in met-enkephalin concentration in
the striatum (Wüster et al., 1980). This suggests that met-enkephalin's anxiolytic effects are actually achieved via the striatum.
The far greater rise in met-enkephalin concentration in predisposed

rats compared with non-disposed animals may be explained by the fact
that the hypothermic action of ethanol is much more effectively
eliminated by naloxone in the former group (Brick et al., 1982).
Met-enkephalin concentration is also reduced in the striatum and
thalamus of the same group, possibly reflecting something of a
deficit in the positive reinforcement system, compensated for by
ethanol consumption during the first two stages of the development of
ethanol dependence. After 10 months of alcohol consumption, decom-
pensation of both the enkephalinergic and other neurotransmitter
systems appears to take place (Sytinsky, 1980) and ethanol dependence
develops.   β-Endorphin concentration, which is raised in the cortex
and the plasma of animals predisposed to ethanol consumption, may be
a factor in the organism's general response to stress (Barta et al.,
1981). It is an interesting fact that treatment with ethanol considerably reduces the concentration of β-endorphin in the thalamus.
Furthermore, although no differences in concentration of the peptide
were seen in that part of the brain until ethanol was administered,
the concentration was more than doubled in predisposed rats following
injection of alcohol solution. Thalamic β-endorphin level rises
above that of light drinkers in heavy drinkers when dependence
develops, and a further rise in its concentration is seen in the
same region during withdrawal. It is difficult to explain such an
unpredicted rise in β-endorphin in the thalamus as distinct from
levels of other peptides. Their concentration, in fact, drops in
heavy-drinking animals over prolonged periods of ethanol consumption.
In our opinion, the rise in thalamic β-endorphin concentration and
increased leu-enkephalin in the cerebral cortex seen during all 3
stages of experimentally-induced alcoholism could be connected with
the development of tolerance to the effects of ethanol produced by a
drop in the concentration of these peptides, as observed following
acute administration of ethanol. Leu-enkephalin is known to react
primarily with δ-opiate receptors, while β-endorphin reacts equally
with all types of opiate receptors (Zukin et al., 1981). When the
agonists react with the δ-receptors of opiates, myoclonic convulsions
develop and high-frequency activity increase, as does locomotion
(Dzolyic et al., 1982). Dysphoria may be produced when opiates react
with these receptors. It has further been shown that leu-enkephalin
increases during convulsions induced by stimulating the amygdala at
the cerebral cortex (Nindrola et al., 1981). Myoclonic convulsions
are produced by administering β-endorphin into the third ventricle
(Beleslin et al., 1982). An increase in the amount of thalamic
β-endorphin and of leu-enkephalin in the cerebral cortex during
withdrawal may mark the genesis of convulsions and of the increased
motor activity occurring during ethanol withdrawal.

We may therefore state that dynamic changes in concentration of
the enkephalins, β-endorphin and DSIP occur following acute treatment
with ethanol and chronic alcohol consumption - an indication that
these central transmitters are involved in shaping ethanol dependence. Furthermore, differences in concentrations of these peptides

between rats disposed and non-disposed to alcohol consumption, as well as pure strains of mice showing varying susceptibility to alcohol, confirm the significance of these factors in the pathogenesis of alcoholism and the determination of genetic predisposition to ethanol consumption. In fact, largely contradictory results were obtained on the functioning of the enkephalinergic system; this system's connection with the development of alcoholism has been discussed by a number of authors (Blum et al., 1977). More particularly, some authors point out that naloxone eliminates the analgesia produced by stress (Booda et al., 1981), and others - that this drug has no effect on ethanol-induced stress (Berkowitz et al., 1977). Studies of ethanol's effects on the development of morphine and opiate dependence and how this affects ethanol dependence have also proved contradictory. It was found that ethanol can eliminate the symptoms of morphine withdrawal (Blum et al., 1976), while morphine alleviates withdrawal when ethanol-dependent animals are actually taken off ethanol (Jones et al., 1977). Other workers observed no connection between morphine and ethanol dependence (Miceli et al., 1980; Goldstein et al., 1971). These widely diverging results make more sense if we bear in mind:

a) the specificity of neuropeptides in different parts of the brain,

b) the existence of different types of opiate receptor and

c) our observation that ethanol affects enkephalin and endorphin concentration differently in different areas of rat brain.

Further clarification of how ethanol changes neuropeptide concentrations is required. We may at least presume that changes in peptide biosynthesis may be occurring, as well as in their secretion and metabolism, as they do in a number of neurotransmitters under the influence of ethanol (Kalan, 1975; Burov, 1982). However, the pathways and methods of peptide metabolism regulation have not yet been studied in sufficient depth to justify any fixed conclusions at this stage. It should be mentioned, however, that even prolonged stimulation of isolated synaptosomes changes enkephalin stores by no more than 30% (Lindberg et al., 1981).

The changes we have observed in peptide concentrations point the way to pathogenetically-based therapy for alcoholism, taking DSIP and the state of the enkephalinergic system into account. Compensatory mechanisms for the pathological changes in concentrations of these peptides could form the basis for correcting homeostatic mechanisms damaged by chronic alcohol consumption.

REFERENCES

Akil, H., Hewlett, W. A., Barchas, J. D., and Li C. H., 1981, Binding
    of ^{3}H-endorphin to rat brain membranes; characterization of
    opiate properties and interaction with ACTH, Eur.J.Pharmacol,
    63:1-8.
Anokhina, I. P., 1979, Neurochemical aspects of pathogenesis of
    alcohol and drug dependence, Drug and Alcohol Dependence,
    4:215-273.
Ashmarin, I. P., and Dovedova, E. I., 1980, Delta-sleep peptide
    influence on the activity of acetylcholinesterase and
    monoamine oxidase in synaptosomes and mitochondria of rabbit
    brain, Rep.Acad.Med.Sci., USSR, 255:1501-1503.
Barta, A., and Yashpal, K., 1981, Regional redistribution of
    β-endorphin in the rat brain: the effect of stress,
    Progr.Neuropsychopharmacol., 5:595-599.
Barbely, A. A., Tobler, J., and Berkowitz, B. A., 1977, Nitrous oxide
    analgesia: reversal by naloxone and development of tolerance,
    J.Pharmacol.Exp.,Ther., 203:539-547.
Beleslin, D. B., Samardzic, R., Krstic, S. K., and Micci, D., 1982,
    Differences in central effects of β-endorphin and enkephalin,
    Neuropharmacology, 21:99-102.
Blum, K., Wallace, J. E., and Schuster, H. A., 1976, Morphine sup-
    pression of ethanol withdrawal in mice, Experientia, 32:79-82.
Blum, K., Hamilton, M. G., and Wallace, J. E., 1977, Alcohol and
    opiates: a review of common neurochemical and behavioural
    mechanism, in: "Alcohol and Opiates. Neural and Behavioral
    Mechanisms," K. Blum, ed., Academic Press, New York,
    pp.203-236.
Booda, I., Feria, M., and Sanz, E., 1981, Inhibitory effects of
    naloxone on the ethanol induced antinociception in mice,
    Pharmacol.Res.Comm., 13:673-679.
Brick, I., and Horowitz, G. P., 1982, Alcohol and morphine induced
    hypothermia in mice selected for sensitivity to ethanol,
    Pharmacol.Biochem.Behav., 16:473-479.
Burov, Yu. V., Zhukov, V. N., and Kampov-Polevoy, A. B., 1980,
    Methodological Recommendations on Experimental Pharmacological
    Study of the Drugs Proposed for Treatment and Prophylaxis of
    Alcoholism, Moscow.
Burov, Yu. V., Absava, G. I., Kampov-Polevoy, A. B., and Klyuev, S.
    M., 1981, Elimination of ethanol in white rats with different
    levels of alcohol motivation, Pharmacol.Toxicol., No.1,
    pp.50-52.
Burov, Yu. V., and Viglinskaya, I. V., 1981, Influence of psycho-
    tropic drugs on sleep disturbances during the period of
    alcohol deprivation in rats, Bull.Exp.Biol.Med., No.6,
    pp.689-691.
Burov, Yu. V., 1982, Searching for pharmacological agents to treat
    alcoholism, Bull.USSR Acad.Med.Sci., No.5, pp.72-77.

Burov, Yu. V., Yukhananov, R. Yu., and Maisky, A. I., 1982, The content of Delta-sleep-inducing peptide in the brain of rats with different levels of alcohol motivation, Bull.Exp.Biol.Med., No.9, pp.67-69.

Carlsson, A., Magnusson, T., Svensson, T. H., and Walder, B., 1973, Effect of ethanol on the metabolism of brain catecholamine, Psychopharmacologia, 30:27-36.

Duncan, P. M., 1981, Ethanol-amphetamine interaction effect on spontaneous motor activity and fixed interval responding, Psychopharmacology, 74:256-259.

Dzolyic, M. R., and Heisterkamp, A. L. P., 1982, Delta opiate receptors are involved in the endopioid-induced myoclonic contractions, Brain Res.Bull., 8:1-6.

Frenk, H., McCarty, B. C., and Liebsking, K., 1978, Different brain areas mediate the analgesic and epileptic properties of enkephalin, Science., 200:335-337.

Glowinsky, I., and Iverson, L. L., 1966, Regional studies of catecholamines in the rat brain, J.Neurochem., 13:655-669.

Goldstein, A., and Judson, B. C., 1971, Alcohol dependence and opiate dependence: lack of relationship in mice, Science., 172:290-292.

Goldstein D. B., and Chia, I. H., 1981, Interaction of ethanol with biological membranes, Fed.Proc., 40:2073-2076.

Graf, M., Christen, M., Tobler, H. I., and Baumann, I. P., 1981, DSIP, a circadian programming substance, Experientia, 47:624-625.

Goodman, R. R., Snyder, S. H., Kuhar, M. I., and Young III, W. S., Differentation of delta opiate receptor localization by light microscopic autoradiography, Proc.Nat.Acad.Sci.,US., 77:6239-6243.

Hiller, J. M., Angel, L. H., and Simon, E. J., 1981, Multiple opiate receptors; alcohol selectively inhibits binding to delta receptors, Science., 214:468-469.

Iwamoto, E. I., and Martin, W. R., 1981, Multiple opiate receptors, Med.Res.Rev., 1:411-440.

Jones, M. A., and Sprato, G., 1977, Ethanol suppression of naloxone induced withdrawal in morphine dependent rat, Life Sci., 20:1549-1557.

Kafi, S., Monnier, M., and Galland, F. M., 1979, The delta-sleep inducing peptide (DSIP) increases duration of sleep in rats, Neurosci.Lett., 13:169-172.

Kalant, H., 1975, Direct effects of ethanol on the nervous system, Fed.Proc., 34:1930-1941.

Kalant, H., and Rengurai, M., 1981, Interactions of catecholamines and ethanol of rat brain (Na$^+$-K$^+$) ATPase, Eur.J.Pharmacol., 70:157-166.

Kastin, A. I., Nissen, C., Schally, A. V., and Coy, D. H., 1978, Radioimmunoassay of DSIP-like material in the rat brain, Braina Res.Bull., 3:691-695.

Kastin, A. I., Olson, G. A., and Schally, A. V., 1981, DSIP - more
    than a sleep peptide, Trends Neurol.Sci., 5:91-93.
De Kloet, E. R., Paekovits, M., Mezey, E., 1981, Opiocortin peptides:
    localization, source and avenues of transport,
    Pharmacol.Ther., 12:325-351.
Lindberg, I., and Dalil, J. L., 1981, Characterization of enkephalin
    release from rat striatum, J.Neurochem, 36:506-512.
Lehtinen, I., Long, A. H., Jänthe, V., and Pukkonen, A., 1981,
    Ethanol induced disturbance in human arousal mechanism,
    Psychopharmacology, 73:223-229.
Lumeng, L., Waller, M. D., Medride, W. I., and Li, T.-K., 1982,
    Different sensitivity to ethanol in alcohol preferring and
    nonpreferring rats, Pharmacol.Biochem.Behav., 16:125-130.
Miceli, D., Marfaing-Jallat, P., and Magnen, I., 1980, Failure of
    naloxone to affect initial and acquired tolerance to ethanol
    rats, Eur.J.Pharmacol., 63:327-332.
Nicoli, R. A., Siggins, G. R., Ling, N., Bloom, F. E., and Guilemin,
    R., 1977, Neuronal actions of endorphins and enkephalins among
    brain regions: a comparative microiontophoretic study.
    Proc.Nat.Acad.Sci.,US., 74:2584-2588.
Olson, I. A., and Olson, R. D., 1978, Endogenous opiates through
    1978, Neurosci.Biobehav.Rev., 3:285-299.
Pollard, H., Llorens, C., and Schwartz, I. C., 1978, Localization of
    opiate and enkephalin in the striatum in relationship with
    nigrostriatal dopaminergic system, Brain Res., 151:392-398.
Poto, H., and Ros, A. S., 1981, Inhibition of PNS and protein
    synthesis by ethanol in regenerative rat liver: evidence for
    transcriptional inhibition of protein synthesis,
    Acta Pharmacol.Toxicol., 49:125-130.
Przewlocki, R., Millan, M. I., and Herz, A., 1980, β-endorphin
    involved in the analgesia generated by stress, endogenous and
    exogenous opiate agonists and antagonists, Proc.Int.Narcotic
    Res., North Falmouth, Mass, 1979, No.4, pp.391-394.
Schoenenberger, G. A., Guam, L. P., Halt, A. M., and Monnier, M.,
    1972, Isolation and physicochemical characterization of a
    humoral sleep inducing substance in rabbit (delta factor),
    Experientia, 28:919-921.
Schvartz, J., Pollard, M., and Lloran, S., 1981, Endorphines and
    endorphine receptors in striatum, connection with dopaminergic
    neurones, Moscow, Mir.
Sitinsky, I. A., 1980, Biochemical principles of ethanol action on
    CNS. Moscow.
Stolerman, J. P., Kumar, R., and Steinberg, H., 1971, Development of
    morphine dependence in rats: lack of effects of previous
    ingestion of other drugs, Psychopharmacologia, 20:321-336.
Tewari, S., Stewart, R., and Fleming E. W., 1981, Ethanol induced
    changes in the transpeptidation reactions on brain ribosomes,
    Res.Comm.Subst.Abuse, 2:93-94.
Windrola, O., Brioness, R., Asal, M., and Fernandez-Guardiola, A.,
    1981, Brain content of Leu-[5] and Met-[5]enkephalins changes

independently during the development of kindling in the rat, Neurosci.Lett., 26:125-130.

Wei, E. T., Tseng, L. F., Loh, H. H., and Lich, H., 1977, Comparison of the behavioural effects of β-endorphin and enkephalin, Life Sci., 21:321-328.

Wüster, M., Duka, T., and Herz, A., 1980, Diazepam effects on striatal metenkephalin levels following long-term pharmacological manipulation, Neuropharmacology., 19:501-505.

Yehuda, S., Kastin, A. J., and Coy, D. H., 1980, Thermoregulatory and locomotor effects of DSIP: paradoxical interaction with d-amphetamine, Pharmacol.Biochem.Behav., 13:895-900.

Zhukov, V. N., Khodorova, N. A., and Burov, Yu. V., 1982, Serotonin content in different parts of brain, in the liver, intestine and blood of rats predisposed and not predisposed to alcohol consumption, Bull.Exp.Biol.Med., No.7, pp.35-37.

Zukin, R. S., and Zukin, S. R., 1981, Multiple opiate receptors: emerging concepts, Life Sci., 29:2681-2696.

Endocrine Factors in the Pathogenesis of Experimentally-induced Alcoholism

Yu.V. Burov and N.N. Vedernikova

Institute of Pharmacology, Academy of Medical Sciences
of the USSR, Moscow

1. INTRODUCTION

The development of experimentally-induced alcoholism is generally attended by pronounced evidence of endocrinopathy. Peptide and
steroid hormones so involved are widely recognized as joint regulators of various forms of animal and human behavior, apart from
their more specific functions. Such considerations go a long way
towards explaining the abundance of etiological and pathogenetic
"endocrine" theories which link a certain illness with ethanol-
induced dysfunction of a) peripheral hormone-producing organs, such
as gonads, adrenals, thyroid and pancreas and b) those acting centrally, such as pituitary and hypothalamus (Sze, 1977; Van Thiel,
1980; Greene and Hollander, 1980). Endocrine research, as applied
to the problem of experimentally-induced alcoholism, follows two
basic lines of approach. The first consists of studying the changing
pattern of concentration and function of hormones with different
chemical structures during the creation of the alcoholism model. The
second involves analyzing the effects of hormones and their derivatives on the differing effects of ethanol - on, for example, the
voluntary ethanol consumption of experimental animals, or the build-
up of tolerance and physical dependence when ethanol is administered
chronically.

These two approaches are evidently closely linked, a fact which
enables us to compare the function of endogenous and exogenously
administered hormones during the creation of the alcoholism model.
This article is intended a) to present findings on hormonal characteristics of rats with varying propensity for alcohol and b) to
assess any potential correlations between the course of alcoholism
and special features of endocrine system function. It also considers

the part played by specific hormonal endowment in the modifying role of hormonal analogs in relation to acute and chronic effects of ethanol.

2. METHODS

Animals Used

Our experiments were performed on white cross-bred rats weighing 250–350 g and white cross-bred mice weighing 20–25 g. (See the 15th article for the conditions under which the animals were kept, the method used for sorting them into those susceptible or not susceptible to alcohol and of simulating experimentally-induced alcoholism, together with criteria used for dividing the rats into heavy and light drinkers.)

Castration

Bilateral orchidectomy was performed under light ether anesthesia. The effectiveness of acute ethanol administration was assessed by the following tests: maintaining sideways position, rotating rod and reduction in rectal temperature.

Its narcotic action was measured using the duration of the sideways position adopted following intraperitioneal administration of 4.75 g/kg ethanol as a yardstick.

Rectal temperature was registered on a "Nihon Kohden" thermometer (Japan), 10 min prior to and 30 min after intraperitoneal administration of 3.5 g/kg 25% ethanol and the difference between the two values obtained was calculated. The number of times rats fell off a rod rotating at a speed of 11 rpm was noted over a period of 1 min, 3.5 g/kg 25% ethanol having been administered 15 min earlier.

The following compounds were used in our investigations: thyroliberin (TRH), L-pyroglutamyl-L-seryl-L-leucineamide (TRH-2) – an analog which affects prolactin but not TSH secretion, and the methyl ester of thyroliberin (TRH-3), with its much lower releasing effect in respect of both pituitary hormones. All peptides used were synthesized at the National Chemical Laboratories for Peptides Hormones, Institute for Endocrine and Hormonal Chemistry Research of the Acad. of Med. Sci.

TRH and its analogs were freshly dissolved in physiological saline, and administered at 0.1 ml solution per 10 g weight, together with ethanol at a dose of 20 mg/kg (IC_{50}TRH) when testing for duration of ethanol-induced sedation and a dose of 5 mg/kg (IC_{50} TRH) when measuring rectal temperature. The highest dose of TRH and its

analogs administered, when using the rotating rod test, did not
exceed 5 mg/kg i.p., since the animals actually began to jump off
the rod if the dose was further increased.

In our investigations into the effects of TRH and its analogs on
the rate of development of tolerance to ethanol, substances were
administered twice per day at a dose of 5 mg/kg, together with 3.5
g/kg ethanol (i.p.)

Assessing Hormone Content of the Blood

Animals were decapitated not later than 11 p.m. In work on rats
inclined (and disinclined) towards voluntary ethanol consumption,
animals were sacrificed not less than 10-14 days after administration
of a test dose of ethanol.

Plasma was separated from blood containing 10 units/ml heparin
(used to determine steroid hormone levels) or 1 mg/ml sodium EDTA
(for measuring pituitary hormone concentration). Concentrations of
all hormones were ascertained by radioimmune assay using the following commercially-produced kits:

RIA kit from "Amersham" which measures total testosterone
plus 5-α-dihydrotestosterone concentration.
"Sorin" kit for measuring 17-β-estradiol and ACTH.
Immuno Nuclear Corp. kit for measuring β-endorphin.

A volume of 200 µl plasma was then taken for radioimmune assay, which
had previously been diluted by 1:4 with buffer for testosterone assay
and a further 200 µl diluted similarly for β-endorphin and ACTH
measurement.

The steroids were extracted from plasma by mixing with 2 ml of
previously purified ester in a round-bottomed test-tube, measuring
11 mm across and 100 mm in height for 2 min. The ethereal layer was
separated off from the steroid-free plasma by freezing at -30°C with
an alcohol-liquid nitrogen mixture. The ester was poured off into a
glass test-tube prior to radioimmuno assay, and was evaporated off by
standing in a water bath at 30°C. Subsequently, steroids were measured according to the RIA kit instructions, as were ACTH and β-
endorphin immunoreactivity. The estrogen/androgen (E/A) index was
expressed as the ratio between 17-β-estradiol and the combined testosterone plus 5-α-dihydrotestosterone level, calculated in pg/ml blood
and multiplied by 100.

Luteinizing hormone (LH) concentration was determined using
highly specific antisera obtained from the National Institute for
Endocrine and Hormonal Chemistry Research of the Acad. of Med. Sci.
For iodination, a standard of rat LH (NIH-LH-SI) was used and for

constructing a calibration curve, 125, 250, 200, 1,000 and 2,000 ng
standard rat LH (NIAMD Rat LH-RP-1, Bethesda, USA) was employed.

Radioimmunoassay was performed in polystyrene test-tubes in
0.01 M Na phosphate buffer, with added merthiolate and bovine serum
albumin, in final concentrations of 0.01 and 1% respectively. The
following were then added: 0.01 ml plasma or spun-down pituitary
homogenate or standard LH (for the calibration curve), 0.1 M LH
antiserum, and following incubation at a temperature of 4°C, 0.1 ml
labelled LH (radioactivity: 20,000 counts/min). After a 24 h incubation period, a second set of antibodies was added (rabbit antibodies against guinea pig globulins at a working dilution of 1:32,
obtained from the Gamalei Institute of Epidemiology and Microbiology
of the Acad. Med. Sci.) Incubation continued for a further 1-2 days
at the same temperature. After incubation the tubes were centrifuged
for 20 min at 2,000 rpm, the upper layer sucked off and the precipitate washed with 0.5 ml distilled water. After repeated centrifugation and removal of supernatant, the radioactivity of the precipitate was measured. Sensitivity of this method: 1.9 ng/ml LH.
Quantitative data were expressed in ng per 1 ml plasma or per 1 mg
pituitary expressed in terms of NIH-LH-SI – the highly purified LH
substance.

The antiserum used to measure prolactin concentration was obtained from the National Institute for Endocrine and Hormonal
Chemistry Research of the Acad. Med. Sci. Prolactin was measured in
freshly-frozen rat anterior pituitary. Activity of this preparation
was 17.0±0.8 IU/mg compared with that of international standard sheep
prolactin from the National Institute for Medical Research, London,
with 22 IU/mg activity. We used 0.001 M Na phosphate buffer, pH 7.6,
containing 0.15 M 1% bovine serum albumin and 0.01% merthiolate to
carry out these reactions and prepare the reagents. ^{125}I-Prolactin
was added (10,000-40,000 counts/sample). Final dilution of antiserum
– 1:125,000. Between 1.5 and 400 ng/ml standard prolactin was added
to the mixture. After a 24 h incubation at room temperature a relative excess of "secondary antibodies" was added. Donkey serum
against rabbit immunoglobulins from the above-mentioned Gamalei
Institute was used for this purpose. After a further 18 h incubation, the precipitate was centrifuged (30 min at 1,000 rpm), washed
twice with buffer and its activity measured on a gamma-counter with a
counting efficiency of ^{125}I of 95%. Under the conditions described,
maximum binding achieved for ^{125}I-prolactin (in the absence of an
unlabelled standard) reached 30% of the total added, while the amount
of non-specific binding did not exceed 7-11% of maximum binding
level. Methods of measuring LH and prolactin are given more fully in
the relevant publications (Babichev et al., 1975; Abramova and
Fedotov, 1982). The findings on LH and prolactin concentrations
included in this present work were obtained jointly with colleagues
from the National Institute of Endocrine and Hormonal Chemistry
Research (V. Ya. lgnatkov and T. I. Ivanenko).

The Pharmacokinetics of Ethanol

Ethanol concentration was measured using a "Svet 152" gas liquid chromatograph with a flame-ionization detector on a steel column, with 15 PEG on a support on NAW-DMCS, 0.200-0.250 mm, 3 m in length. During analysis column temperature was 100°C and that of the detector, 250°C. The flow rate of the carrier gas (high purified nitrogen) was 25 ml/min. Propanol was used as an internal standard. A 50 μl sample of blood was taken by micropipette from the caudal vein and placed in a 15 ml vessel containing 50 μl of a 2 g/liter sodium citrate solution, then an internal standard of $10.4x10^5$ mg in 200 μl was added. The sample was then hermetically sealed and incubated at a temperature of 65°C for 15 min. Next, 2 ml of gaseous phase was analyzed by gas liquid chromatography (Eriksson et al., 1977). A test dose of 1 g/kg 25% ethanol solution was administered i.p. Blood samples were taken after intervals of 15 and 30 minutes and 1, 2, 3, 4 and 5 h. A cone-compartment model (Soloviev et al., 1980) was used for correction of pharmacokinetic parameters with respect to absorption errors. Calculations were performed on a BESM-6 computer, using first order kinetics. Statistical evaluation of data was performed by Student's method; Meddis' non-parametric method was used to compare findings on voluntary water or alcohol consumption.

3. ENDOCRINE CHARACTERISTICS IN RATS WITH VARYING PROPENSITY FOR ALCOHOL

Our knowledge of the etiology and pathogenesis of alcoholism is still little more than speculative. This is because neither the direct study of hormonal patterns in alcoholics nor analogous investigations in animal models of alcoholism have yet enabled us to differentiate a primary endocrine deficit from secondary hormonal defects produced by ethanol consumption.

The fact that we were able to identify animals with a built-in susceptibility to voluntary alcohol consumption from amongst a general rat population helps to answer the following questions: do endocrine defects result from the onset of alcoholism? Or should we look for factors predisposing to this disorder in the hormonal status and its variations from one individual to the next? To this end we picked animals either inclined or disinclined to consume alcohol as models for investigating the function of the pituitary-gonadal and the pituitary-adrenal system, i.e. the elements in the hormonal balance which are most affected by the onset of alcoholism.

Our experiments revealed no significant differences between the two types of animal in plasma concentrations of a) 17-β-estradiol, b) luteinizing hormone and c) total concentration of testosterone plus 5-α-dihydrotestosterone (see Table 1). Nor could any distinction between the same animal categories in luteinizing hormone level

Table 1. Hormone Concentration in Plasma of Rats with and without
          Inclination to Consume Alcohol

Hormones	Inclined rats		Disinclined rats
	Hormone concentration in plasma		
ACTH (ng/ml)	7.88± 1.58		2.58± 0.69
		$p < 0.001$	
β-endorphin (fmol/ml)	207.1 ±25.9		82.1 ±16.3
		$p < 0.01$	
Prolactin (ng/ml)	23.95± 5.8		9.87± 1.37
		$p < 0.05$	
Luteinizing hormone (ng/ml)	30.03± 2.95		34.7 ± 2.83
		$p > 0.05$	
Testosterone + 5-α-dihydrotestosterone (ng/ml)	5.50± 0.98		5.35± 0.25
		$p > 0.05$	
17-β-estradiol (pg/ml)	83.5 ± 1.35		92.95± 5.5
		$p > 0.05$	

measured in the pituitary homogenate (3662.1±133.0 ng/g for alcohol-
inclined and 3869.3±94.2 ng/mg for the disinclined group; $p < 0.05$) be
detected. We therefore presumed that function had not been impaired
in animals with heightened susceptibility to alcohol. It was also
found that ACTH concentration in the plasma of the latter animals was
more than three times as high as the corresponding value in animals
not so disposed. Furthermore, animals with a propensity for alcohol
consumption were characterized by a higher level of immunoreactive
β-endorphin and prolactin. Both are known to occur with raised ACTH
secretion, especially during stress (Höllt et al., 1979; Van Vugt et
al., 1978). Our findings serve as an indication of heightened
pituitary-adrenal system function. This manifests as overproduction
of both ACTH and a number of related hormones. To summarize, we were
led to view raised ACTH concentration partly as a biochemical indicator of susceptibility to alcohol consumption and partly as a potential etiological factor in shaping the need for alcohol by the following considerations:

a) the clear psychotropic properties of ACTH and β-endorphin,
   acting in their dual role of hormones and neurotransmitters,
b) the interaction between these two peptides and central and
   peripheral opiate receptors (De Kloet and De Wied, 1980) and
c) speculation on involvement of ACTH or ACTH fragments in the
   pathogenesis of dysphoric states in man and animals (Reus,
   1980).

4. CHARACTERISTICS OF ENDOCRINE SYSTEM FUNCTION IN RATS WITH VARYING PROPENSITY FOR ALCOHOL DURING THE CREATION OF MODEL ALCOHOLISM

The varying degree of involvement of different hormones in shaping the phenomenon of rats' "predisposition" to alcohol consumption, as demonstrated using special models, gives way to polymorphous generalized reaction of the endocrine system to chronic administration of ethanol. This conclusion was reached during experiments on rats which had been offered the choice of water and 15% ethanol. Experiments comparing ACTH levels in the plasma of heavy and lighter drinking animals after 10 days with access to ethanol failed to establish any difference between them (see Figure 1). Levels were 6.56±1.04 and 6.19±0.34 ng/ml respectively (N.S). A rise in ACTH concentration was, however, observed in light-drinking animals compared with those inherently predisposed to alcoholism during a short period with access to alcohol. ACTH in heavy-drinking animals showed a tendency to decrease as a result of 10 days' ethanol consumption in comparison with rats inclined to alcoholism. We thus established that ethanol is acting as a rather original type of tranquilizer in these instances. Our biochemical data agree with findings on behavioral characteristics of rats inclined to alcohol consumption and those not so inclined (see Chapter 14).

Prolonged consumption of ethanol over an 8-month period causes overproduction of ACTH in both heavy- and light-drinking animals, producing levels of 19.99±3.03 ng/ml and 13.56±1.0 ng/ml respectively. The first group shows this effect more clearly - a reflection of innate hyperfunction of the pituitary-adrenal system in rats inherently inclined towards alcohol. Thus, while ethanol can serve as a tranquilizing agent in the short term, it appears to be one of the factors promoting overproduction of ACTH if consumed over a longer period. The second factor is very typical of endogenous

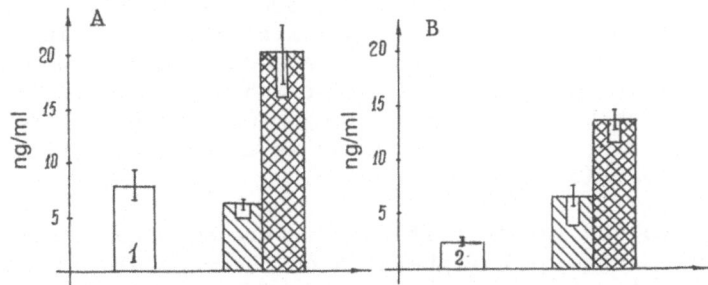

Fig. 1. The changing pattern of ACTH concentration in the plasma of rats varying in their need for alcohol. 1 - predisposed animals; 2 - non-predisposed animals. ▨ 10-day alcohol consumption; ▨ - 8-month alcohol consumption. A - heavy drinking rats; B - light drinking rats.

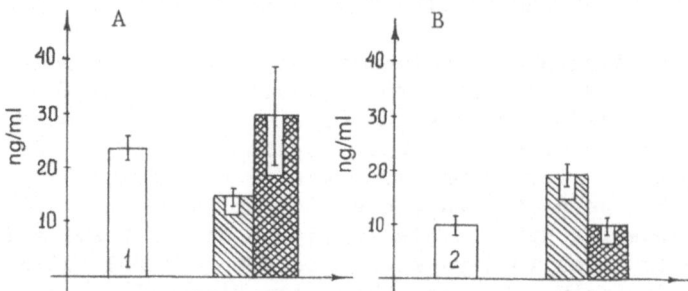

Fig. 2.   The changing pattern of prolactin concentration in the
          plasma of rats varying in their need for alcohol.
          1 - predisposed animals; 2 - non-predisposed animals.
          ▨ - 10-day alcohol consumption; ▩ - 8-month alcohol
          consumption. A - heavy drinking rats; B - light drinking
          rats.

depression, which many workers consider the starting point of alcoholism. We would suggest that, in view of corticotropin's capacity
for producing characteristic behavior patterns indicative of discomfort in animals, high concentrations of this peptide could be
largely responsible for producing the depressive state.

    Changes in both prolactin and ACTH levels in the early stages of
the evolution of alcoholism evidently reflect the effects of ethanol
on heavy-drinking animals physically predisposed to alcoholism. In
the case of long-term alcohol consumption, however, a marked overproduction of this hormone is seen in the heavy drinkers, a tendency
which is reversed in light-drinking animals. Hyperprolactinemia is a
very typical clinical sign of other types of drug addiction apart
from alcoholism (Brambilla et al., 1978). It appears to be a reflection of the particular discoordination imposed by ethanol on neurotransmitters in general and on the dopaminergic system in particular,
since this system controls prolactin release (Van Vugt et al., 1979).
Hyperprolactinemia resulting from experimentally/clinically-induced
alcoholism is probably a secondary phenomenon; raised concentrations
of this hormone can, however, serve as an indicator of both and
severity and specific type of diseases.

    Although no consistent differences were observed in the gonadal
steroid concentration of peripheral blood between alcoholically-
inclined and disinclined rats, testosterone and 5-α-dihydrotesto-
sterone levels dropped abruptly in all experimental animals after
10 days' access to alcohol. This decrease occurred in all experimental animals, irrespective of how prone they were to alcohol, and
amounted to 44.1% of control level for heavy drinkers and 20.6% in
the case of light drinkers (see Figure 3). At the same time ethanol

hardly affected estradiol level in the peripheral blood of rats
varying in their propensity for ethanol, under the same conditions.
When light drinkers had been consuming ethanol from choice over a
10-month period, androgen concentration in their plasma reached
control levels but remained at only 57% of control in heavy
drinkers. The latter group did not, however, show any changes in
plasma estradiol, unlike light drinkers; these had only a low concentration of estrogens, but the level increased by 26% – a statistically significant rise. Figure 4 illustrated the changing pattern
in balance between estrogens and androgens during the evolution of

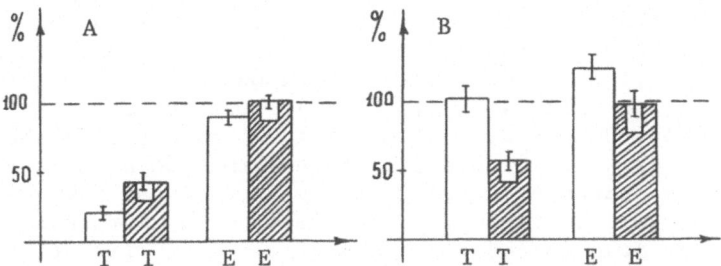

Fig. 3.   Concentration of 17-β-estradiol and concentration of total
          testosterone plus 5-α-dihydrotesterone in rat blood during
          the development of experimental alcoholism. A – 10-day
          consumption; B – 12-day consumption. □ – light drinkers;
          ▨ – heavy drinkers. T – testosterone + 5-α-dihydrotesto-
          sterone; E – 17-β-estradiol.

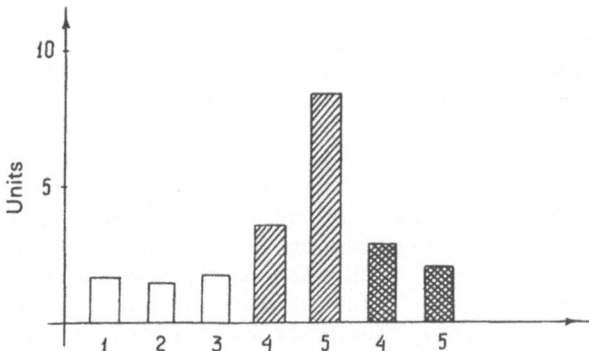

Fig. 4.   Changes in estrogen/androgen index in rats after varying
          periods with access to ethanol. 1 – control; 2 – rats prone
          to alcohol consumption; 3 – rats not prone to alcohol con-
          sumption; 4 – heavy drinking rats; 5 – light drinking rats.
          ▨ – 10-day period of ethanol consumption; ▨ – 12-month
          period of ethanol consumption.

alcoholism in rats. It will be seen that the E/A index amounts to
less than two units (intact animals only) for alcoholically-inclined
and disinclined rats alike. A higher E/A index is observed both
with short and prolonged periods of alcohol consumption. Over a
short period, however, a sharp rise in this interrelationship
occurs in conjunction with a significant drop in testosterone level
in the blood, where estradiol concentration remains unchanged. The
rise in estradiol level (per ml plasma) is associated with some
rise in the index when light-drinking rats consume alcohol over a
12-month period, while androgen concentrations are normal. A like
period of alcohol consumption produces a tendency for the estrogen/
androgen ratio to revert back to normal in heavy drinkers, owing
to a rise in androgen concentrations. It would appear that both
constant (or a rise in) estradiol content of peropheral blood and
a sharp drop in testosterone concentration are typical of the organism's response to long- or short-term exposure to ethanol (Van Thiel
et al., 1979a; Vedernikova et al., 1982). Overproduction of estrogen
is basically viewed as a disruption of the peripheral metabolism of
steroids together with increased biotransformation of androgens into
estrogens. The increase in E/A index typical of the advanced stages
of chronic alcoholism is usually associated with cirrhosis of the
liver. It also accompanies cirrhosis of non-alcoholic origin (Lester
et al., 1979). Some overproduction, which manifests in the light-
drinking group of rats choosing to consume alcohol over a 12-month
period, seems to be connected with impairment of steroid metabolism
arising from the animals' heightened sensitivity to ethanol's toxic
effects.

5. THE ROLE OF ENDOGENOUS TESTOSTERONE IN GOVERNING ETHANOL
    CONSUMPTION AND ETHANOL METABOLISM IN RATS

     Information contained in the previous section dealt with the
significant changes in androgen levels occurring in the peripheral
blood of light-drinking rats, as well as those with a preference for
alcohol.

     Some features of the evolution of alcoholism are also bound
up with sexual differences and changes in concentration of gonadal
steroids and point to an involvement of the latter in the pathogenesis of this condition. Much work has been done recently on
how treatment with exogenous testosterone potentiates alcohol consumption in castrated rats. There is also reason to believe that
administering estradiol causes the opposite effect (Lakoza et al.,
1978; Messiha, 1981). The suggestions that testosterone or 5-α-
dihydrotestosterone levels may have a direct causal relationship
with the amount of ethanol which the rats chose to drink in preference to water disagrees, however, with the data given above; these
findings show how hypoandrogenemia prevails throughout all stages of
experimentally-induced alcoholism. The development of a steady need

for alcohol, in the form of either a stable or increasing voluntary
alcohol consumption, does, in fact, occur against a background of a
significant and continuous decline in plasma androgen levels. While
testosterone may, under certain experimental conditions, be acting
so as to promote ethanol consumption by castrated animals, it apparently fails to mirror precisely the function of the endogenous
hormone in intact mice. We investigated the effects of castration
on elective alcohol consumption and metabolism in rats in order to
ascertain how far endogenous sex steroids govern these animals' need
for alcohol.

From our experiments we were able to establish that castration
led heavy-drinking male rats to consume somewhat more (32%) water
(p<0.05); there was a statistically significant reduction of 18% in
voluntary consumption (p<0.05 - see Table 2). Elimination rate was
only two thirds that of the controls and the half-life considerably
longer in castrated animals than in control. These findings suggest
castration produces a considerable fall in rate of ethanol metabolism, together with reduced consumption of this drug following a
4-month period of voluntary alcohol consumption. Castration has the
opposite effect on light-drinking rats' alcohol consumption, which
tends to increase (p<0.05), while pharmacokinetic studies show a
noticeable fall in the rate of ethanol degradation in comparison
with non-castrated animals (see Table 3). These findings suggest
that animals varying in their need for alcohol may possess several
distinct mechanisms for regulating ethanol consumption. Furthermore,
they suggest that other factors apart from testosterone may exert
their own influence on the phenomenon described (Mezey et al., 1980).
Thus a change in the peripheral androgen pool does not automatically
result in equivalent changes of the same degree in ethanol consumption when a choice is made available. Nor does it alter the pattern
of need for alcohol as seen in light- and heavy-drinkers rats. It
would nevertheless be borne in mind that the peripheral pool of sex
steroids may fail to mirror changes in their concentration in
certain organs and tissues, including the CNS. It is a pity that
insufficient research has been done on brain steroidogenesis to date,
although the presence in this region of a relatively autonomous
system of steroid synthesis and metabolism has been reliably established. The possibility of steroid hormones affecting neurotransmitter uptake and release has also been pointed out - an action
which is not mediated by specific hormonal receptors within the CNS
(McEwen et al., 1979). One suggestion is that sex hormones maintain
their specificity of action in the CNS only with respect to those
effects achieved via the specific receptors, whose function is associated with sexual behavior and differentiation. A detailed study
of how hormones and their analogs act within the CNS in intact
animals and those in which alcoholism has developed might well prove
useful. It could provide the key to discovering how the endocrine
system contributes to the control of behavior in normal subjects, as
well as those under the influence of alcohol.

Table 2.   Effects of Castration on Voluntary Consumption of 15%
           Ethanol by Rats

| | Average 24-hourly consumption of 15% ethanol (ml) per rat | | | | | |
| | Water | | | 15% ethanol | | |
Animals	pre-castration	post-castration	%	pre-castration	post-castration	%
Heavy drinkers	1.74±0.51	2.30±0.49	132	21.24±2.23	17.56±0.47	82
		p>0.05				p<0.05
Light drinkers	16.10±2.65	19.12±1.03	118	1.90±0.21	2.12±0.31	111
		p>0.05				p>0.05

Animals used in this experiment had been offered a choice between
water and 15% ethanol over a 4-month period. The animals were housed
individually during the 10 days preceding the experiment and the same
regime was resumed for a further 10 days, starting on the second day
after castration.

Table 3.   The Pharmacokinetics of Ethanol in Rat Blood Following 4
           Months of Voluntary Ethanol Consumption

Animal group	Elimination constant	Absorption constant	Maximum concentration	Half-life
Heavy drinking non-castrated	0.70±0.28	3.4±1.3	7.6±1.6	1.0±0.4
Heavy drinking castrated	0.42±0.20	4.0±1.8	13.8±2.5	2.3±1.1
Light drinking castrated	0.07±0.02	–	63.9±2.3	10.2±2.5
Intact non-castrated	0.20±0.04	4.5±2.8	17.1±1.2	3.5±0.8

Animals used in this experiment had been offered a choice between
water and 15% ethanol over a 4-month period. The animals were housed
individually during the 10 days preceding the experiment and the same
regime was resumed for a further 10 days, starting on the second day
after castration. After this 10-day interval the animals were given
a test dose of 1.5 g/kg ethanol. Ethanol concentration was then
measured by gas-liquid chromatography (see "Methods" section).

6. HORMONAL MODULATION OF SOME EFFECTS OF ETHANOL

The question of whether endocrine function of the brain is autonomous should not be considered purely theoretical. Even so, the use of hormones in clinical practice to deal with alcoholism and drug addiction is extremely limited, in view of their highly specific action – although much evidence has accumulated concerning the possibility of almost all ethanol's effects being governed hormonally. There is further evidence in the literature relating to the high level of biological activity of a) some hormonally inactive metabolites and b) synthetic analogs of hormones, especially when administered into the cerebral ventricle. Sulfate conjugates of steroids having an opiate-like effect, catecholestrogens and 17-α-estranes with a strong affinity for opiate receptors in rat brain are a few such compounds (Fishman, 1976; La Bella et al., 1978, 1979). The likelihood of extrapituitary activity has also been pointed out in the case of some pituitary and hypothalamic hormones, TRH (thyroliberin) amongst them. We set out to find whether TRH's range of "antialcoholic" action could be divided into hormonal and non-hormonal components. The effects of this hormone and its analogs with modified action on some of ethanol's acute and chronic effects were compared with this end in view.

Figure 5 illustrates the influence of this group of compounds on ethanol's acute effects. As shown in column C, neither TRH itself nor its derivatives affected the animals' ability to stay on the rotating rod after administration of 3.5 g/kg ethanol. The same dose of TRH and TRH-2 did, however, avert the fall in rectal temperature which this amount of ethanol would otherwise have induced (see column B). A drop in rectal temperature occurred following administration of 4.75 g/kg ethanol and an acute dose of TRH, as follows: $28.8 \pm 4.91\%$ of control level ($p < 0.01$), $63.46 \pm 7.25\%$ in the case of TRH-2 ($p < 0.05$) and $95.4 \pm 9.7\%$ with TRH-3, which hardly affected the measurement in question. Both TRH and TRH-2 were equally effective in modifying ethanol-induced sedation (by $35.23 \pm 10.42\%$ – $p < 0.05$, and 36.53 ± 7.47 – $p < 0.05$, respectively, compared with the control group).

None of these substances, however, affected the rate of developing tolerance to ethanol; this was established according to the drop recorded in rectal temperature and performance in the rotating rod test (see Figures 6 and 7). In each case tolerance became apparent during the third day of ethanol administration in both test group and control. The onset of tolerance was signalled by lack of significant changes in certain parameters, taking the previous day of treatment as a yardstick.

It should be stressed, however, than when tolerance is developed to the hypothermic effect of ethanol, especially when administered in combination with TRH, statistically significant differences in temperature in absolute terms, in comparison with the control group,

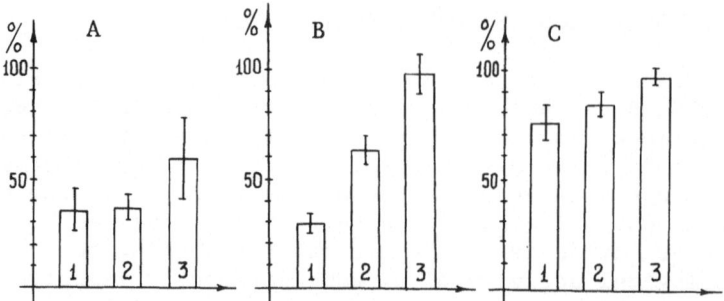

Fig. 5.   Action of thyroliberin and its analogs on the acute effects
          of ethanol. A - duration of ethanol-induced sedation; B -
          change in rectal temperature; C - No. of falls from rotating
          rod. 1 - thyroliberin; 2 - TRH-2; 3 - TRH-3. Values given
          as percentage of control (taken as 100%).

persist throughout the experimental period. A similar tendency was
apparent on observing TRH's effects on the tolerance to ethanol
registered using the rotating rod test (see Figure 6).

     Our findings make it clear that the analeptic properties of TRH
and its analogs scarcely depend on their hormonal activity. This
property is evidently unconnected, in the case of thyroliberin, with
its potentiating effect on TSH or prolactin. TRH-2, which does not
affect TSH secretion but reduces that of prolactin, produces an equal
and similarly directed rousing effect in the case of ethanol-induced
sedation. These results would suggest the involvement of extrapituitary mechanisms in mediating the analeptic effects of TRH and
some of its analog. A considerable number of authors (Cott et al.,
1976; Fukuda et al., 1980) have raised the possibility that the
psychotropic and hormonal properties of the thyroliberin molecule may
be dissociated. Among the group of substances we tested, however, an
interdependence was observed between hormonal activity and ability to
counter a drop in rectal temperature, produced by treatment with
ethanol. None of the compounds, irrespective of their specific
properties, proved effective in countering the effects which ethanol
had been seen to exert on performance in the rotating rod test.
Neither this test nor measurement of rectal temperature showed these
compounds to affect the course of ethanol tolerance. The extent of
TRH's "hormonal" involvement - and that of its analogs - in achieving
its psychotropic effects is thus seen to vary. The different effects
observed could well stem from dissimilar mechanisms within the CNS or
the endocrine system itself.

     We have indicated the possibility of some general properties
persisting parallel with the phasing out of specific action when
hormone molecules have been chemically modified. This would suggest
the feasibility of separating off selected psychotropic and neuro-

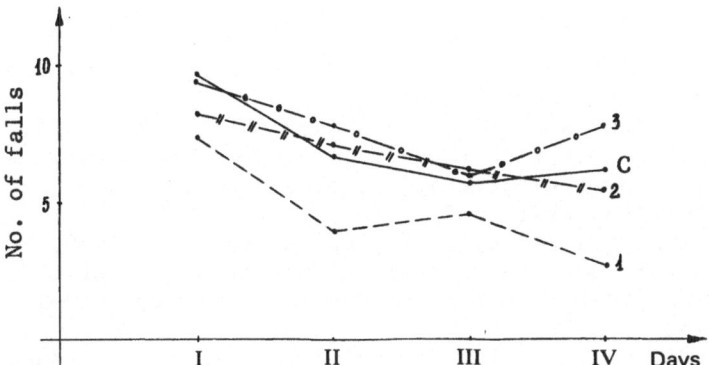

Fig. 6.   The effects of thyroliberin and its analogs on the onset
of ethanol tolerance, as indicated by performance in the
rotating rod test.   C – control.   1 – thyroliberin; 2 –
TRH-2; 3 – TRH-3.

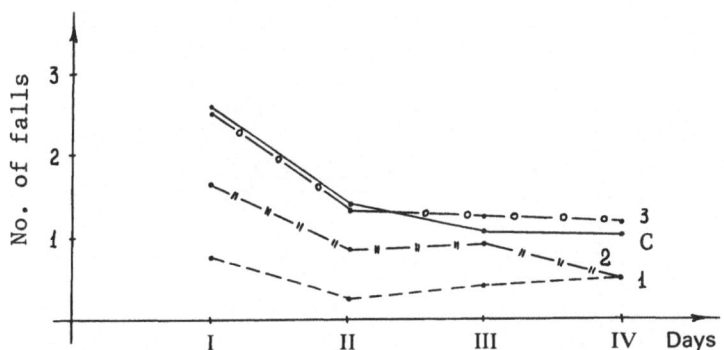

Fig. 7.   The effects of thyroliberin and its analogs on the onset of
tolerance to the hypothermic effects of ethanol.
C – control; 1 – thyroliberin; 2 – TRH-2; 3 – TRH-3.

tropic properties from the overall biological spectrum of hormones
with different chemical structures.   Dissociation between hormonal
and neurotropic properties of hormone molecules could thus present a
promising line of approach in the search for substances effective
against some forms of mental illness, including alcoholism.   These
might well be sought amongst the metabolites of hormones and their
synthetic analogs which are devoid of hormonal activity (Vedernikova,
1982).

REFERENCES

Abramova, V. V., and Fedotov, V. P., 1982, Homologous radioimmuno-
    assay of rat prolactin Probl.Endocrin., 28:68–73.
Babichev, V. N., Adamskaya, E. I., and Samonova, V. M., 1975, Radio-
    immunological determination of luteinizing hormone in the
    hypophysis and the blood in the course of the estral cycle in
    rats, Probl.Endocrin., 21:63–65.
Badr, F. M., Smith, M., Dalterio, S., and Bartke, A., 1979, Role of
    the pituitary and the adrenals in mediating the effects of
    alcohol on testicular steroidogenesis in mice, Steroids,
    34:477–482.
Brambilla, F., Cazzullo, C. L., Bellodi, L., De Maio, D., Zanoboni,
    A., and Zanoboni-Maciacia, W., 1979, Prolactin secretion and
    response to stimuli in male heroin addicts, Neuropsycho-
    biology, 5:294–300.
Cicero, T. Y., Bernard, Y. D., and Newman, K., 1980, Effect of
    castration and chronic morphine administration on liver
    alcohol dehydrogenase and the metabolism of ethanol in the
    male Sprague-Dawley rat, J.Pharmacol.Exp.Ther., 215:317–324.
De Kloet, R., and De Wied, D., 1980, The brain as target tissue for
    hormones of pituitary origin: behavioral and biochemical
    studies, in: "Frontiers in Neuroendocrinology," Vol.6,
    L. Martins and W. Ganong, eds., Raven Press, New York,
    pp.157–201.
Eriksson, C. Y., Sippel, H. W., and Forsander, O. A., 1977, The
    determination of acetaldehyde in biological samples by
    head-space gas chromatography, Anal.Biochem., 80:116–124.
Fisherman, Y., 1976, The catechol estrogens, Neuroendocrinology,
    22:363–374.
Greene, L. W., and Hollander, C. S., 1981, Alcohol and the hypo-
    thalamus, Prog.Biochem.Pharmacol., 18:15–23.
Hölt, A., Müller, O. A., and Fahlbisch, R., 1979, β-Endorphin in
    human plasma: basal and pathologically elevated levels,
    Life Sci., 25:37–44.
Lakoza, G. N., Poliakova, N. B., and Barkov, N. K., 1978, Sex
    hormone influence on the formation of predisposition to
    ethanol, in: "Act. Probl. of Psych. Endocrinology," Moscow,
    pp.122–132.
La Bella, F., Kim, K. S., and Templeton Y. 1978, Opiate receptor
    binding activity of 17-α-estrogenic steroids, Life Sci.,
    23:1797–1804.
La Bella, F. S., Havlicek, V., and Pinsky, C., 1979, Opiate-like
    excitatory effects of steroid sulfates and calcium-complexing
    agents given cerebroventricularly, Brain Res, 160:295–305.
Lester, R., Eagon, P. H., and Van Thiel, D. H., 1979, Feminization of
    the alcoholic: the estrogen testosterone ratio (E/T), Gastro-
    enterology, 76:415–417.
McEwen, B. S., Davis, P. G., Parsons, B., and Pfaft, D. W., 1979, The

brain as a target for steroid hormone action, Ann.Rev.Neuro-
sci., No.2, pp.65-112.

Messiha, F. S., 1981, Steroidal actions and voluntary drinking of
ethanol by male and female rats, in: "Endocrinol. Aspects
Alcohol," 4th Ann Conf. Alcohol, El Paso, Tex., Feb.22-23,
Prog.biochem.Pharmacol., 18:205-215.

Mezey, E., Potter, J. J., Harmon, S. M., and Tsitouras, P., 1980,
Effects of castration and testosterone administration on rat
liver alcohol dehydrogenase activity, Biochem.Pharmacol.,
29:3175-3180.

Reus, V. I., 1980, Neuropeptide modulation of opiate and ethanol
tolerance and dependence, Medical Hypotheses, 6:1141-1148.

Soloviov, V. M., Firsov, A. A., and Filov, V. A., 1980,
Pharmacokinetics, Moscow.

Sze, P. Y., 1977, The permissive role of glucocorticoids in the
development of ethanol dependence and tolerance,
Drug Alcohol Depend, 2:381-396.

Van Thiel, D. H., 1980, Alcoholism and its effect on endocrine functioning, Alcoholism, 4:44-49.

Van Thiel, D. H., Gavaler, G. S., Cobb, C. F., Sherins, R. J.,
and Lester, R., 1979, Alcohol-induced testicular atrophy in
the adult male rat, Endocrinology, 105:888-895.

Van Vugt, D. A., Bruni, Y. F., and Meitis, I., 1978, Naloxone
inhibition of stress-induced increase in prolactin secretion,
Life Sci., 22:85-90.

Van Vugt, D. A., Bruni, Y. F., Sylvester, P. W., Chen, H. T.,Jeiri,
T., and Meites, Y., 1979, Interaction between opiates and
hypothalamic dopamine on prolactin release, Life Sci.,
24:2361-2368.

Vedernikova, N. N., 1982, Contemporary ideas of the role of endocrine factors in pathogenesis and treatment of alcoholism,
in: "Physiologically Active Agents - for Medicine," Abstr.,
All-Union Congress of Pharmacol., Yerevan, 62.

Vedernikova, N. N., Bragin, V. B., and Burov, Yu. V., 1982, Role of
endogenous sex steroids in the formation of experimental
alcoholism in rats, Pharmacol.Toksikol., 15:94-97.

The Role of Reward and Punishment Brain Systems in the Development of Drug Dependence

A.V. Valdman[1] and E.E. Zvartau[2]

[1] Institute of Pharmacology, Academy of Medical Sciences
of the USSR, Moscow
[2] Department of Pharmacology, Pavlov Medical Institute, Leningrad

1. INTRODUCTION

The psychiatric conditions of drug dependence and addiction
belong to the most complex realms of human pathology, where psychological, hereditary, social and cultural factors are closely inter-
woven. This is why the legitimacy of investigating these subjects
using experimental animals may be questioned. Our many years'
experience in psychopharmacological research, however, have shown
that it is indeed possible to study the psychotropic effects of
pharmaceuticals on animals (Valdman, 1972; Valdman et al., 1976,
1979). Investigation of emotional reactions common to man and
animals requires the application of objective methods for evaluating
an animal's "emotional" state. Experimental work on drug taking,
moreover, has special features distinguishing it from classical
psychopharmacological research. It is true to say that research on
a psychotropic compound usually sets out to reveal a special feature
of its action in order to classify it (Ilyuchenok, 1972; Zakusov,
1973; Anichkov, 1974; Rayevsky, 1976). Experimentation with narcotics, on the other hand, is concerned with detecting properties
producing dependence, shared by compounds belonging to different
pharmacological categories. Psychotropic agents of the most diverse
groups are known to cause addiction. One might thus expect to find a
common denominator encompassing the effects of these heterogeneous
substances. Clearly, the physical and mental processes used as
criteria in our studies should take account of the basic principles
of behavioral regulation and how they govern the organism's interaction with its environment.

The past 10 years of research on the physiology of higher
nervous activity have led to the discovery of systems of positive and

negative reinforcement. It is these systems which determine how the
organism reacts to a) stimuli, b) the environment (whether favorably
or unfavorably) and c) forces of attraction and repulsion (Olds and
Milner, 1954; Delgado et al., 1954; Simonov, Valdman and Zvartau,
1975; Makarenko, 1980). The systems of "reward and punishment" may
be indirectly activated. Likewise, conditions under which such activation may in fact be encountered or maintained using instrumental
conditioning methods could be created. If so, the motivational and
emotional forces shaping behavior might be evaluated more objectively. The gap separating the study of human mental processes from
those of animals would thereby be substantially narrowed, and
Pavlov's goal of studying the physiology of higher nervous activity
would be brought one stage nearer.

There appear to be grounds for linking the emotiotropic action
of addictive agents (an action which is central to the phenomenon of
dependence) with their effects on the primary emotional and evaluation systems. It is the reinforcement system which is responsible
for such processes. It was postulated that changes in the functioning of this system could shape both pathological craving for addictive drugs and actual dependence in which this craving finally
culminates. Our standpoint assumes additional importance in the
context of producing methods for identifying pharmacological agents
with abuse potential and chemical compounds suitable for therapeutic
use.

This article combines the results obtained by a number of
research workers over the past 10 years on the role of the reward and
punishment systems of the brain.

2. MATERIAL AND METHODS

Animals

We used a total of 263 male rats weighing 180-250g and 51 cats
weighing 2.5-4 kg. The rats were housed in groups of 3-4 animals or
kept in separate cages at room temperature. The animals were provided with food and water ad libitum, except during experiments
involving administration of substances under a special regimen.

Implantation Technique

Electrodes were implanted into the positive or negative regions
of the hypothalamus following the coordinates of the stereotaxic
atlases of Snieder and Niemer (1961) and König and Klippel (1963).
Electrodes made from nichrome wire, measuring 150-200 µm in diameter,
were insulated along their entire length, except for the tip. A
screw made of stainless steel, let in to the frontal bone, was used

as an electrode. A control panel was attached to the skull with
dental cement and protacryl. The operation was performed under
pentobarbital anesthesia (40-50 mg/kg). The experiments were carried
out 5-7 days later. Cannulae were implanted in the jugular vein
following a modified version of Weeks' method (Zvartau, 1979).

Electrical Stimulation of the Brain

An ESU-2 stimulator (USSR) was used to produce electrical stimulation of the brain. Single-phase rectangular pulses of negative
polarity were generated, lasting 0.5-1 msec; frequency: 40-100
imp/sec; amplitude: 0.05-0.8 mA. Electrical stimulation was monitored by oscilloscope. In the case of experiments performed on cats,
the method of gradual stimulation of the hypothalamic "defense area"
was applied, together with external stimuli intended to produce
offensive-defensive reactions (Valdman et al. 1976). The reaction
of electrical self-stimulation (SS) of the brain in rats was studied
as follows: experiments took place partly in an enclosure measuring
30 x 30 x 40 cm. This was fitted with one or two levers for the
purpose of lever-operated SS. Other experiments were performed in a
shuttle-box enclosure consisting of two compartments, each measuring
25 x 25 x 35 cm, where locomotor SS was monitored. Under the former
system the animals could receive either a bundle of impulses of fixed
duration, lasting 0.2-0.5 sec each time they pressed the lever, or
else stimulation lasting as long as the lever was depressed (free
regimen SS).

A histogram recorded the duration of lever-pressing and intervals indicating operant response using an original type of micro-
computer (Zvartau and Sidorov, 1980). Locomotor SS response
consisted of the animal setting off stimulation by crossing the
"active" half of the enclosure. It was switched off when the animals
crossed back to the "inactive section". The incidence of crossing
over in this way and total length of stimulation was recorded.

The same enclosure was used for investigating active avoidance
response arising from stimulation of "punishment" sites, but the
animal was initially placed in the "active" section. Stimulation was
cut off by the animal changing to the other half of the enclosure.
It was again set up 0.5-3 min later and the animal switched off again
by changing to the other compartment. After a number of repeats,
latency of this avoidance reaction and the number of correct
responses was noted. The avoidance (or cut-off) reaction was studied
using a T-shaped maze in a number of experiments. The "punished"
response, i.e. passive avoidance, consisted of inhibiting a natural
reaction, usually feeding, by electrical stimulation of the brain.
Once the experiments were over, a histological check was made of the
placing of the electrodes by inspecting thionine-stained frozen brain
slices.

Process of Intravenous Self-administration

We used a modified version of the Weeks methods (Zvartau, 1979). Compounds tested were as follows: morphine hydrochloride, diacetyl-morphine, pentobarbital, cocaine hydrochloride, ethanol, chlordiaze-poxide ("Elenium", Polfa), diazepam ("Seduxen", Gedeon Richter, Hungary), pyrroxane, butyroxane, carbidine, haloperidol, lithium chloride and naloxone (Endo Lab, USA). All substances were in the form of aqueous solutions, 0.1 ml/kg body weight, administered i.p. Control groups received an equivalent volume of isotonic saline solution. Statistical evaluation of the results was performed using standard programs on an ES-1030 computer and included general methods of parametric and non-parametric statistics, dispersion analysis and regression analysis.

3. RESULTS

Our experiments showed that drugs with abuse potential (DAP's) elicit positive (or "rewarding") emotional reactions (Zvartau, 1976, 1978; Zvartau et al. 1981). Specificity of action is demonstrated by a lack of correlation between therapeutic activation of SS and a) facilitation of spontaneous motor reaction or b) evidence of persev-eration behavior. The effect of pharmacological activation of the "reward" system is reflected in the basic level of its function and, on balance, fulfills the criterion of a "rise in positive discharge" (see Table 1). The table shows that DAP's a) increase the incidence of self-stimulation, curtail the modal duration of lever-pressings and intervals on histograms showing time in relation to operant reaction and b) increase the correlation between number of lever-pressings and incidence of SS associated with heightening of emotional arousal. They generally reduce the average duration of switching off of current under "free regimen" SS and of the latent period of return to the activation portion of the enclosure (i.e. locomotor SS). DAP's also increase the attraction of lever-pressing for these animals. This was illustrated by their unwillingness for the experiment to cease, as shown by the contracting pauses between each SS and a reduction in SS latency.

It is interesting to note that compounds with a predominantly depressant action could increase the number of prolonged, sometimes greatly extended lever-pressings, as well as their frequency. Opiates, barbiturates, benzodiazepines and ethanol acted in this way, producing asymmetrical movements, discoordination and ataxia, as seen during intoxication.

Table 2 shows the ED_{50} of various DAP's required to promote SS of the lateral hypothalamus in rats. The effects of the substances studied are seen to vary substantially according to how they are administered. Thus, it is hardly surprising that addicts achieve

Table 1. Comparative Change in Parameters of Electrical
Self-stimulation Reaction (SS) Under the Influence of
Addictive Drugs and of Increasing Intensity of "Rewarding"
Stimulation

Level of SS reaction	Change in level of SS reaction occurring with:	
	Increasing strength of current	Treatment with DAP's
Incidence of SS	Increases	Increases
Ratio between incidence of lever-pressing and of SS	Increases	Increases
Strength of pressure on lever	Increases	Increases
Average duration of each press (lever-operated SS) and of closing the circuit (locomotor SS)	Unaffected or reduced	Unaffected or reduced
Overall duration of stimulation	Extended or unchanged	Extended or unchanged
Modal duration of press	Reduced	Reduced
Excessive mode of pressing	Increases	Increases
Incidence of extended presses	Reduced	Reduced or increased (DAP psychodepressants)
Modal duration of pauses	Reduced	Reduced
Excessive mode of pauses	Extended	Reduced
Incidence of extended pauses	Reduced	Reduced
Latent period - "return" response (locomotor SS)	Reduced	Reduced
General behavior during SS: emotional pitch	Heightened	Heightened
alteration in motor activity	Increased or reduced, depending on whether pauses lengthen or shorten	
grooming	id.	
other	Possibility of shaking, biting, sneezing, gnawing, etc.	
Reaction to termination of experiment	Resistance, particularly marked in cats	
Latency of SS	Reduced after "follow-up"	Reduced

maximum psychotropic effect from administering drugs intravenously. This is true of morphine, diacetylmorphine, amphetamine, cocaine and barbiturates, which act as follows: "Positive" feelings are inten- sified as the drug's affect take hold. This in turn enhances the rate of activating the "reward" system.

If we were to compare the maximum facilitation achieved by DAP's using dose-response curves, they could be arranged in the following approximate order of efficacy: diacetylmorphine > morphine > cocaine > amphetamine > pentobarbital > diazepam > ethanol. Although clini- cal practice has no such established ranking of narcotic activity, it recognizes the addiction potential common to these compounds and their ability to promote the action of the "reward" system.

Experiments involving intravenous self-administration revealed that animals use standard DAP's of their own accord in doses suf- ficient to activate SS (see Table 3).

Table 2. Doses (in mol/kg) of Addictive Drugs Producing a 50% Increase in Self-stimulation (control - 100%)

Drug	ED_{50}, administered i.p.	ED_{50}, administered i.v.
Amphetamine	2.6×10^{-3}	0.6×10^{-3}
Diazepam	2.6×10^{-3}	
Diacetylmorphine	5.6×10^{-3}	0.2×10^{-3}
Morphine	7.1×10^{-3}	0.8×10^{-3}
Pentobarbital	9.0×10^{-3}	
Cocaine	2.7×10^{-2}	1.1×10^{-2}
Ethanol	224.0×10^{-1}	151.3×10^{-1}

Table 3. ED_{50} for Activating Self-stimulation Response, and Doses Taken Under Intravenous Self-administration (IS) Regimen

Drug	ED_{50} required to activate SS (mg/kg)	Doses taken over 1 h experiment involving IS (mg/kg)
Diacetylmorphine	0.12	0.13-0.21
Morphine	0.31	0.44
Amphetamine	0.23	0.50-0.80
Cocaine	3.60	3.00-5.00
Ethanol	696.00	480.0-800.0

As experiments proceeded, animals could be divided into three groups - "stable", "labile" and "receding", on the basis of intravenous self-administration response (ISR). In some of the animals ISR was dictated by their own distinctive reaction to a particular substance. When ethanol-induced SS reaction in animals with "stable" ISR was compared with that of rats with "receding" ISR, findings indicated that alcohol produced no significant reproducible promotion of SS in the latter group, whether administered i.v. or i.p. The possibility of a learning deficit occurring should also be envisaged, since instrumental conditioning and reinforcement stimuli, unlike the usual conditioned responses, disappear in time with ISR. Even when using a "stable" group of animals, there is no guarantee that the ISR will be reproduced from one experiment to another over prolonged periods, as the phasing of the reaction may undergo substantial change.

The basic ISR pattern is set up when the action of self-injection first takes place, and is then repeated regularly, after which ISR is usually discontinued. With ISR of opiates, especially during initial experiments, inhibition amounting to catatonia was observed at this point. During self-administration of cocaine or amphetamine, the animals' behavior pattern was distinguished by motor agitation or mild stereotypy. Single or grouped instances of self-administration followed at fairly regular intervals thereafter. The ISR described could be referred to as "controllable", in that animals are, so to speak, "titrating" DSP's effects; their behavior does not get out of control, despite a certain level of intoxication. This is corroborated by the fact that pretreatment with the substance reduces the incidence of self-injection and hence the size of the dose required or, alternatively, completely prevents the reaction occurring. When the concentration of the substance administered is changed, the single dose is seen to be inversely proportional to ISR values. At the same time, repeated experimentation may affect ISR pattern. This was especially noticeable in experiments with diacetylmorphine. In the first half of the experiment, animals which were relatively tolerant to cataleptogenic influences injected themselves with up to 70-80% of the amount required to produce catatonia.

"Uncontrollable" ISR pattern was distinguished by haphazard and unbroken sequences of self-injection and acute intoxication, resulting in complete immobility in the case of opiates. In addition, ethanol induced sleep, while stereotypy was produced by amphetamine and cocaine. The following criteria were used to rate an "uncontrollable" ISR regimen: i) prior development of DAP dependence; ii) frequent and regular experimentation and prolonged experiments; iii) renewed experimentation after a 4-6 interval, or a surprise increase in concentration of test substance and iv) "uncontrolled" self-administration in experiments involving cocaine were noted in a proportion of the animals as an early form of ISR.

When investigating the action of catecholaminergic compounds on morphine's emotionally rewarding effects, it was discovered that 200 mg/kg L-DOPA, the catecholamine precursor, enhances morphine's potentiating effect on SS (I.I., or index of interactions, equals 1.47±0.25)*.

Impaired synthesis of catecholamines (CA) and of noradrenaline (α-methyltyrosine, 100-250 mg/kg, disulfiram, 20 mg/kg) prevents any increase in the incidence of SS after treatment with morphine (I.I. - 0.02±0.04 and 0.08±0.13 respectively). In addition, impaired synthesis of noradrenaline together with a corresponding accumulation of dopamine (disulfiram, 200 mg/kg + DOPA), increase in presynaptic release of dopamine (amantadine, 10 mg/kg) or stimulation of dopamine receptors (apomorphine, 1 mg/kg) also inhibit the emotionally-positive activation produced by morphine (I.I. - 0.12±0.04, 0.55±0.05 and 0.02±0.03 respectively). A similar effect is produced by blocking noradrenaline and dopamine receptors with pyrroxane, butyroxane or haloperidol. However, on studying the accompanying behavioral responses, it became apparent that the inhibitory action of dopaminergic compounds is correlated with potentiation of morphine's effects on behavioral arousal (including sudden bursts of locomotor activity, motor stereotypy and "waltzing"). In fact, treatment with 0.01 mg/kg apomorphine, combined with small doses of morphine insufficient to activate self-stimulation, produced significant activation of SS, unaccompanied by motor disturbances; moreover, changes were observed in operant pauses and behavioral features of the process which are typical of a rise in the "cost of reinforcement".

Serotininergic processes also contribute to achieving morphine's emotionally positive effect. Thus, parachlorophenylalanine (PCPA) at a dose of 300 mg/kg countered activation of SS deriving from the action of morphine (I.I. = 0.12±0.14). Furthermore, increased serotonin concentration at the receptor, due to amine reuptake blockade (fluoxetine, 10 mg/kg), potentiated morphine's effects and promoted opiate-induced SS when sub-threshold doses were used. As the dose of morphine was raised, its catatonic action likewise increased.

Selected studies on the effects of three antidepressants - pyrroxane, butyroxane and carbidine - by Barkov (1971, 1973) and Krylov and Starykh (1973) illustrated their shared property of countering the emotionally positive effects of all DAP's investigated (Table 4). In a number of cases, the antagonism index, which applies

---

*

$$\text{I.I. (index of interactions)} = \frac{\Delta SS_{DAP + T}}{\Delta SS_{DAP + V}} \quad \text{where}$$

$\Delta SS_{DAP + T}$ = % change of SS under the action of DAP (test drug - control, 100%) vs. vehicle (V).

to the ability of the antagonists under study to inhibit SS to below
its baseline level, was greater than one. It should be emphasized
that the compounds used had been approved by addiction treatment
centers.

Pert and Snyder (1973), Simon et al. (1973), Terenius (1973),
Hughes (1975) and Cox et al. (1975) have all helped to develop the
concept of opiate receptors and their endogenous ligands. In this
connection, research into opiate receptors' contribution to the
action of DAP's is of special interest. Our investigations included
the blockade of opiate receptors with naloxone and how this influenced DAP's emotionally positive activating effects. It may be seen
from Table 4 that naloxone counters the action of ethanol, diazepam
and pentobarbital, as well as that of opiates. The effects of
cocaine and amphetamine were uninfluenced by pretreatment with
naloxone (Zvartau, 1979, 1980; Valdman and Zvartau, 1980).

Findings obtained during experiments on the SS reaction were
confirmed using the ISR model - i.e. that pyrroxane, butyroxane,
carbidine and lithium chloride countered the primary reinforcing
effects of DAP's. A time-related study of ISR showed that where the
pharmacological agents had been administered previously, reduction of
both the incidence of self-administration and size of dose required
depends on a reduction in DAPs' primary reinforcing effects. ISR
thus begins with a brisk pattern of self-administration, but it
becomes evident from the animal's behavior that it is experiencing
very little effect and the reaction then fades. Such a pattern of
drug effects could thus be associated with DAPs' ability to reinforce
pharmacological blockade. This phase could, moreover, be preceded by
a stage involving a reduction in DAPs' reinforcing ability, reflecting decreased doses of the agonist, or alternatively manifesting when
the first injections of a course of antagonist treatment are given.

Table 4. Index of Antagonism (1-I.I.) of Some Drugs with Reference
to the Rewarding Activation Produced by Addictive Drugs

Agonists and doses (mg/kg)	Antagonists and doses (mg/kg)			
	Butyroxane (5-10)	Pyrroxane (10-20)	Carbidine (10)	Naloxone (5)
Cocaine (10)	1.17***	0.65*	0.47	0.15
Amphetamine (1)	0.84*	2.05***	0.32	0.01
Morphine (3)	1.72***	1.06**	0.72	1.09**
Diacetylmorphine (5)	1.67***	0.71*	0.60	1.24**
Pentobarbital (3)	1.86***	1.97***	1.33*	0.46*
Diazepam (1)	2.01***	2.14***	1.20*	0.46*
Ethanol (768)	–	2.38***	1.82**	1.12**

$* p < 0.05$; $** p < 0.01$; $*** p < 0.001$.

If DAP's have been administered chronically for some time, changes occur in the rate of SS response when the substance is withdrawn. These are termed type I and II abstinence reactions or AR-1 and AR-2. AR-1 is the response which produces a rise in SS and changes in other parameters indicative of activation of the "reward" system. AR-2 produces the reverse effect of a fall in the incidence of SS, increasing the modal duration of lever-pressing together with other signs of reduced "reward" reaction - sometimes actually culminating in complete suppression of SS.

Table 5 shows approximate intervals after which the two types of AR first appear. The degree of pharmacological loading is to some extent related to how rapidly withdrawal reaction sets in. AR-1 precedes AR-2. The restrained SS produced by withdrawal may be associated with certain signs pointing to a given degree of dependency, such as wet dog shakes, diarrhea, hyperalgesia in the case of opiate taking and convulsions prompted by stimulation of the brain when barbiturates or ethanol are used. Changes in operant response, however, tended to occur earlier than classical signs of dependence. In fact AR-2 can even arise after the 2nd, 3rd and occasionally the first injection of diacetylmorphine or morphine when naloxone is also administered.

The behavior of the reinforcement systems during the withdrawal of DAP's is shown to be a sensitive instrument for rating dependence according to the above data. Hence it may be presumed that deprivation of the habit-forming substance evokes certain responses which go to shape the withdrawal syndrome. Activity levels of the "reward"

Table 5. Timing of Onset of Types I and II Abstinence Reactions in Cases of Chronic Intoxication Caused by Addictive Drugs

Drug	Degree of intoxication (daily dose/LD_{50})	Timing of onset of abstinent reaction (days)	
		Type I	Type II
Morphine	0.06	–	30–50
	0.10	5–7	9
	0.20	–	4
Ethanol	10.47	–	28–30
	0.66	–	16–21
	0.3–1	6–7	–
Pentobarbital	0.04	15–20	50
	0.08	3–5	–
	0.17–0.2	–	12
Diazepam	0.08–0.1	–	12
Amphetamine	0.5–0.7	1–2	2.5–3
Cocaine	0.13–0.16	3–5	10

and "punishment" systems, as described below, react unmistakably when
the DAP is withdrawn, since these systems are apparently directly
related to the realization of a narcotic's pharmacological effects.

Studies of DAP's effects on the processes of perception and
emotional appraisal of negative stress-stimuli were performed on
models of the easily reproducible reactions of "cut-off" and
"punishment" (i.e. active and passive avoidance), applying electrical
stimulation to the brain. Activation of the "punishment" system
occurring during such reactions is termed "aversive", since the
animal is seeking to avoid or prevent contact with the stimulus. It
is possible to control the extent of this aversion, produced by
stimulating the "punishment" region, in such a way that subsequent
stimulation maintains its aversive quality without becoming stressful. Stronger or extended stimulation of these regions may produce
the characteristic changes in behavior which are classed as indications of experimentally-produced neurosis.

The effects of DAP's on stress-induced defensive behavior in
cats were investigated in circumstances permitting assessment of an
animal's goal-directed behavior in response to a threatening test
object. The main distinction of this method, which aimed to provide
stable and reproducible defensive responses, consisted of presenting
the animal with relatively innocuous threatening situations against a
background of drug-induced activation of the nervous substrate of
emotionally stressful reactions. The behavior of the animals in
response to such stimulation has already been described (Waldman et
al., 1976).

Our studies of the effects of DAP's on the above models yielded
the following information (Zvartau and Patkina, 1978):

1. DAP's may be divided into 3 groups on the basis of how they
affect a) the aversive and b) the "punishing" action of hypothalamic
electrical stimulation (as demonstrated during experiments involving
escape reaction and punished response respectively) - see Table 6.
The main characteristics of the first group, comprising morphine,
ethanol, pentobarbital and diacetylmorphine, is their inhibitory
effect during both experimental situations, while the second group
(amphetamine) attenuates "cut-off" as well as "punished" reaction.

2. In 20-50% of experiments, psychodepressant DAP's changed the
effects of brain stimulation from negative to equivocal, as shown by
a switch from "cut-off" to locomotor SS reaction.

3. Qualitative changes were observed in cats' behavior in experiments involving "punishment" using ethanol, pentobarbital, chlordiazepoxide and diazepam. The "anti-punishment" action of these
compounds produced species-specific shows of aggression and threatening behavior.

Table 6. Dose Range (mg/kg ethanol: g/kg) and Mode of Action of Test
          Drugs on "Cut-off" and "Punished" Responses

	Cut-off response Dose range changing (by 50%)		"Punished" response Dose range significantly affecting reaction
	Latency	No.responses	
Morphine	5-10(↑)	10(↓)	5-10(↓)
Ethanol	10.8-1.5(↑)	0.8-3(↓)	0.8-1.5(↓)
Pentobarbital	2-3(↑)	4-5(↓)	2-3(↓)
Chlordiazepoxide	(0.↓)	(0)	3-5(↓)
Amphetamine	(↓)	(0)	1(↑)

↑ - extended, increased, facilitated.
↓ - shortened, reduced, inhibited.
0 - no effect.

4. When electrical stimulation is applied to the hypothalamic
"defense" zone, the cats' response is one of anxiety and threat.
Threatening aspect is followed in turn by a defensive response with
aggressive overtones. Defensive reactions may be either active or
passive. If active, displays of aggression or avoidance may
accompany or dominate the response. If passive, fear will be shown
and action avoided. Type of response (active or passive) will depend
on an animal's inherent characteristics and degree of intensity of
central and externally applied stimulation. Some DAP's, such as
morphine and pentobarbital, do reduce the general pitch of behavioral
defensive reactions (see Table 7). All DAP's possessing a psychodepressant action changed the pattern of defense reactions such as to
reduce display of avoidance and heighten that of aggressive elements.
In the case of "passive" fear, however, they initiated actively
defensive (avoidance or occasionally aggressive) behavior. Amphetamine accentuated both aggressive and avoidance elements; thus, the
nature of their action was determined by the balance between them in
the initial organization of the reaction. Amphetamine also potentiated stenotic features in the case of "passive" fear.

5. Tolerance rapidly develops to the "anti-aversive" action of
morphine, but not to that of ethanol, when these two compounds are
administered repeatedly. This was established using the model of
"cut-off" reaction in rats. The rapid phase - of relative tolerance
- sets in 1.5±0.3 days later, following administration of a twice-
daily dose of 20 mg/kg. Habit formation produced by the analgesic
action of morphine follows a similar time course.

6. Withdrawal of morphine and ethanol after a phase of sub-chronic
intoxication, lasting 5-10 days, produces a reduction in latency of
the cut-off reaction to terminate electrical stimulation of the
brain.

Table 7. The Effect of Test Drugs on Pattern of Induced Defense
Reactions Following Electrical Stimulation of the
Hypothalamic "Defense Area" in Cats

Drug and dose (mg/kg) (ethanol: g/kg)	Initial response pattern					
	Aggressive		Mixed		Avoidance	
	AD	PD	AD	PD	AD	PD
Isotonic solution	+	−	+	+	−	+
Morphine (5-10)	↓	−	0.↓	↓	−	↓
Pentobarbital (3-5)	↓	−	↑	↓	+	↓
Ethanol (0.5)	0.↑.↓(1)	−	↑	↓	+	↓
Chlordiazepoxide (3-10)	0.↓.↑(1)	−	↑	↓	+	↓
Diazepam (1-2)	0.↓.↑(1)	−	↑	↓	+	↓
Amphetamine (1)	↑	−	↓.↑(2)	↑	−	↑

AD: signs of active defense; PD: signs of passive defense for
various kinds of initial defensive response patterns;
+ and −: presence, absence of sign, respectively;
↑ and ↓ facilitating (promotion) and inhibitory effect of drug,
respectively; 0: no effect; (1) effect extremely variable with
defensive response of inherently aggressive type; (2) effect
dependent on intensity of inherent passive defense response displayed.

4. DISCUSSION

Our research shows that DAP's have an unmistakable action on the
functional systems of positive and negative reinforcement. We would
offer the following evidence in support of our original hypothesis
regarding the reinforcement system's role in generating the phenomenon of psychological dependence on pharmacological compounds. These
are DAP's effects of:

- altering the degree of intensity of noxious stimuli,
- modifying behavior patterns in response to stressful influences,
- generally potentiating "rewarding" reactions,
- promoting the taking of DAP's, and finally,
- the response of the reward and punishment systems to the
  withdrawal of the habit-forming substance used.

DAPs' primary reinforcing effect would appear to consist of how
far these substances affect the system of regulation of "rewarding"
reactions. Activation of the reward system produced by DAP's gives
rise to an increase in the "cost of reinforcement" occurring with
electrical stimulation. In other words, the effect of joint stimulation of the system of positive reinforcement by electrical and
pharmacological means is being achieved to a certain extent. Electrical stimulation does not occur in the case of ISR, which may thus

be considered a kind of "chemical" self-stimulation. The way in
which spells of activity and rest alternate are a distinguishing
feature of the operant reaction occurring during ISR in a so-called
"controlled" regimen. If ISR and SS are combined in such a way that
each lever pressing results in brain stimulation and i.v. injection
of a microdose of 1 µl of DAP, the form of operant response will also
produce a wave-like pattern. The dose-effect curve for DAP's effects
on SS usually has a dome-shaped configuration and shows a characteristic reduction in incidence of SS for doses producing catatonia,
stereotypy or motor discoordination. Animals can reach a high level
of intoxication using ISR, however - a feature which radically distinguishes the inhibition of SS produced by neuroleptics from the
restraint placed on operant response by high doses of DAP's.
"Sensitization" of the system of positive reinforcement is reflected
by the effective rise in the level of SS, eventually reaching a
"ceiling" beyond which electrical stimulation is unable to add to the
effects of "chemical" stimulation.

    The interrelatedness of the systems of reward and punishment
suggests that their dependence potential is a factor of changing
balance of function between the two systems. Inhibition of the
perceptive or emotional component of the "punishment" system can in
itself predetermine a switch to dominance of the emotionally positive
system. This disinhibitory means of activating the reward system is
probably highly significant as far as the effects of DAP psychodepressants are concerned, as confirmed by: a) the comparable dose
ranges which are effective in tests for rating the level of function
of these systems b) the effect of switching negative effects to
equivocal ones produced by administering these substances and c) our
own and others' findings on the timing of the onset of tolerance to
"anti-aversive" and euphoriant effects.

    Independent and unmediated factors affecting the workings of the
"reward" system may also contribute to DAPs' effects. The following
considerations bear this out: 1) animals will take opiates, psycho-
stimulants and ethanol of their own accord by intravenous auto-
injection so long as no nociceptive or stressful influence is
involved, 2) amphetamine and cocaine do activate the "reward" system
and help maintain ISR, but also heighten response to adverse stimuli
and 3) activation of SS is observed when electrodes for applying
stimulation are implanted in "equivocal" as well as definitely
"positive" sites.

    Morphine inhibits both the "punished" response and reaction to
painful stimulation, activating the "reward" system simultaneously.
However, tolerance to the "anti-aversive" and analgesic effects of
morphine develops rapidly, while tolerance does not evolve to its
activation of SS when morphine is administered a) chronically, b)
at doses promoting SS (but sub-analgesic and sub-"anti-aversive").
Euphoriant effects are thus seen to emerge superimposed on habitu-

ation to morphine's effect of countering aversion. Dopaminergic
compounds heighten the emotionally positive and attenuate the
analgesic effects of morphine, moreover, while haloperidol does the
opposite. Compounds possessing an antiadrenergic action, such as
pyrroxane and carbidine, dampen the morphine withdrawal syndrome as
displayed in mouse behavior (Patkina and Zvartau, 1980), whereas DOPA
intensifies depressive manifestations of withdrawal in tests involving SS. Furthermore, adrenergic influences attenuate the "rewarding"
effects of opiates. Morphine's "rewarding" action does not appear to
be inseparably linked with its countering of aversion and analgesic
effect, which points to the possibility of separating pain relief and
influencing mood, at least in theory.

AR-1 type response is unconnected with the effects of residual
doses of DAP's since these reactions persist even when abstinence has
continued for 3-10 days. They are produced by treatment with relatively small doses of morphine, pentobarbital and cocaine, but occur
in attenuated form or not at all on raising the daily dose. It
should be remembered that by and large DAP's activating effect on SS
is short-lived. Nor can "rebound" be involved in the genesis of
AR-1, since this phenomenon is produced by administering both stimulants and substances with a primary depressant action.

Activation of the positive emotions is the state from which the
organism, while duly taking into the account the biological function
which they serve, endeavors to derive maximum advantage. We are
aware that sudden withdrawal of narcotics may intensify the activity
of lateral hypothalamic neurones (Olds' "drive-neurones"), while
administering morphine to addicted animals discourages these neurones
from firing. The modulating effect which motivational elements exert
on the "reward" zones could be responsible for the increased effectiveness of positive electrical stimulation of the brain when applied
after the animal has received a few injections of DAP. AR-1 may thus
be viewed as mirroring the evolution of physical and mental mechanisms of craving. A comparable level of prior intoxication produced
by DAP's increases DAP consumption under an ISR regime; interest-
ingly, Burov showed that rats choosing to consume alcohol of their
own accord develop the desire for alcohol over roughly the same
period of time (see preceding articles).

AR-2 offers a particularly striking example of dependence on a
pharmacologically-active substance, in that SS can no longer work
properly at a certain stage of chronic DAP administration in the
absence of the original drug. Depending on the duration of administration, the substance may also maintain its "euphoriant" action or
restore SS to the level of control values only - or ultimately even
to below this level. AR-2 is often associated with tolerance to the
effects of the compounds - although this is not an essential
property, even with morphine administration. If tolerance always
accompanies AR-2, this means that the continuation of SS activation

under the influence of DAP's is associated with AR-1 as well as
AR-2.

The timing of how the reward system's responses to DAP's evolve,
as described above, bears some resemblance to that seen in the course
of human illness.

Substances inhibiting catecholaminergic processes antagonized
the "rewarding" activation evoked by the DAP's tested. Findings on
the two phases in the changing pattern of ISR also confirm that
antagonism occurs. These data are supported by findings on the
important part played by catecholamine release in the mechanisms of
alcoholism and drug addiction.

The involvement of opiate receptors in the "euphoriant" effect
of psychodepressants is interesting to note. The interaction between
ethanol and diazepam on one hand and naloxone on the other may be
partially explained by their effect at the level of the GABA-receptor
or the formation of biogenic morphine-like alkaloids (Davis and
Walsh, 1970; Breuker et al. 1976). However, the suggestion that the
action of these compounds release endorphin-like substances should
also be considered. Indirect evidence in support of this is provided
by the fact that lithium is a successful antagonist (unpublished
findings); it is known to modulate the conformation of opiate receptors - and also the naloxone-sensitive potentiation of diazepam's
antinociceptive action by bacitracin (Wüster et al. 1980).

Compounds possessing an anti-adrenergic action can in themselves
produce some reduction in SS level. This response, which is dose-
dependent, begins spontaneously, but then fades for lack of reinforcement, or the initial spontaneous reaction falters and additional or "follow-up" stimulus is required to reinstate it. The
above compounds thus bring about a reduction in both original motivational tonus and in "reward" effect - this is borne out by the
"drive-creating" effect of electrical stimulation of the hypothalamus.

Naloxone, however, offers an example of more selective antagonism. This compound did not affect original motivational tonus; it
did, however, eliminate opiate induced activation of SS. Hence, two
different approaches may be adopted to the pharmacological control of
addiction (psychological dependence which forms due to the "rewarding" effects of DAP's); one is specific and the other non-specific.
The first approach is physiologically based and involves controlling
the mechanism and achievement of DAPs' effects. The interlinked
sequence of events which involves the positive reinforcement system
in some measure does not apply here. This type of control affects
how the workings of motivation are regulated; and may therefore
extend its action beyond selective suppression of craving for DAP's.
Specific pharmacological control, however, is directed towards the

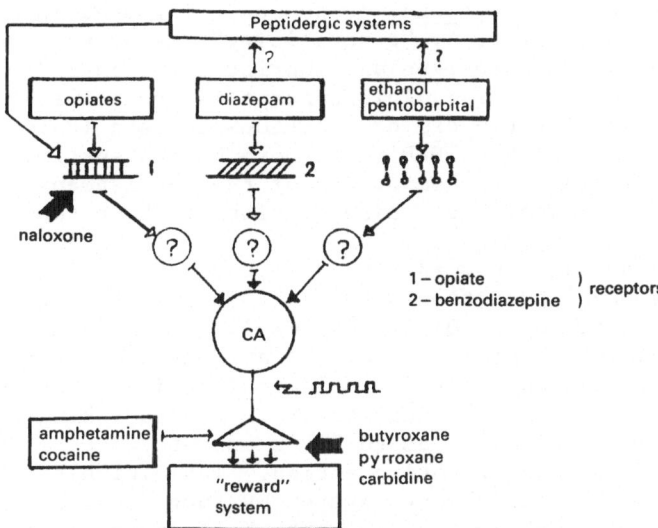

Fig. 1.   Diagram of relationships between substances promoting
           activation of "reward" responses and their antagonists.

original step initiating the sequence of neurophysiological and
neurochemical changes produced by DAP's. The specificity of the
"naloxone-opiate" interaction is probably due to the structural and
functional coinciding of their common target - the opiate receptor.
An antagonist of this type does not alter the tonic motivational
response, but averts the drug of addiction's rewarding effect and the
associated psychological dependence which this brings in its wake.
Evidently, the quest for specific antagonists, especially where the
advances now being made in drug receptor research are concerned,
offers most promise for developing methods of drug addiction control
(Valdman and Zvartau, 1980).

     Our original theoretical point of departure for investigating
the effects of DAP's on "punished" responses was the belief that
stress could be defined as "the organism's response to psychologically aversive (negative, or "punishing") situations" (Valdman et al.
1979). This definition comprises two different aspects; the process
is seen to consist of: a) perception and (negative) appraisal of the
stimulus and b) the organism's response to that stimulus. That is
why models suitable for rating an animal's "punished" state and
investigating the defensive reactions which it produces under stress
were required for this research.

     Our findings enabled us to suggest a twofold composition of the
"punishment" system (Valdman et al. 1976):   a) the perceptual
component of receiving and assigning a "negative" evaluation to the
stimulus; this suffices to accomplish the reaction intended to

suppress it. b) The emotional component, displayed in a "punishing" situation, takes effect by "generalizing" the process sufficiently to suppress conflicting motives and imprint the signal to stop on the emotionally-modal memory. This concept is evocative of the physical and mental patterns typical of reactions to pain (Valdman, 1972). Our research showed that DAP's can both have a specific action on the perceptual or emotional components of the "punishment" system and bring about pharmacological dissociation of these two systems.

It may be asked how far the effects observed are decisive in determining DAPs' addiction potential. It is a fact that stress increases the taking of DAP's, especially ethanol, as shown by Burov in the preceding chapters. From the evidence obtained, we have been able to forge the last link in a logical chain. Even so, we have no scientifically proven explanation of increased taking of DAP's under stress; rather, we have classed DAP's as substances with a stress-protective or stress-provoking action. Inhibition of the perceptual component of the "punishment" system by such drugs as opiates, ethanol or barbiturates may, in fact, be responsible for their dis-adaptive effect in situations requiring coordinated operant activity. Sensitization to the effects of aversive stimulation produced by DAP withdrawal may be looked upon as the root of a number of the emotional signs of withdrawal syndrome.

A few of the predictors of harmfully addictive effects in pharmacological agents provided by our investigations are as follows. These are a compound's ability to:

- activate the "reward" system
- potentiate primary reinforcement effect
- create dependence (as judged by the level of function of the positive reinforcement system)
- depress the function of the negative reinforcement system
- create dependence (as judged by the level of function of the negative reinforcement system)
- alter the pattern of defensive reactions observed in the animal's behavior in a certain way.

One of the tasks now facing experimental pharmacologists has thus become clear - that of detecting such predictors at the stage of preclinical research into new pharmacological products in this country, and of organizing such information into the form of practical recommendations.

REFERENCES

Barkov, N. K., 1973, About the mechanisms of carbidine action, Pharmacol. Toxicol., 2:154-157.
Davis, V. E., and Walsh, M. J., 1970, Alcohol, amines and alkaloids: possible biochemical basis for alcohol addiction, Science, 167:1005-1007.

Delgado, J., Roberts, W., and Miller, N., 1954, Learning motivated by
    electrical stimulation of the brain, Amer.J.Physiol., 179,
    587-593.
König, J. F. R., and Klippel, R. A., 1963, The rat brain, in: "A
    Stereotaxic Atlas of the Forebrain and Lower parts of the
    Brain Stem," Williams and Wilkins, Baltimore.
Krylov, S. S., and Starych, N. T., 1973, Pharmacological properties
    of pyrroxane, Pharmacol.Toxicol., 4:396-399.
Olds, J., and Milner, P., 1954, Positive reinforcement produced by
    electrical stimulation of the septal area and other regions of
    the brain, J.Comp.Physiol.Psychol., 47:419-427.
Patkina, N. A., and Zvartau, E. E., 1980, Influence of some psycho-
    tropic drugs on manifestations of morphine abstinence syndrome
    in mice, Pharmacol. Toxicol., 43:19-24.
Snieder, R. S., and Niemer, W. T., 1961, A stereotaxic atlas of the
    cat brain, University of Chicago Press.
Valdman, A. V., 1972, The neuropharmacology of narcotic analgesics,
    "Meditsina", Leningrad.
Valdman, A. V., Zvartau, E. E., and Kozlovskaya, M. M., 1976, Psycho-
    pharmacology of emotions, "Meditsina", Moscow.
Valdman, A. V., Kozlovskaya, M. M., and Medvedev, O. S., 1979, Phar-
    macological regulation of emotional stress, "Meditsina",
    Moscow.
Valdman, A. V., and Zvartau, E. E., 1982, Systems of reinforcement
    and drug dependence, Drug and Alcohol dependence, in press.
Valdman, A. V., and Zvartau, E. E., 1980, The influence on emotion-
    ally reward brain systems as a way of pathogenetic therapy of
    alcohol dependence, J.Neuropath.Psychiatr., 80:1020-1024.
Weeks, J. R., 1962, Experimental morphine addiction: method for
    automatic intravenous injections in unrestrained rats,
    Science, 138:143-144.
Wikler, A., Martin, W. R., Pescor, P. T., and Eades, C. G., 1966,
    Factors regulating oral consumption of an opioid/etonitazene
    by morphine-addicted rats, Psychopharmacologia, 3:55-78.
Wüster, M., Duka, T., and Herz, A., 1980, Diazepam induced release of
    opioid activity in the rat brain, Neurosci.Lett., 16:335-342.
Zvartau, E. E., 1977, Hypothalamic self-stimulation under the chronic
    morphine treatment, Res.Comm.Chem.Pathol.Pharmacol., 16,
    4:707-720.
Zvartau, E. E., 1979, A simple modification of the method of intra-
    venous drugs, self-administration by experimental animals,
    J.High.Nerv.Activ., 29:877-879.
Zvartau, E. E., 1979, Action of naloxone on emotionally positive and
    anti-nociceptive effects of hypothalamic stimulation in rats,
    Bull.Exper.Biol. Med., 11:569-572.
Zvartau, E. E., 1980, Influence of naloxone on emotionally positive
    reaction activation produced by drugs with a dependence
    liability, Pharmacol.Toxicol., 3:313-317.
Zvartau, E. E., Patkina, N. A., 1974, Motivational components and
    self-stimulation under behavioural reactions produced by

electrical stimulation of the hypothalamus in cats, J.High.
Nerv.Act., 24:529-535.
Zvartau, E. E., and Patkina, N. A., 1978, Changes of hypothalamic
escape response in rats under chronic morphine administration,
J.High.Nerv.Act., 28:1020-1026.
Zvartau, E. E., and Sidorov, V. I., 1980, Technique of automatic
recording of temporal characteristics of brain electrical
self-stimulation reaction, J.High.Nerv.Act., 30:865-867.
Zvartau, E. E., Sheremet, V. V., and Malinina, V. N., 1981, Effect of
euphorigenic drugs on emotionally positive brain systems,
J.High.Nerv.Act., 31:308-314

Characteristics and Mechanisms of Tolerance to Benzodiazepines

T.L. Garibova and T.A. Voronina

Institute of Pharmacology, Academy of Medical Sciences
of the USSR, Moscow

1. INTRODUCTION

Tranquilizers of the benzodiazepine group, which includes diazepam, chlordiazepoxide, nitrazepam, lorazepam and phenazepam, are widely used in various branches of medicine to treat certain types of neuroses. They affect such conditions as fear, anxiety, obsessive states and insomnia (Avrutsky et al., 1974; Aleksandrovsky, 1976, 1979; Greenblatt and Shader, 1974, etc). Benzodiazepines are amongst the most widely used psychotropic agents and are employed for long-term treatment, often in out-patients, without proper medical supervision. A large body of experimental and clinical evidence has shown that they can, like the majority of hypnotics, sedatives and anxiolytics, be habit-forming and produce tolerance or even physical and mental dependence in some instances if taken over long periods. Most workers in this field, however, like Marks in his review article of 1978, consider the risk of benzodiazepine dependence fairly low in comparison with other sedatives and anxiolytics in use. Dependence generally occurs when benzodiazepines have been used in high dosage over long periods by individuals with unstable personalities.

Our understanding of the action of benzodiazepines has been considerably extended during recent years, especially with regard to the involvement of GABA-ergic and serotoninergic systems in these drug's action (Costa et al., 1975; Lippman and Pugsley, 1974; Agarwal et al., 1977; Rastogi et al., 1978). Squires and Braestrup (1977), Mohler and Okada (1977) and Speth et al. (1979) attached considerable importance to the effects of benzodiazepine binding to specific benzodiazepine receptors in animal and human brain. Findings made to date on the neurochemical and molecular mechanisms of

453

benzodiazepines are usually obtained from acute treatment with these
compounds; the field remains wide open for identifying the mechanisms
of benzodiazepine tolerance and dependence during long-term administration.

Contemporary writers such as Vikhlayev et al. (1978), Kalant
et al. (1971) and Garattini (1978) offer a two-fold approach to
understanding the mechanism underlying tolerance and dependence -
pharmacokinetic changes in processes of absorption, distribution,
metabolism and excretion, and adaptational changes in interaction
between the substance and substrate or receptor. The present work
was undertaken with these subjects in view. It describes findings
and concepts regarding the characteristics of and mechanisms underlying tolerance and dependence as developed to compounds of the
benzodiazepine group.

2. MATERIALS AND METHODS

Our investigations on tolerance to tranquilizers of the benzodiazepine group were based on experiments using rats and mice. We
adopted composite methods, with the comparison of the broad range of
the substances' pharmacological effects (acute versus chronic administration) as our point of departure (Vikhlayev et al., 1970). A
comparison of this type enabled us to a) determine whether tolerance
was established or not, b) reveal the strength of each of its effects
and c) find whether the drug's therapeutic range was altered in any
way, according to whether it had been administered acutely or chronically.

We selected methods enabling us to compare the performance of
the tranquilizers, under similar conditions, in achieving several
given effects.

The following products were used: phenazepam [7-bromo-5-ortho-
chlorophenyl(1,2-dihydro-3H-1,4-benzodiazepin-2-one)], developed in
the USSR (Voronina et al., 1982), chlordiazepoxide ("Polfa"), diazepam (Gedeon Richter) and lorazepam (Wyeth). Purified substances in
crystalline form were selected for these experiments, which were
performed on cross-bred male rats and mice. The compounds were
administered orally or intraperitoneally, in suspension with
Tween-80, 1-1½ h and 30-40 min prior to the experiments, respectively. Administration took place daily at 10 a.m. and ended at
15 h over a period of 5-60 days. The animals were kept in a vivarium
and experimental enclosures with water and food (consisting of dry
pellets) freely available. The effect of each dose of test substance
was studied on 8-16 animals. Statistical analysis was performed
calculating the ED_{50} (Litchfield and Wilcoxon, 1949) and the arithmetical mean with standard error of the mean, where p = 0.05
(Belenky, 1959).

The following methods were employed:

1. That of the conflict situation - one of the most selective methods of rating the performance of anxiolytics in the laboratory (Vikhlayev and Kligul, 1973; Cook and Davidson, 1976). This experiment involved training rats to drink water from a trough in the enclosure after thirst had developed. To this end the animals were kept on dry feed for 48 h and then taken, always at the same time of day, to an enclosure where water was provided. This procedure was continued for 3 days. On the day of the experiment, electrical stimulation was given by passing an 80 V electric current through the trough. A conflict situation was thus created by making the two different reflexes of drinking and self-defense clash. The following indices of the rats' behavior were then noted during a 20 min period: incidence of drinking, regardless of the painful electrical stimulation which this involved, the number of approaches made to the drinking trough and overall level of motor activity. All this information was automatically recorded and indices of behavior fed into a computer in numerical form (Kligul and Krivopalov, 1966).

2. Maze test and drinking reflex. The effects of these substances on conditioned drinking reflex were studied using the T-maze device. The rats had been previously trained to find the trough of water in the maze, while the time taken to cross from the starting enclosure to the water trough and the number of mistakes made (e.g. turning round and taking the wrong turning to the trough) were recorded. These investigations were performed using animals with a strongly-developed reflex. A total of 5 out of 5 successful attempts to locate the trough with a latent period not exceeding 10 sec over 2-3 days was taken as the criterion of reliability for this reflex.

3. Outside interference with the above reflex. Both this and the conflict situation device were used to assess the drugs' tranquilizing effects. Rats' fixed drinking reflex pattern in a T-maze was inhibited by beaming a bright light two seconds before the 2nd or 3rd conditioned reflex took place; the ensuing 5th, 6th and 7th reflexes followed the normal pattern. Performance of the conditioned reflex was judged according to the number of errors made and the latent period of passing through the maze. Switching on the stimulus (light) before the reflex occurred produced pronounced inhibition of fixed conditioned reflex in control animals with a developed pattern marked by a great increase (of 500%) in the latent period preceding running through the maze.

4. Aggressive behavior in rats. This was analyzed according to the changes in aggression threshold observed in rats subjected to increasing sporadic stimulation via strengthening burst of current passed through an electrode floor. Aggression threshold was taken as the minimum strength of current at which "fighting" broke out amongst the animals (Tedeschi et al., 1959).

5. <u>Antagonism to pentylenetetrazole</u>. Antagonism to pentylenetetrazole was assessed in mice according to the drugs' ability to counter the tonic-clonic element of convulsion. A dose of 110–130 mg/kg pentylenetetrazole (ED_{95} or the dose producing convulsions in 95% of the animals) was administered subcutaneously 10 min before the maximum effect of the test drugs was seen (Swinyard et al., 1952).

6. <u>Antagonism to thiosemicarbazide</u>. Interaction between these tranquilizers and thiosemicarbazide was studied in experiments on mice. It was measured according to the drugs' ability to prevent an i.p. dose of thiosemicarbazide from killing the animals. The drugs were administered 30 min before injecting thiosemicarbazide and observations on how the animals were affected continued for 90 min.

7. <u>Antagonism to strychnine</u>. The drugs' ability to antagonize strychnine was judged by their success in preventing strychnine's convulsant and ultimately lethal effects in animals. A 2.5 mg/kg dose of strychnine (ED_{95}, or the dose giving rise to convulsions and death in 95–100% of the animals) was administered subcutaneously 40 min after tranquilizer administration. Animals which had survived for an hour after treatment with strychnine were considered "protected" against its effects.

8. <u>The prevention of maximum electroconvulsant fits in mice</u>. Maximum electroconvulsant fits were applied by running an alternating current of 50 Hz and 50 mA strength lasting 0.2 sec through the animal's head by means of corneal electrodes. The drug's anticonvulsant effect was judged by its ability to prevent tonic-extensor fits (Toman, 1946).

9. <u>Loss of coordination in movement (rats and mice) as shown in the rotating rod test</u>. Loss of coordination in movement, gait and balance was measured by the rotating rod test. Rats and mice were placed on a horizontal rod measuring 2 cm across, revolving at the rate of 5 rpm. The animals' inability to stay balancing on the rod for two minutes after treatment with the drug was interpreted as loss of coordination (Dunham and Mija, 1957).

10. <u>Pentylenetetrazole sensitivity threshold</u>. The intravenous titration method was used in order to investigate drugs' effects on pentylenetetrazole sensitivity threshold (Orloff et al., 1949). A 1% solution was administered into the caudal vein of mice at a constant rate of 0.01 ml/sec and the sequence of the three stages of this substance's action were measured, viz: clonic and tonic convulsions and clonic twitching. The corrected volume of pentylenetetrazole administered in mg/kg body weight was recorded during the onset of these stages. Pentylenetetrazole sensitivity threshold was set according to the quantity administered. Degree of change in sensitivity to this substance was also estimated compared with the control, which was taken as one.

11. **Means of investigating "withdrawal syndrome"**. Withdrawal syndrome was assessed according to an animal's impairment of conditioned
reflex. A conditioned maze reflex was previously developed in the
animals together with positive reinforcement (in the form of water)
in the T-maze (fully described in Section 2). These animals were
then given 2 mg/kg phenazepam daily for 30 days. Speed and accuracy
in performance of this reflex were monitored in rats receiving
phenazepam. Withdrawal syndrome first set in 24 and 48 h after the
drug was withdrawn, whereupon monitoring of successful performance of
conditioned reflex began. Changes of an autonomic and emotional
nature were also taken into account.

3. RESULTS

3.1. BENZODIAZEPINE TOLERANCE

Our laboratory investigations on tolerance were carried out by
administering the drugs over long periods while monitoring their main
observable effects continuously. Changes in a drug's action as a
tranquilizer were identified using "conflict situation" and "outside
interference" devices; similarly, it was tested for:

> sedative action by how long the animal spent running through the
> maze,
> anti-aggressive action by effects on aggression threshold, as
> provoked by painful electrical stimulation applied to pairs of
> rats,
> anti-convulsive action by antagonism to the convulsant effects
> of pentylenetetrazole, thiosemicarbazide and strychnine,
> anti-convulsive action by performance in averting convulsions at
> ceiling electric shock levels,
> myorelaxant action by discoordinating effects on movement.

Long-term oral administration of phenazepam or lorazepam over
30 days produced no substantial change in tranquilizing effect in a
conflict situation compared with the effects of acute administration
(see Table 1). Our findings show that phenazepam maintains its
ability to influence emotional and behavioral effects in a conflict
situation and reduce the general level of "neurosis" in animals'
behavior, even after a four-week period of daily administration.
Lack of tolerance to the tranquilizing effects of diazepam, chlordiazepoxide, oxazepam and nitrazepam is also typical (Vikhlayev et
al., 1970).

In the context of outside interference with trained rats' "maze
behavior" by a beam of light, the pattern of separate changes produced by long-term administration of phenazepam and lorazepam differs. An acute i.p. dose of 1 mg/kg of both substances extends the
length of time spent running through the maze in the absence of

Table 1. The Effects of a) Phenazepam (2 mg/kg, i.p.), b) its
         Metabolites and c) Lorazepam (2 mg/kg i.p.) on Rat Behavior
         Tested in a Conflict Situation. Acute and Long-term
         Administration

Substance	No. days treatment	No. approaches made to drinking trough	Water-taking (counts)	Motor activity
Control (suspension with Tween-80)	1	6.70	1.70	14.0
		4.20÷ 9.20	1.08÷ 2.32	7.73÷20.27
Control (suspension with Tween-80)	30	3.16	1.66	10.66
		0.96÷ 5.36	0.76÷ 2.56	3.46÷17.86
Phenazepam	1	4.70	9.85	16.57
		2.86÷ 6.56	4.41÷14.29	9.17÷23.97
Phenazepam	30	5.62	10.87	10.62
		1.23÷10.01	5.54÷16.20	5.29÷15.95
Lorazepam	1	12.33	10.50	15.00
		5.97÷18.69	5.96÷15.04	6.37÷23.63
Lorazepam	30	14.10	13.00	7.00
		10.43÷17.77	9.09÷16.91	3.58÷10.42

Substance	Dose in mg/kg i.p.	Incidence of drinking notwithstanding painful electrical stimulation
Control	Suspension with Tween-80	2.2(1.8-2.6)
Phenazepam	0.5	7.5(5.5-9.5)
3-Hydroxyphenazepam	0.5	9.4(7.8-11)
Chinazolinone	20	6.5(3.8-9.2)
Benzophenone	80	2.6(1.8-3.4)

outside interference, thereby showing their inhibitory effect on the
maze reflex. Treatment at the same dosage over an 8-day period,
however, produces a considerable reduction in the drugs' sedative
effect. This is further brought out by a shortening of the reflex
to control levels seen in intact animals (see Figure 1). The tranquilizing effect, on the other hand, viewed as the drugs' ability to
suppress outside interference, continues unchanged, even when lorazepam and phenazepam administration has lasted a month. A clear-cut
stratification of tranquilizing and sedative effects is thus observed
during long-term administration of benzodiazepines, as illustrated by
the " outside interference" set-up described.

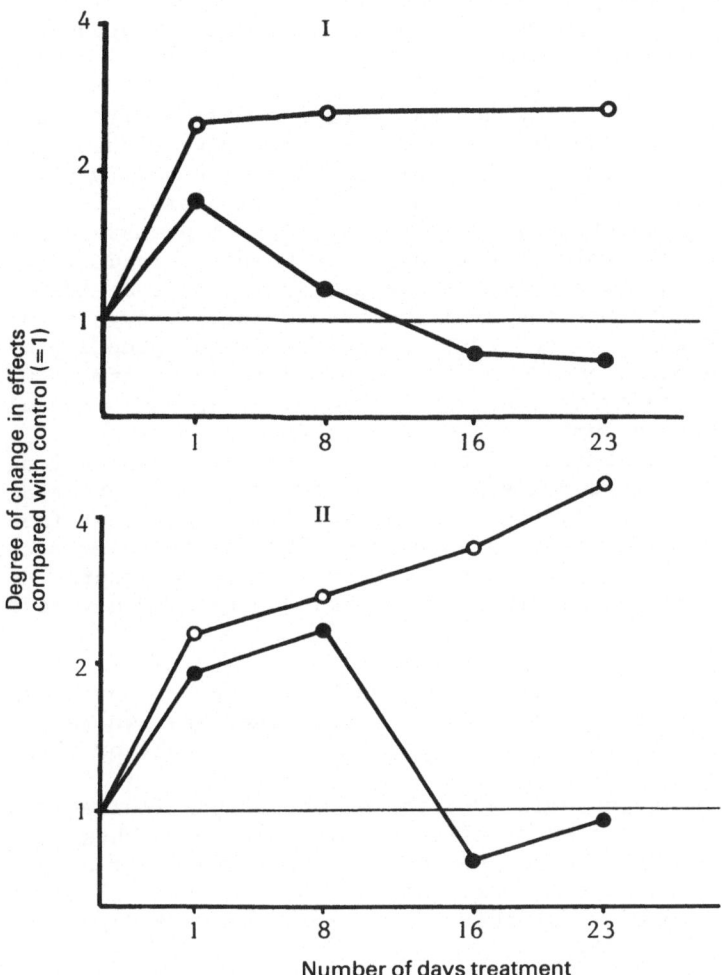

Fig. 1.   "Layering" of benzodiazepines' sedative and tranquilizing
          (anxiolytic) effect during long-term administration to rats
          in a context of outside interference. Abscissa - No. of
          days of treatment; ordinate - degree of change in effects
          compared with control (=1).   -•-•- sedative effects; -o-o-
          tranquilizing effects.   I - phenazepam at a dose of 1 mg/kg,
          i.p.; II - lorazepam at a dose of 1 mg/kg, i.p.

     Substantial changes in anti-aggression effects were noted during
long-term treatment with phenazepam, expressed by the drop in aggression threshold produced by applying painful electrical stimulation to
pairs of rats. Hence, a 1 mg/kg dose of phenazepam, administered for

2 weeks, reduced rats' aggression threshold by a half to
1.1(0.7-1.4) mA as compared with the 2.2(1.8-2.6) mA produced by
acute administration.

Benzodiazepines' anticonvulsant effects were contrasted
with those of phenobarbital and diphenylhydantoin in the course of
a long-lasting experiment. It had already been established, using
various devices, that different changes in convulsant response had
occurred during long-term i.p. administration of phenazepam to
mice. Unmistakable tolerance to phenazepam was noticed from this
drug's ability to prevent tonic-extensor convulsions at maximal
electro-convulsant fits and from its countering of strychnine's
convulsant effects. No substantial attenuation of phenazepam's
ability to prevent clonic convulsions produced by treatment with
pentylenetetrazole or of countering thiosemicarbazide's convulsant
action was observed during long-term treatment (see Figure 2).
Unlike phenazepam, phenobarbital, when administered for a long
period (18 days), produced signs of tolerance judging by antagonism
to pentylenetetrazole and the maximum electric shock test (see
Figure 2). Diphenylhydantoin's anticonvulsant effect, as recorded
in the maximum electric shock test, underwent no substantial changes
when this compound was administered to mice over a 30-day period
(see Figure 2).

It was thus seen that each of the test drugs - phenazepam,
diphenylhydantoin and phenobarbital - possessed its own range of
anticonvulsant activity, whether administered acutely or chronically.

During long-term administration of phenazepam indications of
tolerance to its myorelaxant effects were observed. When acute
doses were given of 2 mg/kg phenazepam or lorazepam, or alternatively
5 mg/kg diazepam or nitrazepam, the movements of 80-100% of the rats
became discoordinated, but this effect completely disappeared in all
animals by the 14th day.

Mice developed tolerance to myorelaxant effects even more rapidly. Discoordination persisted in the movements of only 20% of the
animals 24 h after repeated doses of 10 mg/kg phenazepam. Similar
changes were observed when lorazepam, diazepam and oxazepam were
administered repeatedly (see Figure 3). Such an abrupt loss of
effect shows that a substantial degree of tolerance has developed.
The extent of the myorelaxant effect produced in mice largely depends
on how frequently the drug is administered. Hourly treatment with
30 mg/kg phenazepam i.p. produced maximal myorelaxant effect, with
50% of the animals adopting the sideways position. With two-hourly
treatment myorelaxant effects stayed at roughly the same level, while
4-hourly administration progressively reduced these effects, which
completely disappeared in all animals when phenazepam was given on an
8-hourly schedule.

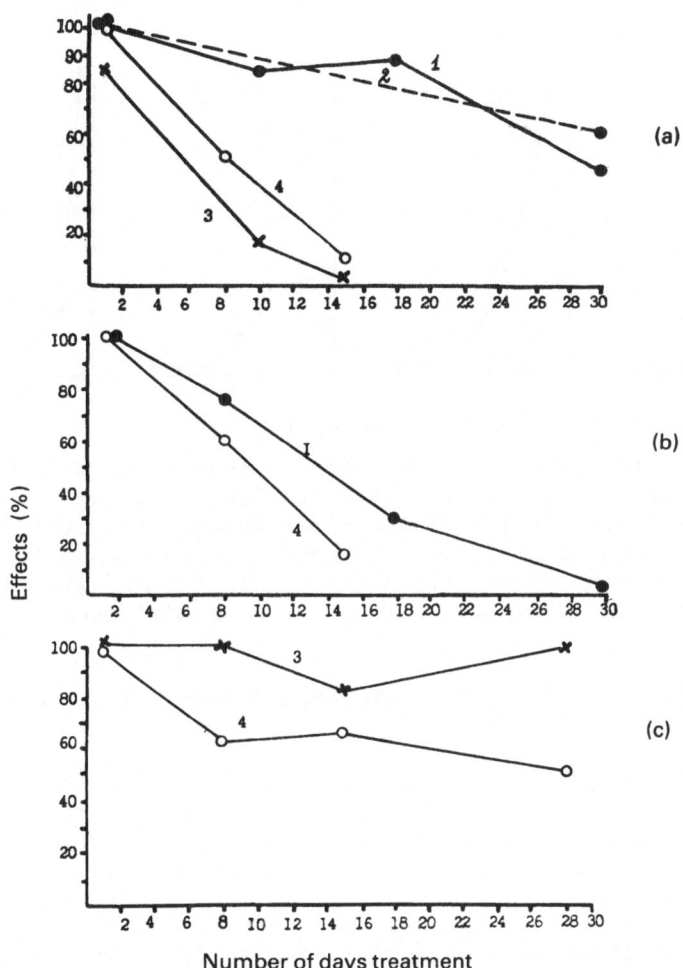

Fig. 2. Anticonvulsant effects of phenazepam, phenobarbital and
diphenylhydantoin during long-term administration to mice.
Abscissa - no. of days of treatment; ordinate - effect (%).
(a) - range of phenazepam; (b) - range of phenobarbital;
(c) - range of diphenylhydantoin. 1 - antagonism to
pentylenetetrazol (phenazepam - 0.5 mg/kg, phenobarbital -
15 mg/kg); 2 - antagonism to thiosemicarbazide (phenazepam
0.5 mg/kg); 3 - antagonism to strychnine (phenazepam
25 mg/kg, phenobarbital 15 mg/kg, diphenylhydantoin -
80 mg/kg); 4 - prevention of convulsions at maximum
electric shock level (phenazepam - 25 mg/kg, phenobarbital -
25 mg/kg, diphenylhydantoin - 25 mg/kg).

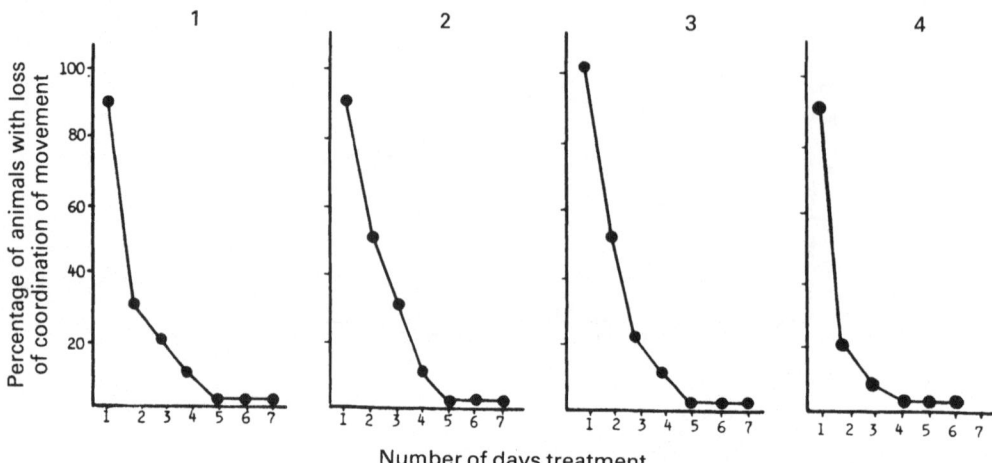

Fig. 3.   Changes in the myorelaxant effect produced in mice by
          long-term tranquilizer administration. Abscissa – no.
          days of treatment; ordinate – percentage of animals with
          loss of coordination in movement. 1 – lorazepam (30 mg/kg;
          2 – diazepam (10 mg/kg); 3 – oxazepam (10 mg/kg); 4 –
          phenazepam (10 mg/kg). All substances administered i.p.

3.2.   BENZODIAZEPINE WITHDRAWAL – MECHANISMS OF TOLERANCE AND
       WITHDRAWAL SYNDROME

       Termination of long-term benzodiazepine administration produces
substantial changes in animal behavior. The most marked of these is
the behavioral disruption noted in the pattern of an animal's conditioned reflex actions during the course of phenazepam's sedative
effects. As mentioned earlier, an acute dose of 2 mg/kg phenazepam
inhibits performance of conditioned "maze reflexes". Prolonged
administration of phenazepam extending over 30 days results in habituation, associated with reduced activity, while termination of treatment with the drug produces "withdrawal syndrome" with characteristic
distortion of its effects. Between 24 and 48 h after the last treatment with phenazepam general inhibition (sluggishness) sets in,
response patterns to test devices become inadequate and conditioned
reflex performance (time spent in maze-running) takes 20 times longer
(see Table 2). When placed in the maze, the animal tenses, squeaks
and shows signs of tachycardia and accelerated respiration rate.

       For the purpose of studying the anti-aggression aspects of with-
drawal syndrome's effects, aggression threshold was monitored 24 h
after the 28th-32nd administration of diazepam and nitrazepam at doses
which had raised aggression threshold considerably when administered
acutely to rats. It was established that termination of phenazepam
treatment made animals more aggressive, as expressed in a drop

in aggression threshold to below that of untreated control animals, while aggression had set in spontaneously in 80% of the animals (see Figure 4).

An examination of the anticonvulsant effects of withdrawal syndrome showed that phenazepam and clonazepam, having been administered for 7 days and subsequently withdrawn, produced a statistically significant drop in sensitivity threshold to all effects induced by pentylenetetrazole - clonic twitching, clonic convulsions and tonic convulsions - to below control levels calculated for untreated animals (see Figure 4).

Behavior is thus seen to be evidently disrupted when long-term phenazepam administration is discontinued. This is illustrated by rats' impaired performance in the conditioned "maze reflex", their suppressed exploratory behavior in "open field" situations, heightened aggressiveness and a reduction in their convulsive threshold. All these effects are easily rectified. Administering phenazepam to the animals completely eliminated withdrawal syndrome in itself. Supposedly, the disturbances seen after drug withdrawal represent a kind of compensatory response, which had been held in check by the drug during the period of its administration. Such effects can probably be considered signs of the withdrawal syndrome occurring after treatment with some medicament has ended.

The power of various psychotropic substances and putative endogenous benzodiazepine receptor ligands to replace benzodiazepines under schedules involving repeated administration was investigated, to help explain the mechanisms of benzodiazepine tolerance.

A single high dose of 40 mg/kg phenazepam produced marked disruption of a myorelaxant type in all animals - i.e. discoordinated movement and inhibited stretching reflex. When phenazepam was administered repeatedly, its myorelaxant effect was completely suppressed. On the 6th day, when tolerance to phenazepam had set in, it was possible to replace phenazepam, preserving the tolerance effects, with either lorazepam or diazepam, while treatment with grandaxin produced myorelaxant effects in only 20% of the animals. Chlorpromazine, tryptazine, ethanol, phenobarbital, non-benzodiazepine tranquilizers such as meprobamate and atarax, or muscimol (a GABA-ergic receptor agonist), when used after tolerance to phenazepam had developed, all suppressed this tolerance and produced pronounced myorelaxant effects in all animals. These effects distinguished them from benzodiazepines. Lonetil, calcium valproate, sodium hydroxybutyrate, inozine and nicotinamide were all able to replace phenazepam in some measure and preserve the tolerance developed to its effects.

This experiment was reversed in the second series; phenazepam was administered after tolerance had set in to other substances, which had already been administered for 10 days. Under these

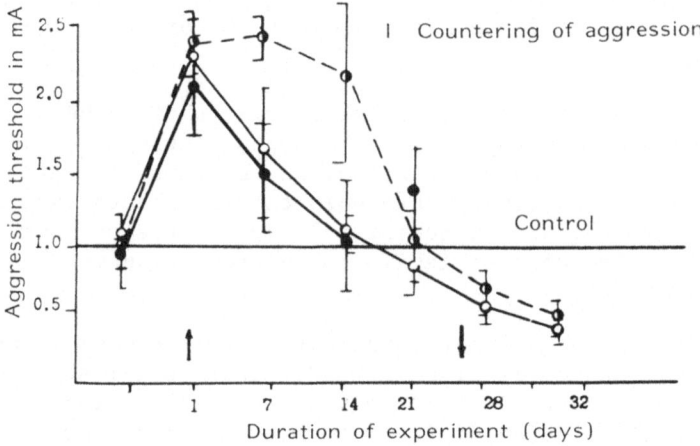

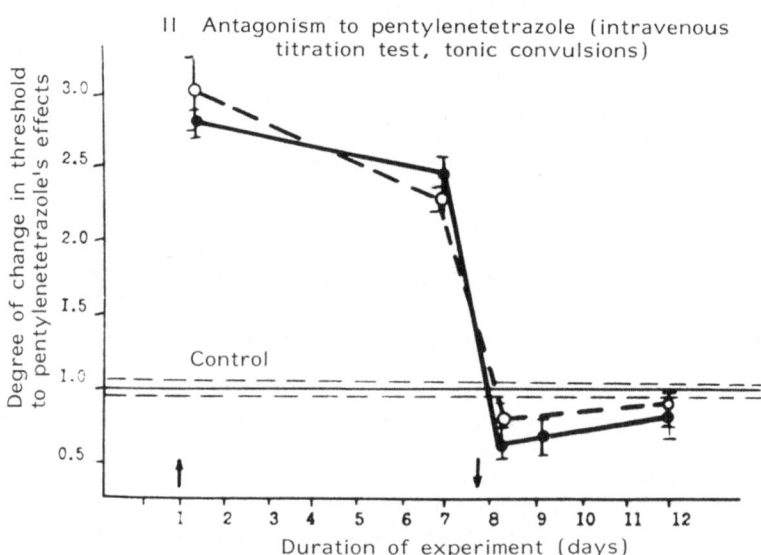

Fig. 4. Changing pattern of benzodiazepines' effects during long-
         term administration and after withdrawal. I. Countering of
         aggression. Abscissa - duration of experiment (days);
         ordinate - aggression threshold in mA. -●-●- nitrazepam
         (5 mg/kg); -o-o- phenazepam (1 mg/kg); -●-●- lorazepam
         (1 mg/kg); ↑ - start of daily administration of drug to
         rats; ↓ - end of daily administration of drug to rats.
         II. Antagonism to pentylenetetrazole (intravenous titration
         test, tonic convulsions). Abscissa - duration of experi-
         ment; ordinate - degree of change in threshold to pentylene-
         tetrazole's effects. -●-●- phenazepam (2.5 mg/kg); -o-o-
         clonazepam (3.5 mg/kg); ↑ - start of daily administration of
         drug to mice; ↓ - end of daily administration of drug to mice.

circumstances it was found that phenazepam was able to replace
nicotinamide in part and diazepam in full. Treating animals which
had built up tolerance to meprobamate or phenobarbital with phenazepam, moreover, produced myorelaxation in all the mice.

The third set of experiments involved administering phenazepam
and other compounds alternately. A "sandwiching" of the various
substances was achieved by administering phenazepam alternately with
meprobamate, atarax, chlorpromazine or phenobarbital (one day on
phenazepam, one on "substitute"), each substance independently from
the others, irrespective of their common features. It was thus seen
that the myorelaxant effect of phenazepam when alternated in this way
had disappeared by the 2nd-3rd day of treatment, while the effects of
chlorpromazine, meprobamate, atarax and phenobarbital remained steady
in all animals, even when treatment had been continuing for as long
as 10 days.

Quite the opposite occurred when phenazepam was alternated with
members of the benzodiazepine group, such as diazepam or lorazepam.
In this situation, the drugs interreacted with each other and the
tolerance pattern was just the same as for phenazepam given alone.
Nicotinamide and calcium valproate were the only non-benzodiazepines
which showed even partial cross-tolerance with phenazepam when the
test substances were administered alternately (as tolerance first
began to develop). The only substances which showed complete cross
tolerance with phenazepam as regards myorelaxant action were benzodiazepines. Neither neuroleptics, ethanol, tranquilizers with a
different chemical structure - e.g. meprobamate and atarax - nor the
GABA-ergic receptor agonist, muscimol, were able to replace benzodiazepines. Nicotinamide, chlorpromazine, calcium valproate and
lonetil all showed partial cross-tolerance with phenazepam.

The chances of arresting the onset of withdrawal syndrome which
arises when long-term administration of phenazepam terminates were
considered, in order to throw light on the specificity of withdrawal
syndrome. Other psychotropic agents resembling phenazepam's effects
in some way were administered to animals withdrawn from phenazepam.
These included tranquilizers differing in their action - diazepam,
meprobamate, trioxazin and lonetil, phenobarbital and the neuroleptic, chlorpromazine. These substances were used at dosages which
had produced a similar inhibitory effect on stable conditioned reflex
when administered acutely to control animals.

It was found that diazepam, one of the benzodiazepine group, was
the most successful in substituting for phenazepam and completely
eliminating withdrawal syndrome effects. When phenazepam was replaced by diazepam, complete recovery of the conditioned "maze reflex" (disrupted by withdrawal) and righting of emotional disturbances was observed. Tranquilizers belonging to other groups of
chemical compounds, such as meprobamate and trioxazine, were unable

to arrest the onset of withdrawal syndrome, nor did the neuroleptic chlorpromazine. When the latter compounds were administered to rats, their state of inhibition and fear persisted and their conditioned reflex remained unimpaired. A partial (25%) recovery in the performance of conditioned reflex was observed, while some animals emerged from a stuporose state when phenazepam was replaced by the new Bulgarian tranquilizer lonetil (a chinazoline derivative). Phenobarbital produced a similar effect; it prevented withdrawal syndrome developing in some of the drug-deprived rats (see Table 2).

As with crossed tolerance, it was thus seen that only benzodiazepines (phenazepam and diazepam) were fully able to eliminate withdrawal syndrome arising after discontinuation of long-term administration of phenazepam. Tranquilizers and sedatives belonging to other chemical groups were either unable to replace benzodiazepine, as was the case with meprobamate, trioxazine and chlorpromazine, or were only partially successful in restoring reactions which had been disrupted by phenazepam withdrawal (e.g. lonetil and phenobarbital).

The neurochemical mechanisms of tolerance and withdrawal syndrome were examined during long-term treatment with phenazepam to throw light on the part played by different transmitter systems in shaping these phenomena. Tolerance and onset of withdrawal syndrome were studied in the course of experiments on impairment of conditioned reflex performance and the countering of aggression in rats. The following chemical tools were used to examine the role of GABA-ergic transmission in the onset of withdrawal syndrome produced by withdrawal of phenazepam administration: depakin (the calcium salt of dipropylacetic acid - calcium valproate) - a GABA agonist which raises GABA concentration in the brain, thiosemicarbazide - an inhibitor of GABA synthesis and bicuculline, which blocks GABA-ergic receptors. It was found that depakin corrects the behavioral disturbances produced by discontinuing long-term phenazepam administration. In the majority of animals (60%) depakin's effects abolished stupor, restored conditioned reflex performance, abolished spontaneous aggression and normalized aggressive response threshold. Improved performance in maze-running is not connected with any stimulating effect of depakin, though, since it had hardly affected reflex performance in control animals. The "GABA negative" substances thiosemicarbazide and bicuculline, however, have the reverse effect of reinforcing the effects of withdrawal syndrome, while countering phenazepam's corrective action in cases of withdrawal syndrome (see Table 2).

The compound 5-hydroxytryptophan, the serotonin precursor, hardly affects the withdrawal syndrome pattern arising after discontinuation of long-term administration of phenazepam. Treatment with α-methyl DOPA, a substance which tends to prevent catecholamine synthesis by inhibiting DOPA decarboxylase, is largely successful in

Table 2. Drugs' Effects on Withdrawal Syndrome as Observed 24 h
          after Discontinuation of Long-term Phenazepam Adminis-
          tration (2 mg/kg per 24 h)

Substance	Dose in mg/kg	Interval between drug administration and experiment	Duration of "maze reflex" (sec)
Suspension with Tween-80		30 min	104.2(96.1-112.3)
			"withdrawal syndrome"
Phenazepam	2	30 min	2.4 (1.6-3.2)
Diazepam	5	30 min	1.5 (1.3-1.7)
Meprobamate	100	40 min	87.6(60.7-114.5)
Trioxazine	100	40 min	107.3(95.7-118.9)
Lonetil	100	40 min	77.9(65.2- 90.6)
Phenobarbital	40	40 min	60.7(31.6- 89.8)
Chlorpromazine	1	1 h	98.4(70.1-126.7)
Calcium valproate	200	1 h	45.3(37.1- 53.5)
α- Methyl DOPA	200	24 h	62.9(50.3- 75.5)
5-Hydroxytryptophan	100	4 h	93.7(81.6-105.8)
3,4-DOPA	200	30 min	120
Thiosemicarbazide	4	20 min	120
Disulfiram	230	20 h	101.8(89.9-113.7)
Phenazepam	2	30 min	
+	+		79.3(54.2-94.4)
Bicuculline	1	10 min	
Phenazepam	2	30 min	
+	+		90.0(71.1-108.9)
Thiosemicarbazide	4	50 min	
3,4-DOPA	200	1 h	
+	+		36.6(21.5-51.7)
Phenazepam	2	30 min	
α-Methyl DOPA	200	24 h	
+	+		4.2 (3.6-4.8)
Phenazepam	2	30 min	
SKF-525A	25	1 h	
+	+		5.4 (1.6-9.2)
Phenazepam	2	30 min	

suppressing the effects of withdrawal syndrome. Like phenazepam and
depakin, it restores loss of balance. The effects of α-methyl DOPA,
however, are somewhat less pronounced than those of depakin. DOPA,
or 3,4-dihydroxyphenylalanine, which, as the precursor of dopamine
and noradrenaline raises the concentration of these catecholamines in
the brain, reinforces withdrawal syndrome (see Table 2). A similar
effect on this phenomenon is also produced by disulfiram - a powerful

dopamine β-hydroxylase inhibitor, which reduces or hardly affects
noradrenaline concentration and increases that of dopamine within the
brain (Ngai, 1978).

Our findings appear to suggest that GABA- and dopaminergic
mechanisms may be involved in the build-up of tolerance and withdrawal syndrome following long-term administration of benzodiazepines. It was shown how the effects of withdrawal syndrome are
regularly suppressed by substances which reinforce the inhibitory
action of endogenous GABA, while GABA-negative ones have the reverse
effect of heightening this syndrome. The effects of withdrawal are
also attenuated by treatment with α-methyl DOPA, whereas a rise in
dopamine concentration serves to reinforce withdrawal syndrome.

This work also includes an examination of the pharmacological
activity of phenazepam metabolites, generated in the human or animal
organism, in order to survey metabolic aspects of tolerance. Signs
were noted of the build-up of tolerance in the context of inhibition
of microsomal enzyme systems of the liver.

By comparing phenazepam's pharmacological range of action with
that of its metabolites we showed that the effects of the latter
differ considerably in performance. For instance, the tranquilizing
effects of the metabolite 3-hydroxyphenazepam do not fall below those
of the parent compound, but sometimes exceed it - in the conflict
situation test, for example. It also has similar action as a sedative and an anti-convulsant, but slightly lower myorelaxant properties (Table 1) (Figure 5).

Another phenazepam metabolite - 6-bromo-4-phenylchinazolinone,
which is formed by reduction of a 7- to a 6-membered diazepine ring,
maintains the properties of the 1,4-benzodiazepines and has the same
types of effect as phenazepam. This metabolite is between 10 and 100
times less active than the original compound, however. Aminochloro-
benzophenone, one of the final metabolites of phenazepam, showed no
tranquilizing effect and in general was relatively inactive (Table 1,
Figure 5).

This enables us to state that the process of phenazepam's metabolism in the animal organism gives rise to the formation of less
active degradation products.

The compound SKF-525A, and spasmolytin, which resembles SKF-525A
in its chemical structure and action, were used to investigate benzodiazepine tolerance when microsomal enzyme systems were inhibited.
When this was brought about by giving SKF-525A to rats which had
become tolerant to phenazepam's sedative effects (impaired conditioned reflex performance in a T-maze), it was found that the tranquilizer's effects had changed substantially. When tolerant animals
received the usual dose of phenazepam following pre-treatment with

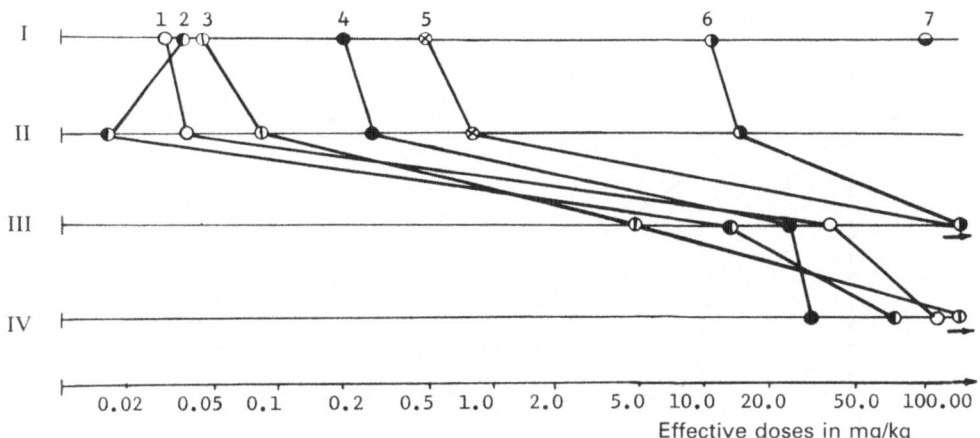

Fig. 5. Comparative activity of phenazepam metabolites. Abscissa -
effective doses in mg/kg. I - phenazepam; II - 3-hydroxy-
phenazepam; III - 6-bromo-4-phenylchinazolinone; IV - benzophenone. 1 - antagonism to thiosemicarbazide; 2 - antagonism to pentylenetetrazole; 3 - potentiation of hexenal; 4 -
countering of aggression; 5 - loss of coordination in movement; 6 - electric shock; 7 - sideways position.

SKF-525A, they spent 2.2 times longer in maze-running, thus providing
evidence of partial return of the compound's sedative effect, once
enzymatic activity has been suppressed (see Table 2). When spasmolytin was administered to mice which had developed tolerance to phenazepam and lorazepam's myorelaxant effects, as illustrated by the
rotating rod test, myorelaxant effects were seen to return in 40%　of
the animals, as against 100%　during SKF-induced enzymatic inhibition.
(see Figure 6). When enzymatic activity had been inhibited by
SKF-525A or spasmolytin in mice which had become tolerant to phenazepam, however, this caused no impairment of coordination in movement.
These results contrasted with findings obtained from rat experiments,
thereby indicating that involvement of metabolic mechanisms in the
development of benzodiazepine tolerance varies from one species to
another.

4. CONCLUSION

     Investigations into phenazepam, lorazepam, nitrazepam, clonaz-
epam and other members of the benzodiazepine group over prolonged
trial periods enabled us to confirm the tolerance which had set up to
their effects - as myorelaxants, sedatives and anticonvulsants which
counter aggression and the build-up of tolerance to tranquilizing
action. Toxicity is also substantially reduced during long-term

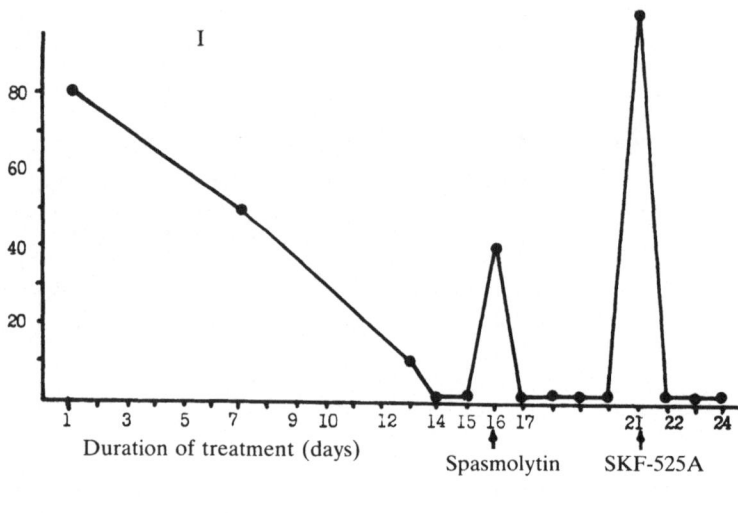

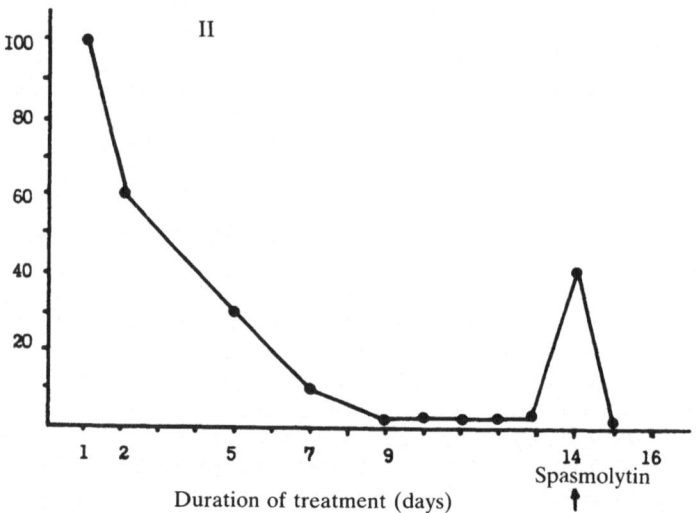

Fig. 6. Influence of SKF-525A and spasmolytin on the discoordinating
         effect of tolerance to tranquilizers in rats. Abscissa –
         duration of treatment (days); ordinate – effect (as %). I –
         phenazepam at a dose of 2 mg/kg; II – lorazepam at a dose of
         2 mg/kg. SKF – used at a dose of 25 mg/kg; spasmolytin used
         at a dose of 50 mg/kg. All substances administered i.p. to
         rats.

benzodiazepine administration (Vikhlayev et al., 1970). This
accounts for the mixed nature of benzodiazepines' range of pharmacological action and their extended therapeutic usefulness as tranquilizers and sedatives.

The above contrasts with morphine's characteristic development
of tolerance during long-term treatment, which is associated with a
reduction in its main effects of inducing euphoria and analgesia, as
well as toxicity (Seevers and Deneau, 1963; Cox et al., 1970). We
would add that even when morphine is used over long periods, at
considerably increased doses, no change in the scope of its therapeutic action is observed.

When barbiturates are used extensively, tolerance develops to
their sleep-inducing, sedative and anticonvulsant action, while their
toxic and lethal effects are unaffected; their therapeutic role as
sleep-inducing and anticonvulsant agents is thereby substantially
reduced (Frazer et al., 1954; Marlow, 1970, etc).

Long-term treatment with meprobamate builds up tolerance to its
main action (as a tranquilizer); thus meprobamate's tolerance effects
are seen to differ from those of benzodiazepine tranquilizers.

On the basis of our findings on the tolerance which develops to
morphine, barbiturates, meprobamate and benzodiazepine tranquilizers,
we would suggest that habituation to benzodiazepine derivatives may
be seen as a newly-revealed aspect of tolerance. The main feature
distinguishing the effects of benzodiazepine tolerance from tolerance
to like substances is that their tranquilizing ability remains virtually unchanged, while their side-effects are actually reduced.
This increases the scope of their therapeutic application.

In accordance with current ideas on the ways in which tolerance
is initiated and developed, there are several explanations for our
findings on habituation to tranquilizers. Firstly, tolerance may be
produced by central nervous system cells adapting to the permanent
presence of the tranquilizer within the organism; i.e. receptor
sensitivity towards the drug may alter, or compensatory changes occur
within the transmitter system. Secondly, an explanation of the benzodiazepine tolerance phenomenon could be provided by changes occurring in metabolic processes during long-term administration, connected with accelerated metabolism and excretion of these substances.

Evidence is supplied for the first explanation (involving
adaptational mechanisms) principally by our findings on the degree
of withdrawal syndrome observed in animals following severance of
long-term treatment with phenazepam. It is customary to consider
withdrawal syndrome a compensatory response, linked to heightened
arousal of receptor systems which had been inhibited during the
period of administration (Seevers and Deneau, 1963, 1964; Kalant,
1972; Collier, 1973; Rastogi, 1978; Pereu et al., 1978). Withdrawal
syndrome is marked by the animals becoming prone to convulsions and
spontaneous aggression, and losing their powers of orientation and of
conditioned reflexes. Other characteristics include fear, anxiety
and stupor. These features resemble clinically observed withdrawal

syndrome, to be seen in some patients after discontinuation of long-
term treatment with tranquilizers (Lees et al., 1978).

An extremely high degree of specificity is an important feature
of benzodiazepine withdrawal syndrome; the restoration of withdrawal-
induced conditioned reflex impairment is observed almost exclusively
when compounds with benzodiazepine structure are used. Compounds
belonging to other chemical groups, such as meprobamate, trioxazine
and chlorpromazine are either unable to replace phenazepam and thereby cancel out the emotional and behavioral effects of withdrawal, or
only partially replace this benzodiazepine, as in the case of phenobarbital or depakin.

Benzodiazepines' action is known to be connected with the different neurotransmitter systems of the brain. This work, having
examined the role of some neurotransmitters in the development of
tolerance to and withdrawal effect arising from benzodiazepines, has
illustrated the involvement of the GABA-ergic and catecholaminergic
systems in this process. Their involvement is shown by our finding
that depakin, which reinforces the action of endogenous GABA, and
α-methyl-DOPA, which inhibits catecholamine synthesis, both attenuate
the effects of benzodiazepine withdrawal. In addition, noradrenaline
and DOPA (which, as the precursor or dopamine, raises the concentration of these two catechalamines within the brain), as well as
bicucculine and thiosemicarbazide ("GABA-positive" compounds) all
reinforce withdrawal syndrome.

Current literature also contains information on specific receptor sites common to man and animals which bind substances with a
benzodiazepine structure (Squires and Braestrup, 1977; Mohler and
Okada, 1977; Tallam et al., 1979; Speth et al., 1979, etc). The
degree of specific binding depends on the concentration of benzodiazepine administered, moreover, and tends towards saturation.
Benzodiazepine-receptor binding is mainly distinguished by its high
stereospecificity and selectivity almost exclusively for compounds
with a benzodiazepine structure. Neither pharmacologically inactive
benzodiazepine stereoisomers nor any of the well-known neurotransmitters, whether noradrenaline, dopamine, serotonin, GABA, glycine or
others - nor, indeed, any of the substances like meprobamate, bar-
biturate, etc., whose action resembles that of benzodiazepines, are
able to bind with benzodiazepine receptors.

It was found that various compounds with a benzodiazepine structure differ in their affinity for these receptors. The strong correlation between the tranquilizing, myorelaxant and anticonvulsant
action of benzodiazepines and degree of receptor binding was also
revealed. During long-term benzodiazepine administration the pattern
of drug-receptor binding is impaired, with a change in affinity
(dissociation constant) and number of binding sites (B_{max}) (Braestrup
et al., 1979; Rosenberg and Chin, 1979).

It might be suggested, by analogy with the way in which morphine
tolerance develops, that impaired drug-receptor interaction also
lies at the root of tolerance to benzodiazepines - more specifically
the occurrence of adaptational changes in receptor sensitivity and
activation of so-called receptors and changes in the processes of
receptor occupancy. Taking these factors into account can help us
understand the selectivity and specificity of benzodiazepine-induced
cross tolerance and withdrawal. Substances like nicotinamide and
inosine, which bind fairly readily with benzodiazepine receptors,
possess the ability to displace benzodiazepines (Braestrup and
Nielsen, 1980; Möhler et al., 1979; Skolnick et al., 1979). The
reverse also applies, that substances which do not bind, like non-
benzodiazepine tranquilizers, neuroleptics, ethanol, phenobarbital
and muscimol, do not show cross-tolerance (Möhler and Okada, 1977;
Squires and Braestrup, 1977). Our data support the role attributed
to nicotinamide and inosine as endogenous ligands (Möhler and Okada,
1977; Skolnick et al., 1979). Benzodiazepine-induced withdrawal
effects are also averted by giving depakin - a substance which
reinforces the effects of endogenous GABA in the brain; this, too,
requires special explanation. As findings have recently shown,
benzodiazepines do not have the ability to bind with GABA-ergic
receptors; likewise, nor can GABA or muscimol bind with benzodiazepine receptors (Squires and Braestrup, 1977; Möhler and Okada,
1977). There is also evidence to show that GABA and muscimol
reinforce benzodiazepine binding to their specific receptors, and
benzodiazepines, in turn, strengthen GABA's specific binding
(Gallager et al., 1978; Iversen, 1978). This serves as evidence of
a bond indirectly linking GABA-ergic and benzodiazepine receptors.
Only a short time ago a protein with a molecular weight of 15,000
was separated from human brain which had the capacity of modulating
GABA-ergic receptor function (Tallam et al., 1978; Guidotti et al.,
1978). When added to synaptosomal membranes, this compound, which
was named GABA-modulin, produced substantial reduction in the number
of GABA binding sites and their affinity for GABA. It was further
shown that benzodiazepines inhibit GABA-modulin's effects competitively, thereby producing modifications in GABA receptors and reinforcement of GABA action. GABA-modulin itself is also seen to
act as a modulator of benzodiazepine receptors, since it affects the
binding of benzodiazepines to their receptors when added to synaptosomal membranes (Tallam et al., 1978; Iverson, 1978; Baraldi et al.,
1979). These findings suggest that the specific effect of depakin
in benzodiazepine-induced withdrawal syndrome occurs neither ac-
cidentally nor symptomatically, but is bound up with the effects
produced on receptor sites via GABA-modulin.

Our findings using spasmolytin and SKF-525A (inhibitors of
hepatic microsomal enzyme activity) provided evidence of changes
occurring in the process of phenazepam metabolism during long-term
administration. In view of the fact that SKF-525A and spasmolytin
inhibit the activity of enzymes involved in benzodiazepine hydroxy-

lation and glucuronide conjugation, it may be supposed that administering these inhibitors can produce a) inhibition of tranquilizer metabolic processes (which had been accelerated by long-term treatment) and b) a return to initial concentrations of these substances. These changes result in: a) the return of myorelaxant and sedative effect on rats and b) changed anti-convulsant effects observed when phenazepam is administered over long periods in combination with SKF-525A and spasmolytin.

Our work using SKF-525A and spasmolytin suggest that habituation arising during long-term benzodiazepine administration contributes to the observed changes in pharmacokinetic parameters or, more specifically, to speedier inactivation of tranquilizers within the organism due to enhanced activity of hepatic enzymes. This suggestion is supported by the findings of Jablonska et al. (1975) and Fukusama et al. (1975) concerning the potentiating effect of certain well-known benzodiazepines on the activity of metabolizing enzymes.

It might be supposed, by analogy with the effects of other psychotropic agents, that benzodiazepine tolerance is associated with a redistribution of its degradation products and an accumulation of less active metabolites, as well as altered rate of its inactivation and excretion. Our results would support this. It was seen that 3-hydroxyphenazepam, a phenazepam metabolite, equals phenazepam both as a tranquilizer and in countering pentylenetetrazole's effects in a conflict situation. It also has less pronounced myorelaxant effects then phenazepam. This tolerance to myorelaxant but not to tranquilizing influence could thus be attributed to accumulation of this metabolite within the organism during long-term phenazepam administration.

We have given some of the reasons why the way in which tolerance to myorelaxant, sedative and anticonvulsant effects develop could be explained by a) accelerated metabolic processes and b) the changing relationship between metabolites so formed. It should still be borne in mind, however, that a pharmacokinetic explanation is not universal in benzodiazepine tolerance development; it is only involved under certain experimental conditions and in particular animal species. Restoration of myorelaxant effects, for instance, is observed in phenazepam-tolerant rats when treated with microsomal enzyme inhibitors, but not in mice. The lack of correlation which we established between drug concentration values confirms that metabolic mechanisms are not involved in tolerance developed to phenazepam as far as its myorelaxant effects are concerned. Such effects completely disappear soon after repeated administration of phenazepam however. The concentrations of this drug and 3-hydroxyphenazepam, its active metabolite, stay at a high value in the blood; 3.55 and 1.43 µg/ml respectively (acute administration) and 3.75 and 1.58 µg/ml respectively in the case of long-term administration.

These results are confirmed by Christensen's (1973) findings of high concentrations of chlordiazepoxide and its active metabolites in mice which had developed tolerance to the drug's myorelaxant effect.

We would thus conclude that as yet we have insufficient information at our disposal to provide a comprehensive solution as to how tolerance forms to these drugs. Mechanisms of benzodiazepine tolerance appear to be mixed; they are associated both with changes in metabolism of the substances within the organism and with adaptation of central nervous system cells. Which of these factors prevails in the development of tolerance to benzodiazepines depends on how the individual substance is metabolized, which animal species is used and the effects produced.

REFERENCES

Agarwal, K. A., Lapierre, Y. D., Rastogi, R. B., and Singhal, K. L., 1977, Alterations in brain 5-hydroxytryptamine metabolism during the "withdrawal" phase after chronic treatment with diazepam and bromazepam, Brit.J.Pharmac., 60:3-9.

Alexandrovsky, Yu. A., 1976, States of psychic disadaptation and their compensation, in: "Frontiers of Neuropsychic Dysfunctions," Moscow, pp.124-157.

Alexandrovsky, Yu. A., 1979, "Clinical Pharmacology", Results of science and technology, Pharmac.Chemotherap.Agents, Moscow, II, 3-46.

Avrutsky, G. J., Gurovitch, I. J., and Gromova, V. V., 1974, Pharmacology of Mental Diseases, Moscow, "Medicine", p.472.

Baraldi, M., Guidotti, A., Schwartz, J., and Costa, E., GABA receptors in clonal cell lines, a model for study of benzodiazepine action at molecular level, Science, 205:281-283.

Baulu, R. B., Labas, M., and Ngoma-Jengo, P., 1978, Some neurochemical correlates of "rebound" phenomenon observed during withdrawal after long-term exposure to 1,4-benzodiazepines, Progr.Neuropsychopharmacol., 2:43-54.

Belenky, M. L., 1959, "Elements of Quantitative Valuation of Pharmacological Effect," Riga, pp.9-28.

Biswas, B., and Carlsson, A., 1977, The effect of intracerebroventricularly administered GABA on brain monoamine metabolism, Naunyn Schmiedeberg's Arch.Pharmacol., 299:41-46.

Braestrup, C., Nielsen, M., and Squires, R. F., 1979, No changes in rat benzodiazepine receptors after withdrawal from continuous treatment with lorazepam and diazepam, Life Sci., 24:347-350.

Braestrup, C., and Nielsen, M., 1980, Benzodiazepine receptors, Arzneim.Forsch., 30:852-857.

Christensen, J. D., 1973, Tolerance development with chlordiazepoxide in relation to the plasma levels of the parent compound and its main metabolites in mice, Acta Pharmacol.Toxicol., 33:262-272.

Cook, L., and Davidson, A. B., 1978, Psychopharmacological characterization of anxiolytics, J.Neuropsychopharmacol., 1:891-897.

Costa, E., Guidotti, A., Mao, C., and Suria, A., 1975, New concepts
    on the mechanism of action of benzodiazepines, Life Sci.,
    17:167-185.
Cox, B. M, and Osman, O. H., 1970, Inhibition of development in
    rats by drugs which inhibit ribonucleic acid or protein
    synthesis, Brit.J.Pharmacol., 38:157-170.
Crawley, J. N., Marangos, P. J., Stivers, J., and Goodwin, G. K.,
    1982, Chronic clonazepam administration induces benzodiazepine
    receptor subsensitivity, Neuropharmacology, 21:85-89.
Dunham, N. W., and Miya, T. S., 1957, A note on a simple apparatus
    for detecting neurological deficit in rats, J.Amer.Pharm.Ass.,
    46:208-209.
Fraser, H. F., Isbell, H., Eisenman, A. E., Wikler, A., and Pescor,
    F., 1954, Chronic barbiturate intoxication; further studies,
    Arch.Int.Med., 94:34-41.
Fukasama Hideo, Iwase Hiroaki, Ichishita Hiroko, Tokizawa Tsuyoshi,
    and Shimizu Hirotoshi, 1975, Effects of chronic administration
    of bromazepam on its blood level profile and on the hepatic
    microsomal drug metabolizing enzymes in the rat. Drug Metab.
    and Disposition: Biol.Fate Chem., 3:235-244.
Gallager, D. M., Thomas, J. W., and Tallman, J. F., 1978, Effect of
    GABA-ergic drugs on benzodiazepine binding site sensitivity in
    rat cerebral cortex, Biochem.Pharmacol., 27:2745-2749.
Garattini, S., 1978, Some examples of resistance to psychotropic
    drugs in animals and man, Neuropsychopharmacol., 1:311-319.
Greenblatt, D., and Shader, R., 1974, Benzodiazepines in Clinical
    Practice, Raven Press, New York.
Guidotti, A., Toffano, G., and Costa, E., 1978, An endogenous protein
    modulates the affinity of GABA and benzodiazepine receptors in
    rat brain, Nature, 275:553-555.
Iversen, L. L., 1978, Gaba and benzodiazepine receptors, Nature,
    275:477.
Jablonska, K., Knobloch, M. J., and Wisniewska, K. V.,1975, Stimu-
    latory effect of chlordiazepoxide, diazepam and oxazepam on the
    drug metabolizing enzymes in microsomes, Toxicology, 5:103-111.
Kalant, H., Le Blanc, A. E., and Gibbins, R. J., 1971, Tolerance to,
    and dependence on some non-opiate psychotropic drugs, Pharmacol.
    Rev., 23:135-136.
Kalant, H., 1972, Tolerance phenomena in drug dependence, Proceedings
    of the 30th International Congress on Alcoholism and Drug
    Dependence, A. Tongue and E. Tongue, eds., International Council
    on Alcohol and Addictions, Amsterdam, Sept., pp.49-56.
Klygul, T. A., and Krivopalov, V. A., 1966, Installation with auto-
    matic registration of the rat behaviour for experimental valu-
    ation of the action of minor tranquilizers, Pharmac.Toxicol.,
    No.2, 241-244.
Lippman, W., and Pugsley, T., 1974, Effects of benzodiazepine and
    chlordiazepoxide in the brain, Brit.J.Pharmac., 51:571-575.
Litchfield, J., and Wilcoxon, F., 1949, A simplified method of
    evaluating dose-effect experiments, J.Pharmacol.exp.Ther.,
    96:99-114.

Marlow, H. F., 1970, The pharmacological basis of drug dependence, S.Afr.Med.J., 44:610-615.

Marks, J., 1978, The Benzodiazepines. Use, Overuse, Misuse, Abuse, MTP Press Limited, Lancaster, England.

Möhler, H., and Okada, T., 1977, Benzodiazepine receptor: demonstration in the central nervous system, Science, 198:849-851.

Möhler, H., Polc, P., Cumin, R., Pieri, L., and Kettler, R., 1979, Nicotinamide is a brain constituent with benzodiazepine-like actions, Nature, 278:563-565.

Orloff, M., Williams, P., and Pfeiffer, C., 1949, Timed intravenous infusion of metrazol and strychnine for testing anticonvulsant drugs, Proc.Soc.exp.Biol., 70:254.

Pereu, G., Gasamenti, F., and Pedata, F., 1978, Neurotransmitters and opiate addiction, Proc.Eur.Neurochem., 2nd Meet. ESN, Göttingen Weinhein - New York, pp.425-438.

Rastogi, R. B., Lapierre, J. D., and Singhal, R. L., 1978, Some neurochemical correlates of "rebound" phenomenon observed during withdrawal after long-term exposure to 1,4-benzodiazepines, Progr.Neuropsychopharmacol., 2:43-54.

Rosenberg, H. C., and Chin, T. H., 1979, Benzodiazepine binding after in vivo elevation of GABA, Neuroscience, 15:277-281.

Seevers, M. H., and Dencau, G. A., 1963, Physiological aspects of tolerance and physical dependence, in: "Physiological Pharmacology," W. S. Root and F. G. Hofman, eds., Academic Press, New York - London, 1A:565-640.

Seevers, M. H., and Deneau, G. A., 1964, Drug dependence, in: "Evaluation of Drug Activities. Pharmacometrics," D. Laurence and A. L. Bacharach, eds., Academic Press, London, 1:167-179.

Skolnick, P., Marangos, P. J., Syapin, P., Goodwin, F. K., and Paul, S. M., 1979, CNS benzodiazepine receptors: physiological studies and putative endogenous ligands, Pharmacol.Biochem.Behav., 10:815-823.

Speth, R. C., Wastek, G. J., and Yamamura, H. I., 1979, Benzodiazepine receptors: temperature dependence of (^{3}H) flunitrazepam binding, Life Sci., 24:351-358.

Squires, R. F., and Braestrup, A., 1977, Benzodiazepine receptors in rat brain, Nature, 266:732-734.

Swinyard, E., Brown, W., and Goodman, L., 1952, Comparative assays of antiepileptic drugs in mice and rats, J.Pharmacol.Exper.Ther., 106:313-314.

Tallman, J. F., Thomas, J. W., and Gallager, D. W., 1978, GABA-ergic modulation of benzodiazepine binding site sensitivity, Nature 274:383-385.

Tallman, J. F., Thomas, J. W., and Gallager, D. W., 1979, Identification of diazepam binding in intact animals, Life Sci., 24:873-880.

Tedeschi, R., Tedeschi, D., Mucha, A., Cook, H., Mattis, P., and Fellow, E., 1959, Effects of various centrally acting drugs on fighting behaviour of mice, J.Pharmacol.Exp.Ther., 125:28-34.

Toman, J., Swinyard, E., and Goodman, L., 1946, Properties of maximal
    seizures and their alteration by anticonvulsant drugs and other
    agents, J.Neurophysiol., 9:231-271.
Vikhlyaev Yu, I., Klygul, T. A., and Jagatspanian, I. A., 1970,
    Experimental characteristics of the development of tolerance to
    diazepam and chlordiazepoxide, J.Neuropath.Psych. S.S. Korsakov,
    No.12, 1867-1872.
Vikhlyaev, Yu, I., and Klygul, T. A., 1973, Influence of the benzo-
    diazepine series tranquilizers on various forms of the operant
    animal behavior, Pharmac.Toxic., No.6, 657-663.
Vikhlyaev, Yu, I., Klygul, T. A., Garibova, T. L., Jagatspanian, I.
    A., Zhilina, Z. I., and Andronati, S. A., 1978, Physiological
    activity of diazepam metabolites and their role in the mani-
    festation of its effect and development of tolerance, in:
    "Physiol. Active. Drugs," Naukova dumka, Kiev, No.10, pp.47-53.
Voronina, T. A., Vikhlyaev, Yu, I., Nerobkova, L. N., Garibova, T.
    L., Tozhanova, N. M., and Kosoy, M. Yu., 1982, Characteristics
    of pharmacological properties of phenazepam, in: "Phenazepam,"
    Naukova dumka, Kiev, pp.87-145.

Index